NEW COURSE

ENGINEERING CHEMISTRY

(Theory and Practicals)

By

Prof. (Dr.) S.K. BHASIN
Director, Asian Institute of
Management & Technology,
Dhaurang, Yamuna Nagar
(Haryana)

PUNIT BHARDWAJ
HOD, Applied Sciences,
Asstt. Prof. in Chemistry
Himalayan Group of Professional
Institutions, KALA-AHB, (H.P.)

S.K. KATARIA & SONS®
Publisher of Engineering Books
4885/109, Prakash Mahal, Dr. Subhash Bhargav Lane,
Opposite Delhi Medical Association, Daryaganj, New Delhi-110002
Phone: +91-11-43551243, Mobile: +91-9871775858, 7678425209
e-mail: katariabook@gmail.com, katariabooks@yahoo.com
Website: www.skkatariaandsons.com
(e-books, subscriptions and bulk buying also available on our website)

Published By :

S.K. KATARIA & SONS®

4885/109, Prakash Mahal, Dr. Subhash Bhargav Lane,
Opposite Delhi Medical Association, Daryaganj, New Delhi-110002
Phone: +91-11-43551243, Mobile: +91-9871775858, 7678425209
e-mail: katariabook@gmail.com, katariabooks@yahoo.com

First Edition: January, 2013, Reprint: 2025

Price: ₹ 595/-

Laser Typesetting at: Goswami Associates, Delhi.

Printed at: Arora Printing Press, Delhi-110051

Preface to the First Edition

This book has been written strictly according to the latest syllabus (in theory as well as practicals) of the engineering students of Himachal Technical University, Dharamsala (H.P.). The book has grown out of the vast teaching experience and interaction with the students of chemistry and applied chemistry. Every care has been taken to develop the topics from the fundamentals, so that it can be easily understood and used by the students for self-study also. We have tried our best to give a balanced account of the topics in the syllabus along with the latest problems for engineering students. No doubt the market is flooded with number of good books for engineering students, but this book contains the subject matter strictly according to the latest examination pattern of Himachal Technical University. The pattern of examination has been kept well in view and every effort has been made to make the subject matter simple.

It is no exaggeration to mention that this publication is a **three-in-one book**. The present book is comprised of three parts namely;

Part–I: Text strictly in accordance with latest H.P. Tech. Univ. syllabus.

Part–II: Laboratory Manual as per H.P. Tech. Univ. syllabus.

Part–III: Solved H.P. Tech. Univ. Question Paper Dec., 2012 along with Model Question Papers.

The special features of the book have made the book— ***A book with a difference***. Here is a book which caters to all your queries and doubts. It gives you the fundamentals and the basics of applied chemistry required by you as an engineer. The book will not only help you to score better in the examination, but would develop your skills to apply your knowledge of applied chemistry in solving the problems that you may face during your technical career.

We express our sincere thanks to the authors and publishers whose works we have consulted in framing this book. Our acknowledgements are due to M/s S.K. Kataria and sons, who have left no stone unturned in bringing out the book well in time. Although care has been taken while preparing and typing the manuscript of the book, yet errors and misprints might have crept in. We shall be grateful to the teachers and students who are kind enough to send their suggestions for further improvement of the book.

We shall consider our efforts amply rewarded if all those for whom the book is intended are benefited by it.

January, 2013

S.K. BHASIN
PUNIT BHARDWAT

Preface to the First Edition

Contents

PART I

SECTION A

2 PHASE RULE 71–94

SECTION B

3 WATER TREATMENT 97–129

6 FUELS AND COMBUSTION 225–255

SECTION D

7 POLYMERS 259–286

8 COMPOSITE MATERIALS 287–307

PART II

PRACTICALS

PART III

PART I

Section A

Chapter 1: Electrochemistry

Chapter 2: Phase Rule

1

Electrochemistry

INSIDE THIS CHAPTER

1.1 Introduction; **1.2** Electrical Conductance; **1.2.1** Metallic and Electrolytic Conductors; **1.2.2** Differences Between Metallic Conductor and Electrolytic Conductor; **1.2.3** Types of Electrolyte; **1.2.4** Factors Affecting Electrolytic Conduction; **1.3** Specific Conductance; **1.3.1** Equivalent Conductance and Molecular Conductance; **1.3.2** Relation Between Specific Conductance and Equivalent Conductance; **1.4** Experimental Measurement of Conductance; **1.4.1** Calculations of Specific Conductance; **1.5** Effect of Dilution on Conductance; **1.5.1** Conductance Behaviour of Strong Electrolytes; **1.5.2** Conductance Behaviour of Weak Electrolytes; **1.6** Migration of Ions; **1.6.1** Hittorf's Theoretical Device; **1.7** Transport Number; **1.7.1** Factors Controlling Transport Number; **1.7.2** Determination of Transport Number by Hittorff Method; **1.8** Kohlrausch Law; **1.8.1** Relation between Ionic Conductance and Transport Number; **1.8.2** Applications of Kohlrausch's Law; **1.9** Interionic Attraction Theory of Conductance or Modern Theory of Strong Electrolytes; **1.10** Hydration of Ions; **1.11** Electrochemical Cell or Galvanic Cell; **1.11.1** Functions of Salt Bridge; **1.11.2** Representation of Galvanic Cells; **1.12** Electrode Potential; **1.12.1** Standard Electrode Potential; **1.12.2** Measurement of Electrode Potential; **1.13** Cell Potential or EMF of a Cell; **1.13.1** Calculation of EMF of a Cell; **1.13.2** EMF of the Cell and Free Energy Change; **1.13.3** Criteria for Spontaneity in Terms of Cell EMF **1.13.4** Calculation of Equilibrium constant K_C from $\Delta G°$; **1.14** Nernst Equation or Concentration Dependence of Electrode Potential; **1.14.1** Applications of Nernst Equation; **1.15** Reference Electrodes; **1.15.1** Primary Reference Electrodes; **1.15.2** Secondary Reference Electrodes; **1.16** Electrochemical Series; **1.16.1** Applications of Electrochemical Series; **1.17** Concentration Cells; **1.18** Batteries; **1.18.1** Metal-air Batteries; **1.19** Fuel Cell

1.1 INTRODUCTION

Electrochemistry is the study of the processes involved in the interconversion of electrical energy and chemical energy. In other words, *electrochemistry is that branch of chemistry which deals with chemical changes produced by electricity and production of electricity by chemical changes.*

Electrochemical changes are of two types:

(*i*) A chemical change is effected by passing electric current through electrolytes in the molten state or through their aqueous solutions. In these changes, electric current can be employed to bring about a chemical reaction. The phenomenon of decomposition of an electrolyte with the help of electric current is called ***electrolysis*** and a device in which electrolysis is carried out is called an ***electrolytic cell.***

(*ii*) A chemical change which produces electric current, because of the electron transfer from one substance to another, is a redox reaction. When a redox reaction is carried out directly, chemical energy in the form of heat is produced. However, when a redox reaction is carried out indirectly, chemical energy is produced in the form of electrical energy. A device employed to convert the chemical energy of a redox reaction into electrical energy is called an ***electrochemical cell***. The study of the principles involved in utilizing chemical changes for producing electrical energy is referred to as ***electromotive chemistry***.

The flow of current in the external circuits, *i.e.*, through metallic conductors is nothing but the flow of electrons. However, the flow of electricity through solutions is due to the flow of ions. The passage of electricity through electrolytes in the molten state or through their solutions due to the movements of ions is called ***electrolytic conduction.*** The knowledge of electrolytic conduction being of primary importance in the study of electrochemistry.

Besides electrolytic conduction, we will discuss in this chapter a very important law (Kohlrausch law) dealing with the amount of current carried in solution and transport number which relates to the current carried by an ion and its mobility.

1.2 ELECTRICAL CONDUCTANCE

The term conductance of any conductor (metallic or electrolytic solution) is related to the resistance of the conductor. The resistance of the conductor can be understood through Ohm's law. The law states that,

'*The strength of current* (I) *passing through a conductor is directly proportional to the potential difference* (E) *applied across the conductor and is inversely proportional to the resistance* (R) *of the conductor*'.

Mathematically,

$$R \propto \frac{1}{C} \quad \text{or} \quad R = k\frac{1}{C} \qquad \ldots(1.1)$$

The current strength is measured in amperes, the potential difference in volts and the resistance in ohms. The three units of electrical measurements are, therefore, related to each other by Ohm's law, as

$$R = \frac{E}{I} \quad \text{or} \quad \text{Ohms} = \frac{\text{Volts}}{\text{Amperes}}$$

Thus, if 1 ampere of current flows through a conductor when a voltage of 1 V is applied to it, the resistance of the conductor is taken as 1 ohm. Every substance offers resistance to electricity to a small or large extent. Ohm's law gives us the exact value of resistance. It holds good for metallic as well as electrolytic conductors.

In electrochemistry, the term conductance (C) is more commonly used. The term "C" implies the ease with which the current flows through the conductor.

It is a common observation that a substance with little resistance must possess a high conductance and vice versa. Therefore, if C is the conductance of a solution and R the resistance it offers to the passage of current, then

$$R \propto \frac{1}{C} \qquad \text{or} \quad R = k\frac{1}{C}$$

where k is constant of proportionality.

when $\qquad k = 1, \quad R = \frac{1}{C} \quad \text{or} \quad C = \frac{1}{R}$

Hence, conductance is defined as the reciprocal of resistance.

Units: The units of electrical conductance are also the reciprocal of ohm, *i.e.*, **Ohm^{-1}** or **mho.** Its SI unit is siemens represented by symbol S. For example, if the resistance of an electrolyte is 10 ohms, its conductivity will be 1/10 mhos or ohm^{-1} or siemens or Ω^{-1}.

1.2.1 Metallic and Electrolytic Conductors

A substance which allows the electric current to flow through it is called a conductor while another which does not allow any electric current to flow through it is called a non-conductor or an insulator.

Conductors are of two kinds:

(*i*) **Metallic conductors** *which allow the electric current to flow through them without undergoing any chemical decomposition.* They remain unchanged even after the flow of electricity through them. Examples are copper, silver, etc. The conduction is due to the flow of electron.

(*ii*) **Electrolytic conductors (Electrolytes)** *are substances whose molten or aqueous solutions conduct the electric current and which are decomposed by the passage of electric current.* Examples are acids, bases and salts. The conduction is due to the movement of ions.

A compound whose aqueous solution does not conduct electric current is called a **non-electrolyte.** Solutions of cane sugar, glycerine and alcohol do not conduct electricity and hence are examples of non-electrolytes.

1.2.2 Differences Between Metallic Conductor and Electrolytic Conductor

The main differences between metallic conductor and electrolytic conductors are as follows:

S.No.	*Metallic Conductor*	*Electrolytic Conductor*
1.	The flow of electricity through metallic conductor is due to the flow of electrons.	The flow of electricity through electrolytic conductor is due to the flow of ions.
2.	Flow of electricity does not cause the decomposition of metallic conductor.	Flow of electricity causes the decomposition of electrolytic conductor.
3.	Metallic conductance decreases with the rise in temperature.	Electrical conductance increases with the rise in temperature.
4	Resistance of metals increases with rise in temperature.	Resistance of electrolytes decreases with rise in temperature.
5.	Flow of electricity does not cause the flow of matter.	Flow of electricity causes the flow of matter.
6.	Metals usually have high conductivity.	Electrolytic solutions have lower conductivity than metals.

1.2.3 Types of Electrolytes

There are two types of electrolytes:

Strong electrolyte. *A substance which dissociates completely into ions in aqueous solution* is called *strong electrolyte.* For example mineral acids such as, HCI, H_2SO_4, etc; alkalies like NaOH and KOH, salts like NaCl, NH_4Cl, etc.

On the other hand, a **weak electrolyte** is one *which dissociates only partially into ions in aqueous solution.* For example, organic acids like HCOOH, CH_3COOH, and bases like C_6H_5 NH_2, etc.

Strong electrolytes have higher value of conductance than those of weak electrolytes. This is due to the fact that the number of ions in solution in case of strong electrolytes (which are completely ionised) is larger than in case of weak electrolytes (which are only partially ionised). Electric current in solutions is conducted by ions. So greater the number of ions in solution, the higher would be the conductance of the solution. In other words, the conductance of any solution depends upon the degree of ionisation of the electrolyte in solution.

Note: The above given definition of strong and weak electrolytes is ambiguous, since in very dilute solutions all electrolytes are almost equally ionised. To remove the ambiguity, a convention is adopted. The criterion of dilution is fixed as one molar solution. If the substance is highly dissociated in 1M solution, it is considered as strong electrolyte. If in 1 M solution it is slightly dissociated it is considered as a weak electrolyte.

1.2.4 Factors Affecting Electrolytic Conduction

The factors which influence the free movement of ions in solution and thus the conductance of the solution are given below:

(*i*) **Nature of the electrolyte:** *Strong electrolytes,* A substance that dissociates completely into ions in aqueous solution is called a strong electrolyte. For example, mineral acids such as, HCl, H_2SO_4, etc. alkalies such as NaOH and KOH; salts such as NaCl, NH_4Cl, etc.

On the other hand, a weak electrolyte is one, which dissociates only partially into ions in an aqueous solution. For example, organic acids like HCOOH, CH_3COOH, and bases like $C_6H_5NH_2$, etc.

Strong electrolytes have higher value of conductance than those of weak electrolytes. This is because the number of ions in solution in the case of strong electrolytes (which are completely ionized) is larger than in the case of weak electrolytes (which are only partially ionized.)

(*ii*) **The interionic interactions:** The forces existing between the oppositely charged ions of the electrolyte (solute) in solution are also called solute—solute interactions. In dilute solutions, the ions are far apart so that the forces of attraction between the oppositely charged ions are almost negligible. The ions are free to move and they make maximum contribution towards conductance.

However, in concentrated solutions, ions are close to each other and ionic forces of attraction are no longer negligible. Consequently, the mobility of ions falls so that the conductance of the solution decreases.

(*iii*) **The solvation of ions:** The ions of the electrolyte in solution get associated with the molecules of the solvent. The hydration of Na^+ ions and Cl^- ions from NaCl in water is shown in Fig. 1.1.

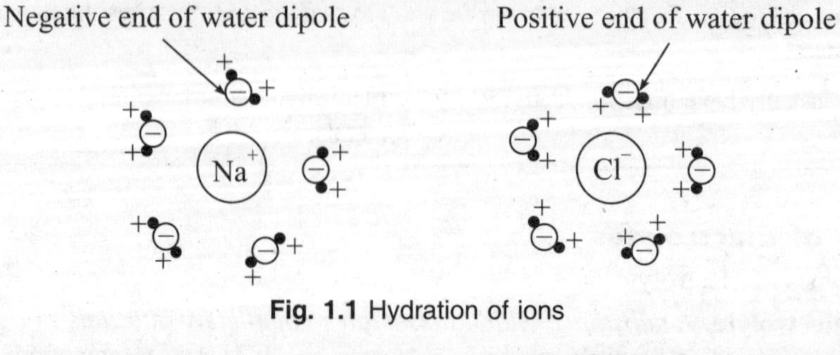

Fig. 1.1 Hydration of ions

Thus, there exists an interaction between the ions of the solute and solvent molecules due to solvation. It affects mobility of the ions and hence conductance of the solution.

(*iv*) **Viscosity of the solvent:** The viscosity of the solvent depends upon the interaction amongst the molecules of the solvent, *i.e.*, solvent-solvent interactions. Since the ions have to pass through the solvent, their mobility is affected by the viscosity of the solvent. Greater the viscosity of the solvent, lower would be the mobility of the ions passing through it and lower would be the conductance of the solution.

(*v*) **Temperature:** With increase in temperature, the conductance of an electrolytic solution also increases, whereas the conductance of metallic conductors decreases with rise in temperature. The increase in conductance of solutions with temperature is due to increase in the average kinetic energy of ions carrying the current. Moreover, the viscosity of the solvent decreases with rise in temperature and thus the speed of ions increases.

It can be concluded that the conductance of an electrolyte in solution depends upon the following:

(*i*) The number of ions per unit volume of the solution.

(*ii*) The charge on the ions in solution: Greater the charge higher would be the conduction. The conductance of $MgCl_2$ solution is greater than that of NaCl solution of the same concentration.

(*iii*) The mobility (speed) of the ions. Greater the mobility of the ions in solution, the higher would be its conductance.

1.3 SPECIFIC CONDUCTANCE

It is observed that resistance R of a conductor is

(*i*) directly proportional to its length (I) and

(*ii*) inversely proportional to its area of cross-section (a)

i.e.

$$R \propto \frac{1}{a} \text{ or } R = \rho\frac{1}{a} \qquad ...(1.2)$$

where ρ is a constant of proportionality called **specific resistance or resistivity.** Its value depends upon the material of the conductor.

If $l = 1$ cm and $a = 1$ sq. cm, then $\rho = R$, then specific resistance of a material is defined as the *resistance of material of a specimen of a unit length with unit area of cross-section. In other words, specific resistance is the resistance of one centimeter cube of material.*

"The reciprocal of specific resistance is known as conductivity or specific conductance." It is denoted by k (*kappa*). Thus, if k is the specific conductance and ρ is the specific resistance, then the two terms are related as

$$k = \frac{1}{\rho}$$

We know

$$R = \rho \times \frac{1}{a} \quad \therefore \quad \rho = \frac{a}{l} \times R$$

or

$$k = \frac{1}{\rho} = \frac{1}{a} \times \frac{1}{R} \quad \therefore \quad k = \frac{1}{a} \times \text{Conductance}$$

Now, if $l = 1$ cm and $a = 1$ sq. cm, then $C = k$.

Hence, conductivity or specific conductance of a solution may be defined as the conductance of a solution of unit length with unit area of cross-section. Alternatively, it may be defined as conductance of one centimetre cube of the solution of electrolyte or it may be defined as the conductance of one centimetre cube of the material.

1.3.1 Equivalent Conductance and Molecular Conductance

Specific conductance (k) of a metallic conductor (of a fixed length and fixed area of cross-section) has a definite value at a given temperature. However, this is not so in the case of electrolytic solutions. In the electrolytic solution, the current is carried by ions. Thus, conductance of a solution depends upon the number of ions present in the given volume. The number of ions present per cubic centimetre of the solution varies with the concentration of the electrolyte. Therefore, the specific conductance of the electrolytic solution, amongst other factors, depends on its concentration also.

For comparison of the conductances of different electrolytic solutions, it is essential that equal volume of these solutions contain such a definite amount of the electrolytes which give ions carrying the same total charge of electricity. In other words, it is necessary to take solutions of equivalent concentrations.

For example, one mole of potassium chloride (KCl) in its aqueous solution gives one mole of K^+ and one mole of Cl^- ions. One mole of potassium sulphate K_2SO_4 will give two moles of K^+ ion and one mole of SO_4^{2-} ions. One mole of potassium chloride gives ions carrying a total charge of 2 faradays; one mole of potassium sulphate gives ions carrying a total charge of 4 faradays. Thus, in order to get comparable results, one-half mole of K_2SO_4 should be taken for every one mole of KCl. Therefore consider the conductance of a certain volume of the solution in which definite amount of the solid (g eq. wt. or g mol. wt.) is dissolved. The conductance is called *equivalent conductance or molecular conductance* according as the weight dissolved in gram equivalent mass or gram molecular mass (Fig. 1.2)

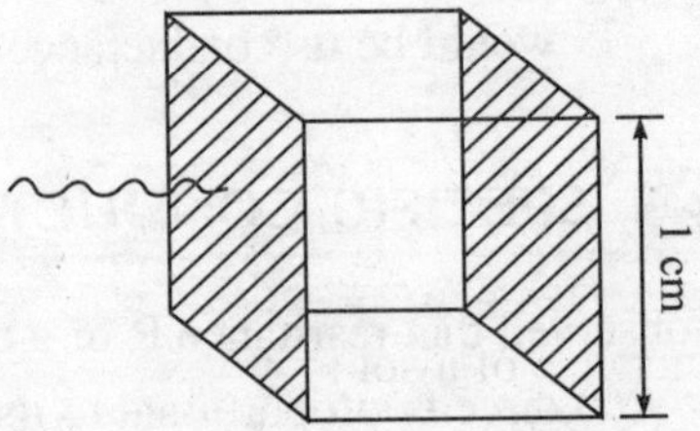

Fig. 1.2 Diagramatic illustration of specific conductance

Hence, **equivalent conductance** *is defined as the conductance of all the ions produced by 1 g equivalent of electrolyte dissolved in the solution.* It is represented by Λ_{eq}.

Its value is given by the product of the specific conductance k and the volume V in c.c. containing 1 g equivalent of the electrolyte.

1.3.2 Relation Between Specific Conductance and Equivalent Conductance

Consider 1 c.c. of a solution containing 1 g equivalent of an electrolyte placed between two large electrodes 1 cm apart. The area of cross-section of the solution will obviously be 1 sq. cm. Therefore, in such a case

$$\text{Conductance (C)} = \text{Specific conductance } (k)$$
$$= \text{Equivalent conductance } (\Lambda_{eq})$$

If the solution is now diluted, say 9 c.c., the conductance of the solution will be nine times the specific conductance as there will be now nine cubes (Fig. 1.3)

However, even now, as the solution contains gram equivalent of the electrolyte between the electrodes, the conductance measured will be the equivalent conductance. Thus, in this case:

Equivalent conductance = 9 × Specific conductance.

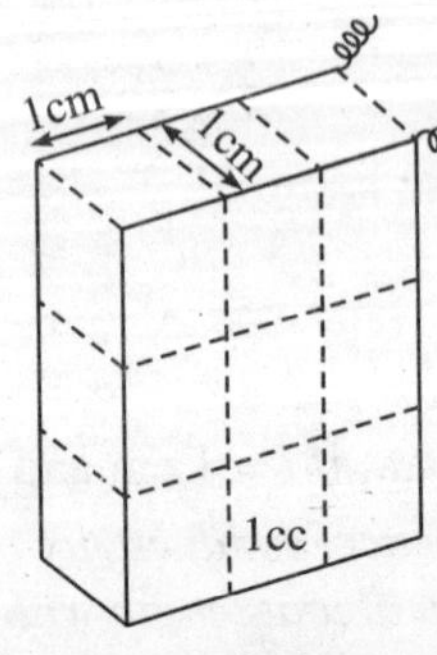

Fig. 1.3 Illustration of Equivalent conduct

If the solution is now diluted to 1,000 c.c., the conductance of the solution will be 1,000 times the specific conductance as there will be now 1,000 cubes of the material. However, even now, as the solution contains only 1 g equivalent of the electrolyte between the electrodes, the conductance measured will be the equivalent conductance.

Hence, in general,

$$\Lambda_{eq} = k \times V \quad \text{...(1.3)}$$

where V is the volume of the solution in c.c. containing 1 g equivalent of the electrolyte.

If C gram equivalents of an electrolyte are present in 1 L of the solution or if C gram equivalents per litre are the concentration of the solution, the volume containing 1 g equivalent of the electrolyte will be

$$\frac{1}{C} \text{ litre} \quad \text{or} \quad \frac{1,000}{C} \text{c.c.}$$

Therefore, the equivalent conductance, $\Lambda_{eq} = k \times \frac{1000}{\text{Normality}} = k \times \frac{1,000}{C_{eq}}$ ...(1.4)

Molar conductance *may be defined as the conducing power of all the ions produced by dissolving one mole of an electrolyte in a solution. It is denoted by* Λ_m. Like equivalent conductancc, the molar conductance of a solution is related to its specific conductance as

$$\Lambda_m = k_v \times V \quad \text{...(1.5)}$$

where V is the volume in c.c., containing one mole (g mol. mass) of the electrolyte.

or

$$\Lambda_m = k_c \times \frac{1,000}{\text{molarity}} = k_c \times \frac{1,000}{C_m} \quad \text{...(1.6)}$$

where C_m is the concentration of solution in moles per litre.

1.4 EXPERIMENTAL MEASUREMENT OF CONDUCTANCE

The electrical conductance of a solution can be determined by measuring its resistance because conductance is just the reciprocal of resistance. The **Wheatstone Bridge method** is employed for the purpose. The apparatus used is shown in Fig. 1.4. To determine the resistance of an electrolytic solution, specially designed cells known as conductivity cells are used.

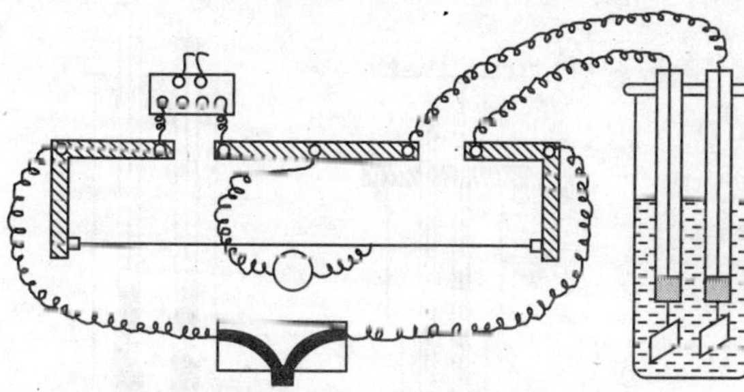

Fig. 1.4 Circuit for finding the conductance of the solution

AB is a uniform wire and a sliding contact point H moves over it. S represents the source of AC current and C is the conductivity cell containing the solution whose resistance is to be measured. R is the resistance box and P is the phone head to detect the current. Galvanometer can not be used to detect the null point because we are using AC and not DC in the circuit. Direct current can not be used because it will lead to electrolysis resulting into polarisation and change in concentration of the solution under test. In case of AC, the direction of the current is reversed about 500-1000 times per second and polarisaton effect is destroyed.

From the resistance box a known resistance R, which is of the same order (approximately) as that of solution under examination, is introduced. The sliding contact H is moved along the wire AB to detect the point of minimum sound. Under these circumstances

$$\frac{\text{Resistance of solution (X)}}{\text{Resistance R}} = \frac{\text{Length HB}}{\text{Length HA}}$$

The length HA and HB can be read from the scale, R is known, therefore, resistance of the solution X can be calculated.

Reciprocal of this resistance gives the conductance of the solution

$$\left[C = \frac{1}{X}\right]$$

For accurate results, the solutions should be prepared in *absolutely pure water* which has no conductance due to dissolved impurities. For this purpose, ordinary distilled water is not suitable, So, *conductivity water* is used for preparing solution. Conductivity water is prepared by distilling the water a number of times (at least twice) to which a few crystals of $KMnO_4$ is added, so that the conductance of water is as low as possible.

Conductivity Cells

The conductivity cells are of various forms, some of which are shown in Fig. 1.5. They are made of Pyrex glass and are fitted with platinum electrodes. The electrodes usually consist of stout platinum plates sealed into glass tubes. The electrodes are connected to the circuit by means of mercury placed in the tubes. The relative positions of the electrodes is fixed by fixing the tubes to the ebonite cover, as in A and B. The electrodes are freshly platinised by putting a solution of chloroplatinitic acid in the cell and passing alternating current for sometime. The electrodes get blackened due to a coating of finely divided platinum. This fine deposit gives larger surface to the electrodes and improves the sharpness of the sound-minimum is the telephone head.

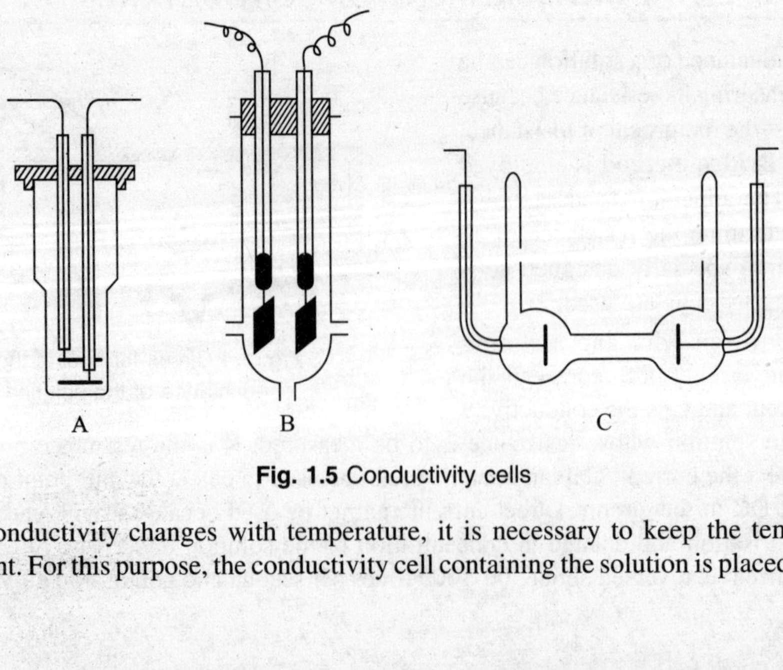

Fig. 1.5 Conductivity cells

Since conductivity changes with temperature, it is necessary to keep the temperature of the solution constant. For this purpose, the conductivity cell containing the solution is placed in a thermostat.

1.4.1 Calculations of Specific Conductance

We know that the specific conductance (k) is the reciprocal of specific resistance (ρ):

$$\text{Specific conductance} = \frac{1}{a} \times \text{Conductance}$$

or
$$k = C \times \frac{1}{a}$$

where C is the conductance of the solution, 1 is the length of solution column, *i.e.*, distance between the electrodes and a is the area of cross-section of electrons. The factor l/a is constant for a given conductivity cell. This factor is called **'cell constant'** and is expressed in cm^{-1}. Thus

∴ Specific conductance (k) = Conductance (C) × Cell constant

The value of the cell constant is not calculated from the value of l and a, as these are difficult to measure for a given cell from its dimensions. It is determined from the relation:

$$\text{Cell constant} = \frac{\text{Specific conductance } (k)}{\text{Observed conductance (C)}}$$

For this purpose, the conductance of a standard solution (say N/50 KCl) is determined at 25°C by using the given conductivity cell. The specific conductance of N/50 KCl, solution at 25°C has been found to be 0.002765 mhos. Therefore, cell constant can be determined from the following expression once for all:

$$\text{Cell constant} = \frac{0.002765}{\text{Observed conductance of N/50 KCl}} \quad \ldots(1.7)$$

By knowing the cell constant, the specific conductance of any solution can be calculated from the measured conductance.

It may be noted that for electrolytic solution direct current cannot be used because it causes electrolysis because of which the concentration of the solution changes. Moreover, the products of electrolysis may get accumulated at the electrodes and back electromotive force (EMF) may be set up. The resistance of the solution thus changes.

NUMERICAL PROBLEMS

Example 1: ***The resistance of a decinormal solution of an electrolyte in a conductivity cell was found to be 245 ohm. Calculate the equivalent conductance of the solution if the electrodes in the cell were 2 cm apart and each area of 3.5 sq. cm.***

Solution: We know that conductivity is the reciprocal of resistance.

Therefore, conductivity, $C = \frac{1}{R} = \frac{1}{245}\ ohm^{-1}$

Now specific conductance (k) = Observed conductance × Cell constant

$$= \text{Observed conductance} \times \frac{1}{a} = \frac{1}{245} \times \frac{2}{3.5}\ ohm^{-1}cm^{-1}$$

Now, equivalent conductance $\Lambda_v = k \times V$

where V is the volume containing 1 g equivalent of the electrolyte.

As the solution is decinormal (*i.e.*, 0.1 N), it means that 0.1 g equivalent of the electrolyte is present in 1,000 c.c. of the solution. Therefore, volume containing 1 g of equivalent of the electrolyte = 1,000/0.1 = 10,000 c.c.

Equivalent conductance $(\Lambda_v) = \frac{1}{245} \times \frac{2}{3.5} \times 10{,}000 = \mathbf{23.32\ ohms^{-1}cm^2eq^{-1}}$

Example 2: ***Specific conductance of N/50 KCl at 25°C is 0.002768 ohms and it has a resistance of 520 ohms. An N/25 solution of a salt kept in the same cell was found to have a resistance of 300 ohms at 25°C. Calculate (i) the cell constant and (ii) the equivalent conductance of salt solution.***

Solution:

(*i*) We know that

Specific conductance (k) = Observed conductance × Cell constant

or $$\text{Cell constant} = \frac{\text{Specific conductance}}{\text{Observed conductance}}$$

$$= \text{Specific conductance} \times \text{Observed resistance}$$

Now, we for KCl solution, we are given that

Specific conductance = 0.002768 mhos

and observed resistance = 520 ohms

Therefore, cell constant = 0.002768 × 520 = **1.43936**

(*ii*) As before, we have for the salt solution

Conductivity (k) = Observed conductance × Cell constant

$$= \frac{\text{Cell constant}}{\text{Observed resistance}}$$

But, cell constant = 1.43936

and observed resistance = 300 ohms (given)

$$\therefore \quad \text{Conductivity } (k) = \frac{1.43936}{300} \text{ ohms}$$

As the solution is N/25, therefore, volume containing 1 g equivalent of the salt will be

= 25 × 1,000 = 25,000 c.c.

Hence, equivalent conductance,

$$\Lambda_v = k - V = \frac{1.43936}{300} \times 25{,}000 \text{ mhos}$$

$$= \mathbf{119.945 \text{ mhos cm}^2 \text{g eq}^{-1}}$$

Example 3: ***Calculate equivalent conductivity of 1 MH_2SO_4 solution if its specific conductivity is 2.6×10^{-1} ohm^{-1} cm^{-1}.***

Solution: Equivalent mass of $$H_2SO_4 = \frac{\text{Molar mass}}{\text{Basicity}} = \frac{98}{2} = 49$$

Therefore, $1\ MH_2SO_4 = 2\ NH_2SO_4$

or concentration of the solution,

$$C_{eq} = 2 \text{ eq L}^{-1}$$

Now, $$\Lambda_{eq} = \kappa \times \frac{1{,}000}{C_{eq}} = 2.6 \times 10^{-1} \times \frac{1{,}000}{2}$$

$$= \mathbf{130 \text{ ohm}^{-1} \text{ cm}^2 \text{ eq}^{-1}}$$

Example 4: ***0.05 M NaOH solution offered a resistance of 31.6 ohm in a conductivity cell at 298 K. If the cell constant of the cell is 0.367 cm^{-1}, calculate the molar conductivity of the sodium hydroxide.***

Solution: It is given that resistance of the solution = 31.6 ohm

$$\text{Therefore, conductance} = \frac{1}{31.6} \text{ ohm}^{-1}$$

Therefore, specific conduction (k) = Conductance × cell constant

$$= \frac{1}{31.6} \text{ohm}^{-1} \times 0.367 \text{ cm}^{-1}$$

$$= 0.0116 \text{ ohm}^{-1} \text{ cm}^{-1}$$

$$\text{Molar conductance } (\Lambda_m) = \frac{k \times 1{,}000}{C}$$

$$= \frac{0.0016 \times 1000}{0.05}$$

$$= \mathbf{232 \ ohm^{-1} \ cm^2 \ mol^{-1}}$$

1.5 EFFECT OF DILUTION ON CONDUCTANCE

Equivalent conductance of an electrolyte increases with dilution. This is due to the fact that on dilution, the same amount of an electrolyte furnishes a large number of ions in solution. However, the increase in the number of ions on dilution is much less as compared to the increase in the volume of the solution. As a result, the total number of ions present per c.c., decreases on dilution; thus, the specific conductance decreases while equivalent conductance increases.

On progressive dilution, the values of equivalent conductance go on increasing and ultimately it becomes constant due to the fact that dissociation of electrolyte is complete. The maximum (limiting) value of equivalent conductance at infinite dilution (or zero concentration) is represented by Λ_∞ or Λ_0.

The ratio of the equivalent conductance at dilution V(Λ_v) to the equivalent conductance at infinite dilution (Λ_∞) gives the degree of dissociation *a*, *i.e.*,

$$\alpha = \frac{\Lambda_v}{\Lambda_\infty} \quad \text{or} \quad \frac{\Lambda_c}{\Lambda_0}$$

Thus, when $\alpha = 1$, *i.e.*, when dissociation is complete, then $\Lambda_c = \Lambda_v$.

The manner in which the equivalent conductance varies with dilution is shown in Fig. 1.5 in which the values of equivalent conductances of some electrolytes are plotted against the square root of concentration, $\sqrt{c}$. It follows from the curves obtained that there is a wide difference in the conductance behaviour of strong and weak electrolytes.

1.5.1 Conductance Behaviour of Strong Electrolytes

It is seen from Fig. 1.5 that for strong electrolytes like KCl, $BaCl_2$, etc., there is only a small increase in conductance with dilution. This is due to the fact that a strong electrolyte is completely dissociated in solution and hence there is no change in the number of ions. However, if the concentration is high, the movement of the ions is retarded by greater interionic attractive forces so that the conductance falls. With increase in dilution the ions are far apart so that the interionic attractive forces are almost negligible and hence the conductance increases. It is noted that in case of strong electrolytes, the equivalent conductance has a tendency to attain a certain limiting value as the concentration approaches zero near infinite dilution.

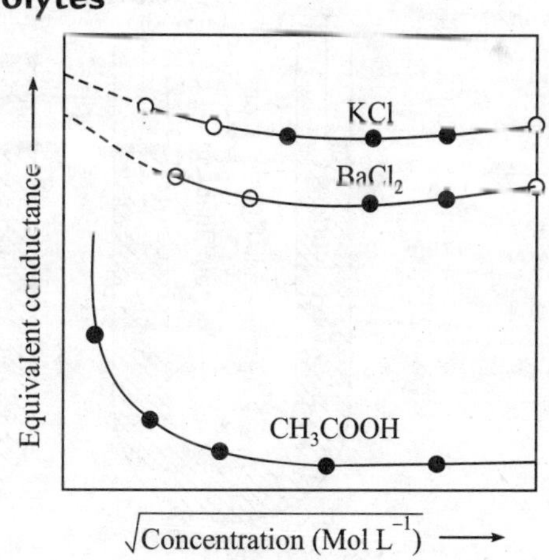

Fig. 1.6 Vatiation of equivalent conductance with concentration

This is called equivalent conductance at infinite dilution and is commonly denoted by Λ_0. This limiting value represents complete dissociation of the electrolyte. In case of strong electrolytes, it is obtained by extrapolating the equivalent conductance graph to zero concentration as shown in Fig. 1.6.

It has been observed that the variation of molar conductance with concentration in case of strong electrolytes at low concentrations is given by the expressions:

$$\Lambda_m = \Lambda_0^{\infty} - b\sqrt{C}$$

where C is the molar concentration of the electrolyte and b is a constant depending upon the nature of the solvent and temperature. Thus, a infinite dilution or zero concentration, $\Lambda_m = \Lambda_m^{\infty}$.

1.5.2 Conductance Behaviour of Weak Electrolytes

It follows from Fig. 1.5 that in case of weak electrolytes (*e.g.*, CH_3COOH) there is a sharp increase in equivalent conductance near infinite dilution.

This is due to the fact that at infinite dilution, the ionization of the electrolyte is complete and hence the increase in conductance is due to an increase in the number of ions in solutions. Further, in case of weak electrolytes, the nature of the plot is a curve and not a straight line. Thus, it is not possible to obtain the limiting value of equivalent conductance (Λ_0) by the curve to zero concentration.

So, Λ_0 cannot be obtained by extrapolation in case of weak electrolytes. The equivalent conductance at infinite dilution in such cases is obtained by an indirect method based on Kohlrausch law.

1.6 MIGRATION OF IONS

The migration of ions with different velocities can be demonstrated by a simple experiment devised by Noyes.

The U-tube shown in Fig. 1.7(*a*) is filled three-fourths with a dilute solution of agar-agar containing potassium chloride and a little phenolphthalein. Potassium chloride is added to increase the conductance of the solution. A drop of a dilute soluton of sodium hydroxide is added in the right limb and a drop of dilute hydrochloric acid is added in the left limb. As a result, the solution of agar-agar is coloured pink in the right limb and remains colourless in the left limb.

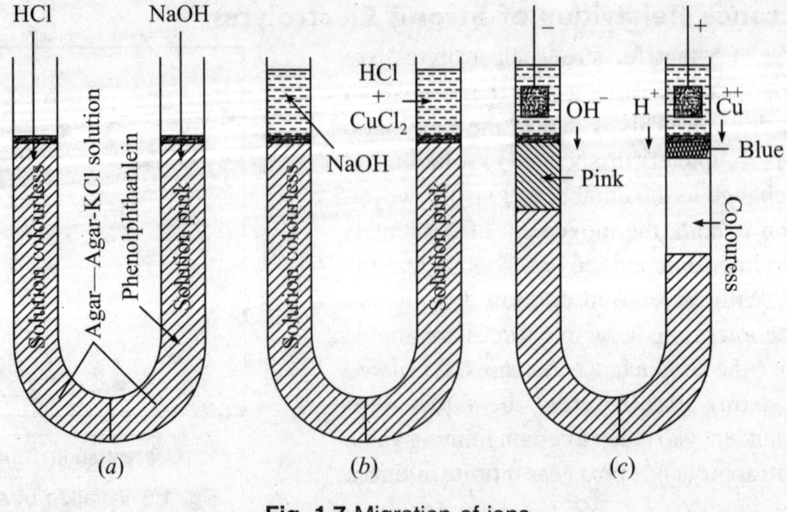

Fig. 1.7 Migration of ions

The agar-agar solution is then allowed to set into a jelly by placing the U-tube in an ice bath for a few minutes. A small amount of powdered charcoal is then scattered on the surface of the solution on each side to mark the position of the boundary.

A solution of hydrochloric acid and cupric chloride is poured gently over the charcoal boundary on the right side and a solution of sodium hydroxide is poured over the charcoal boundary on the left side as shown in Fig. 1.7(*b*). Current is allowed to flow by inserting a metal electrode in each limb, the anode in the right limb and the cathode in the left limb. As a result, the hydrogen and cupric ions move towards the left limb while the hydroxyl ions move towards the right limb. The movement of the hydrogen ions will be indicated by the disappearance of the pink colour and that of the cupric ions by the development of blue colour in the solution on the right side. The movement of the hydroxyl ions will be indicated by the appearance of pink colour of the solution on the left side. These changes are shown in Fig. 1.7(*c*).

If the experiment is performed carefully, it will be found that if, in a given time, the hydrogen ions move through a distance of 5 cm, the cupric ions will move through a distance of 0.91 cm, while the hydroxyl ions will move in the opposite direction through a distance of 2.83 cm.

1.6.1 Hittorf's Theoretical Device

It looks surprising at first sight that although most of the ions differ largely in their mobilities, *their equal numbers are discharged,* on electrolysis, at the appropriate electrodes (cf. Faraday's second law of electrolysis). Hittorf explained this anomaly by means of a theoretical device represented in Fig. 1.8. The electrolytic vessel contains equal number of positive and negative ions of the *same* valency. The two metal electrodes A and B represent the anode and cathode respectively. The vessel is supposed to be divided into three compartments by putting imaginary partitions at C and D which are permeable to ions. The three imaginary chambers AC, CD and DB are termed anodic, central and cathodic chambers, respectively.

Before electrolysis, the position of the solution is represented as at 1 in the figure. There are, say, 8 pairs of positive and negative ions in the anodic and cathodic chambes and 6 pairs in the central chamber.

Now, suppose that on applying potential, only cations are capable of moving and that in a given time, two of the cations, move towards the cathode. The condition will then be as represented in the figure at II.

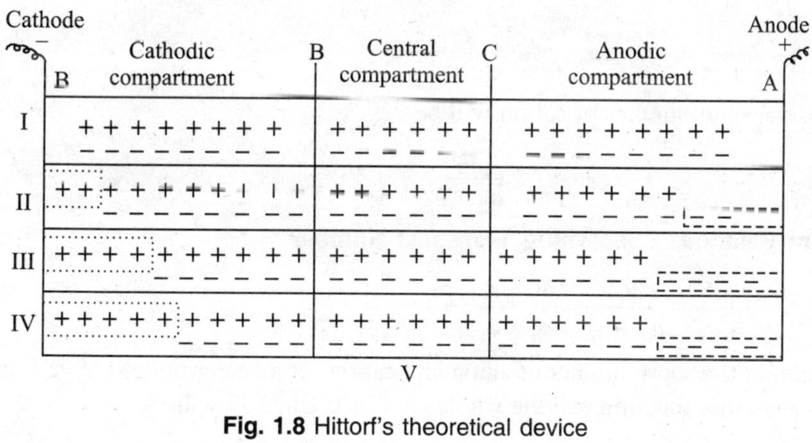

Fig. 1.8 Hittorf's theoretical device

There will be two unpaired cations at the cathode. There will also be two unpaired anions left at the anode. As unpaired ions always get discharged at the respective electrodes (by the gain or loss of electrons as the case may be), two cations will be discharged at the cathode and two anions will be discharged at the anode. Thus, the *same number of cations and anions will be discharged at the cathode and the anode respectively, even though only cations were able to move.*

Now, suppose that both cations and anions can move and that both have the same velocity. Thus, if two cations move towards the cathode, two anions will move towards the anode in the same time. The state of affairs will now be represented as at III. Now four cations and four anions, *i.e.,* the *same number*, will be discharged at the respective electrodes.

Finally, suppose that cations and anions move with different velocities so that if in a given time two cations move towards the cathode, three anions move towards the anode. In this case, too, the *same number, i.e.,* five unpaired cations and five unpaired anions will be discharged at the respective electrodes, as shown at IV.

In the above illustration, the concentration of the central compartment has remained constant. This condition is realised in practice provided the current strength is small and there are no temperature variations.

1.7 TRANSPORT NUMBER

As seen earlier, electric current is carried by ions. However, each ion carries a definite fraction of the total current depending upon its speed. The fraction of the total current carried by an ion is called its **transport number or Hittorf's number or Transference number.**

The greater the speed of the ion, the greater the amount of current it will carry with it and hence greater will be the transport number. Let u_a and u_c be the speed of the anion and the cation. Then

Current carried by the anion $\alpha u_a = ku_a$

Current carried by the cation $\alpha u_c = ku_c$

Since current is carried only by ions.

Total current carried = Current carried by anion + current carried by cation

$$= ku_a + ku_c = k(u_a + u_c).$$

Transport number of the anion, $t_a = \dfrac{ku_a}{k(u_a+u_c)}$

or
$$t_a \text{ or } t^- = \frac{u_a}{u_a+u_c} \qquad \text{...(1.9)}$$

Similarly, transport number of cation will be

$$t_c \text{ or } t_+ = \frac{u_c}{u_a+u_c} \qquad \text{...(1.10)}$$

Important Relations Concerning Transport Number

(*i*) $t^- + t^+ = \dfrac{u_a}{u_a+u_c} = \dfrac{u_c}{u_a+u_c} = \dfrac{u_a+u_c}{u_a+u_c} = 1$

That is, sum of transport number of anion and cations of an electrolyte is always unity. Thus, no current is lost is any other function and the whole current is carried by the ions.

(ii) $$t^- = \frac{u_a}{u_a + u_c} = \frac{1}{1 + \frac{u_c}{u_a}} = \frac{1}{1+r}$$

Thus, $$t^- = \frac{u_a}{u_a + u_c} = \frac{1}{1+r} \quad ...(1.11)$$

$$t^+ = \frac{u_c}{u_a + u_c} = \frac{r}{1+r} \quad ...(1.12)$$

1.7.1 Factors Controlling Transport Number

The transport number of an ion is directly proportional to the speed of the ion, which in turn depends upon the following factors.

(i) **Nature of the ion:** The smaller size of the ion the greater will be its speed and higher would be its transport number.

(ii) **Nature of the counter ion present:** The transport number of an ion depends not only upon the mobility of the ion alone, but also on the mobility of counter ion associated with it in the electrolyte as is evident from the following example.

For example, transport number of chloride ion in solution chloride is 0.604 and in hydrochloric acid, it is 0.166. This can be understood from the fact that the mobility of hydrogen ion is greater than that of sodium ion.

$$t_{Cl^-} \text{ (in NaCl)} = \frac{u_{Cl^-}}{u_{Cl^-} + u_{Na^+}}$$

$$t_{Cl^-} \text{ (in HCl)} = \frac{u_{Cl^-}}{u_{Cl^-} + u_{H^+}}$$

Obviously, the t_{Cl^-} in NaCl is greater than t_{Cl^-} in HCl as in the latter case, the denominator $(u_{Cl^-} + u_{H^+})$ is much greater than that in the former case $(u_{Cl^-} + u_{Na^+})$ while numerator u_{Cl^-} *i.e.*, the mobility of the chloride ion, is same in each case.

(iii) **Hydration of the ions:** Hydration of the ion affects the speed and thus the transport number of that ion. For instance, the size of Li^+ ions is smaller than Na^+ ion, but it is more prone to hydration. The transport number of Li^+ is expected to be higher based on its size, but in practice, the value has been found to be less. This is due to its greater tendency towards hydration.

(iv) **Concentration:** Although the effect of concentration on transport number is very small, yet sometimes it becomes significant due to the formation of complex ion with change in concentration.

It has been observed that for some salts the values of the transport numbers are abnormal in the sense that they are even greater than unity for cation and so are negative for anion. A simple case is that of the solution of cadmium iodide in which the transport number of cations is greater than unity and for the anions a negative value is obtained. This is because of the formation of the complex ions, *e.g.*,

$$CdI_2 + \text{aqueous} \longrightarrow Cd^{+2}(aq) + 2\ I^-(aq)$$

$$Cd^{2+}\ (aq) + 4I\ (aq) \longrightarrow Cd\ I^{+2}(aq)$$

At high concentration, cadmium exists as Cd^{-2} as well as CdI_4^{-2} so cadmium migrates towards both cathode and anode giving negative values for its transport numbers.

(*iv*) **Temperature:** With increase of temperature, the ions tend to approach a 0.5 value of transport number. It has been observed that ions having transport number less than 0.5 at ordinary temperature show an increase in the value, whereas the other ion which has transport number more than 0.5 shows a decrease in the value, with rise of temperature. It implies that the cation and the anion tend to equalize the value of transport number with increase in temperature. The effect of temperature is very small when the value of transport number of the cation and the anion is already 0.5 as in the case of KCl.

(*vi*) **Interionic effects:** The speed of the ions and consequently their transport numbers are affected due to interionic effects as discussed.

1.7.2 Determination of Transport Numbers By Hittorff Method

Principle: The method is based upon the principle that the fall in concentration around an electrode is proportional to the speed of the ion moving away from it.

$$\frac{\text{Fall in concentration around cathode}}{\text{Fall in concentration around anode}}$$

$$= \frac{\text{Speed of anion}}{\text{Speed of cation}} = \frac{u_a}{u_c} \qquad \text{...(1.13)}$$

Since transport number of cation

$$t^+ = \frac{\text{Speed of the cation}}{\text{Speed of the cation + Speed of the anion}}$$

Transport number of cation,

$$t^+ = \frac{\text{Fall around anode}}{\text{Fall around the anode}} + \text{Fall around the cathode}$$

$$= \frac{\text{Fall around anode}}{\text{Total fall around both the electrodes}}$$

If the concentration are measured in terms of gram equivalent, then

Transport number of cation, t^+

Number of gram equivalents lost from the anodic compartment = Total number of gram equivalent lost from both the compartment.

Further, the total number of gram equivalent lost from both the compartments is equal to the gram equivalent deposited on each electrode. This is clear from Hittorf's theoretical device since in each case considering the total number of ions from both anodic and cathodic compartments is equal to the number of ions discharged or deposited on each electrode.

It is more convenient to use a copper coulometer in series with the experimental solution in the determination of transport number since the number of gram equivalents deposited shall be equal to the number of gram equivalents of copper deposited in copper coulometer.

Therefore, the transport number of cation,

$$= \frac{\text{Number of gram equivalents lost from the Anode compartment}}{\text{Number of gram equivalents of copper deposited in the copper coulometer}}$$

Now, $t^+ + t^- = 1$ ...(1.16)

Transport number of anion = $1 - t^+$

Thus, transport number of the cation (t^+) can be determined by measuring fall in concentrations around anode and total amount deposited on an electrode.

It may be pointed out here that in order to apply the above relationship, the electrolysis must be carried out so that intermixing of the solution around the two electrodes is avoided completely.

The above equation is valid only if the electrodes are not attacked by the ions in solutions. If the anode is attacked by the anions, there will be an increase in the concentration around the anode instead of decrease. However, this increase in concentration at anode is equal to decrease in concentration that would have taken place if the anode were inert. Examples are as in the case of electrolysis of $AgNO_3$ using silver anode.

Experimental Procedure

The most widely applicable form of Hittorf's apparatus is shown in Fig. 1.9. It consists of a two-limbed vessel connected by a narrow U-tube in the middle. Both limbs as well as the U-tube are provided with stop cocks at the bottom for the withdrawal of the solution. Two limbs are provided either with Pt electrodes (non-attackable) or with suitable attackable electrodes. Suppose we want to determine the transport number of Ag^+ in $AgNO_3$. Then, the electrodes are either of Pt or pure Ag.

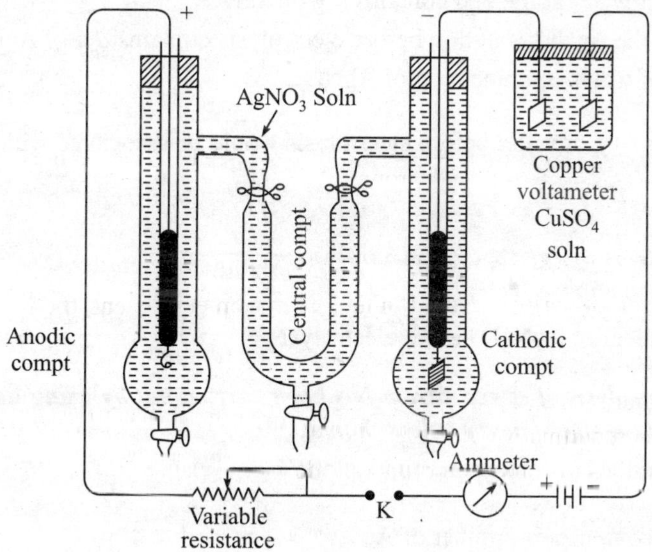

Fig. 1.9 Hittorf's apparatus of measuring transport number

The apparatus in filled with a standard solution of $AgNO_3$ (usually 0.IN) and the cell is connected in series with a source of direct and a milliammeter as shown. A low current (10–15 mA) is passed for 2–3 h so as to produce an appreciable (but not too large change in concentrations). Low current is used to minimize thermal and resulting diffusion effects.

After the electrolysis has taken place, the solutions from anodic as well as central compartments are withdrawn separately by opening stopcock at the bottom and each analyzed by weighing and titration with standard ammonium thiocyanate (NH_4CNS) solution. The change in the amount of electrolyte in the anode compartment is thus obtained. There should be no change in the concentration of the solution taken out from the central compartment (*i.e.*, U-tube). It is important that the amount of solution withdrawn from anode for analysis should include all the solution whose composition has changed. This can be checked by withdrawing a further portion of solution from the anode limb and finding whether it has the same composition as the original $AgNO_3$ solution or not. The weight of Cu deposited in the voltameter is accurately determined.

Observations and Calculations

With the change in concentration of a solution, the density of the solution also undergoes a change and consequently the volume also changes. The change in concentration is thus determined with reference to a definite weight of the solvent. Therefore, while determining the transport number of ions we take into account the known weight of the solution and not the known volume of the solution before and after the electrolysis.

(*i*) Let the weight of copper deposited on cathode of copper voltameter = W g

the equivalent of the copper deposited in copper voltameter = $\dfrac{W}{31.75} = z$

(*ii*) Let *a* g of anode solution after electrolysis contains *b* g of $AgNO_3$.

Wt. of water = $(a - b)$ g

$(a - b)$ g of water at the end contains *b* g of $AgNO_3$

(*iii*) Let *c* g of the $AgNO_3$ solution before electrolysis contains *d* g of $AgNO_3$.

$(c - d)$ g of water contained *d* g of $AgNO_3$.

Hence $(a - b)$ g of water before electrolysis would be associated with $\dfrac{d}{(c-d)} \times (a-b)$ g of $AgNO_3$

$\therefore$ *Equivalents of* $AgNO_3$ *after electrolysis* = $\dfrac{b}{170} = y$

and equivalent of $AgNO_3$ before electrolysis = $\dfrac{d}{(c-d)} \times \dfrac{(a-b)}{170} = x$

(*a*) ***If the electrolysis of silver nitrate has been carried out by using unattackable electrodes, i.e., Platinum electrodes, y is less than x.***

$\therefore$ Fall in the concentration round anode = $(x - y)$

Hence transport number of Ag^+, $t^+_{Ag} = \left(\dfrac{x-y}{z}\right)$

(*b*) ***In case the electrodes are attacked by the ions discharged*** *i.e.*, the electrolysis of $AgNO_3$ is carried out using Ag electrodes, the nitrate ions react with the Ag anode dissolving Ag^+ ions in the anodic compartment. Therefore, the concentration of anodic compartment increases and so $y > x$.

$\therefore$ The increase in concentration round the anode = $(y - x)$ equivalents. The fall in concentration of $AgNO_3$ in the anodic compartment due to the migration of the Ag^+ ions may be calculated as follows:

The amount of $AgNO_3$ formed by the discharge of NO_3^- ions at the anode

= Number of the NO_3^- ions discharged

= Total $AgNO_3$ electrolysed = z (say)

If no silver ion had migrated, then increase is the anodic compartment should have been = z equivalent *i.e.,* the theoretical increase in the anodic compartment = z equivalent

But the observed increase in the concentration of = $(y - x)$ equivalent. Therefore, the fall in concentration round the anode due to the migration of Ag^+ ions from anodic to the cathodic compartment = $z - (y - x)$ equivalent.

Therefore, the transport number of silver, $t_{Ag^+} = \dfrac{z-(y-x)}{z}$

Transport number of NO_3^-, $t_{NO_3^-} = 1 - t_{Ag^+}$.

NUMERICAL PROBLEMS

(A) Attackable Electrodes

Example 5: ***An aqueous solution of copper sulphate containing 5.3 g of the salt in 500 g solution was electrolyzed between copper electrodes. At the end of the experiment 54.47 g of anode solution was found to contain 5.726 g of the salt. During the experiment, 0.5101 g silver was deposited in a silver coulometer placed in series. Calculate the transport number of cupric ions.***

Solution: After electrolysis, weight of anodic solution = 54.57 g

Weight of $CuSO_4$ in it = 5.726 g

Weight of water = 54.57 – 5.726 = 48.844 g

$$\text{Weight of } CuSO_4 \text{ per gram of water} = \frac{5.726}{48.844} = 0.1172 \text{ g}$$

Before electrolysis, weight of solution = 500 g

Weight of $CuSO_4$ = 5.3 g

Weight of water = 500 – 5.3. = 494.7 g

$$\text{Weight of } CuSO_4 \text{ per gram of water} = \frac{5.3}{494.7} = 0.0107 \text{ g}$$

$$\text{Increase in the conc. of } CuSO_4 \text{ around anode} = 0.1172 - 0.0107 = 0.1065 \text{ g } \frac{0.1065}{79.75}$$

$$= 0.00136 \text{ g eq.}$$

$$\text{Equivalent wt of } CuSO_4 = \frac{63.5 + 32 + 64}{2} = 79.75$$

$$\text{Weight of Ag deposited in coulometer} = 0.5101 \text{ g} = \frac{0.5101}{108} = 0.00472 \text{ g eq.}$$

Fall in conc. around anode due to migration of Cu^{+2} ions

$$= 0.00472 - 0.00136 = 0.00336 \text{ g eq.}$$

$$\text{Transport No. of } Cu^{+2} \text{ ions} = \frac{\text{g eq. of } Cu^{+2} \text{ ions lost from anodic solution}}{\text{g eq. of } Ag^{+} \text{ ions deposited}}$$

$$= \frac{0.00336}{0.00472} = \mathbf{0.712}$$

$$\text{Transport No. of } SO_4^{2-} \text{ ions} = 1 - 0.712 = \mathbf{0.288}$$

Example 6: ***A solution of $AgNO_3$ containing 0.00739 g of the solute per gram of water was electrolyzed between Ag electrodes. During the experiment, 0.078 of Ag got deposited at the cathode. At the end of the experiment, the anode portion contained 23.14 of water and 0.236 of $AgNO_3$. What are the transport numbers of Ag^+ and NO_3^- ions?***

Solution: After electrolysis, 23.14 g of water contains 0.236 g of $AgNO_3$

$$1 \text{ g of water contains} = \frac{0.236}{23.14} = 0.0102 \text{ g of } AgNO_3$$

Before electrolysis, 1 g of water contains

$$= 0.0074 \text{ g of } AgNO_3$$

Increase of $AgNO_3$ around anode = 0.0102 – 0.0074

$$= 0.0028 \text{ g}$$

$$= \frac{0.0028}{170} = 0.0000165 \text{ g eq.}$$

Weight of Ag deposited at the cathode

$$= 0.078 \text{ g}$$

$$= \frac{0.078 \text{ g}}{170} = 0.0000458 \text{ g eq.}$$

Fall in conc. around anode due to migration of Ag^+ ions

$$= 0.000458 - 0.0000165 = 0.0000293 \text{ g eq.}$$

$$\text{Transport No. of } Ag^+ \text{ ions} = \frac{\text{g eq. of } Ag^+ \text{ ions from anodic solution}}{\text{g eq. of Ag deposited at the cathode}}$$

$$= \frac{0.0000293}{0.0000458} = \mathbf{0.6397}$$

$$\text{Transport number of } NO_3^- \text{ ions} = 1 - 0.6397 = \mathbf{0.3603}$$

(B) Unattackable Electrodes

Example 7: ***During electrolysis of a solution of potassium chloride using platinum electrodes 0.0137 g of the chloride ion was lost from the anodic compartment and 0.0857 g of silver was deposited in a silver coulometer attached is series with the cell. Determine the transport number of K^+ and Cl^- ions.***

Solution: Wt. of Cl^- ions lost from the anodic compartment = 0.0137 g.

$$\text{Number of g eq. of } Cl^- \text{ ions lost} = \frac{0.0137}{35.7} = 0.0003859 \qquad \text{(Eq. wt. of Cl = 35.5)}$$

Number of g eq. of Ag deposited in the coulometer

$$= \frac{0.0857}{108} = 0.0007935 \qquad \text{(Eq. wt. of Ag = 108)}$$

$$\text{Transport number of } K^+ = \frac{\text{g eq. lost from anodic compartment}}{\text{g eq. deposited in the coulometer}}$$

$$= \frac{0.0003859}{0.0007935} = \mathbf{0.0486}$$

$$\text{Transport number of } Cl^- = 1 - \text{Transport number of } K^+$$

$$= 1 - 0.0486 = \mathbf{0.0514.}$$

Example 8: ***A solution of silver nitrate containing 12.14 g of silver in 50 ml of solution was electrolysed between platinum electrodes. After electrolysis, 50 ml of the anode solutions was found to contain 11.55 g of Ag, while 1.25 g of metallic silver was deposited on the cathode. Calculate the transport number of Ag^+ and NO_3^- ions.***

Solution: Wt. of Ag is 50 ml of the solution before electrolysis = 12.14 g

Wt. of Ag is 50 ml of the solution after electrolysis = 11.55 g

Fall in the concentration of Ag = 12.14 – 11.55 = 0.59 g

$$= \frac{0.59}{180} = 0.0055 \text{ g eq.} \qquad \text{(Eq. wt. of Ag = 180)}$$

$$\text{Wt. of Ag deposited in the cathode} = 1.25 \text{ g} = \frac{1.25}{180} = 0.0116 \text{ g eq.}$$

$$\text{Hence, transport number of } Ag^+ \text{ ion} = \frac{\text{Fall in conc. around anode}}{\text{Number of g. eq. of Ag deposited}} = \frac{0.0055}{0.0116}$$

$$= \mathbf{0.474}$$

$$\text{Transport number of } NO_3^- \text{ ion} = 1 - 0.474 = \mathbf{0.526}$$

1.8 KOHLRAUSCH LAW

We know that equivalent conductance of the electrolytes increases with dilution until at a limiting value it becomes constant. This limiting value (or maximum value) is called equivalent conductance at infinite dilution.

Kohlrausch observed a regularity in the equivalent conductance at infinite dilution of pairs of salts having a common ion. Table 1.1 shows equivalent conductances at infinite dilution of electrolytes at 298 K. Table 1.1 also shows that the difference in conductance at infinite dilution of potassium and sodium salts having a common anion gives the same value (23.41). It is independent of the anion. In the same manner, the difference in equivalent conductances at infinite dilution of pairs of salts having the anions bromide and chlorides (2.06) is independent of the cations. Kohlrausch observed that difference between the Λ_0 value for two salts, (which were strong electrolytes) having the same cation, is always constant.

Table 1.1 Equivalent conductances at infinite dilution at 298 K (in $ohms^{-1}\ cm^2\ eq^{-1}$).

Pairs of Electrolytes Having Same Anions			*Pairs of Electrolytes Having Same Cations*		
Electrolytes	Λ_0	*Difference*	*Electrolyte*	Λ_0	*Difference*
KBr	151.92	23.41	NaBr	128.51	2.06
NaBr	128.51	—	NaCl	126.45	—
KCl	149.86	23.41	KBr	151.92	2.06
NaCl	126.45	—	KCl	149.86	—
KNO_3	144.96	23.41	HBr	428.22	2.06
$NaNO_3$	121.55	—	HCl	426.16	—

From these observations, Kohlrausch concluded that Λ_0 value of an electrolyte is the sum of two terms, one for the anion and other for the cation. These terms are known as ionic conductances of the ions concerned. This was first noticed by Kohlrausch and the following law was put forward by him. *At infinite dilution when electrolyte is fully dissociated, each ion makes a definite contribution towards equivalent conductance of the electrolyte, irrespective of the nature of the other ion with which it is associated. In other words, the value of equivalent conductance at infinite dilution of an electrolyte is equal to the sum of two values, one depending upon cation and other upon anion at a constant temperature.* Thus, mathematically

$$\Lambda_0 = \lambda_\alpha^0 + \lambda_c^0$$

where λ_α^0 and λ_c^0 signify the ionic conductance of the anion and cation respectively at infinite dilution and are expressed in $ohm^{-1}\ cm^2$ or mhos cm^2. The law is also known as law of independent migration of ions or the law of independent ionic mobilities.

It may be emphasised that *ionic conductance is not the same thing as ionic mobility.* They are, however, proportional to each other. Thus, at infinite dilution,

$$\lambda_c^\circ = kU_c^\circ$$

and
$$\lambda_a^\circ = kU_a^\circ$$

where U_c° and U_a° are the mobilities of cationa nd anion, respectively, at infinite dilution.

Ionic conductance is measured in terms of mhos while ionic mobility is expressed in cm. per sec. under a unit potential gradient, as already defined. The value of k is known to be 96500, *i.e.,* 1 faraday.

$$U^\circ = \frac{\lambda^\circ}{F} = \frac{\lambda^\circ}{96500}$$

1.8.1 Relation between Ionic conductance and Transport Number

Ionic conductance is directly proportional to the absolute velocities of ions, *i.e.,*

$$\Lambda_a \propto u_a \text{ or } \Lambda_a = Ku_a.$$

and
$$\Lambda_c \propto u_c \text{ or } \Lambda_c = Ku_c.$$

By Kohlrausch's Law;

$$\Lambda_0 = \Lambda_a + \Lambda_c = K(u_a + u_c)$$

Hence
$$\frac{\Lambda_a}{\Lambda_0} = \frac{Ku_a}{K(u_a + u_c)} = \frac{u_a}{u_a + u_c} = t$$

or $$\Lambda_a = t^- \times \Lambda_0 \quad \text{...(1.11)}$$

Again $$\frac{\Lambda_c}{\Lambda_0} = \frac{Ku_c}{K(u_a + n_c)} = \frac{u_c}{u_a + n_c} = t^+$$

or $$\Lambda_c = t^+ \times \Lambda_0 \quad \text{...(1.12)}$$

Example 9: ***The transport number of silver ion is 0.466 and that of nitrate is 0.534 at 18°C. The equivalent conductance of a very dilute solution of $AgNO_3$ is 115.8 at the same temperature. Calculate the ionic conductance of silver and nitrate ion.***

Solution: We know that

Ionic conductance = Transport no. of the ion × Λ_0 of strong electrolyte containing that ion

i.e., $$\Lambda_a = t^- \times \Lambda_0$$

or $$\Lambda_{NO_3^-} = 0.534 \times 115.8 = 61.837 \text{ mhos.}$$

and $$\Lambda_c = t^+ \times \Lambda_0$$

or $$\Lambda Ag^+ = 0.466 \times 115.8 = 53.962 \text{ mhos.}$$

$$\text{Absolute velocity } (u) = \frac{\text{Ionic mobility } (\Lambda)}{96500}$$

$$\text{Absolute velocity of } Ag^+ = \frac{53.962}{96500} = 0.000559 \text{ cm/sec.}$$

$$\text{and absolute velocity of } NO_3^- = \frac{61.837}{96500} = 0.000641 \text{ cm/sec.}$$

1.8.2 Applications of Kohlrausch's Law

(1) Calculation of equivalent conductance at infinite dilution (Λ_0) for weak electrolytes: Weak electrolytes do not possess sufficient conductance and moreover it does not become constant even at very high dilutions. Hence practically it is not possible to determine their values directly. Kohlrausch's law helps in the determination as follows:

$$\Lambda_0 = \Lambda_a + \Lambda_c \quad \text{...(Kohlrausch's Law)}$$

Thus the equivalent conductance of the weak electrolytes at infinite dilution can be known by knowing the ionic conductances of the anion (Λ_a) and the cation (Λ_c) of that weak electrolyte. These can be found by making use of the relationship:

Ionic conductance of a cation = Equivalent conductance at infinite dilution of any strong electrolyte containing that ion × Transport number of that ion.

For example, in case of acetic acid (which is a weak electrolyte)

$$\Lambda_0 \text{ for } CH_3COOH = \Lambda_{CH_3COO^-} + \Lambda_{H^+}$$

To know $\Lambda_{CH_3COO^-}$ and Λ_{H^+} ions, the equivalent conductance of strong electrolytes CH_3COONa (containing CH_3COO^- ion) and HO (containing H^+ ion) at infinite dilution are determined and then

$$\Lambda_{CH_3COO^-} = \Lambda_0 \text{ for } CH_3COONa \times t_{CH_3COO^-}$$

and

$$\Lambda_{H^+} = \Lambda_0 \text{ (for HCl)} \times t_{H^+}$$

Example 10: ***At 25°C eq. conductance at infinite dilution of HC and CH_3COONa are 400 and 85 mhos respectively. If the transport number of H^+ and CH_3COO^- are 0.8 and 0.4 respectively, calculate the eq. conductance of acetic acid at infinite dilution.***

Solution: We have,

$$\Lambda_{H^+} = \Lambda_0 \text{ of HCl} \times t_{H^+} = 400 \times 0.8 = 320 \text{ mhos.}$$

$$\Lambda_{CH_3COO^-} = \Lambda_0 \text{ for } CH_3COONa \times t_{CH_3COO^-}$$

$$= 85 \times 0.4 = 34 \text{ mhos.}$$

$$\Lambda_0 \text{ for } CH_3COOH = \Lambda_{H^+} + \Lambda_{CH_3COO^-} \quad \text{[Kohlrausch's Law]}$$

$$= 320 + 34 = \mathbf{354 \text{ mhos.}}$$

Another method of calculating the equivalent conductance at infinite dilution (Λ_0) for weak electrolytes.

Another way to calculate the equivalent conductance at infinite dilution of the weak electrolytes is from the conductance data only. An example of acetic acid is taken to illustrate it. Three strong electrolytes are properly chosen:

(*i*) HCl which contains the H^+ and Cl^-

(*ii*) Sodium acetate which contains the CH_3COO^- and Na^+

(*iii*) Sodium chloride containing Na^+ and Cl^-.

The equivalent conductances at infinite dilution of the above three strong electrolytes HCl, CH_3COONa and NaCl are experimentanally determined.

Then on adding these values of first and second and subtracting that of the third gives the values of Λ^0 of acetic acid.

Thus

$$\Lambda^0_{HCl} = \Lambda_{H^+} + \Lambda_{Cl^-} \quad ...(i)$$

$$\Lambda^0_{CH_3COONa} = \Lambda_{Na^+} + \Lambda_{CH_3COO^-} \quad ...(ii)$$

$$\Lambda^0_{NaCl} = \Lambda_{Na^+} + \Lambda_{Cl^-} \quad ...(iii)$$

Adding (*i*) and (*ii*) and subtracting (*iii*) we have value ;

$$\Lambda^0_{HCl} + \Lambda^0_{CH_3COONa} - \Lambda^0_{NaCl}$$

$$\Lambda_{H^+} \Lambda_{CHCOO^-} = \Lambda^0 \text{ for } CH_3COOH$$

Example 11: ***If the equivalent conductance of HCl, CH_3COONa and NaCl are 426.16, 91.0 and 126.45 repectively, what will be that of CH_3COOH?***

Solution:

$$\Lambda^0_{HCl} = 426.16 = \Lambda_{H^+} + \Lambda_{Cl^-} \quad ...(i)$$

$$\Lambda^0_{CH_3COONa} = 91.0 = \Lambda_{Na^+} \Lambda_{Na_2COO^-} \quad ...(ii)$$

$$\Lambda^0_{NaCl} = 126.45 = \Lambda_{Na^+} + \Lambda_{Cl^-} \quad ...(iii)$$

Adding (*i*) and (*ii*) and subtracting from (*iii*), we have

$$\Lambda_{H^+}\Lambda_{CH_3COO^-} = \Lambda^0 \text{ of acetic acid}$$
$$= 426.16 + 91.00 - 126.45 = 390.71 \quad \text{mhos}$$

(2) Calculation of absolute ionic mobilities. (velocities): The ionic mobilities of Kohlraush are not the velocities of ions. The absolute ionic mobility of an ion represent the velocity of an ion in cm./sec. under a potential gradient of 1 volt per cm. [Potential gradient = E.M.F./distance between electrodes]. As said above,

$$\Lambda_a = K \times u_a$$

and

$$\Lambda_c = K \times u_c$$

It is found that under a potential gradient of 1 volt per cm the value of K, is the charge on one gram equivalent of ion *i.e.,* one Faraday or 96,500 Coulombs.

$$\text{Therefore, absolute ionic mobility} = \frac{\text{Ionic conductance}}{96500} \quad ...(1.13)$$

Example 12: ***The ionic conductance of Na^+ and Cl^- ions are 43.4 and 65.6 respectively. Calculate their absolute ionic mobilities.***

Solution: Absolute ionic mobility of Na^+ (u_{Na^+}) will be

$$u_{Na^+} = \frac{\Lambda_{Na^+}}{96500} = \frac{43.4}{96500} = 0.000449 \text{ cm/sec.}$$

Absolute ionic mobility of Cl^- (u_{Cl^-}) will be

$$u_{Cl^-} = \frac{65.6}{96500} = 0.000679 \text{ cm/sec.}$$

(3) Degree of dissociation and ionisation constant: An equilibrium exists between the undissociated and dissociated ions of a weak electrolyte and the degree of dissociation (α) of a weak electrolyte is given by the relationship,

$$\alpha = \frac{\Lambda}{\Lambda_0},$$

where Λ is equivalent conductance at a given dilution or concentration obtained by adding the ionic conductances of the electrolyte.

The dissociation constant K for a weak acid or for a weak base is given by equation:

$$K = \frac{Ca^2}{1-a}$$

where (α) represents degree of ionisation and is given by the relation: $\alpha = \frac{\Lambda_c}{\Lambda_0}$

Λ_0 can be calculated by Kohlrausch's law $\Lambda_0 = \Lambda_a + \Lambda_c$. The values of ionic conductances can be looked from tables.

Example 13: ***The ionic conductance of Na^+ and Cl^- at 15°C is 43.4 and 65.5 respectively. The eq. conductance at some particular dilution 96.8 mhos. Calculate the degree of ionisation at that dilution.***

Solution: According to Kohlrausch's law,

$$\Lambda_0 \text{ for NaCl} = \Lambda_{Na^+} + \Lambda_{Cl^-} = 43.4 + 65.5 = 108.9$$

But $\Lambda_c = 96.8$ (given)

$\therefore$ α, the degree of ionisation $= \frac{\Lambda_c}{\Lambda_0} = \frac{96.8}{108.9} = 0.889$

(4) Calculation of ionic product of water: Ionic product of water (K_w) is defined as the product of the H^+ is concentration and OH^- ion concentration in moles per litre of an aqueous solution *i.e.,*

$$[H^+][OH^-] = K_w$$

At a given temperature, the value of K_w is constant. Its numerical value can be calculated by conductivity measurements applying Kohlrausch law.

The specific conductance of pure water at 25°C has been found out experimentally to be $5.54 \times 10^{-8}\ ohm^{-1}\ cm^{-1}$. The conductance of one litre of water would be $5.54 \times 10^{-8} \times 5.54 \times 10^{-5}\ ohm^{-1}$.

The equivalent conductance at infinite dilution (λ_α) of water can be calculated by Kohlrausch law.

$$\lambda_\alpha \text{ of water} = \lambda_{H^+} + \lambda_{OH^-} = 349.8 + 198.5 = 548.3\ ohm^{-1}$$

This is the conductance of water due to one gram equivalent of H^+ and one gram equivalent of OH^- ions. The conductance is proportional to the number of ions. So the gram equivalents H^+ ions per litre of waer of conductivity $5.54 \times 10^{-5}\ cm^{-1}$ can be known.

$$\text{Gram equivalents of } H^+ \text{ ion per litre of water} = \frac{5.54 \times 10^{-5}}{548.3} = 1.01 \times 10^{-7}$$

The number of gram equivalents of OH^- ions will also be the same.

Hence $K_w = [H^+][OH^-] = 1.01 \times 10^{-7} \times 1.01 \times 10^{-7}$

$= 1.02 \times 10^{-14}$ at 25°C $= 1 \times 10^{-14}$

(5) Calculation of solubility of sparingly soluble salts: Substances such as AgCl, $BaSO_4$, $CaSO_4$, $CaCO_3$ etc., which are known as insoluble substances possess some definite solubility though very small. By Kohlrausch's law, it has been possible to calculate the solubility by determining experimentally the specific conductance of its saturated solution.

Let k be sp. conductance of the salt containing C equivalent per litre of the solution.

$\therefore$ Equivalent conductance at this dilution $= k \times \frac{1000}{C}$.

because volume of the solution containing 1 equivalent $= \frac{1000}{C}$.

As the solubility of the salt is very small, therefore, it is supposed to be completely dissociated into ions even in saturated solution. In such a case the equivalent conductance at this dilution is equal to equivalent conductance at infinite dilution (Λ_0)

or $\Lambda = \Lambda_0$

By Kohlrausch's Law,

$$\Lambda_0 = \Lambda_a + \Lambda_c$$

Now $\Lambda_0 = k \times \frac{1000}{C}$

$$\therefore \qquad C = \frac{k \times 1000}{\Lambda_0} = \frac{k \times 1000}{\lambda_a + \lambda_c} \qquad ...(3.14)$$

Thus value of C is obtained in terms of equivalents per litre. Multiplying C equivalents litre by equivalent weight we get solubility in terms of g/litre. We may calculate the solubility in moles per litre from which solubility product can be calculated.

Example 14: ***Specific conductance of a saturated solution of AgCl at 25°C was found to be 3.41 × 10^{-6} ohm^{-1} cm^{-1}. The sp. conductance for water used to make up the solution was 1.60 × 10^{-6} ohm^{-1} cm^{-1}. Determine the solubility of AgCl in water. Ionic mobilities of Ag^+ and Cl^- at 25°C are 60.3 and 78.0 ohm^{-1} cm^{-1}.***

Solution: Let V be the volume of saturated solution in which one g eq. of AgCl, *i.e.*, 143.5 g has been dissolved.

Now,
$$\Lambda_0 = \Lambda_{Ag^+} + \Lambda_{Cl^-} \text{ (Kohlrausch's law)}$$
$$= 60.3 + 78.0 = 138.3$$
$$k \text{ for AgCl} = k \text{ for solution} - k \text{ for } H_2O$$
$$= (3.41 - 1.60) + 10^{-6} = 1.81 \times 10^{-6}$$
$$\therefore \qquad \Lambda_0 = 1.81 \times 10^{-6} \times \frac{1000}{C}$$
$$\therefore \qquad 1.81 \times 10^{-6} \times \frac{1000}{C} = 138.3$$
$$C = \frac{1.81 \times 10^{-6} \times 1000}{138.3} \text{ equivalents of AgCl/litre}$$
$$= \frac{1.81 \times 10^{-6} \times 1000}{138.3} \times 143.5 = 0.0116 \text{ g/litre.}$$

1.9 INTERIONIC ATTRACTION THEORY OF CONDUCTANCE OR MODERN THEORY OF STRONG ELECTROLYTES

A satisfactory explanation for the behaviour of strong electrolytes at different concentration was dealt with quantitatively by ***Debye-Huckesl-Onsagar***. Here, we shall deal with only the qualitative aspects of interionic attraction theory of conductance. This theory is based on the realization of the following facts:

(*i*) Although the strong electrolytes are completely ionized at all dilutions, but at present it is believed that there is a very small amount of ionised electrolyte also present. Therefore, instead of believing that strong electrolytes are completely ionised, we should believe that they are *almost completely ionised.*

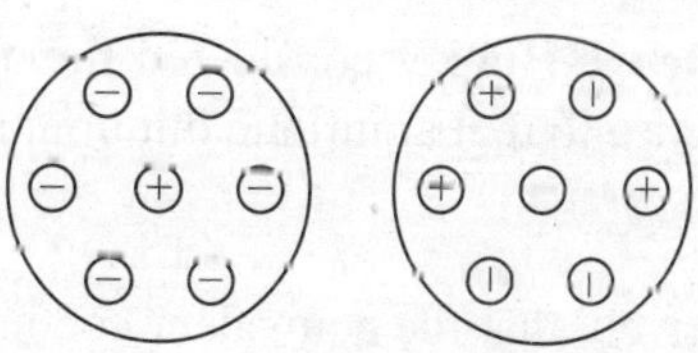

Fig. 1.10 Representation of situation for the ions in solution at rest *i.e.*, when no applied field. (*a*) ion atmosphere for positive ion; (*b*) ion atmosphere for negative ion

(*ii*) Because of attraction between positive and negative ions, each ion in solution is surrounded by ions of opposite charge, whose net charge is equal to that of the central ion. Thus, *there is present an ionic atmosphere of opposite charge around any ion*. In the absence of any external field applied, the ionic atmosphere has a central symmetry, *i.e.*, atmosphere of opposite charges is spherically and symmetrically distributed about any ion (Fig. 1.10).

(*iii*) When the potential is applied across two electrodes immersed in the solution of the electrolyte for conductance measurement, the above situation of the ions in solutions gets disturbed, thus giving rise to two-fold effects, namely (*a*) *the relaxation or asymmetry effects*, and (*b*) *the electrophoretic effect*. Both these effects finally result in the decrease of the speed of the ions and so their conducting power.

(*a*) **Relaxation effect:** This arises due to occurrence of the following situation. When the current is passed through the solution, positive ions will tend to move towards cathode and, the negative ions towards the negative electrode, *i.e.*, anode. Now, both the ions are surrounded by an atmosphere of opposite charge, so that under an applied potential, the central ion will tend to move in the direction opposite to the movement of the atmosphere. That is to say, when the positive central ion moves towards the cathode, its ionic atmosphere will move towards the anode. During its motion, the central ion will build up its ionic atmosphere again whereas the old one will die out. If this formation of the new atmosphere and the destruction of the previous one occur simultaneously, there will be no retarding force affecting the movement of the central ion. However, there is a time gap or the so-called time of relaxation between the formation of new ionic atmosphere and the breaking up of the old.

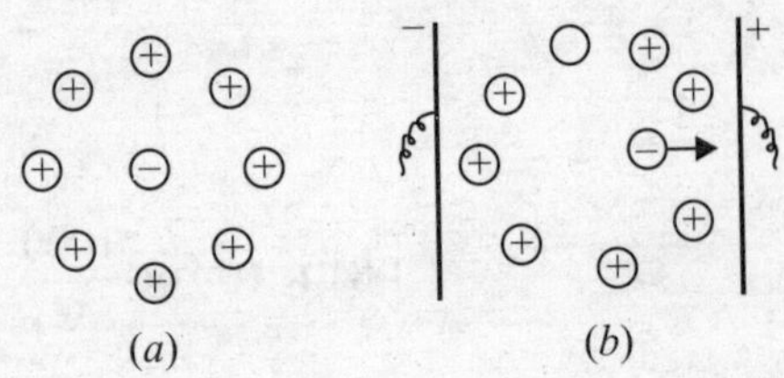

Fig. 1.11 (*a*) Symmetrical ionic atmosphere at rest; (*b*) Asymmetrical ionic atmosphere under the influence of applied field.

Thus, the symmetry of the ionic atmosphere will be disturbed and a new situation arises as shown in Fig. 1.11. The ionic atmosphere is no longer uniform around the central ion and consequently the ion experiences a retarding force against the direction of its motion and the ion slows down by these interionic attractions.

(*b*) **Electrophoretic effect:** This is another factor, which also oppose the motion of the moving ions. As the ions are solvated in solution and when the ions move, they carry these solvent molecules with them. In this way, any ion under the influence of applied field does not move through a stationary medium. Thus, a positive ion moving to the cathode has to make its way through the stream of the solvent molecules with the negative ions migrating towards the anode in the opposite direction. Similarly, the negative ions have to make their way through the solvent molecules carried by the positive ions in the opposite direction.

Debye-Huckel considered the above factors and obtained an equation relating the observed conductance to that at infinite dilution, thus showing

$$\Lambda_c = \Lambda_0 - A\sqrt{c} \qquad \text{...(1.15)}$$

quantitatively that the equivalent conductance depends on the square root of the concentration of the electrolyte. A is a constant and depends on the dielectric constant and viscosity of the medium. However, when the slopes of the straight lines in the plot of Λ vs. $\sqrt{c}$ were calculated based on the above equation, it agreed with the observed behaviour only up to concentrations of about 0.1 M. Onsagar modified the Debye-Huckel equation, which is known as **Deby Huckel-Onsagar equation or simply as Onsagar equation.** For a univalent electrolyte such as KCl, it is

$$\Lambda_c = \Lambda_0 - \left[\frac{82.4}{(DT)^{1/2}\,n} + \frac{8.20\times10^5}{(DT)^{3/2}}\right]\sqrt{c} \qquad \text{...(1.16)}$$

where D is the dielectric constant, T is the temperature in Kelvin scale, Λ_c is the equivalent conductance at concentration c in moles/litre and Λ_0 is the equivalent conductance at infinite dilution.

D and n are constants for a particular solvent, so at constant temperature the above equation is reduce to

$$\Lambda_c = \Lambda_0 - [A + B\Lambda_0]\sqrt{c} \qquad \text{...(1.17)}$$

where A and B are constants depending on the temperature and the nature of solvent.

Significance of Debye-Huckel Theory:

1. This equation explains the fact that the equivalent conductance of a strong electrolyte at any concentration (Λ_c) is less than the limiting value of Λ_0 at infinite dilution in spite of the almost complete ionization of strong electrolyte.
2. Decrease in equivalent conductance with increase in concentration is due to fall is mobilities of the ions due to greater inter-ionic attractions and vice versa.
3. The ratio Λ_c/Λ_0 does not correctly represent the degree of ionisation (α) for strong electrolytes, but it is only a conductivity ratio.
4. Inspite of almost complete dissociation of strong electrolytes, λ_0 is much less than λ_α.

1.10 HYDRATION OF IONS

It has been observed that ions in any solution, aqueous or non-aqueous are not present as such, they are always associated with molecules of the solvent in which the substance is dissolved. *The process by which ions are surrounded by the molecules of the solvent through weak Vander Waal's forces between the ions of the solute and molecules of the solvent is called* ***Solvation***. In general, this term is applied in case of non-aqueous solution. In case of aqueous solution, the term used is **hydration** of ions. *Hydration implies that the ions in solution with water as a solvent are always surrounded by the molecules of water through weak Vander Waal's interaction between the ions and the molecules of water.* We know that each ion is a charged particle having either positive or negative charge on them. The positively charged ions are called **Cations** and negatively charged ions are called **anions.** Water is a polar molecule having a slight positive charge with the two H atoms at the ends and negative charge with the central oxygen atom. Oxygen being more electronegative than H, the shared pair of electrons is displaced towards the oxygen atom. It may be represented as

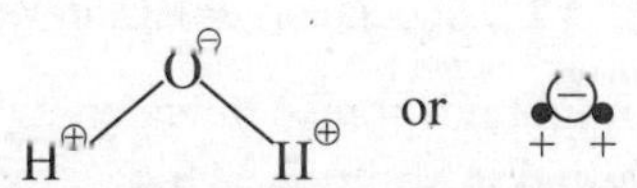

On dissolving any ionic solid (say NaCl, KNO_3), it breaks into charged particles called ions in solution. According to Arrhenius theory of ionisation, *the mere process of dissolution of any electrolyte breaks it up into the oppositely charged ions held together in the crystal of that solid.* These ions get surrounded by the molecules of the solvent and there are interactions between them. The hydration of Na^+ ions and Cl^- ions from NaCl has been depicted in Fig. 1.12.

As shown in the figure, the negatively charged oxygen atom of water molecules surround the positively charged Na^+ ion. Similarly, the positively charged hydrogen atoms of the water molecule are directed towards the negatively charged Cl^- ion. These hydrated ions move about in all possible directions [random motion] in the solution. On passing electric current, they move towards the oppositely charged electrodes. The positively charged cations move towards the cathode and negatively charged anions move towards the anode. On reaching their respective electrodes, they are discharged and are said to be electrolysed.

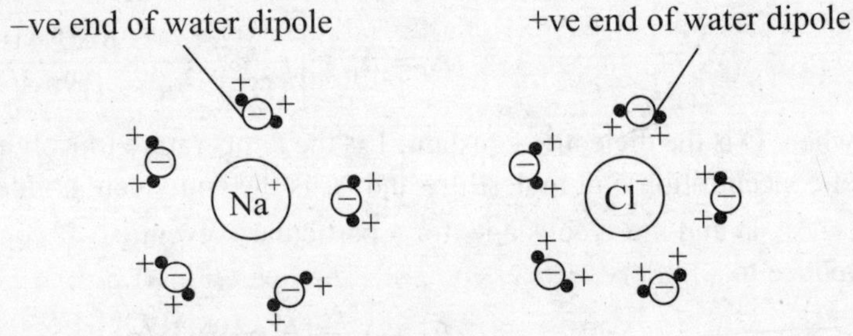

Fig. 1.12 Hydration of Ions

It is clear that *current is conducted in solutions of the electrolytes through the movement of ions.* Due to hydration of the ions, their mobility is affected which affects the conductance of current in the solution.

To cite another example of hydration of ions, let us consider the dissociation of hydrochloric acid [a covalent compound] in water. When HCl gas is dissolved into water, HCl molecules collide with water molecules. Collisions of sufficient energy and proper orientation produce hydrated protons called ***hydronium ions*** and chloride ions.

All the acidic properties shown by HCl acid are infact the properties of H^+ ions that HCl gives in solution. Dry HCl gas or when it is dissolved in non-aqueous solvents like benzene does not behave as an acid. Fig. 1.13 shows the dissociation of HCl molecule by H_2O.

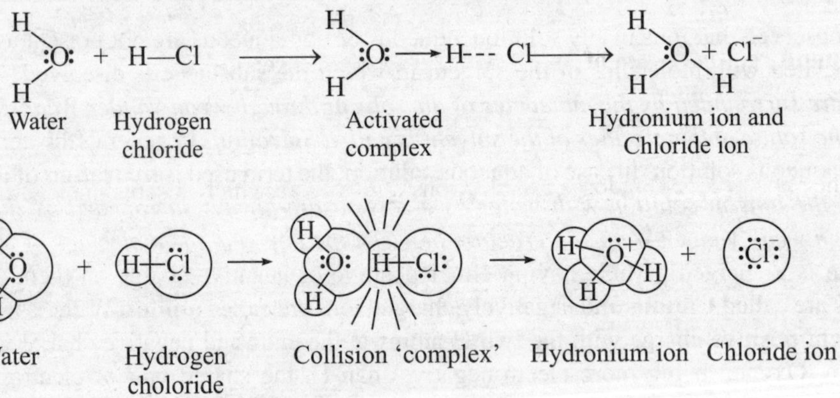

Fig. 1.13 Dissociation of HCl in water.

It means that not only ionic compounds ionise in water, even covalent molecules are dissociated in water. But the ions are always hydrated in solution, what so ever be their source.

Let us consider another example of *self ionisation of water or autoprotolysis of water.* We know water is feebly ionized into $H^{\oplus}$ and $OH^{\ominus}$ ions. But the H^+ ions exist as hydronium ions (H_3O^+) in water. The dual behaviour of water *i.e.,* it can act as an acid as well as a base is very will explained on the basis of Bronsted and Lowery concept of acids and bases.

According to this concept, *an acid is a proton doner and a base is a proton acceptor.* Water is amphoteric in nature. In the presence of acids (say HCl or CH_3COOH), it behaves as a base and in presence of bases (say NH_3), it acts as an acid. According to this theory, one molecule of water acts as a Bronsted acid and the other water molecule acts as a base as shown below.

$$\underset{\text{Acid}_1}{H_2O} + \underset{\text{Base}_2}{H_2O} \rightleftharpoons \underset{\text{Acid}_2}{H_3\overset{+}{O}} + \underset{\text{Base}_1}{\overset{-}{O}H}$$

In addition to hydronium ion, the hydration of proton gives rise to other ions such as $H^+ \cdot 2H_2O$ or $[H_5O_2^+]$, $H^+ \cdot 3H_2O$ or $[H_7O_3^+]$ and $H^+ \cdot 4H_2O$ or $[H_9O_4^+]$. For the sake of simplicity, they are represented as **hydronium ion $[H_3O^+]$**. The structure of these ions are given below in Fig. 1.14.

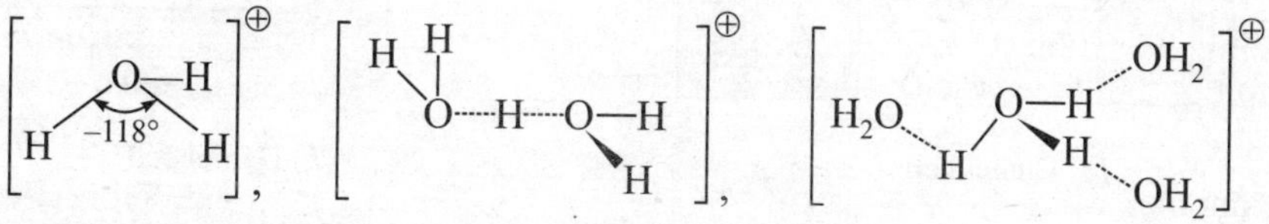

Fig. 1.14 Structures of hydrated proton

1.11 ELECTROCHEMICAL CELL OR GALVANIC CELL

A device employed to convert the chemical energy of a redox reaction into electrical energy is called an electrochemical cell or simply a chemical cell. It is also commonly referred to as a galvanic or ***voltaic cell*** *after the names of Galvani (1780) and Volta (1800), the pioneer in this field.*

The main requirements of an electrochemical cell are:

(*i*) A suitable redox reaction is carried out **indirectly in the two separate half cells.** The electrons are lost in one cell and gained in the other.

(*ii*) The substance which loses the electrons and the one which accepts the electrons should not be in direct contact with each other. The electron transfer must take place through an **external circuit.**

Let us illustrate the principles of an electrochemical cell by taking example of **Daniel cell**, *i.e.*, an electrochemical cell in which Zn – $CuSO_4$ reaction takes place.

$$Zn(s) + CuSO_4(aq) \longrightarrow ZnSO_4(aq) + Cu(s)$$

The ionic equation for this reaction is:

$$Zn(s) + Cu^{2+}(aq) \longrightarrow Zn^{2+}(aq) + Cu(s)$$

Construction: It consists of two breakers one containing solution of $ZnSO_4$ (say 0.1 M) and zinc plate is dipped in it. The other beaker contains $CuSO_4$ solution (say 0.1 M) and a copper plate is dipped in it. The two solutions are connected by means of a salt bridge. A salt bridge is U-shaped tube containing concentrated solution of an electrolyte like KCl, KNO_3, NH_4NO_3 etc, or solidified solution of an electrolyte in Agar-Agar and gelatine. The ends of the bridge are plugged with glass wool or cotton. The salt in the bridge is such that it does not react chemically with either of the two solutions connected by it (Fig. 1.15(*a*)).

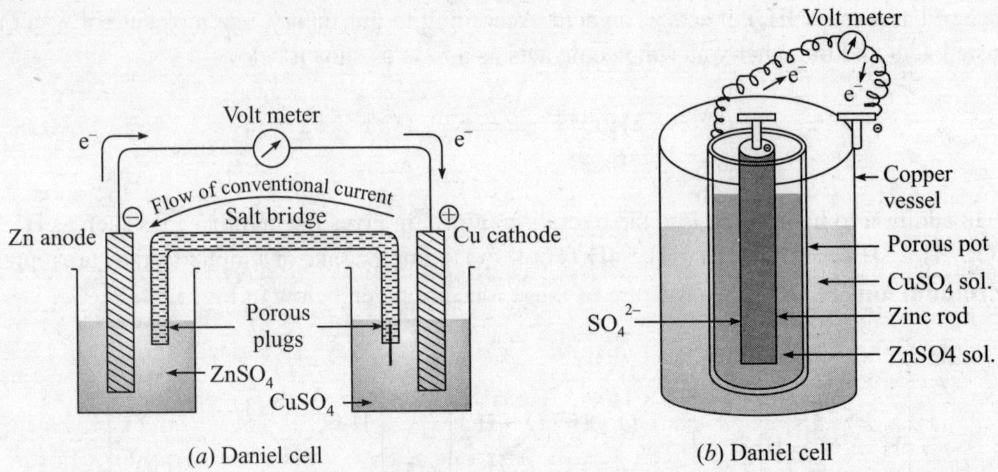

Fig. 1.15 Structures of hydrated proton

Working and Reactions: When the two metallic plates are joined through a connecting wire, the reaction takes place due to the fact that electrons flow from zinc to the copper plate. This flow of electrons constitutes electric current, since electrons, *i.e.*, negative electricity flows from zinc to the copper plate, **the conventional current,** *i.e.*, positive electricity flows from copper to zinc plate. Thus zinc plate serves as **negative electrode** and copper plate serves as **positive electrode.**

The following observations are recorded:

(*i*) ***Zinc plate loses its weight:*** It is due to the fact the zinc gets oxidized to zinc ions which go into the solution.

$$Zn\,(s) \longrightarrow Zn^{2+}(aq) + 2e^-$$

Such an electrode at the surface of which oxidation, *i.e.*, loss of electrons takes place is known as **anode.**

(*ii*) ***Copper plate gains weight:*** It is due to the fact that copper ions gain electrons at the copper plate.

$$Cu^{2+}(aq) + 2e^- \longrightarrow Cu(s)$$

Such an electrode at the surface of which reduction *i.e.*, gain of electrons takes place is known as **cathode.**

(*iii*) ***The solution remains electrically neutral:*** For details see functions of salt bridges.

(*iv*) The loss in weight of zinc and gain in weight of copper are in the ratio of their equivalent weights, *i.e.*,

$$\frac{\text{Loss in wt. of Zn}}{\text{Gain in wt. of Cu}} = \frac{\text{Eq. wt. of Zn}}{\text{Eq. wt. of Cu}} = \frac{32.70}{31.75}$$

(*v*) ***Electrical energy is obtained due to conversion of chemical energy:*** It is due to the fact that chemical energy gets transformed to electrical energy as evident from the deflection in the voltmeter.

(*vi*) The *reaction and working of cell, i.e., flow of electric current stops after sometimes.*

Reason: It is due to the fact that reduction potentials of both zinc and copper plates become same after sometime. Thus the e.m.f. of the cell, *i.e.*, the driving force becomes zero.

The e.m.f. of the cell is made up of two half cell is each of which a half reaction takes place as shown below:

$$Zn(s) \longrightarrow Zn^{2+}(aq) + 2e^- \text{ (Oxidation half-cell reaction)}$$
$$Cu^{2+}(aq) + 2e^- \longrightarrow Cu(s) \text{ (Reduction half-cell reaction)}$$

1.11.1 Functions of Salt Bridge

(i) To complete the circuit.

(ii) To maintain electrical neutrality of the solutions in both the half cells.

As the reaction progresses, in the oxidation half cell, Zn will lose electrons and would change to Zn^{2+} ions and positive charge may accumulate in this cell due to excess of Zn^{2+} ions. *This will prevents the release of electrons from zinc plate.*

Similarly, in the reduction half cell, Cu^{2+} ions would gain electrons and would be deposited as Cu on the electrode. So, the negative charge would accumulate due to the excess of SO_4^{2-} ions in this half cell. *This prevents the flow of electrons upto the copper plate.* To maintain the electrical neutrality in the two half cells, the negative ions (anions) of the electrolyte in the salt bridge would move to the oxidation half cell in order to neutralize excess of zinc ions. Similarly the positive ions (cation) move to the reduction half cell, in order to neutralize the excess of sulphate ions.

In the absence of salt bridge, the current would have stopped flowing in the circuit due, accumulation of opposite charge in the two half cells.

(iii) To prevent intermixing of the solutions of both half cells.

1.11.2 Representation of Galvanic Cells

The galvanic cells are represented with the help of certain notations and conventions, which are given below:

1. The interfaces across which a potential difference exists are shown by a semi-colon (;) or a single vertical line (|). For example, in the half-cell zinc in contact with Zn^{2+} ion is shown as

$$Zn; Zn^{2+} \text{ or } Zn \mid Zn^{2+}$$

This indicates that the potential difference exists at the zinc and Zn^{2+} ion interface. In the formulation of another half-cell

$$Ag, AgCl \mid Cl^-$$

The comma or the plus sign between Ag and AgCl indicates that these two materials taken together constitute the electrode.

2. The contact between the solutions with the help of a salt bridge is indicated by a double vertical line (||) between them. For example, two solutions containing Zn^{2+} ions and Cu^{2+} ions respectively connected by a salt bridge or a porous barrier are shown by

$$Zn^{2+}/\text{salt bridge}/Cu^{2+}; \quad Zn^{2+} \parallel Cu^{2+}$$

3. The oxidation half-cell is written on the left hand side and the reduction half-cell on the right-hand side with the metal electrodes always on the outside extremes. For example, Daniel cell is formulated as:

$$Zn; Zn^{2+} \parallel Cu^{2+}; Cu$$

or

$$Zn \mid Zn^{2+} \parallel Cu^{2+} \mid Cu$$

Sometimes negative and positive signs are put on the electrodes to show that they are negative and positive electrodes as shown below:

$$Zn \overset{\ominus}{\mid} Zn^{2+} \parallel Cu^{2+} \overset{\oplus}{\mid} Cu$$

4. An arrow when drawn below the cell formulation gives the direction of the positive current inside the cell while an arrow drawn above the cell formulation gives the direction in which the electrons flow in the outer circuit.

$$Zn \mid Zn^{2+} \parallel Cu^{2+} \mid Cu$$

5. The concentration of solutions, pressures of gases and physical state of the solid or liquid involved are indicated by suitable signs within brackets as in the following cells:

(*a*) Daniel cell: $Zn \mid Zn^{2+}$ (0.1 M) $\parallel Cu^{2+}$ (0.1 M) Cu

(*b*) Copper-silver nitrate cell: Cu; Cu^{2+} (0.1 M) $\parallel Ag^+$ (0.1 M) | Ag

(*c*) Zinc-hydrochloric acid cell: Zn; Zn^{2+} (0.1 M) $\parallel H^+$ (0.1 M); (1 atm.) H_2, Pt

(*d*) Zinc-silver nitrate cell: Zn; Zn^{2+} (0.1 M) $\parallel Ag^+$ (0.1 M); Ag

1.12 ELECTRODE POTENTIAL

When a metal is immersed in an aqueous solution containing the metal ions, the metal atoms have some tendency to pass into solution as cations or metals ion start depositing on the metal surface (Fig. 1.16).

In the beginning, both these changes occur with different speeds but soon an equilibrium is established.

$$M \underset{\text{Reduction}}{\overset{\text{Oxidation}}{\rightleftharpoons}} M^{n+} + ne^-$$

"The equilibrium potential difference between the metal electrode and the solution surrounding it" is called the ***electrode potential.*** *It is also defined as the tendency of an electrode to lose or gain electrons.*

The electrode potential is further termed oxidation potential, if oxidation takes place at the electrode.

Oxidation involves loss of electrons. So, the tendency of an electrode to lose electrons is called ***oxidation potential.***

$$M \longrightarrow M^{n+} + ne^-;$$

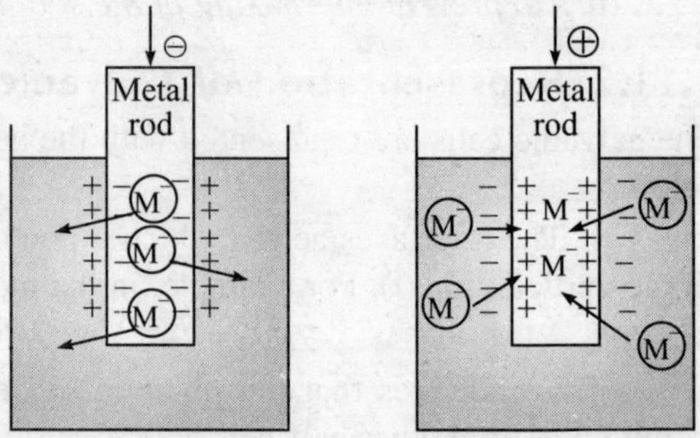

(*a*) Electrode potential oxidation (deelectronation)

(*b*) Electrode potential reduction (electronation)

Fig. 1.16 Electrode Potential

The metal ions go into solution and electrons remain on the metal rod making it negatively charged. The tendency of the metal to change into ions is also known as **electrolytic solution pressure.**

The electrode potential is called **reduction potential** if reduction takes place at the electrode.

Reduction involves gain of electrons so, the tendency of an electrode to gain electrons is called is ***reduction potential.***

$$M^{n+} + ne^- \longrightarrow M;$$

Since oxidation is just a reverse of reduction, therefore, reduction potential is obtained from the oxidation potential by simply changing the sign.

In general, for an electrode,

$$\textit{Oxidation potential} = -\ \textit{Reduction potential}$$

For example, Zinc electrode, standard oxidation potential is represented as

$$E^\circ_{Zn/Zn^{2+}} = 0.76 \text{ V}$$

and standard reduction potential as

$$E^\circ_{Zn^{2+}/Zn} = -0.76 \text{ V}$$

It is a common practice to express all the electrode potentials as reduction potentials.

The differences between oxidation potential and reduction potential are summarized as:

Oxidation Potential	*Reduction Potential*
1. The tendency of an electrode to lose electrons when in contact with the solution of its own ions in a half cell is known as oxidation potential.	The tendency of an electrode to gain electrons when in contact with solution of its own ions in a half cell is known as reduction potential.
2. It is inversely proportional to the concentration of ions of the electrode.	It is directly proportional to the concentration of ions of the electrode.

1.12.1 Standard Electrode Potential

The electrode potential of a half cell, in which the metal is suspended in a solution of one molar (1 M) concentration and the temperature is kept at 25°C (298 K) is called standard electrode potential. It is usually represented by E°.

It is not possible to determine the **absolute value** of electrode potential of a single electrode. It is because a single electrode constitutes only a half cell. A half cell, can not lose or gain electrons by itself. The loss or gain of electrons can take place only in a complete circuit containing two half cells connected to each other. So, the electrode potential can only be measured by using some electrode as reference electrode. The reference electrode commonly employed is the standard or normal hydrogen electrode. The electrode potential of NHE is taken as zero.

1.12.2 Measurement of Electrode Potential

To measure the relative value of electrode potential of any given electrode, a cell is set up which consists of the given electrode as one of the electrode and the second electrode is NHE. The e.m.f. of the cell is measured. The measured e.m.f. of the cell would give the electrode potential of the electrode under investigation because the electrode potential of NHE is taken as zero. The direction of the flow of current further indicates whether oxidation takes place or the reduction take place at the electrode under investigation with respect to hydrogen electrode. Accordingly the electrode potential is termed as oxidation or reduction potential.

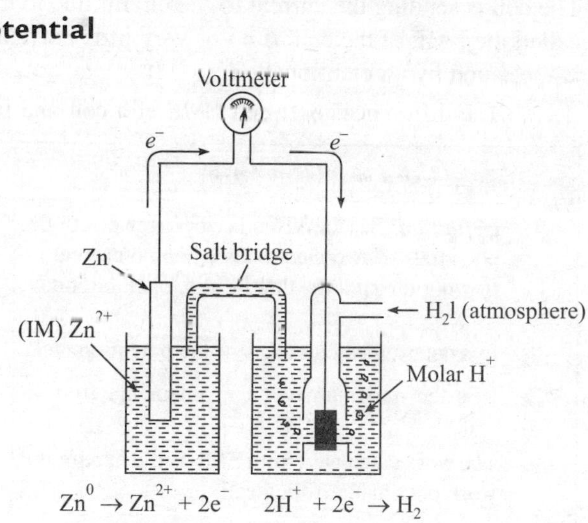

Fig. 1.17 Zinc electrode acts as anode when it is connected to NHE

For measurement of standard electrode potential, a cell is set up in which one-half of the cell consists of metal and its ion at 1 M concentration and the other half of the cell consists of the hydrogen electrode. The temperature in both half cells is kept at 25°C.

(*a*) **Measurement of electrode potential of Zn^{2+}/Zn electrode:** When coupled with NHE, some electrodes undergo spontaneous oxidation and permit loss of electrons. Zinc electrode when combined with NHE is oxidized to Zn^{2+} ions. In other words, the zinc electrode pushes electrons into the external circuit and forms the anode. By convention the anode is shown on the left (Fig. 1.17) and it is said to possess a negative value of *reduction electrode potential* (– 0.76 volts).

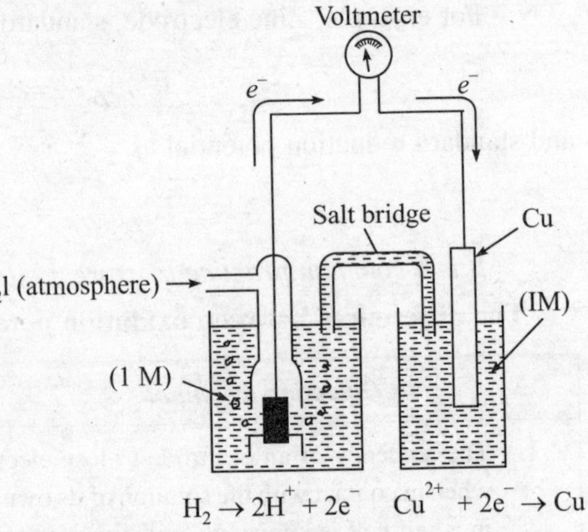

Fig. 1.18 Copper electrode acts as cathode it is connected to NHE

(*b*) **Measurement of electrode potential of Cu^{2+}/Cu electrode:** In case the electrode coupled with NHE undergoes spontaneous reduction, it allows oxidation of NHE. When copper electrode dipped in Cu^{2+} solution coupled with NHE, Cu^{2+} ions are reduced to Cu(*s*) and at NHE, H_2 is oxidized to H^+ ion by pushing electrons into the external circuit (Fig. 1.18). Copper electrode acts as cathode and it is said to have positive value of reduction electrode potential (0.34 volts).

1.13 CELL POTENTIAL OR EMF OF A CELL

An electrochemical cell consists of two electrodes, which have different electrode potentials. *The difference in electrode potential, which causes the current to flow from the electrode at a higher potential to the electrode at lower potential, is known as the cell potential.*

The differences of potential between the electrodes of a cell can be measured by using a voltmeter. The cell is sending the current to the circuit due to the potential difference. This potential difference is called the EMF of the cell, if no or very little current is allowed to flow at the time of measurement. It is measured by a potentiometer.

The differences between EMF of a cell and potential difference are summarized below:

Cell Potential or EMF	*Potential Difference*
1. ***Definition:*** It is the difference between electrode potentials of two electrodes when no current is flowing through the circuit, *i.e.*, open circuit.	1. ***Definition:*** It is the difference between electrode potentials of two electrodes when current is flowing through the circuit, *i.e.*, closed circuit.
2. It is determined by means of the potentiometer.	2. It is determined by means of a simple voltmeter.
3. It is the maximum voltage obtainable from a cell.	3. It is less than maximum voltage obtainable from a cell.
4. The work calculated from EMF is the maximum work obtainable from a cell.	4. The work calculated from the potential difference is less than maximum work obtainable from a cell.
5. It is responsible for the flow of steady current through the circuit.	5. It is not responsible for the flow of steady current through the circuit.

1.13.1 Calculation of EMF of a Cell

An electrochemical cell is always based upon a redox reaction, which can be split into two half-reactions namely the oxidation half-reaction and the reduction half-reaction. The standard EMF of a cell (E°_{cell}) is given by the relation

$$E^\circ_{cell} = \begin{bmatrix}\text{Standard oxidation potential}\\ \text{of oxidation half-reaction}\end{bmatrix} - \begin{bmatrix}\text{Standard oxidation potential}\\ \text{of reduction half-reaction}\end{bmatrix}$$

Since oxidation potential = – reduction potential
The above expression may be written as

$$E^\circ_{cell} = \begin{bmatrix}\text{Standard reduction potential}\\ \text{of reduction half-reaction}\end{bmatrix} - \begin{bmatrix}\text{Standard reduction potential}\\ \text{of oxidation half-reaction}\end{bmatrix}$$

Notation and Convention

In the representation of a cell, the electrode at which oxidation takes place is shown on the left-hand side and the electrode at which reduction takes place is shown on the right-hand side. Hence, the above expression is written as

$$E^\circ_{cell} = E^\circ_{(\text{right electrode})} - E^\circ_{(\text{left electrode})} = E^\circ_{(\text{cathode})} - E^\circ_{(\text{anode})}$$

[both the electrode potentials being reduction potentials]

Thus, the EMF of a galvanic cell (E°_{cell}) is calculated through the following steps:

Step I: The overall electrochemical cell reaction is split into two half-reaction, *i.e.*, oxidation half-reaction and reduction half-reaction.

Step II: Equate the number of electrons in the two half-reactions by multiplying with a suitable number, if necessary. However, E° values are not multiplied.

Step III: Formulate the galvanic cell in such a way that the electrode on which oxidation takes place and acts as anode is shown on the left-hand side while the electrode on which reduction takes place and acts as cathode is shown on the right-hand side.

Step IV: The standard EMF of the cell is given by the relationship:

$$E^\circ_{cell} = E^\circ_{cathode} - E^\circ_{anode}$$
$$= E^\circ_{right} - E^\circ_{left}$$

The electrode potentials of the cathode as well as anode should be in terms of reduction potentials.

Note: On calculation, if E°_{cell} has a positive value, the cell reaction is feasible. On the other hand, if E°_{cell} has a negative value, the cell reaction is not feasible.

NUMERICAL PROBLEMS BASED ON THE CALCULATION OF EMF

Example 15: ***Determine the standard EMF of the cell:***

$$Cu + 2Ag^+ \longrightarrow Cu^{2+} + 2Ag$$

Given $\quad E^\circ_{Ag/Ag^+} \longrightarrow$ ***0.80 (oxidation potential)***

$E^\circ_{Cu^{2+}/Cu} \longrightarrow$ ***0.34 (reduction potential)***

Solution: The given cell reaction

$$Cu + 2Ag^+ \longrightarrow Cu^{2+} + 2Ag$$

The cell for above cell reaction is represented as

$$Cu \mid Cu^{2+} \parallel Ag^+ + 2Ag$$

$$E^\circ_{cell} = E^\circ_{(right)} - E^\circ_{(left)} = E^\circ_{(cathode)} - E^\circ_{(anode)}$$

$$= 0.80 - 0.34 = 0.46$$

Therefore, the EMF of the given cell reaction = 0.46 V.

Example 16: ***Predict the feasibility of the given reaction.***

$$2Ag + Cu^{2+} \longrightarrow 2Ag^+ + Cu$$

Solution: Cell involving the given reaction is

$$Ag \mid Ag^+ \parallel Cu^{2+} \mid Cu$$

We know that $E^\circ_{cell} = E^\circ_{(right)} - E^\circ_{(left)}$

$$= E^\circ_{Cu^{2+}/Cu} - E^\circ_{Ag^+/Ag}$$

$$= 0.34 - (0.80) = -0.46 \text{ V}$$

Since the value E°_{cell} comes out to be negative, the given reaction is not feasible.

Example 17: ***Can a solution of 1 M $CuSO_4$ be stored, in a vessel made of Ni metal? Given that***

$$E^\circ_{Ni/Ni^{2+}} = +0.25 \text{ V and } E^\circ_{Cu^{2+}/Cu} = -0.34 \text{ V.}$$

Solution: The reduction potentials of the two electrodes,

i.e., $E^\circ_{Ni/Ni^{2+}} = +0.25$ V and $E^\circ_{Cu^{2+}/Cu} = -0.34$ V

suggest that nickel will act as an anode and copper will act as a cathode. Now, we wish to know whether the following reactions take place or not.

$$Ni + CuSO_4 \longrightarrow NiSO_4 + Cu$$

or

$$Ni + Cu^{2+} \longrightarrow Ni^{2+} + Cu$$

The cell involving this reaction can be represented as

$$Ni \mid Ni^{2+} \parallel Cu^{2+} \mid Cu$$

We know that $E^\circ_{cell} = E^\circ_{(right)} - E^\circ_{(left)} = 0.34 - (-0.25) = 0.59$ V

Since E°_{cell} is positive, it means that $CuSO_4$ reacts with Ni. Hence, $CuSO_4$ cannot be stored in a Ni vessel.

Example 18: ***Predict whether silver reacts with 1 M H_2SO_4 to give out hydrogen gas or not. Given that standard reduction potential of Ag is 0.80 V.***

Solution: If silver reacts with sulphuric acid, the following reaction should take place.

$$Ag + H_2SO_4 \longrightarrow Ag_2SO_4 + H_2$$

or

$$2Ag + 2H^+ \longrightarrow 2Ag^+ + H_2$$

The cell involving the above reaction is represented as

$$Ag \mid Ag^+ \parallel H^+ \mid H_2$$

$$E^\circ_{cell} = E^\circ_{(right)} - E^\circ_{(left)}$$

$$= E^\circ_{H^+, H} + E^\circ_{Ag^+, Ag}$$

$$= 0 - 0.80 = -0.80 \text{ V}$$

Since E°_{cell} of the supposed reaction comes out to be negative, this reaction does not take place. Hence, silver cannot liberate hydrogen from the acid.

Example 19: ***State whether the following representation of a galvanic cell is correct or not.***

$$Pb/Pb^{2+}\ (1\ M)\ ||\ Mg^{2+}\ (1\ M)/Mg$$

Given that $E^\circ_{Pb^{2+},\,Pb} = -1.126$ ***and*** $E^\circ_{Mg^{2+}/Mg} = -2.363\ V$

Solution: The representation is correct if E°_{cell} comes out to be positive.

$$E^\circ_{cell} = -2.363 - (-0.126) = -2.363 + 0.126 = -2.237\ V$$

Since E°_{cell} is negative the position of the cathode and anode must be interchanged, *i.e.*, the representation is not correct.

Example 20: ***Calculate*** ΔG° ***for the cell reaction at 25°C:*** $Zn + FeSO_4 \rightarrow Fe + ZnSO_4$. ***The standard reduction potentials are:***

$$E^\circ_{(Zn^{2+},\,Zn)} = -1.76 \quad \text{and} \quad E^\circ_{(Fe^{2+},\,Fe)} = -0.44\ V$$

Predict whether the reduction is spontaneous or not.

Solution: For the given cell reaction, the cell can be represented as

$$Zn\ |\ Zn^{2+}\ ||\ Fe^{2+}\ |\ Fe$$

Standard EMF of the cell

$$(E^\circ_{cell}) = E^\circ_{cathode} - E^\circ_{anode} = -0.44\ V - (-0.76\ V) = 0.32\ V$$

Now, $\Delta G^\circ = -nFE^\circ = -2 \times 96500\ C \times 0.32\ V = -61760\ CV = -61760\ J$

Since the free energy change (ΔG°) is negative, the reaction is spontaneous.

1.13.2 EMF of the Cell and Free Energy Change

The cell potential is simply related to the free energy change (ΔG) for the reaction. Therefore, measurement of the cell potential is often a convenient way to obtain free energy changes (useful work done) and consequently equilibrium constants.

In an electrochemical cell, the system does work by transferring electrical charge through an external circuit. For an electrochemical cell to work reversibly, it is ensured that while operating the cell only an infinitesimal current from the cell is to be drawn. The free energy change, ΔG is equal to electrical work done which is written as:

$$\Delta G = W_{elect}$$

If free energy change, ΔG, of the system is negative, the system does electrical work on its surroundings. If ΔG is positive, the change can be affected only if the surroundings does electrical work on the system.

For a reaction occurring in an electrochemical cell, the work done when amount of charge, nF, is transferred is given by $-nFE$. Thus:

$$\Delta G = -nFE_{cell}$$

Here F is the Faraday Constant and n is the number of moles of electrons transferred in the reaction. For comparing cell voltage on a common basis, standard cell potential (E°_{cell}) is used.

The free energy associated with standard cell potential is the standard free energy for the reaction denoted by

$$\Delta G^\circ = -nFE^\circ_{cell}$$

1.13.3 Criteria for Spontaneity in Terms of Cell EMF

The value of standard free energy change ($\Delta G°$) can be evaluated knowing the standard e.m.f (E°) of the cell experimentally by applying the above relation. Thus, the criteria of spontaneity is given as:

(i) If $\Delta G < 0$ i.e., negative or $E > 0$ i.e., positive, the reaction will be spontaneous.

(ii) If $\Delta G = 0$ or $E = 0$, the reaction will be in equilibrium.

(iii) If $\Delta G > 0$ i.e., positive or $E < 0$ i.e., negative, the reaction will be non-spontaneous. However, it will be spontaneous in the reverse direction.

1.13.4 Calculation of Equilibrium Constant K_C from $\Delta G°$

Knowing the value of standard free energy $\Delta G°$, the equilibrium constant can also be calculated by the relation,

$$E°_{cell} = \frac{RT}{nF} \ln K_C$$

$$-nF\,E°_{cell} = -nF \times \frac{RT}{nF} \ln K_C = -RT \ln K_C \text{ i.e. } \Delta G° = -RT \ln K_C$$

Substituting the value $\Delta G°$, R and T the value of equilibrium constant K_C is calculated. In case of Daniel cell $K_C = 2 \times 10^{-7}$ which shows that very little of Cu^{2+} will be in solution at equilibrium.

Example 21: ***Calculate the standard free energy change of the cell reaction***

$$Fe^{2+}(aq) + Ag^{+}(aq) \longrightarrow Fe^{3+}(aq) + Ag(s)$$

How is it related to the equilibrium constant of the reaction?

(Given $E°_{Fe^{3+}/Fe^{2+}} = 0.77$ V; $E°_{Ag^{+}/Ag} = 0.80$ V; $F = 96500$ C mol^{-1})

Solution: The electrochemical cell for the given reaction is

$$Fe \mid Fe^{3+} \parallel Ag^{+} \mid Ag$$

$$E°_{cell} = E°_{Cathode} - E°_{Anode}$$

$$= E°_{Ag^{+},Ag} - E°_{Fe^{3+}/Fe^{2+}} = 0.80\text{ V} - 0.77\text{ V} = 0.03\text{ V}$$

We know $\Delta G° = -nFE°$

$$= -1 \text{ mol} \times 96500 \text{ C mol}^{-1} \times 0.03 \text{ V}$$

$$= \mathbf{-2895}$$

$\Delta G°$ is related to the equilibrium constant K as under:

$$\Delta G° = -2.303\, RT \log K.$$

Example 22: ***For the cell reaction Ni | Ni^{2+} || Ag^{+} | Ag. Calculate the equilibrium constant at 25°C. How much maximum work would be obtained by operation of this cell?***

$$(E°_{Ni^{2+}/Ni} = -0.25 \text{ V},\ E°_{Ag^{+}/Ag} = 0.80 \text{ V})$$

Solution:

$$E°_{cell} = E°_{right} - E°_{left}$$

$$= 0.80 - (-0.25) = 1.05 \text{ V}$$

Now $$\log K_c = \frac{E^\circ_{cell} \times n}{0.059} = \frac{1.05 \times 2}{0.059} = 35.59$$

$$K_c = 3.89 \times 10^{35}$$

Maximum work obtained, $\Delta G^\circ = -nFE^\circ_{cell} = 2 \times 96500 \times 1.05 = 202650 \text{ J} = 202.650 \text{ kJ}$

Note: ΔG° and K_c are also related as: $\Delta G^\circ = -2.303 \text{ R} \log K_c$
Thus K_c can be calculated from ΔG° also.

1.14 NERNST EQUATION OR CONCENTRATION DEPENDENCE OF ELECTRODE POTENTIAL

The e.m.f. of a cell depends both on (i) Standard e.m.f. of cell, i.e., its nature (ii) the concentrations of the species involved in the cell reaction and (iii) temperature.

For a general cell reaction of the type,

$$aA + bB \longrightarrow lL + mM$$

In thermodynamics, the free energy (ΔG) accompanying the above general equation is given as:

$$\Delta G = \Delta G^\circ + RT \ln \frac{[L]^l [M]^m}{[A]^a [B]^b}$$

where ΔG° is the free energy change of the reaction under standard conditions.

Since $$\Delta G = -nFE \quad \text{and} \quad \Delta G^\circ = -nFE^\circ$$

$$\therefore \quad -nFE = -nFE^\circ + RT \ln \frac{[L]^l [M]^m}{[A]^a [B]^b}$$

or $$E = E^\circ - \frac{RT}{nF} \ln \frac{[L]^l [M]^m}{[A]^a [B]^b}$$

$$E = E^\circ - \frac{2.303 RT}{nF} \log \frac{[L]^l [M]^m}{[A]^a [B]^b}$$

where E is the e.m.f. of the cell, E° is the standard e.m.f. of the cell:

[A], [B], [L] and [M] are the molar concentration of the species A, B, L, and M respectively.

a, b, l and m are the stoichiometric co-efficients in the chemical equation, n is the number of electrons involved, R is the gas constant, T is the absolute temperature, F is one faraday (96500 coulombs).

This equation is known as **Nernst Equation.**

At 25°C, R = 8.314 Joules/kelvin/mole, T = 25°C + 273 = 298 K (at standard conditions) and F = 96500 coulombs

$$\therefore \quad \frac{2.303RT}{F} = \frac{2.303 \times 8.314 \times 298}{96500} = 0.0591$$

Substituting the value of $\frac{2.303RT}{F}$ in equation above, we get

$$E = E^\circ - \frac{0.0591}{n} \log \frac{[L]^l [M]^m}{[A]^a [B]^b} \quad ...(1.22)$$

or

$$E = E^\circ + \frac{0.0591}{n} \log \frac{[A]^a [B]^b}{[L]^l [M]^m}$$

It is seen from Nernst equation, that if the product of the molar concentration of products is more than the product of the reactants, e.m.f., E is less than the standard e.m.f., E°. On the other hand, if the product of the molar concentration of the reactants is greater than that of the molar concentration of products, E is greater than E°.

1.14.1 Applications of Nernst Equation

(*i*) **Calculation of the e.m.f. of a cell.** Consider for example the cell:

The cell reaction may be written as:

$$Zn \mid Zn^{2+} \parallel Cu^{2+} \mid Cu$$

Applying Nernst equation (4.1):

$$E = E^\circ - \frac{0.0591}{2} \log \frac{[Cu][Zn^{2+}]}{[Zn][Cu^{2+}]}$$

Here, number of electrons involved in the half cell reaction is 2.

In the standard state, the activity of a solid is taken as unity.

$$E = E^\circ - \frac{0.0591}{2} \log \frac{[Zn^{2+}]}{[Cu^{2+}]}$$

From this equation, it is obvious that E may be equal to E° or may be greater or less than E°. E depends on the ratio of the concentration of Zn^{2+} ions and Cu^{2+} ions.

(*ii*) **Calculation of the Single Electrode Potential:** Although the standard electrode potential is fixed, the concentration of the ion in the half-cell affects its single reduction electrode potential. This is also given by Nernst's equation.

$$Zn^{2+} + 2e^- \longrightarrow Zn$$

$$E = E^\circ - \frac{RT}{nF} \ln \frac{[Zn]}{[Zn^{2+}]} = E^\circ - \frac{RT}{nF} \ln \frac{1}{[Zn^{2+}]} = E^\circ + \frac{RT}{nF} \ln [Zn^{2+}]$$

[∵ Zn being solid, its concentration is taken as unity]

$$E = E^\circ + \frac{RT}{nF} \ln [Zn^{2+}] = E^\circ - \frac{2.303RT}{nF} \log [Zn^{2+}]$$

$$= E^\circ + \frac{0.0591}{n} \log [Zn^{2+}]$$

Where E and E° are the single electrode potential and standard potential, *n*, R and F have the usual significance.

The general expression for electrode potential of a metal in contact with M^{n+} ions reaction involving the electrode may be written as

$$M^{n+} + ne^- \rightarrow M$$

$$E = E° + \frac{0.0591}{n} \log [M^{n+}]$$

(*iii*) **Calculation of equilibrium constant:** At the equilibrium concentration of the two metal ions, there is no potential difference between the two electrodes and the e.m.f. of the cell becomes zero.

Consider a reaction at equilibrium

$$aA + bB \rightleftharpoons cC + dD$$

$$K_C = \frac{[C]^c [D]^d}{[A]^a [B]^b}$$

The Nernst equation can be represented as

$$E_{cell} = E°_{cell} - \frac{0.0591}{n} \log \frac{[C]^c [D]^d}{[A]^a [B]^b}$$

At the stage of equilibrium, $E_{cell} = 0$

Therefore, the above equation can be written as

$$\text{Zero} = E°_{cell} - \frac{0.0591}{n} \log K_c$$

or
$$E°_{cell} = \frac{0.0591}{n} \log K_c,$$

where K_c is the equilibrium constant for the reaction.

Knowledge of standard cell potential can thus be employed for calculating the equilibrium constant for the reactions.

Example 23: ***Calculate the standard electrode potential of Cu^{2+}/Cu, if the electrode potential at 25°C is 0.296 V, when***

$$[Cu^{2+}] = 0.015 \text{ M. } Cu^{2+}_{(aq)} + 2e^- \longrightarrow Cu(s)$$

Solution: According to Nernst equation.

$$E = E° + \frac{0.0592 \text{ V}}{n} \log [M^{n+}] = E° + \frac{0.0592 \text{ V}}{n} \log [Cu^{2+}]$$

$$E° = E - 0.0296 \text{ V} \log [Cu^{2+}]$$

$$= 0.296 \text{ V} - 0.0296 \text{ V} \log 0.015$$

$$= 0.296 \text{ V} - 0.0296 \text{ V} (-1.8239)$$

$$= 0.296 \text{ V} - 0.054 \text{ V} = \mathbf{0.350 \text{ V}}$$

Example 24: ***What is the emf of the following cell at 25°C?***

$$Zn(s) \mid Zn^{2+}(0.2 \text{ M}) \parallel Ag^+(0.002 \text{ M}) \mid Ag(s)$$

The standard emf of the cell is 1.54 V.

Solution: According to Nernst equation, we have

$$E_{cell} = E°_{cell} + \frac{0.0592 \text{ V}}{2} \log \frac{[Ag^+]^2}{[Zn^{2+}]}$$

$$= 0.154 + 0.0296 \text{ V} \log \frac{(0.002)^2}{(0.2)}$$

$$= 1.54 \text{ V} + 0.0296 \text{ V} \log 2 \times 10^{-5} = 1.54 \text{ V} + 0.096 \text{ V} [0.301 - 5]$$

$$= 1.54 \text{ V} + 0.0296 \text{ V} (-4.699) = 1.54 - 0.139 \text{ V} = 1.401 \text{ V}$$

Example 25: ***Calculate the voltage generated in the following cell:***

$$Fe \mid Fe^{2+} \parallel Mn^{2+} \mid Mn$$

When iron rod is immersed in 1.0 M $FeSO_4$ solution and Mn rod is immersed in 0.1 M $MnSO_4$ Solution.

Given E° for Fe^{2+}/Fe is – 0.40 V and Mn^{2+}/Mn is – 1.18 V. **(VT, March 01)**

Solution:

$$E^\circ_{cell} = E^\circ_{Cathode} - E^\circ_{Anode} = E^\circ_{right} - E^\circ_{left}$$

$$= (-1.18 + 0.40) \text{ V} = -0.78 \text{ V}$$

According to Nernst equation:

$$E_{cell} = E^\circ_{cell} + \frac{0.0591}{2} \log \frac{[Mn^{2+}]}{[Fe^{2+}]}$$

$$= -0.78 \text{ V} + 0.0296 \log \frac{0.1}{1.0}$$

$$= -0.78 \text{ V} + 0.0296 (-1) = \mathbf{-0.8096 \text{ V}}$$

Minus sign shows that the actual cell should be represented as

$Mn \mid Mn^{2+} \parallel Fe^{2+} \mid Fe$, whose E.M.F. is 0.8096 V.

Example 26: ***Calculate the equilibrium constant for the reaction:***

$$Fe^{2+}(aq) + Ag^{+}(aq) \mid Fe^{3+}(aq) + Ag(s) \text{ at } 25°C$$

If $E^\circ_{Fe^{3+}/Fe^{2+}} = 0.77$ V and $E^\circ_{Ag^+/Ag} = 0.80$ V **(Sindri, May 01)**

Solution: The cell for the given cell reaction is represented as:

$$Fe^{2+} \mid Fe^{3+} \parallel Ag^{+} \mid Ag$$

Now $E^\circ_{cell} = E^\circ_{Cathode} - E^\circ_{Anode} = 0.80 \text{ V} - (0.77 \text{ V}) = 0.03 \text{ V}$

We know that, $E^\circ_{cell} = \frac{0.059}{n} \log K_C$ or $\log K_C = \frac{n \times E^\circ_{cell}}{0.059}$

or $\log K_C = \frac{1 \times 0.03 \text{ V}}{0.0591 \text{ V}} = 0.50847$

or $K_C = \text{Antilog } 0.50847 = \mathbf{3.2245}$

Example 27: ***What is concentration of Ni^{2+} in the cell at 25° C, if the e.m.f. is 0.601 V?***

$$Ni(s) \mid Ni^{2+}(a = ?) \parallel Cu^{2+}(0.75 \text{ M}) \mid Cu(s)$$

Given $E^\circ_{Ni/Ni^{2+}} = 0.25$ V, and $E^\circ_{Cu^{2+}/Cu} = 0.34$ V **(Sindri May, 01)**

Solution: We know that $E^\circ_{cell} = E^\circ_{right} - E^\circ_{left}$, where E°_{right} and E°_{left} are reduction potentials of cathode and anode respectively.

Given $E^\circ_{Ni/Ni^{2+}} = 0.25$ V is the oxidation potential. Therefore, the reduction Potential

$$E^\circ_{Ni^{2+}/Ni} = -0.25\text{ V}$$

$$E^\circ_{cell} = 0.34\text{ V} - (-0.25\text{ V}) = 0.59\text{ V}$$

According to Nernst equation:

$$E_{cell} = E^\circ_{cell} + \frac{0.0592}{2}\log\frac{[Cu^{2+}]}{[Ni^{2+}]}$$

$$0.601\text{ V} = 0.59\text{ V} + 0.0296\log\frac{0.75}{[Ni^{2+}]}$$

$$0.011 = 0.0296\log\frac{0.75}{[Ni^{2+}]}$$

or $$\log\frac{0.75}{[Ni^{2+}]}\frac{0.011}{0.0296} = 0.37162$$

∴ $$\frac{0.75}{[Ni^{2+}]} = \text{Antilog } 0.37162 = 2.3529$$

or $$[Ni^{2+}] = \frac{0.75}{2.3529} = \mathbf{0.3188\ M}$$

Example 28: ***Calculate the equilibrium constant for the reaction.***

$$Sn^{++}(aq) + Pb(s) \rightleftharpoons Sn(s) + Pb^{++}(aq) \text{ at } 298\text{ K.}$$

Given ***E°(Sn⁺⁺/Sn) = – 0.14, and E°(Pb⁺⁺/Pb) = – 0.13 V*** **(PTU. Dec, 2004)**

Solution: According to Nernst equation:

$$E_{cell} = E^\circ_{cell} - \frac{2.303RT}{nF}\log K_C$$

For the given equation:

$$Sn^{2+}(aq) + Pb(s) \rightleftharpoons Sn(s) + Pb^{2+}(aq)$$

$$E^\circ_{cell} = E^\circ_{Sn^{2+}/Sn} - E^\circ_{Pb^{2+}/Pb}$$

(as Pb is being oxidised in the given equation)

$$= -0.14 - (-0.13) = -0.01\text{ V}$$

As, at equilibrium, E_{cell} becomes zero:

$$E^\circ_{cell} = \frac{0.059}{n}\log K \text{ or } -0.01\text{ V} = \frac{0.059}{n}\log K = \frac{0.0591}{2}\log K$$

$$-0.3389 = \log K \quad \text{or} \quad \mathbf{0.4581 = K}$$

Example 29: ***Write the electrode reaction and calculate the oxidation potential of each of the following at 298 K.***

(i) Ag, AgCl(s) Cl^-(a = 0.0001); E°Ag/AgCl = – 0.225 V

(ii) Pt, H_2(1 atm) | H^+(a = 0.15)

(iii) Calculate the potential of the cell by combining (i) and (ii) above **(P.T.U. May 2006)**

Solution: For the electrode; Ag, AgCl(s)/Cl^- (a = 0.0001)

$$E°Ag, Cl^- \mid AgCl = -0.225 \text{ V}$$

Oxidation half reaction

$$Ag(s) + Cl^- \longrightarrow AgCl_{(s)} + e^-$$

$$E_{electrode} = E^0_{0.p} - \frac{0.0591}{n} \log \frac{[AgCl(s)]}{[Ag(s)][Cl^-]}$$

$$= -0.225 - \frac{0.059}{1} \log \frac{1}{0.0001}$$

$$= -0.225 - 0.059 \times 4 = -0.225 - 0.236 = -0.461 \text{ V}]$$

(ii) For the electrode, pt, H_2(1 atm) | H^+(a = 0.15)

$$E°_{H_2/H^+} = 0 \qquad \text{(by convention)}$$

Oxidation half reaction: $H_2 \longrightarrow 2H^+ + 2e^-$

$$E_{electrode} = E^0_{0.p} - \frac{0.059}{n} \log \frac{[H^+]^2}{p[H_2]}$$

$$= 0 - \frac{0.059}{2} \log \frac{(0.15)^2}{1}$$

$$= -0.059 \log 0.15 = -0.059 \times -0.823 = 0.0486 \text{ V}$$

(iii) If the above two half cells are combined then the cell will be represented as

pt, H_2 | H^+(a = 0.15) || Cl^-(a = 0.0001) | AgCl(s), Ag

At anode: oxidation half reaction

$$\frac{1}{2} H_2 \longrightarrow H^+_{(aq)} + e^-$$

At cathode: Reduction half reaction

$$AgCl(s) + e^- \longrightarrow Ag(s) + Cl^-(aq)$$

The overall reaction is:

$$\frac{1}{2} H_2 + AgCl(s) \longrightarrow H^+ + Ag(s) + Cl^-$$

$$\therefore \qquad E_{cell} = E°_{cell} - \frac{0.059}{n} \log \frac{[H^+][Ag(s)][Cl^-]}{p[H]^{1/2}\,[AgCl(s)]}$$

$$= (0.225 - 0) - \frac{0.059}{1} \log .15 \times 0.0001$$

[$E°_{cell} = E_R - E_L$ when both electrode potentials are measured in reduction potential]

$$= 0.225 - 0.059 \times -4.823$$

$$= 0.225 + 0.2846 = 0.5096 \text{ V}$$

E_{cell} can also be checked from (*i*) and (*ii*) as:

$$E_{cell} = E_R - E_L = 0.461 - (-0.0486) = 0.5096 \text{ V}$$

Example 30: ***For a cell reaction A(s) + 2B(aq) ⟶ A + 2(aq) + 2B(s) at 298 K, the equilibrium constant is 1.0 × 104. Calculate E°cell.*** **(P.T.U. Dec. 2006, May 2007)**

Solution: The half cell reactions are:

$$A(s) \qquad A^{2+}_{(aq)} + 2e^-$$

$$2B^+_{(aq)} + 2e^- \qquad 2B(s)$$

$$A(s) + 2B^+(aq) \qquad A^{2+}_{(aq)} + 2B(s)$$

According to Nernst equation

$$E_{cell} = E^\circ_{cell} - \frac{0.0591}{n} \log \frac{[A^{2+}_{(aq)}][B(s)]^2}{[A(s)][B^{2+}_{(aq)}]^2}$$

$$E_{cell} = E^\circ_{cell} - \frac{0.0591}{2} \log K$$

At equilibrium $E_{cell} = 0$

$$\therefore \quad E^\circ_{cell} = \frac{0.0591}{2} \log(1 \times 10^4) = \frac{0.0591}{2} \times 4 = 0.1182 \text{ V}$$

1.15 REFERENCE ELECTRODES

An electrode whose electrode potential is accurately known or whose electrode potential has been arbitrarily fixed is called a reference electrode. In other words, it is an electrode of standard potential with which we can compare the potentials of other electrodes.

It is combined with any electrode whose electrode potential has to be determined to form an electrochemical cell through a salt bridge. The EMF of the cell thus formed is noted and the electrode potential of the electrode under test can be known. In the case of Normal Hydrogen Electrode, the electrode potential is taken as zero. Therefore, the EMF of the cell itself becomes the electrode potential of the electrode concerned. A reference electrode should satisfy the following conditions.

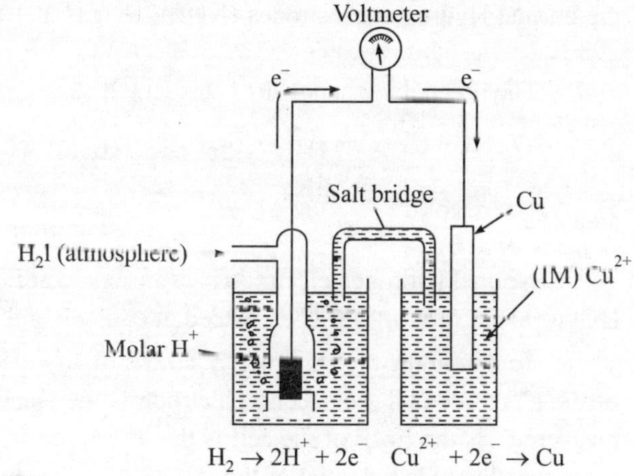

Fig. 1.19 Copper electrode acts as cathode when it is connected to NHE

(*i*) It should be easily set up.

(*ii*) It should not be bulky or space occupying.

(*iii*) It should not be easily poisoned by the impurities.

(*iv*) It should be usable even when the solutes are volatile.

(*v*) It electrode potential should not change even by change in pH of the solution.

Reference electrodes are divided into two types, which we discuss in the following sections.

1.15.1 Primary Reference Electrodes

Standard hydrogen electrode: It is the primary reference electrode employed for cell potential measurements. Its potential is taken as zero. It is combined with the electrode whose electrode potential has to be measured to form a cell. The EMF of the cell so constituted is measured, which gives the potential of the electrode since potential of NHE is taken as zero. It is also known as Standard Hydrogen Electrode (SHE). However, it is quite difficult to set up a hydrogen electrode. Moreover, it cannot be used in solutions containing redox solution. It is also affected by oxidizing agents such as MnO_4^-, $Cr_2O_7^{2-}$ and Fe^{3+} and some compounds of Hg, As and S. Consequently, this electrode cannot be used for solutions containing these ions. In such cases, some other electrodes are employed which are standardized using SHE. These electrodes are called as secondary reference electrodes.

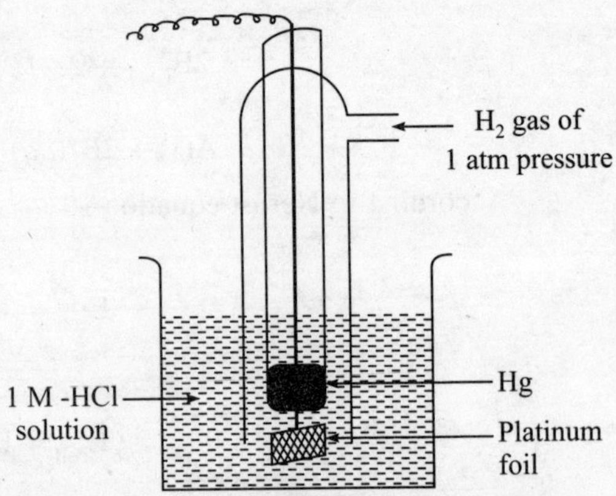

Fig. 1.20 Normal Hydrogen Electrode (NHE) or Slandard Hydrogen Electrode (SHE)

Hydrogen electrode: It is a familiar example of gas electrode. It consists of a Platinized platinum strip dipped in an acid solution through which hydrogen gas can be bubbled as shown in Fig. 1.20. In the normal Hydrogen Electrodes (NHE), H_2 gas at 1 atmospheric pressure is passed into 1 M- HCl at 298 K.

The reversible change are represented as

$$2\,H^+_{(aq)} + 2e^- \rightleftharpoons H_2(g)$$

and $$E^\circ_{H^+/H_2} = 0;\ E^\circ_{H_2/H^+} = 0;$$

Normal hydrogen electrode is represented at Pt; H_2(1 atm/HCl (1 M)) when placed at anode and HCl (1 M)/H_2(1 atm); Pt when placed at cathode end.

To determine the electrode potential of any electrode, a cell is set up using this electrode as one of the electrode and the second electrode is the standard hydrogen electrode. The EMF of the cell is measured. As the EMF of the cell is the difference in the electrode potentials of the two half cells and since the electrode potential of the standard hydrogen is taken to be zero, therefore the EMF of such a cell will directly give the electrode potential or the cell under investigation. The direction of flow of current further indicates whether oxidation or reduction takes place at the electrode under investigation with respect to the hydrogen electrode. **According to IUPAC convention, potential is given a positive**

sign if reduction occurs and a negative sign if oxidation occurs at the electrode under investigation with respect to the hydrogen electrode.

1.15.2 Secondary Reference Electrodes

They include calomel electrode, Ag—AgCl electrode, glass electrode, quinhydrone electrode, etc.

(*i*) **Calomel electrode:** It consists of a tube, at the bottom of which there is a layer of mercury and mercurous chloride, Hg_2Cl_2 (Fig. 1.21). The remaining portion of the cell is filled with a solution of normal or decinormal or saturated solution of KCl. A platinum wire dipping into the mercury layer is used for making electrical contact. The electrode is connected with the help of a side tube through a salt bridge with other electrode to make a complete cell. A compact form of the electrode is shown in Fig.1.21(*b*).

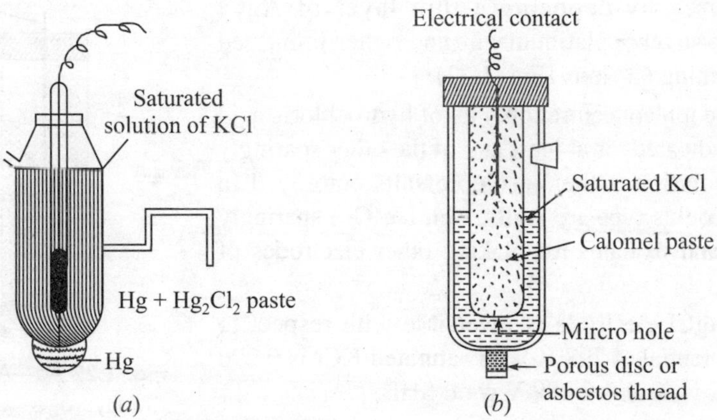

Fig.1.21 The calomel electrode

The electrode may be represented as Hg; $Hg_2Cl_2(s)$ KCl (solution)

When oxidation occurs at this electrode, the following reaction takes place

$$
\begin{aligned}
2Hg(l) &\longrightarrow Hg_2^{2+} + 2e^- \\
Hg_2^{2+} + 2Cl^- &\longrightarrow Hg_2 + Cl_2(s) \\
\hline
2Hg(l) + 2Cl^- &\longrightarrow Hg_2Cl_2(s) + 2e
\end{aligned}
$$

When reduction occurs at this electrode, the following reactions take place

$$
\begin{aligned}
Hg_2Cl_2(s) &\longrightarrow Hg_2^{2+} + 2Cl^- \\
Hg_2^{2+} + 2e^- &\longrightarrow 2Hg(l) \\
\hline
Hg_2Cl_2(s) + 2e^- &\longrightarrow 2Hg(l) + 2Cl^-
\end{aligned}
$$

The electrode is thus, reversible with respect to chloride ion (anion). The equilibrium reaction may be represented as

$$2Hg + 2Cl^- \longrightarrow Hg_2Cl_2(s) + 2e^-$$

The potential of the calomel electrode depends upon the concentration of the potassium chloride solution used in the calomel electrode. If potassium chloride used is as saturated solutions, the electrode is known as **saturated calomel electrode (SCE)** and if potassium chloride solution is IN, the electrode is known as **normal calomel electrode (NCE)** while for 0.1 N potassium chloride solution the electrode is referred to as **decinormal calomel electrode (DNCE).** The reduction potentials of calomel electrodes on hydrogen scale at 298 K are as follows:

Saturated KCl:	**0.2422 V**
1.0 KCl:	**0.2800 V**
0.1 N KCl:	**0.3338 V**

Calomel electrode is preferred to NHE as a reference electrode. Because it can be set up conveniently and once set, it can be used for long. In the case of NHE, it is difficult to maintain the supply of H_2 gas at 1 atm pressure for long.

(*ii*) **Silver-Silver chloride electrode:** The electrode is represented as

Ag: AgCl(*s*) KCl solution; or Ag + AgCl/Cl^-($AgCl^-$)

(*ii*) Its electrode reaction is

$$Ag(s) + Cl^- \longrightarrow AgCl(s) + e^-$$

It is prepared by deposing a thin layer of AgCl electrolytically on a silver or platinum wire and is then immersed in a solution containing Cl^- ions (Fig. 1.22).

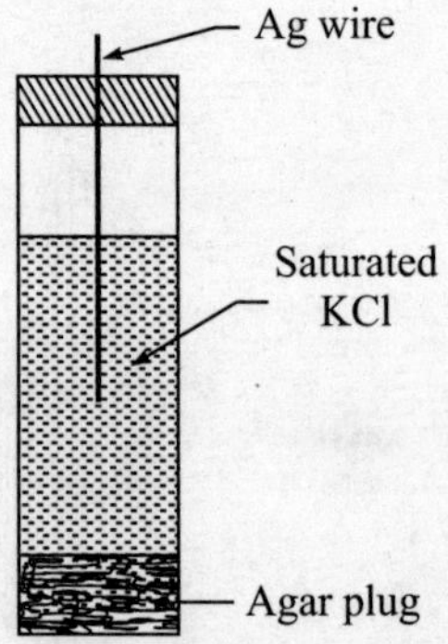

Fig. 1.22 Ag—AgCl Elects Electro

The chloride ion concentration, *i.e.*, of hydrochloric acid solution must be indicated, or at least any of the other sparingly soluble chloride solution. Other insoluble salts employed in making electrode of this type are AgBr, Agl, Hg_2Cl_2, sparingly soluble sulphates and oxalates for making other electrodes of Ag—AgCl type.

The Ag—AgCl electrode is reversible with respect to chloride ion. Its potential at 298 K with saturated KCl is 0.290 V and that with 0.1 N KCl is 0.199 V w.r.t. SHE.

(*iii*) **Glass electrode:** When two solutions of different pH are separated by a glass membrane, a potential is established across the membrane whose magnitude depends on the difference in pH of the two solutions, a potential is established between the glass and the solution. The value of the potential is a function of the particular glass used and is given by the equation:

$$E_g = E_g^\circ + \frac{RT}{F} \ln (H^+) = E_g^\circ - \left(\frac{2.303RT}{F}\right)$$

The electrode reaction is assumed to involve the reduction of H^+ ions. The glass electrode thus functions in the same manner as a reversible hydrogen electrode.

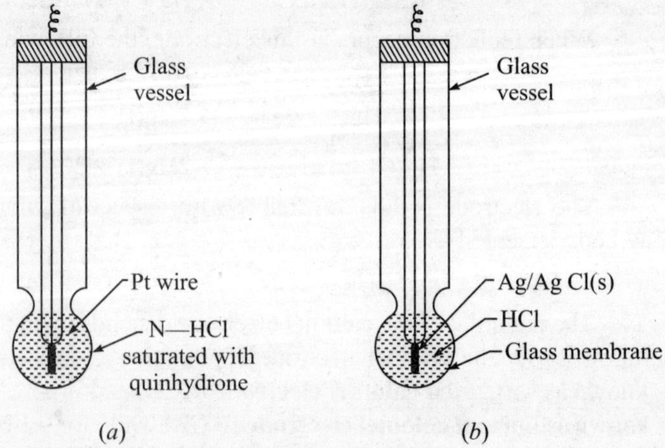

Fig. 1.23 Glass electrode

Construction of glass electrode: It consists of a thin walled glass bulb with a long neck and is made up of a special type of glass of low melting point and comparatively high conductivity [Figs. 1.23(*a*) and 1.23(*b*)]. A solution of known pH (usually 1 N HCl is taken in the bulb. A platinum electrode is introduced into the solution and a pinch of quinhydrone is added to it. A glass electrode is used for determining pH of the solution, especially for coloured solutions and solutions containing oxidizing and reducing agents. The combination representing the half-cell for glass electrode is represented as

$$Pt \mid 1.0\ N\ HCl \mid Glass$$

Sometimes a silver-silver chloride electrode instead of platinum is also employed. It is then represented as: Ag/AgCl(*s*) | 0.1 M HCl | Glass.

Advantages of Glass Electrode

The advantages of using a glass electrode are listed below:

(*i*) It is the most simple and convenient to use and gives accurate results.

(*ii*) It can be used in turbid, coloured and even colloidal solutions.

(*iii*) It can be used in alkaline as well as in solutions containing oxidizing agents.

(*iv*) It is not easily poisoned by any kind of cations and anions present in the solutions.

(*v*) It can measure pH value of solutions available in small quantities (only a few millilitres of the solutions are required).

(*vi*) Equilibrium is attained rapidly.

(*vii*) It does not require the use of any salt bridge.

Limitations of Glass Electrode

(*i*) For EMF measurement using glass electrodes, ordinary potentiometers cannot be employed because of high-resistance glass. Either quadrant electrometers or vacuum tube voltameters (VTVMs) that require practically no current for their operations must be employed.

(*ii*) It can be used up to a pH of 13 but being sensitive to Na^+ ions above a pH of 9, the linear relationship between glass electrode and pH is altered. Therefore, an alkaline error is introduced beyond a pH of 9, which should be taken into consideration.

(*iii*) In strongly acidic solutions of pH < 1, the salt effect due to anions is observed.

(*iv*) It does not function properly in some organic solvents such as pure alcohol.

1.16 ELECTROCHEMICAL SERIES

The standard reduction potentials of a number of electrodes along with their corresponding half cell reaction are given in Table 1.2. The table in which the reduction potentials of various electrodes have been arranged in the increasing order is called electrochemical series or activity series or electromotive series.

1.16.1 Applications of Electrochemical Series

(*i*) **To predict cathode and anode:** In setting up of electrochemical cell, the electrode with higher reduction potential will act as cathode and with lower will act as anode.

(*ii*) **To predict the relative ease of oxidation and reduction:** In the electrochemical series, F_2 has the highest reduction potential with +ve sign (or reduced most easily) and Li^+ has the lowest reduction potential with –ve sign (or reduced with greatest difficulty). In other words, F_2 is the strongest oxidizing agent or Li^+ ion the weakest oxidizing agent *i.e.*, strongest reducing agent. Thus the substances which have lower reduction potential have the tendency to be oxidized easily and act as reducing agents while those which have higher reduction potential have the tendency to be reduced easily and act as oxidizing agents.

Table 1.2 Standard reduction electrode potentials at 298 K.

Electrode	*Electrode Reaction*	*E°(Volts)*	*Electrode*	*Electrode Reaction*	*E°(Volts)*
Li^+/Li	$Li^+ + e^- \rightleftharpoons Li$	– 3.045	Sn^{2+}/Sn	$Sn^{2+} + 2e^- \rightleftharpoons Sn$	– 0.136
K^+/K	$K^+ + e^- \rightleftharpoons K$	– 2.925	Pb^{2+}/Pb	$Pb^{2+} + 2e^- \rightleftharpoons Pb$	– 0.126
Cs^+/Cs	$Cs^+ + e^- \rightleftharpoons Cs$	– 2.923	Pe^{3+}/Pe	$Pe^{3+} + 3e^- \rightleftharpoons Fe$	– 0.036
Ba^{2+}/Ba	$Ba^{2+} + 2e^- \rightleftharpoons Ba$	– 2.906	$H^+/H_2/Pt$	$H^+ + 2e^- \rightleftharpoons H_2$	0.0
Ca^{2+}/Ca	$Ca^{2+} + 2e^- \rightleftharpoons Ca$	– 2.866	$Cu^{2+}, Cu^+/Pt$	$Cu^{2+} + 2e^- \rightleftharpoons Cu^+$	+ 0.153
Na^+/Na	$Na^+ + e^- \rightleftharpoons Na$	– 2.71	Cu^{2+}/Cu	$Cu^{2+} + 2e \rightleftharpoons Cu$	+ 0.337
Mg^{2+}/Mg	$Mg^{2+} + 2e^- \rightleftharpoons Mg$	– 2.36	$I_2/I/Pt$	$I_2 + 2e^- \rightleftharpoons 2I^-$	+ 0.534
Al^{3+}/Al	$Al^{3+} + 3e^- \rightleftharpoons Al$	– 1.66	Ag^+/Ag	$Ag^+ + e^- \rightleftharpoons Ag$	+ 0.799
Zn^{2+}/Zn	$Zn^{2+} + 2e^- \rightleftharpoons Zn$	– 0.76	Hg^{2+}/Hg	$Hg^{2+} + 2e^- \rightleftharpoons Hg$	+ 0.854
Cr^{3+}/Cr	$Cr^{3+} + 3e^- \rightleftharpoons Cr$	– 0.74	Br_2/Br^-Pt	$Br_2 + 2e^- \rightleftharpoons 2Br^-$	+ 1.066
Fe^{2+}/Fe	$Fe^{2+} + 2e^- \rightleftharpoons Fe$	– 0.44	Cl_2/Cl^-pt	$Cl_2 + 2e^- \rightleftharpoons 2Cl^-$	+ 1.359
Cd^{2+}/Cd	$Cd^{2+} + 2e^- \rightleftharpoons Cd$	– 0.403	Au^{3+}/Au	$Au^{3+} + 3e^- \rightleftharpoons Au$	+ 1.498
Co^{2+}/Co	$Co^{2+} + 2e^- \rightleftharpoons Co$	– 0.28	F_2/F^-Pt	$F_2 + 2e^- \rightleftharpoons 2F$	+ 2.870
Ni^{2+}/Ni	$Ni^{2+} + 2e^- \rightleftharpoons Ni$	– 0.25			

(*iii*) **To calculate the standard e.m.f. of an electrochemical cell:** It is possible to calculate the e.m.f. of a cell from the standard electrode potentials of the half cell reactions from the electrochemical series.

An electrochemical cell is always based upon a redox reaction which can be split into two half reactions namely the oxidation half reaction and the reduction half reaction. The standard e.m.f. of a cell ($E°_{cell}$) is given by the relation.

$$E°_{cell} = \begin{bmatrix}\text{Standard oxidation potential}\\ \text{of oxidation half reaction}\end{bmatrix} - \begin{bmatrix}\text{Standard oxidation potential}\\ \text{of reduction half reaction}\end{bmatrix}$$

Since oxidation potential = – reduction potential

The above expression may be written as:

$$E°_{cell} = \begin{bmatrix}\text{Standard reduction potential}\\ \text{of reduction half reaction}\end{bmatrix} - \begin{bmatrix}\text{Standard reduction potential}\\ \text{of oxidation half reaction}\end{bmatrix}$$

In the representation of a cell, the electrode at which oxidation takes place is shown on the left hand side and the electrode at which reduction takes place is shown on the right hand side. Hence, the above expression is written as

$$E°_{cell} = E°_{(Right\ electrode)} - E°_{(Left\ electrode)} = E°_{(cathode)} - E°_{(anode)}$$

[Both the electrode potentials being reduction potentials].

Example 31: ***Determine the standard e.m.f. of the cell:***

$$Cu + 2Ag^+ \longrightarrow Cu^{2+} + 2Ag$$

Given $E°_{Ag;\,Ag^+} \longrightarrow -0.80$ ***(oxidation potential)***

$E°_{Cu^{2+};\,Cu} \longrightarrow 0.34$ ***(reduction potential)***

Solution: The given cell reaction

$$Cu + 2Ag^+ \longrightarrow Cu^{2+} + 2Ag$$

The cell for the above cell reaction is represented is

$$Cu \mid Cu^{2+} \parallel Ag^{+} \mid Ag$$

$$E^\circ_{cell} = E^\circ_{(right)} - E^\circ_{(left)} = E^\circ_{(cathode)} - E^\circ_{(anode)}$$
$$= 0.80 - 0.34 = 0.46 \text{ V}$$

So, the e.m.f. of the given cell reaction = 0.46 V

(*iv*) **To predict the feasibility of a redox reaction:** A reaction is feasible only when the value of standard free energy change (ΔG) for the cell reaction is negative. Now decrease in standard free energy ($-\Delta G^\circ$) is equal to the electrical energy (nFE°), where n is the number of electrons gained or lost, F is Faraday (96500 Coulombs) and E° is the standard cell potential. ΔG° is negative only if E° has a positive value.

Hence standard e.m.f. of the cell is calculated based upon the given redox reaction.

E°_{cell} comes out to be positive; the redox reaction under observation is feasible. In case E°_{cell} comes out to be negative, the reaction cannot take place. It may takes place in the reverse direction.

Example 32: ***Predict the feasibility of the given reaction:***

$$2Ag + Cu^{2+} \longrightarrow 2Ag^{+} + Cu$$

Solution: Cell involving the given reactions is: $Ag \mid Ag^{+} \parallel Cu^{2+} \mid Cu$

We know

$$E^\circ_{cell} = E^\circ_{(Right)} - E^\circ_{(Left)}$$
$$= E^\circ_{Cu^{2+},Cu} - E^\circ_{Ag^{+},Ag}$$
$$= 0.34 - (0.80)$$
$$= -0.46 \text{ volt.}$$

Since the value E°_{cell} comes out to be negative, the given reaction is not feasible.

Example 33: ***Can a solute 1 M $CuSO_4$ be stored, in a vessel made of Ni metal? Given that***

$$E^\circ_{Ni, Ni^{2+}} = +0.25 \text{ volt} \quad \text{and} \quad E^\circ_{Cu, Cu^{2+}} = -0.34 \text{ volt.}$$

Solution: The reduction potentials of the two electrodes

i.e., $E^\circ_{Ni^{2+}/Ni} = -0.25$ and $E^\circ_{Cu^{2+}/Cu} = +0.34$ V.

Suggest that Nickel will act as anode and copper will act as cathode. Now

We wish to know whether the following reactions take place or not.

$$Ni + CuSO_4 \longrightarrow NiSO_4 + Cu$$

or

$$Ni + Cu^{2+} \longrightarrow Ni^{2+} + Cu$$

The cell involving this reaction can be represented as:

$$Ni \mid Ni^{2+} \parallel Cu^{2+} \mid Cu$$

We know: $E^\circ_{cell} = E^\circ_{(Right)} - E^\circ_{(Left)} = 0.34 - (-0.25) = 0.59$ volt.

Since E°_{cell} is positive. It means that $CuSO_4$ reacts with Ni. Hence $CuSO_4$ cannot be stored in Ni vessel.

(*v*) **To predict whether a metal reacts with acid to give H_2 gas or not:** A metal can react with an acid only if it has a greater tendency to lose electrons than hydrogen, so that the electrons lost by the metal are gained by H^+ ions from the acid.

$$M \longrightarrow M^{+} + e^{-} \quad \text{(oxidation half reaction)}$$

$$H^{+} + e^{-} \longrightarrow \frac{1}{2} H_2 \quad \text{(reduction half reaction)}$$

The overall reaction is:

$$M + H^+ \longrightarrow M^+ + \frac{1}{2}H_2$$

A metal which lies above hydrogen in electrochemical series can form positive ions more readily than hydrogen. This means that such a metal can displace hydrogen from acids. For example,

$$Zn + 2H^+ \longrightarrow Zn^{2+} + H_2$$

Obviously metals like Cu and Ag which are below hydrogen in electrochemical series cannot liberate H_2 gas from acids.

Alternatively, the e.m.f. of the concerned redox reaction, $M + H^+ \longrightarrow M^+ + \frac{1}{2}H_2$ can be calculated as above.

If the value of e.m..f. comes out to be positive, the metal will liberate H_2 from the acid, otherwise not.

Example 34: ***Predict whether silver reacts with IM H_2SO_4 to give out hydrogen gas or not. Given that standard reduction potential of Ag is 0.80 volt.***

Solution: If silver reacts with sulphuric acid, the following reaction should take place.

$$Ag + H_2SO_4 \longrightarrow Ag_2SO_4 + H_2$$

or

$$2Ag + 2H^+ \longrightarrow 2Ag^+ + H_2$$

The cell involving the above reaction is represented as:

$$Ag \mid Ag^+ \parallel H^+ \mid H_2$$

$$E^\circ_{cell} = E^\circ_{(Right)} - E^\circ_{(Left)}$$

$$= E^\circ_{H^+, H} - E^\circ_{Ag^+, Ag}$$

$$= \text{Zero} - 0.80 = -0.80 \text{ volt.}$$

Since E°_{cell} of the supposed reaction comes out to be negative, this reaction does not take places. Hence silver cannot liberate hydrogen from the acid.

(*vi*) **Thermal stability of metallic oxides:** The thermal stability of the metal oxides depends on its electropositive nature. As electropositivity decrease from top to bottom, the thermal stability of the oxide also decrease from top to bottom.

$$Ag_2O \xrightarrow{\text{heat}} Ag + \frac{1}{2}O_2$$

$$2HgO \xrightarrow{\text{heat}} 2Hg + O_2$$

$$\left.\begin{array}{l} BaO \\ Na_2O \\ Al_2O_3 \end{array}\right] \xrightarrow{\text{heat}} \text{No decomposition}$$

The metal which fall below copper form unstable oxides.

Example 35: ***State whether the following representation of a galvanic cell is correct or not.***

$Pb/Pb^{+2}(1\ M) \parallel Mg^{2+}(1\ M)/Mg$

Given that $E^\circ_{Pb^{2+}, Pb} = -1.126$ ***V and*** $E^\circ_{Mg^{2+}, Mg} = -2.36$ ***V***

Solution: The representation is correct if E° cell comes out to be positive.

$$E^{\circ}_{cell} = -2.363 - (-0.126) = -2.363 + 0.126 = -2.237 \text{ V}$$

Since E° is negative, the position of the cathode and anode must be interchanged *i.e.*, the representation is not correct.

Example 36: ***Calculate ΔG° for the cell reaction at 25°C:***

$$Zn + FeSO_4 \longrightarrow Fe + ZnSO_4$$

The standard reduction potentials are:

$$E^{\circ}_{(Zn^{2+}, Zn)} = -0.76 \text{ V and } E^{\circ}_{(Fe^{2+}, Fe)} = -0.44 \text{ V.}$$

Predict whether the reduction is spontaneous or not. **(Anna IBE, Nov. 01)**

Solution: For the given cell reaction, the cell can be represented as: $Zn \mid Zn^{2+} \parallel Fe^{2+} \mid Fe$

Standard e.m.f. of the cell

$$(E^{\circ}_{cell}) = E^{\circ}_{cathode} - E^{\circ}_{anode}$$
$$= -0.44 \text{ V} - (-0.76 \text{ V}) = 0.32 \text{ V}$$

Now $$\Delta G^{\circ} = -nFE^{\circ} = -2 \times 96500 \text{ C} \times 0.32 \text{ V} = -61760 \text{ CV} = -61760 \text{ J}$$

Since the free energy change (ΔG°) is negative, so the reaction is spontaneous.

1.17 CONCENTRATION CELLS

There are different kinds of cells as shown below:

We have already discussed the details about electro-chemical cells. We will now take up the details of concentration cells.

A concentration cell is a galvanic cell in which electrical energy is produced by the transference of a substance from a system of high concentration to that of low concentration and there is no net chemical reaction involved.

In general, there are two types of concentration cells.

1. ***Those having an electrolyte of two different concentrations but identical electrodes are called Electrolyte concentration cells.***

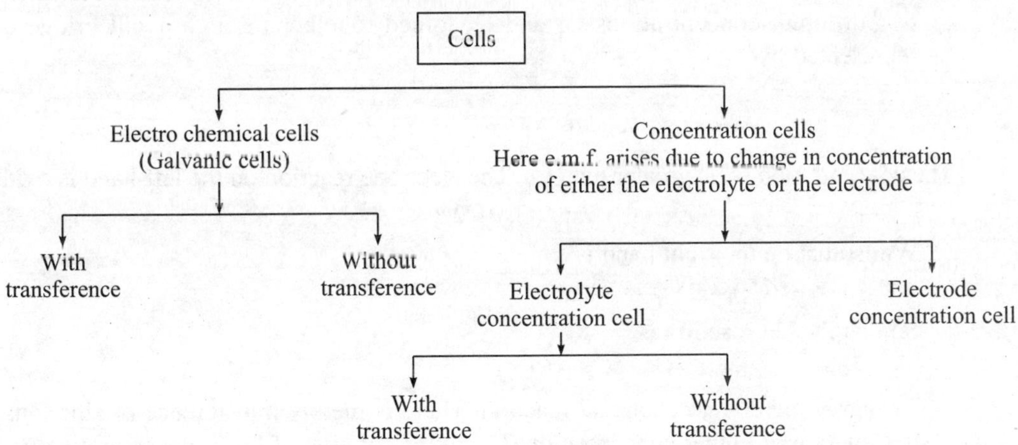

2. ***Those having electrodes of two different concentrations but the electrolyte of same strength are called Electrode concentration cells.***

Of these the **first type** is by far the most important.

(A) Electrolyte concentration cell without transference: When the two electrolytic solutions are not in direct contact with each other and transference of ions from one solution to other does not take place directly. The two solutions are separated by means of salt bridge.

Concentration cells with same electrodes but having different concentrations of electrolytes and are connected by a salt bridge.

When the concentration of an electrolyte is altered there occurs corresponding change in the potential of the electrodes in contact with the electrolyte. Thus a cell can be constructed by dipping two identical electrodes in solutions of different concentration of the same electrolyte. The source of electrical energy in the cell is the tendency of the electrolyte to diffuse from a solution of higher concentration to that of lower concentration. Thus with passage of time the two concentrations tend to become equal. For this reason, the e.m.f. of a concentration cell is maximum at the start and it gradually falls to zero as the two concentrations become equal.

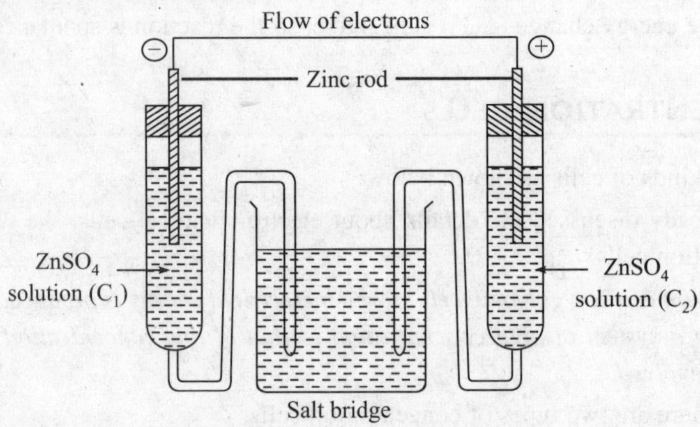

Fig. 1.24 Typical concentration cell

(*i*) A concentration cell (Fig. 1.24) with zinc electrodes and the solution of zinc sulphate of two different concentrations C_1 and C_2 joined together through a salt bridge can be represented by

$$\overset{\ominus}{Zn} \mid Zn^{2+}(C_1) \| Zn^{2+}(C_2) \mid \overset{\oplus}{Zn}$$

The concentration C_2 is greater than C_1. The electrode reaction on the left-hand is oxidation.

$$Zn \longrightarrow Zn^{2+}(C_1) + 2e^-$$

Whilst that on the right hand is:

$$Zn^{2+}(C_2) + 2e^- \longrightarrow Zn$$

So that the net reaction is

$$Zn^{2+}(C_2) \longrightarrow Zn^{2+}(C_1)$$

Evidently there is no chemical reaction. There is merely transference of zinc ions from the solution of higher concentration (C_2) to the solution of lower concentrating (C_1).

$$E_{el}(\text{R.H.S.}) = E^\circ_{el} + \frac{2.303RT}{nF}\log C_2$$

$$E_{el}(\text{L.H.S.}) = E^\circ_{el} + \frac{2.303RT}{nF}\log C_1$$

The e.m.f. of such a cell is given by the expression

$$E_{cell} = E_{el}(\text{R.H.S.}) - E_{el}(\text{L.H.S.})$$

$$= \frac{2.303RT}{nF}\log\frac{C_2}{C_1}$$

By substituting the value of R = (8.314 Joules) and F = (96500 coulombs). We can calculate the e.m.f. of the cell at the given temperature, 25°C. We have

$$E = \frac{2.303 \times 8.314 \times (273 + 25)}{n \times 96500}\log\frac{C_2}{C_1}$$

$$E = \frac{0.0591}{n}\log\frac{C_2}{C_1}$$

(*ii*) **Another example of this type of concentration cell is given below in which both the electrodes are hydrogen electrodes (Fig. 1.25).**

In each electrode H_2 gas at one atmospheric pressure is passed but the activity of H^+ in the two same electrolytes is different. These electrodes are joined by a salt bridge.

$$\text{Pt, } H_2\,(1\text{ atm}) \mid H^+(a_1) \parallel H^+(a_2) \mid H_2\,(1\text{ atm}),\text{ Pt } (a_2 > a_1)$$

On the L.H.S. electrode oxidaticn takes place and the electrode reaction is

$$\frac{1}{2}H_2\,(\text{atm}) \longrightarrow H^+(a_1) + e$$

To apply Nernst equation, the reduction half of the above reaction is written as:

$$H^+(a_1) + e^- \longrightarrow \frac{1}{2}H_2\,(1\text{ atm})$$

Electrode potential of this electrode will be

$$E_1 = E^\circ_{(H^+/H_2)} - \frac{RT}{F}\ln\frac{1}{a_1}$$

$$(\because aH_2 = 1)$$

$$E_1 = E^\circ_{(H^+/H_2)} + \frac{RT}{F}\ln a_1$$

$$E_1 = \frac{RT}{F}\ln a_1$$

$$(\because E^\circ_{H^+/H_2} = 0)$$

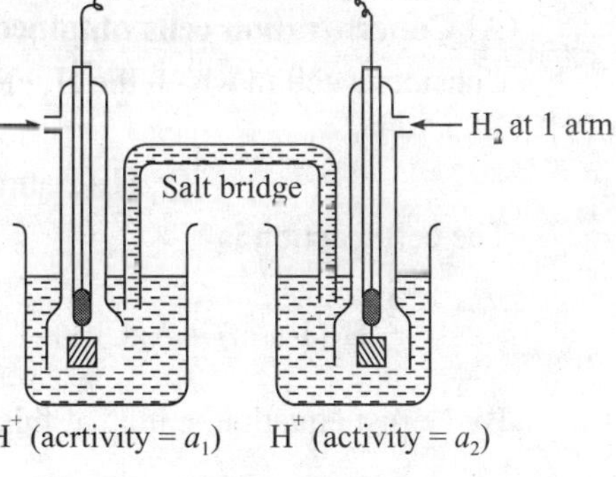

Fig. 1.25 Concentration cells consisting of hydrogen electrode

On the RHS, electrode reduction takes place.

The electrode reaction is

$$H^+_{(a_2)} + e^- \longrightarrow \frac{1}{2} H_2 \text{ (1 atm)}$$

By Nernst equation, the e.m.f. of this electrode will be

$$E_2 = \frac{RT}{F} \ln a_2$$

The overall reaction is obtained by adding two half reaction. This gives

$$H^+_{(a_2)} \longrightarrow H^+_{(a_1)}$$

The e.m.f. of cell is given by

$$E = E_2 - E_1$$

$$= \frac{RT}{F} \ln a_2 - \frac{RT}{F} \ln a_1$$

$$= \frac{RT}{F} \ln \frac{a_2}{a_1}$$

(B) Electrolyte concentration cell with transference: *Concentration cells with same electrodes but having different concentrations of electrolytes which are directly in contact with each other. In these cases the cells are represented as:*

(*i*) $Zn \mid ZnSO_4(a_1) \vdots ZnSO_4(a_2) \mid Zn$

(*ii*) Pt, H_2 (1 atm) | $HCl(a_1) \vdots HCl(a_2)$ | H_2 (1 atm), Pt

(The double line has been replaced by a dotted line showing the two solutions are in direct contact).

These cells can be set up by putting two solutions in a U-tube very carefully so that there is no mixing of the two solutions (Fig. 1.26). The junction at which the two solutions meet is called the liquid junction potential. The expression for e.m.f. of these concentration cells have already been given.

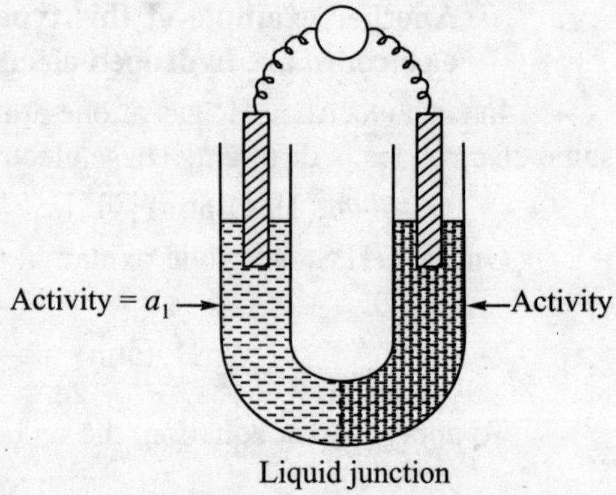

Fig. 1.26 Concentration cell with a liquid junction

(A) Concentration cells obtained by the combination of two chemicals cells

Consider a cell in which the H_2 electrode and Ag, AgCl electrode are dipped in the same solution of HCl of activity a_1

$$\text{Pt, } H_2\text{(1 atm)} \mid HCl(a_1) \mid AgCl(s), Ag$$

The cell reaction is

$$\frac{1}{2} H_2\text{(1 atm)} + AgCl(s) \longrightarrow Ag(s) + HCl(a_1)$$

By Nernst equation, e.m.f. of this cell is

$$E_1 = E° - \frac{RT}{F} \ln a_1 \quad (n = 1)$$

The activities of solids Ag, AgCl and that of H_2(1 atm) is unity.

Let us have another similar cell but with different activity of HCl(a_2). The e.m.f. of this cell is

$$E_2 = E^\circ - \frac{RT}{F} \ln a_2 \quad (n = 1)$$

If the two cells are connected so as to oppose each other and then make one concentration cell.

Pt, H_2(1 atm) |' HCl(a_1), AgCl(s), Ag | Ag, AgCl(s), HCl(a_2) |' H_2(1 atm) Pt

The overall reaction of this cell will be the difference of the cell reaction which will come out to be

$$HCl(a_2) \longrightarrow HCl(a_1)$$

Similarly e.m.f. is given by

$$E = E_1 - E_2 = \left(E^\circ - \frac{RT}{F} \ln a_1\right) - \left(E^\circ - \frac{RT}{F} \ln a_2\right)$$

$$= -\frac{RT}{F} \ln a_1 + \frac{RT}{F} \ln a_2 - \frac{RT}{F} \ln \frac{a_2}{a_1}$$

(B) Chemical cell with transfer: The common example of this type of cells is Daniel cell.

$$Zn \mid Zn^{2+} \parallel Cu^{2+} \mid Cu$$

There is liquid junction potential in the cell because of two different electrolytes in the cells.

Electrode Concentration cell

The electrodes may have different concentration. Mostly these electrodes are gas electrodes operating at different pressure or these are amalgams (Hg + metals) with different concentration.

For example, $Cd_{(Hg)}(C_{Cd} = C_1) \mid Cd^{2+} \mid Cd_{(Hg)}(C_{Cd} = C_2)$

Electrode Reaction

Oxidation $Cd_{(C_1)} \longrightarrow Cd^{2+} + 2e^-$

Reduction $Cd^{2+} + 2e^- \longrightarrow Cd_{(C_2)}$

Overall Reaction $Cd_{(C_1)} \longrightarrow Cd_{(C_2)}$

e.m.f. of cell $E = E_R - E_L$

$$E = \frac{0.0591}{2} \log \frac{C_1}{C_2}$$

If $C_2 < C_1$ then e.m.f. will be positive and the reaction will be spontanesous *i.e.,* Cd^{2+} will go from higher to low concentration of amalgam.

In case of gas electrode concentration cell, for example:

$$Pt \mid H_2(P_1) \mid HCl \mid H_2(P_2) \mid Pt$$

Reduction $2H^+ + 2e^- \longrightarrow H_2(P_2)$

Oxidation $H_2(P_1) \longrightarrow 2H^+ + 2e^-$

Overall $H_2(P_1) \longrightarrow H_2(P_2)$

$$E = E^\circ - \frac{0.0591}{2} \log \frac{P_2}{P_1}$$

If $P_2 < P_1$ then e.m.f. is +ve, so process will be spontaneous.

Example 37: ***Determine the e.m.f. of the concentration cell consisting of zinc electrodes, one immersed in a solution of 0.01 M concentration and the other in a solution of 0.1 M concentration of its ions at 25°C.***

Solution: The cell may be shown as

$$Zn; Zn^{2} (C = 0.01) \| Zn^{2+} (C = 0.1); Zn$$

e.m.f. of the cell,

$$E = \frac{2.303RT}{2F} \log \frac{0.1}{0.01} = \frac{0.0591}{2} \log 10 = \mathbf{0.0295\ volt.}$$

Example 38: ***Calculate the e.m.f. of the following cell at 25°C.***

$$Cd \mid CdSO_4(m = 0.01, \gamma = 0.383) \| CdSO_4(m = 1.00, \gamma = 0.042) \mid Cd$$

Solution: It is given than $m_1 = 0.01$, $\gamma_1 = 0.383$, $m_2 = 1.00$, $\gamma_2 = 0.042$

And for a Cd/Cd^{-2} electrode, $n = 2$

We know $$E = \frac{RT}{nF} \ln \frac{a_2}{a_1} = \frac{0.059}{2} \log \frac{m_2 \gamma_2}{m_1 \gamma_1}$$

We know that $a = \gamma m$,

where a = activity, γ = activity coefficient and m = molality of the electrolyte

$$= \frac{0.059}{2} \log \frac{1 \times 0.042}{0.01 \times 0.383} = \mathbf{0.0307\ volt}$$

Example 39: ***Write equation for each half reaction and calculate E_{cell} for the following concentration cell. $Zn/Zn^{++}(1.0\ M) \| Zn^{++}(0.15\ M)/Zn$ as the cell discharges, does the concentration of two solution becomes smaller or larger.*** (P.T.U. Jan. 2006)

Solution: The given concentration cell is represented as:

$$\underset{C_1(\text{say})}{Zn/Zn^{2+} (1.0M)} \| \underset{C_2(\text{say})}{Zn^{2+} (0.1\ M)/Zn}$$

This is a case of electrolytic concentration cell without transference of ions as the two half cells are connected by salt bridge. The half cell reactions as per given concentration cell will be:

Oxidation half cell (Left Half cell):

$$Zn(s) \longrightarrow \underset{(C_1)}{Zn^{2+}} + 2e^-$$

Reduction half cell (Right half cell):

$$\underset{(C_2)}{Zn^{2+}} + 2e \longrightarrow Zn(s)$$

$$E_{cell} = \frac{2.303RT}{nF} \log \frac{C_2}{C_1}$$

After substituting standard values of R = 8.314 $JK^{-1}\ mol^{-1}$, F = 90500 C mol^{-1} and T = 298 K, we get

$$= \frac{0.059}{2} \log \frac{0.15}{1} = 0.0295 - 0.823 = -0.0243\ V$$

Since E_{cell} is negative therefore this concentration cell is not feasible. However to make the cell work to give electricity, the above cell should be represented as:

$$Zn/Zn^{2+} (0.15\ M) \| Zn^{2+} (1.0\ M)/Zn$$
$$C_1(say) \qquad C_2(say)$$

$$E_{cell} = \frac{0.059}{n} \log \frac{C_2}{C_1}$$

$$= \frac{0.059}{n} \log \frac{1}{0.15} = 0.295 - 0.823 = 0.0243\ V$$

Since E_{cell} is positive, the cell will work and concentration of more dilute solution will increase and whereas that of more concentrated solution will decrease.

1.18 BATTERIES

A commercial cell, is used as a source of electrical energy is basically an electrochemical cell there oxidizing and reducing agents are made to react by using a suitable device. In principle any redox reaction can be used as the basis of an electrochemical cell, but there are some limitations to the use of many reactions. Two or more electrochemical cells connected in series constitute a battery. The term battery should not be used while referring to a single cell. A battery should be reasonably light and compact and its voltage should not vary appreciably during the use.

Types of cell: There are mainly two types of cells:

(*i*) primary cells and (*ii*) secondary cells

(*i*) In primary cells, the reaction occurs only once and the battery then becomes dead over a period of time and cannot be used again. (For example, dry cell, mercury cell).

(*ii*) Secondary cells can be recharged by passing current through them so that they can be used again and again. (For example lead storage battery, nickel-cadmium storage cell).

(A) Primary cells:

(*a*) **Dry cell:** The most familiar type of battery is the dry cell, which is a compact version of Laclanche cell (Fig. 1.27). In this cell, the anode consists of a zinc container and the cathode is a graphite rod surrounded by powdered MnO_2 and carbon. The space between the electrodes is filled with a moist paste of NH_4Cl and $ZnCl_2$. The electrode reactions can be written approximately as.

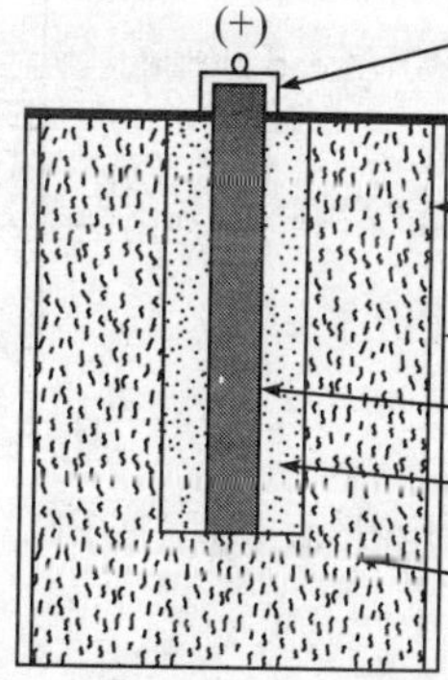

Fig. 1.27 A dry cell

At anode: $Zn(s) \longrightarrow Zn^{2+} + 2e^-$

At cathode: $MnO_2 + NH_4^+ + e^- \longrightarrow MnO(OH) + NH_3$

In the cathode reaction, manganese is reduced from the + 4 oxidation state to the + 3 state. Ammonia is not liberated as a gas but combines with Zn^{2+} to form $[Zn(NH_3)_4]^{2+}$ ion. Dry cells do not have an indefinite life as acidic NH_4Cl corrodes the zinc container even while not in use. Dry cells have a potential of approximately 1.25 to 1.5 V. An improved form of dry cell in which NH_4Cl is replaced by KOH as an electrolyte avoids corrosion of zinc container since zinc does not dissolve readily in a basic medium in this cell, zinc is powered form is mixed with KOH to get a gel. This cell is called **alkaline cell.**

As the life of alkaline cell is longer and maintains better voltage than dry cell, it finds application in use of calculators, watches camera, etc.

(*b*) **Mercury cell:** A new type of dry cell, which has found use in small electrical circuits (such as hearing aids, digital watches and cameras), is the mercury cell. Here zinc mercury-amalgam is the anode; a paste of HgO and carbon is the cathode. The electrolyte is a paste of KOH and ZnO. The reaction of the cell is as follows:

At Anode: $Zn(amalgam) + 2OH^- \longrightarrow ZnO(s) + H_2O + 2e^-$

At Cathode: $HgO(s) + H_2O + 2e^- \longrightarrow Hg(l) + 2OH^-$

Overall reaction: $Zn(amalgam) + HgO(s) \longrightarrow ZnO(s) + Hg(l)$

Since the overall cell reaction does not involve any ion solution whose concentration can change the cell shows constancy in potential through out its life. The cell potential is approximately 1.35 V.

(B) Secondary cells: The most important secondary cell is the **lead storage battery** containing six cells of 2 V each connected in series. It consists of a lead grid filled with spongy lead as anode and a lead grid packed with lead dioxide as cathode (Fig. 1.28). A solution of sulphuric acid (38% by mass or density of 1.3 g cm^{-3}) is used as an electrolyte. The reactions taking place in the cells during use or discharge are:

At Anode:

$$Pb(s) \longrightarrow Pb^{2+} + 2e^-, \quad Pb^{2+} + SO_4^{2-}(aq) \longrightarrow PbSO_4 \downarrow$$

At Cathode:

$$PbO_2(s) + 4H^+(aq) + 2e^- \longrightarrow Pb^{2+} + 2H_2O, \quad Pb^{2+} + SO_4^{2-} \longrightarrow PbSO_4 \downarrow$$

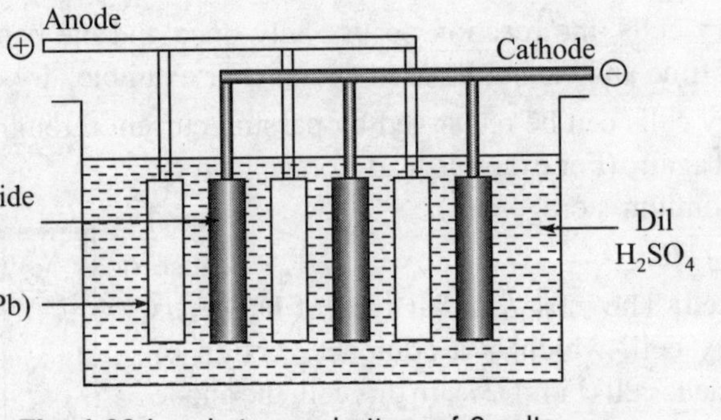

Fig. 1.28 Lead storage battery of 6 volts

Overall reaction: $Pb(s) + PbO_2(s) + 4H^+(aq) + 2SO_4^{2-}(aq) \longrightarrow 2PbSO_4(s) + 2H_2O$

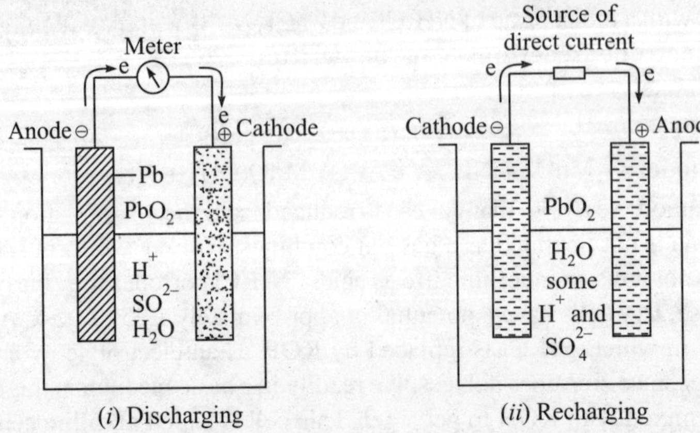

Fig. 1.29 Discharging and recharging of a lead storage cell

During discharge, H_2SO_4 is used and its density falls. When the density falls below 12 g cm^{-3}, the battery can be recharged after use by passing electric current. During recharging, the cell reaction gets reversed at the respective electrodes and operates like an electrolyte cell (Fig. 1.29) as under:

At cathode: $PbSO_4(s) + 2e^- \longrightarrow Pb + SO_4^{2-}(aq)$

At anode: $PbSO_4(s) + 2H_2O^- \longrightarrow PbO_2(s) + SO_4^{2-}(aq) + 4H^+(aq) + 2e^-$

Overall reaction: $2PbSO_4 + 2H_2O \longrightarrow Pb(s) + PbSO_2(s) + 4H^+(aq) + 2SO_4^{2-}(aq)$

This is possible because solid $PbSO_4$ formed during the working of the cell sticks to the electrodes and can gain or lose electrons during recharging.

Nickel-Cadmium Storage Cell

Another secondary cell is the **nickel-cadmium storage cell,** which has a longer life than the lead storage cell but is more expensive to manufacture.

It consists of Cd as anode and metal grid containing Ni(IV) oxides as cathode immersed in KOH solution. The cell reaction is given below:

At anode: $Cd(s) + 2OH^-(aq) \longrightarrow Cd(OH)_2(s) + 2e^-$

At cathode: $NiO_2(s) + 2H_2O + 2e^- \longrightarrow Ni(OH)_2(s) + 2OH(aq)$

Overall reaction: $Cd + NiO_2 + 2H_2O \longrightarrow Cd(OH)_2(s) + Ni(OH)_2(s)$

In this cell, the reaction products adhere to the electrodes and the reaction can be reversed by charging. Since no gases are evolved during charging or discharging, so this cell has very long life and its potential is about 1.4 V.

It is widely used in cordless phones, cordless electronic shavers and electronic flash units.

1.18.1 Metal-Air Batteries

An important example of Metal-Air batteries is that of zinc-air battery discussed below. Zinc-air battery is at the fore front of modern battery technology because one of the active materials (air) does not contribute to the mass of the battery and hence offers a very high energy density. It has an energy density of about 100 Wh kg^{-1} which is three times the classical lead acid battery or twice that of Ni-Cd battery. It finds applications in railways and military radio recovers.

Zinc-air Battery

In this type of cell, the zinc anode and the porous carbon cathode are immersed in 20% solution of NaOH.

Anode reaction: $Zn + 4(OH^-) \longrightarrow ZnO_2 + 2H_2O + 2e^-$

Cathode reaction: $H_2O + [O] + 2e^- \longrightarrow 2(OH^-)$

Overall cell reaction: $Zn + 2(OH^-) + [O] \longrightarrow ZnO_2 + H_2O$

Since OH^- ions take part in the electrode reaction, the alkali gets diluted. The voltage of the cell is 1.45 V. The zinc-air batteries offer the possibility of obtaining very high energy densities. In place of a cathode material placed in the battery during manufacture, oxygen from the atmosphere is used as cathode material, thus allowing for a more efficient design. The cutaway view of miniature air cell is shown in Fig. 1.30.

It resembles a typical miniature cell, except for the air access holes in the can. The anode occupies more internal volume of the cell. However, in place of a thick cathode pellet, it is provided with a thin layer containing the cathode catalyst and air distribution passages. Air enters the cell through the holes in the can and the oxygen reacts at the surface of the cathode catalyst. The air access holes are generally covered with a protective tape, which can be removed when the cell is used.

Small amounts of metal oxides can also be used as catalysts. If the reaction is well catalyzed, the battery can operate at a higher and more useful voltage.

Air cell are ideal for use in such applications where the usage is largely continuous. They are not general purpose cells. They have to be carefully designed for particular application. They are mainly used in hearing aids, in railway and the locations where the electric supply by cables is either expensive or impossible.

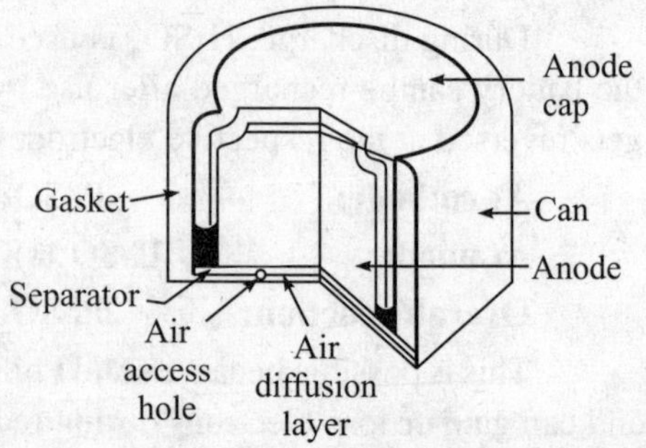

Fig. 1.30 Cross-sectional view of a miniature air cell battery.

The miniature, button type cells are small, button shaped cells, which use zinc anodes, alkaline (NaOH or KOH) electrolyte and a variety of cathodic materials *e.g.*, MnO_2, HgO, Ag_2O or even air (in case of Zn-air batteries). These batteries are used in watches, calculators cameras, hearing aids and other miniature devices.

1.19 FUEL CELLS

A fuel cell is an electrochemical cell which can convert the chemical energy contained in a readily available fuel-oxidant system into electrical energy by an electrochemical process, in which the fuel is oxidised at the anode. Fuel cell consists of an electrolyte and two electrodes, similar to any electrochemical cell. However, the fuel and the oxidising agents are continuously and separately supplied to the cells, at which they undergo reactions. These cells are capable of supplying current as long as the reactants are supplied. Electrical cells that are designated to convert the energy from the combustion of fuels such as hydrogen, carbon monoxide or methane directly into electrical energy are called fuels cells.

Hydrogen-Oxygen Fuel Cell

One of the most successful fuel cell uses the reaction of hydrogen with oxygen to form water (Fig. 1.29). This cell was used for electric power in the Appollo space programme. The water vapour produced was condensed and added to the drinking water supply for the astronauts.

In this cell (Fig. 1.29) Hydrogen-Oxygen fuel cell is used as an electric power source for space vehicles, mobile power systems and military. Hydrogen and oxygen gases are bubbled through a porous carbon anode and a cathode compartiments respectively. The gases diffuse through the electrode pores into concentrated aqueous sodium hydroxide. Catalysts [Pt or Ni] are incorporated in the electrode. The electrode reactions are:

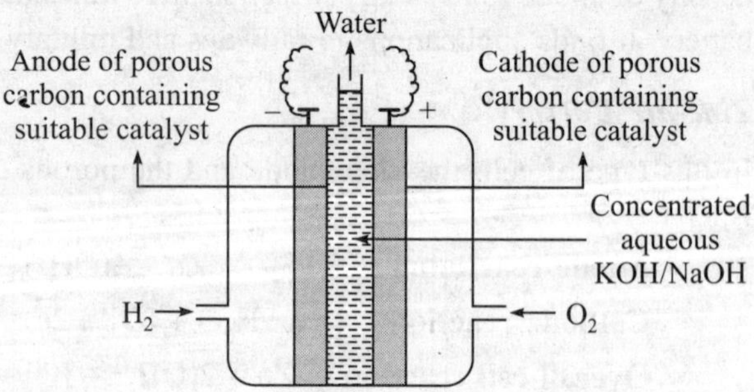

Fig. 1.29 A simple fuel cell

At anode: $H_2(g) + 2OH^-(aq) \longrightarrow 2H_2O(l) + 2e^-$

At cathode: $\frac{1}{2}O_2(g) + H_2O(l) + 2e^- \longrightarrow 2OH^-(aq)$

Overall reaction: $H_2(g) + \frac{1}{2}O_2(g) \longrightarrow H_2O(l)$

This cell runs continuously as long as the reactants are supplied. It should be noted that only product discharged by the cell is water. Water should be removed at the same rate as it is formed. Otherwise, water dilute the NaOH and the cell becomes inoperative. Therefore, electrolyte is kept hot, so that water evaporates as fast as it formed.

Since fuel cells converts the energy of a fuel directly to electricity, they are potentially more efficient than the conventional methods of generating electricity at a large scale by burning hydrogen, carbon or fuels or by using a nuclear reactor. Although theoretically we expect 100% efficiency in fuel cells, so far as efficiency of only 60–70 has been attained. Since fuel cells are efficient and free from pollution, efforts are being made to produce better commercially practicable fuel cells.

This cell consists of two electrodes made of porous carbon [graphite]. The electrodes are placed in a concentrated solution of NaOH or KOH in water.

Other important fuel cells are: Propane-oxygen fuel cell and methyl alcohol-oxygen fuel cell.

In case of propane-oxygen fuel cell: the half reaction are:

At anode: $C_3H_8(g) + 6H_2O(l) \longrightarrow 3CO_2(g) + 20H^+ + 20e^-(aq)$

At cathode:

$$5O_2 + 20H^+ + 20e^- \rightarrow 10H_2O(l)$$

Overall reaction $$C_3H_8(g) + 5O_2(g) \rightarrow 3CO_2(g) + 4H_2O(l)$$

REVISION QUESTIONS

1. Define the following terms:
(*a*) Specific conductance (*b*) Equivalent conductance
(*c*) Molar conductance

2. How is the specific conductance of an electrolyte solution determined? Described the experimental method.

3. (*a*) Write a short note on variation of equivalent conductance with dilution.
(*b*) Write down the methods for determination of equivalent conductance at infinite dilution for strong and weak electrolytes.
(*c*) How would you measure the conductivity of an aqueous salt solution?

4. Define specific conductance, equivalent conductivity and molar conductivity. What is the relationship between them? What effect do they produce upon dilution?

5. What do you understand by the terms: transport number and ionic mobility of an ion? How are these related to each other? Write their units.

6. Describe Hittorf's method for determining the transport number of Ag^+ and NO_3^- ions in a solution of $AgNO_3$ when the electrodes are made of silver.

7. Write a short notes on:
(*a*) Transport number
(*b*) Kohlrausch law of independent migration of ions

8. How does Kohlrausch's law helps in determining the dissociation constants of a weak acid by conductance measurements?

9. (*a*) Derive the expression for the determination of transport number of Hittorf's method when electrodes are not attacked?

(*b*) Explain the factors on which the transport number depends.

(*c*) Give two important differences between EMF and potential difference.

10. (*a*) How will you experimentally show that ions move with different speeds under the influence of electric current?

(*b*) Justify the following:

(*i*) Fall in concentration around an electrode is proportional to the speed of ion moving away from it.

(*ii*) Although ion move with different speeds, yet the number of ions discharged at the two electrodes remains the same.

(*iii*) Why equivalent conductance of hydrogen and hydroxyl ions are much greater than those other ions.

11. (*a*) What are the assumptions on which the theory of strong electrolytes based? How does it explain the low values of ionic mobilities in concentrated solution of strong electrolytes?

(*b*) Explain the term transport number of an ion. Explain the factors on which it depends.

12. (*a*) What is meant by transport number of an ions. Describe Hittrof's method of determination of transport number.

(*b*) Give details of calculations of determining the transport number using attackable as well as unteackable electrodes.

13. (*a*) State and explain Kohlrausch Law of independent migration of ions.

(*b*) Define the terms: Ionic conductance, Ionic mobilities and Transport number. How are they related?

14. Apply Kohlrausch law in determining the following:

(*i*) The equivalent conductance of a weak electrolyte at infinite dilution.

(*ii*) The degree of dissociation and dissociations constant of a weak electrolyte

(*iii*) The ionic product of water.

15. How are the conductance measurement used for determining the solubility of sparingly soluble salt? Briefly explain taking one suitable example. Name and state the law that has been used in such determinations.

16. Describe the construction and working of the following electrodes:

(*a*) hydrogen electrode (*b*) calomel electrode

17. (*a*) Explain the meaning of an electrode potential being positive or negative with respect to hydrogen electrode.

(*b*) What is a reference electrode? Describe briefly.

18. Explain briefly how does Debye-Huckel theory explain the increase in equivalent conductance with delution in case of strong electrolytes.

19. Explain in detail the Inter-ionic theory of strong electrolytes and discuss its significance.

20. Write short notes on the following:

(*a*) Relaxation effect and Electrophoretic effect in case of strong electrolytes

(*b*) Hydration of ions

(*c*) Inter-ionic attraction theory

21. What are concentration cells? Explain in detail the electrolyte concentration cells with transference and without transference.
22. (*a*) How is glass electrode constructed? Write the cell reactions.
(*b*) What is the effect of electrolyte concentration on electrode potential?
23. (*a*) What do you understand by electrochemical series?
(*b*) Write down the cell for which the overall reaction is

$$Cd + Cu^{2+}_{(a=1)} \longrightarrow Cd^{2+}_{(a=2)} + Cu$$

24. (*a*) Derive Nernst equation for the calculation of cell's EMF.
(*b*) What is the cell reaction and cell EMF at 298 K of the cell?

$$Zn/Zn^{2+}_{(a=1)} \,||\, Pb^{2+}_{(a=1)} \,|\, Pb$$

25. (*a*) Explain ionic conductance, transport number of an ion and absolute ionic velocity.
(*b*) What is an electrochemical cell? *(Anna, Dec. 2002, May, 2003)*
(*c*) What is the role of salt bridge in a voltaic cell?
(*d*) Zinc reacts with dil H_2SO_4 to give H_2, but Ag does not. Why? *(Anna, Dec. 2005)*
26. (*a*) Explain the construction and working of a calomel electrode. Indicate the role of calomel electrode.
(*b*) What are ion-selective electrodes? *(VTU, B.E., August, 2002)*
27. (*a*) Define single electrode potential and standard electrode potential? Derive the Nernst equation for electrode potential.
(*b*) Explain the construction and working of a calomel electrode. Mention its advantages.
(*c*) Write a short note on ion-selective electrode. *(VTU, B.E., Feb., 2003)*
28. (*a*) Discuss the origin of electrode potential and derive Nernst equation for electrode potential.
(*b*) What are ion-selective electrodes? Give the construction of glass electrode and explain the experimental method of determining pH glass electrode.
(*c*) An electrochemical cell consists of iron electrode dipped in 0.1 M $FeSO_4$ and silver electrode dipped in 0.05 M $AgNO_3$. Write the cell representation, cell reaction and calculate EMF of the cell at 298 K. Given that the standard reduction potentials of iron and silver electrodes are – 0.44 and + 0.80 V, respectively. *(VTU, B.E., July, 2004)*
29. Describe the various points of difference between a primary and a secondary battery. Discuss the various characteristics of a battery.
30. Describe the construction and working of Lead-Acid battery with the reactions occurring during discharge and charging. *(VTU, March 1999)*
31. What are the electrode reactions involved in a zinc-air battery? Mention the various characteristics and applications of Zn-air battery.
32. (*a*) Why zinc-air battery is at the fore front of modern battery technology?
(*b*) Write the charge-discharged cell reactions of zinc-air battery.
33. Write short note on fuel cell. How is it different from commercial cell? Mention the advantages of fuel cells. *(VTU, Aug, 2K, Aug, 2001, Jan., 2010)*
34. Define fuel cell. Explain the construction and working H_2—O_2 fuel cell. What are the advantages and limitations of fuel cell? Write the reactions involved. Why is water formed in this fuel cell removed continuously?

35. (*i*) Explain the following battery characteristics (*a*) capacity (*b*) cycle life.

(*ii*) Explain the construction and working of Pb-acid battery

(*iii*) Explain the construction and working of H_2—O_2 fuel cell. Why is water formed in this cell removed continuously. *(VTU, Dec 2009 and June 2010)*

36. Describe the construction and working of Zn-MnO_2 battery with relevant reactions occurring during discharge. *(VTU, Aug. 2K, Aug. 2001)*

37. (*a*) Write briefly about: (*i*) primary cells, (*ii*) secondary cells. *(SK, Aug. 2001)*

38. (*a*) What is standard electrode potential? Give its importance. *(Pt, Dec. 2K)*

(*b*) Derive Nernst's equation for single electrode potential and explain the terms involved in it. Write its applications.

(MD, May 2K; Mg, Dec. 1998; Amravati Dec. 2K; Tripura, 2K; MK, April 2001; VT, March 1999, Aug., 2001; SK, June 2001; PT, Dec. 2K)

39. Differentiate between the characteristics of an electrolytic cell and those of a galvanic cell. What do you mean by standard electrode potential? How can it be measured? *(Dibrugarh, June 2K)*

40. What are reference electrodes? Describe the construction of calomel electrode.

(VT, Aug. 2001; PT, Dec. 2K; Anna, Jan. 2001)

41. (*a*) Derive Nernst's equation for the calculation of cell e.m.f. *(PT, May 04)*

(*b*) Write short note on electrode potential.

(*c*) Why do electrochemical cells stop working after some time?

42. (*a*) Derive Nernst's equation.

(*b*) Describe the following electrodes giving the diagram, electrode notation and electrode reaction: (*i*) standard hydrogen electrode, (*ii*) calomel electrode. *(Anna, Dec. 2003)*

43. (*a*) Explain the function of salt bridge. *(Calicut, June 2001)*

(*b*) Derive Nearnst equation giving the e.m.f. of a cell in terms of concentrations of reactants and products of the cell reaction. *(MD, Oct. 1999; Surat 2K)*

44. (*a*) The electrode potential of Zn is assigned a 'negative' value; whereas that of Cu is assigned a 'positive' value. Give reasons.

(*b*) Write short note on calomel electrode. *(PT, Dec. 2002)*

45. (*a*) Describe the following electrodes, giving the diagrams, electrode notation and reaction: (*i*) standard hydrogen electrode; (*ii*) calomel electrode. *(Anna, Dec. 2003)*

(*b*) Derive Nernst's equation for e.m.f. of a cell. Write Nernst's equation for e.m.f. of Daniel cell. *(Anna, Dec. 2004)*

(*c*) What is single electrode potential? How is single electrode potential determined?

(Anna, Dec. 2006)

❑❑❑

2

Phase Rule

2.1 INTRODUCTION

J. Willard Gibbs, in 1875, as a result of his mathematical and thermodynamic studies, put forward a rule known as **phase rule.** This rule is applicable to all **heterogeneous systems** in equilibrium without any exception. With its help we can predict *qualitatively the effect of temperature, pressure and concentration on a heterogeneous system in equilibrium by means of phase diagrams.*

It is assumed that the equilibrium is not influenced by gravity, electrical or magnetic forces or by surface action and is influenced only by *temperature, pressure and concentration.* The maximum number of degree of freedom is taken as three.

Mathematically, Gibbs phase rule may be stated as:

$$F + P = C + 2 \text{ or } F = C - P + 2$$

where, F = number of degrees of freedom.

C = number of components

and P = number of phases.

2.2 EXPLANATION OF THE TERMS INVOLVED

Before proceeding to the statement of phase rule, it will be necessary to define and explain in detail some terms which are employed frequently in this connection. These terms are *(i) true and metastable equilibrium (ii) number of phases (iii) number of components and (iv) degrees of freedom of a system.*

1. True and Metastable Equilibrium: A state of true equilibrium is said to exist in a system when the same state can be attained by approach from either direction. Thermodynamically speaking, *true equilibrium is attained when the free energy content of the system is at the minimum* for the given values of the variables.

Example: An instance of true equilibrium is ice and liquid water at 1 atm pressure and 0°C (273 K). This state can be achieved from either of the two directions. Water at room temperature can be cooled upto 0°C when the solid phase (ice) will appear and both the phases solid ice and liquid water exist in equilibrium at 0°C. SImilarly ice at 0°C or at lower temperature can be slightly warmed to get both the phases

$$\text{Solid Ice} \xrightleftharpoons{0°C} \text{Liquid water}$$

Another example may be that of water in equilibrium with its vapour at 30°C. This equilibrium can be attained from either direction *i.e.,* either by cooling water above 30°C or by warming water below 30°C.

$$\text{Ice} \rightleftharpoons \text{Water}$$

Water at –5°C is said to be in a state of metastable equilibrium. Such a state can be achieved only by careful approach from one direction. In other words water at room temperature can not be cooled up to –5°C because it will normally freeze to ice at 0°C and below this temperature it will exist as solid. But water at –20°C can be brought to water at –5°C by carefully warming water existing at a lower temperature (say –20°C).

It may be preserved provided the system is not subjected to a sudden 'shock', 'stirring', or 'seeding' by the solid phase. As soon as a crystal of ice is introduced, solidification starts and the temperature rises to 0°C when the liquid water changes to solid ice.

2. Phase: A heterogeneous system consists of various homogeneous parts in contact with each other but with distinct boundaries. The *various homogeneous parts* which have distinct physical and chemical properties and which may (if necessary) be mechanically separated from each other are known as phases.

Hence a phase may be defined "as any part of a system which is homogeneous in itself physically distinct from others, and mechanically separable from other parts of the system".

The number of phases (P) in a heterogeneous system is given by the number of such physically distinct and homogeneous parts constituting the system.

Essential Conditions for a Phase

(*i*) It should be homogeneous within itself. It may be a homogeneous part of a heterogeneous system.

(*ii*) It should be *physically distinct.*

(*iii*) It is separated from other parts of system by *well defined boundary.*

(*iv*) There must be *dynamic equilibrium* between different phases of the system due to exchange of chemical species.

Examples:

(*i*) **Gaseous Phases:** *Gases are completely miscible in all proportions* and therefore, *mixtures of two* or *more gases* are homogenous and constitute only *one phase.* There is no boundary between the gas or vapours. Air, which is a mixture of nitrogen oxygen, carbon dioxide, water vapour, etc., is homogenous and thus constitute only one phase.

(*ii*) **Liquid Phases:** The number of liquid phases in a system is given by the number of liquids present depending on their mutual solubilities.

(*a*) *Two or more completely miscible liquids* present in a system such as water, alcohol, acetone, etc., constitute only *one liquid phase* because it form homogenous mixture.

(*b*) *Two immiscible liquids* such as water and mercury, water and carbon disulphide, water and carbon tetrachloride in contact, form *two different phases.*

(*c*) In case of *two partially miscible liquids such as water and ether, the number of liquid phases, depends on the composition, i.e., the ratio in which the two liquids are present,* When the amount of one liquid is very large and the amount of the other is very small, the system is a homogeneous solution and hence only *one phase.*

At *intermediate compositions,* the system is composed of two homogeneous solutions—one rich in ether and the other rich in water and therefore, *two phases.*

In such systems composed of partially miscible liquids, the number of phases depends on temperature as their mutual solubility varies with temperature.

(*iii*) **Solid Phases:** A heterogeneous mixture of solid substances consists of *as many phases as there are substances present.* Different crystalline forms of a substance or allotropic forms of a substance constitute different phases and these are considered different solids for this purpose. But a solid solution is regarded as a single phase as it is considered as one solid truely dissolved in another solid. For example solid solution of lead and silver is considered as one phase.

For example, Decomposition of $CaCO_{3(s)} \rightleftharpoons CaO_{(S)} + CO_{2(g)}$ is a system with three phases two solids and one gaseous (Because any number of gases will constitute one phase only).

Remember:

(*i*) *Miscible gases and miscible liquids constitute a single phase.*

(*ii*) *Solid + liquid, solid + gas, liquid + gas, two solids and two immiscible liquids constitute two phases.*

(*iii*) *Different solids (except in solutions) are regarded as different phases, howsoever, intimately they may be mixed.*

3. Components: In phase rule, the term component does not have its literal meaning *i.e. does not represent the number of constituents*. The composition of a phase in a heterogeneous system is expressed quantitatively in terms of the molecular species constituting it. Some of these molecular species present in the system may be in chemical equilibrium and hence the concentration of these species is not independent of each other. By fixing the concentration of some of the species, the concentrations of other species which are in equilibrium with them are automatically fixed due to equilibrium between them.

The number of components (C) of a system at equilibrium is defined as "the smallest number of independently variable constituents by means of which the composition of each phase present can be expressed either directly or in the form of a chemical equation".

It may also be defined as 'the minimum number of molecular species in terms of which the compositions of all the phases can be quantitatively expressed'.

It is important to note that the number of components of a system may or may not be the same as the actual number of substances present in the system *i.e.* the constituents of the system. Only those constituents of an equilibrium mixture which can undergo *independent* variation are regarded as components.

Example:

***(i)* Water system consists of the equilibrium**

$$\underset{\text{(Solid)}}{\text{Ice}} \rightleftharpoons \underset{\text{(Liquid)}}{\text{Water}} \rightleftharpoons \underset{\text{(Gaseous)}}{\text{Vapour}}$$

This system consists of one component only *i.e.,* it is a *one component system* because the composition of each of the three phases present can be directly expressed as H_2O. Hydrogen and Oxygen are not taken as components because their ratio in water is fixed (1 : 8) and they not independently variable.

***(ii)* Sulphur System Consisting of Four Phases:** Monoclinic (*s*), rhombic (*s*), liquid (*l*) and vapours (*g*) is also *one component system* because the composition of each phase can be expressed in terms of composition of one constituents—sulphur only.

***(iii)* Solution of Sugar in Water:** The composition of solution of sugar in water is described by specifying that sugar and water are present in the solution phase. If the system has vapour as well as solid sugar in addition to solution, the composition of the three phases can be expressed again by sugar and H_2O, and in all two species are required to express the composition of the phases. The system has two chemical species and the number of components is also two.

***(iv)* Decomposition of $CaCO_3$** by heat can be expressed according to the equilibrium

$$\underset{\text{(Solid)}}{CaCO_3} \rightleftharpoons \underset{\text{(Solid)}}{CaO} + \underset{\text{(Gas)}}{CO_2}$$

Thus there are three different constituents forming three different phases. But the composition of each phase of the system can be expressed by chosing any two constituents and not all the constituents. So the system is a two component system although there are three constituents.

(*a*) ***If the components chosen are CaO and CO_2 then,***

Phase		**Composition**
$CaCO_3$	$\longrightarrow$	$CaO + CO_2$
CaO	$\longrightarrow$	$CaO + 0\,CO_2$
CO_2	$\longrightarrow$	$0\,CaO + CO_2$

(*b*) ***If the components chosen are $CaCO_3$ and CO_2 then***

Phase		**Composition**
$CaCO_3$	$\longrightarrow$	$CaCO_3 + 0\,CO_2$
CaO	$\longrightarrow$	$CaCO_3 - CO_2$
CO_2	$\longrightarrow$	$0\,CaCO_3 - CO_2$

(*c*) ***If the components chosen are $CaCO_3$ and CaO, then***

Phase		**Composition**
$CaCO_3$	$\longrightarrow$	$CaCO_3 + 0\,CaO$
CaO	$\longrightarrow$	$0\,CaCO_3 + CaO$
CO_2	$\longrightarrow$	$CaCO_3 - CaO$

Thus in all these cases, the smallest number of constituents which can fix the composition of phases present at equilibrium is two, hence *"dissociation of $CaCO_3$ by heat" is a two component system.*

(v) Dissociation of Ammonium Chloride in a Closed Vessel: One component system does not always mean that only one chemical species is present. For example, in a solid vapour system of ammonium chloride.

(*a*) $$NH_4Cl(s) \rightleftharpoons NH_3(g) + HCl(g)$$

the solid phase is crystalline ammonium chloride and the vapour phase consists of mostly of hydrochloric acid gas and ammonia gas with small amounts of ammonium chloride. It is a one component system though there are three molecular species in the vapour phase because all the three species are related the following equations.

(*b*) $$NH_4Cl(s) \rightleftharpoons NH_3(g) + HCl(g) \rightleftharpoons NH_4Cl_{(g)}$$

In this case concentration of NH_3 and that of HCl in the solid phase as well as in the gase of phase is same.

It may be emphasized here that number of components need not be the same as the number chemical species *i.e.,*

As long as NH_3 and HCl are present in equivalent amounts, the composition of each phase may be represented as NH_4Cl. Hence it is *one component system.*

However, if an excess of either NH_3 or HCl is introduced, the composition of the gaseous phase can no longer be the same and the composition of this phase can no longer be represented by NH_4Cl alone but one more component is required. When NH_3 is in excess in the vapour phase, its amount has to be expressed. Hence it becomes *two-component system.* Similarly when HCl is in excess in the vapour phase, it becomes a two component system.

(vi) A System of NaCl, KCl and H_2O is three component system because the composition of all five phases *i.e.,* solid NaCl, solid KCl, solid H_2O, solution and vapours can be given by three species. NaCl, KCl and H_2O.

(vii) A System Consisting of NaCl, KBr and H_2O is a four-component system (Note the difference in these last two examples. The first one has two salts with common chloride ion). This system can yield maximum seven phases under varying conditions, solid KBr, solid KCl, solid NaBr, solid NaCl, solid H_2O, solution and vapour. Obviously, the composition of NaBr, and KCl cannot be given by only NaCl, KBr and H_2O. If however, one more species, say KCl, is included the composition of NaBr can be expressed as follow:

$$\text{Phase NaBr} - \text{NaCl} + \text{KBr} - \text{KCl} + 0\ H_2O$$

and thus the system is a four-component system.

(viii) In Case of Ions Present in the System: If ions are also present in the system then the condition of electro neutrality should be maintained, *i.e.* number of positively and negatively charged ions should be equal. Hence the number of components may be calculated by the modified relation is:

$$C = S - (R + 1)$$

where S stands for number of species and R for number ionic relations.

For example,

(*i*) In the system NaBr – KCl – H_2O, the number of component is calculated as follow:

Here numbers of species are nine NaBr, KCl, NaCl, H_2O, Na^+, Br^-, K^+, Cl^- (The dissociation of water is ignored). The number of independent ionic reactions, *i.e.* relations is four.

$$NaBr \rightleftharpoons Na^+ + Br^-$$

$$KCl \rightleftharpoons K^+ + Cl^-$$
$$Na^+ + Cl^- \rightleftharpoons NaCl$$
$$K^+ + Br \rightleftharpoons KBr$$

$\therefore$ Number of components C = S – (R + 1) = 9 – (4 +1) = 4.

Hence it is a four components system.

(*ii*) A dilute solution of sulphuric acid in water may be represented as:

$$H_2SO_4 + H_2O \rightleftharpoons H_3O^+ + HSO_4^-$$
$$HSO_4^- \rightleftharpoons H^+ + SO_4^{2-}$$

Here the number of species is five H_2SO_4, H_2O, H_3O^+, HSO_4^- and SO_4^{2-}

The number of relations is two

$\therefore$ Number of components C = S – (R + 1) = 5 – (2 + 1) = 2

Hence, it is a two component system.

4. Degrees of Freedom or Variance

The number of degrees of freedom or variance of a system (F) is "the smallest *number of variable factors i.e.,* temperature, pressure and concentration of the components which must be arbitrarily fixed in order that the condition of the system may be completely defined."

Examples:

(*i*) **Ice-Water-Vapour System:** In this system there are three phases of one component (H_2O). These three phases can coexist in equilibrium only at one particular temperature and under one particular pressure. Any variation of these factors will lead to the ultimate disappearance of one or more of the phase and, therefore, to the destruction of the system. Hence this system has no degrees of freedom or in other words it is *invariant.*

(*ii*) **Water-Vapour-System:** This system consists of two phases of one component. The vapour pressure of water is definite at a definite temperature independent of the amount of water present. It follows, therefore, that if the temperature is fixed, the vapour pressure is also fixed and conversely, fixing the pressure, fixes the temperature. Thus the system has only one degree of freedom or in other words, it is *univariant*.

(*iii*) For a gas, temperature, pressure and concentration (volume) are related according to equation PV = RT. Thus, in order that the system may be completely defined, any two factors, out of P, V and T must be fixed. Hence it follows that a gas has two degrees of freedom or it is *bivariant.*

(*iv*) The system consisting of **saturated sodium chloride solution** is contact with solid salt is completely defined if we mention only temperature, since, the solubility of sodium chloride has a definite value at a certain temperature. The system is, therefore, *univariant*.

But the unsaturated solution of sodium chloride in water in equilibrium would be bivariant. In this case, besides temperature, the composition of the solution has to be fixed to define the system completely.

Thus, system may be described as *nonvariant, univariant, bivariant,* and *trivarial* depending upon the number of its degrees of freedom *i.e., zero, one, two,* or *three* respectively.

2.3 MATHEMATICAL STATEMENT OF PHASE RULE

The phase rule gives relationship between the number of phases, components and degree of freedom of a heterogeneous system.

The Gibb's **phase rule** may be stated as follows:

"In a heterogeneous system in equilibrium, the number of degree of freedom plus phase is equal to the number of components plus two."

Mathematically,

$$F + P = C + 2 \text{ or } F = C - P + 2$$

where, F = number of degrees of freedom.

C = number of components

and P = number of phases.

It is assumed that the equilibrium is not influenced by gravity, electric field and magnetic field and by surface action and is influenced only by temperature pressure and concentration. Hence, the number of degree of freedom is taken as three.

The phase rule is valid only under the following assumptions:

(*i*) *The system is in thermal and mechanical equilibrium. Consequently the pressure and the same in all the phase of system.*

(*ii*) *Surface contributions as well as contributions from any electric or magnetic field to properties of the system, are negligible.*

(*iii*) *Interphase surfaces are deformable and permeable to components.*

2.4 UTILITY OF PHASE RULE

The following are some of the advantages from the study of phase rule:

(*i*) *This rule helps in classifying various equilibrium states of the systems in terms of the number of components, the number of phases or the number of degrees of freedom.*

(*ii*) Different systems having the same degrees of freedom according to phase rule would behave in a similar fashion.

(*iii*) With its help we can predict the behaviour of any system with the changes in the variables such as temperature, pressure and concentration.

(*iv*) Phase rule applies to physical as well as chemical equilibria irrespective of the nature or amounts of the substances.

(*v*) By applying this rule we can predict whether a number of substances taken together would remain in equilibrium as such or would involve inter—conversion or elimination of some of them under a given set of conditions.

(*vi*) It finds extensive use in the study of heterogeneous systems. It has been found quite useful in *metallurgy* and conveys useful information about the *complex formation* among different components.

2.5 LIMITATIONS OF PHASE RULE

Although phase rule gives the conditions at which the different phases of a heterogeneous system can exist in equilibrium, yet it suffers from the following limitations:

1. Phase rule applies only to a single equilibrium state and does not tell us about the number of other equilibria possible in the system. It is not of much use for those systems *which attain the equilibrium state very slowly.*

2. Phase rule deals with macroscopic systems only and it does not tell us anything about *the molecular structure*.
3. The phase rule equation (F = C – P + 2) is applicable only to *system having three variables*. When the number of variables changes the equation shall have to be changed accordingly.
4. The phase rule does not give any information regarding *the time taken* for the system to attain equilibrium.
5. It does not give any information regarding *the amount of any phase* under equilibrium. It takes into account the number of phase and not their amounts. Even the small amount of phase or impurity, if present, counts to the number of phase present in the system in equilibrium.

PRACTICE PROBLEMS

1. Give the number of phases, components and degrees of freedom of following:
(*a*) Mixture of N_2 and H_2 contained in a vessel.
(*b*) Ice, Water and Vapour in equilibrium.
(*c*) An unsaturated sugar solution.
(*d*) A system consisting of NaCl, KCl and H_2O.
(*e*) Dissociation of NH_4Cl in a closed vessel.
(*f*) Dissociation of NH_4Cl in a closed vessel containing NH_3 also.

2. Classify the following into true and metastable equilibria:
(*a*) Ice and liquid water at atmospheric pressure and 0°C.
(*b*) Water at – 3°C.
(*c*) Water and its vapours at 20°C.
(*d*) $S_{(Rhombic)}$ and $S_{(Monoclinic)}$ at 95.5°C.
(*e*) $S_{(Rhombic)}$ and $S_{(Monoclinic)}$ below 95.5°C.

3. Give a mathematical statement of phase rule in the following cases:
(*a*) When temperature, pressure and concentration are variables.
(*b*) When temperature and concentration are variables.
(*c*) When concentration and pressure are variables.
(*d*) When concentration, pressure, temperature and magnetic force are variables.

4. Give the degrees of freedom of one component system in the following cases:
(*a*) When a single phase exists.
(*b*) When two phases are in equilibrium.
(*c*) When three phases are in equilibrium.

5. What is the minimum number of phases that can co-exist in a one component system?

Answers

1. (*a*) P – 1, C = 2, F = 3 (*b*) P = 3, C = 1, F = 0
(*c*) P = 2, C = 2, F = 2 (*d*) P – 5, C – 3, F = 0
(*e*) P = 2, C = 1, F = 1 (*f*) P = 2, C = 2, F = 2

2. (*a*), (*c*), (*d*) true equilibrium; (*b*) and (*e*) metastable equilibrium.

3. (*a*) F = C – P + 2, (*b*) F = C – P + 1,
(*c*) F = C – P + 1, (*d*) F = C – P + 3
4. (*a*) Two, (*b*) One,
(*c*) Zero
5. Three.

2.6 ONE-COMPONENT SYSTEMS

The general shape of the phase diagram for one component system is shown below in Fig. 2.1(*a*) CO_2 and that in case of water system in Fig. 2.1(*b*).

The inclination of curve OB representing solid-liquid equilibrium in Fig. 2.1(*a*) in general is always away from the Y-axis, whereas for water it is inclined towards Y-axis. This explains why the effect of pressure on melting point of ice is different from that of other solids in general. It is clear from the curves that melting point of ice decreases, while that of other solids increases with increase of temperature.

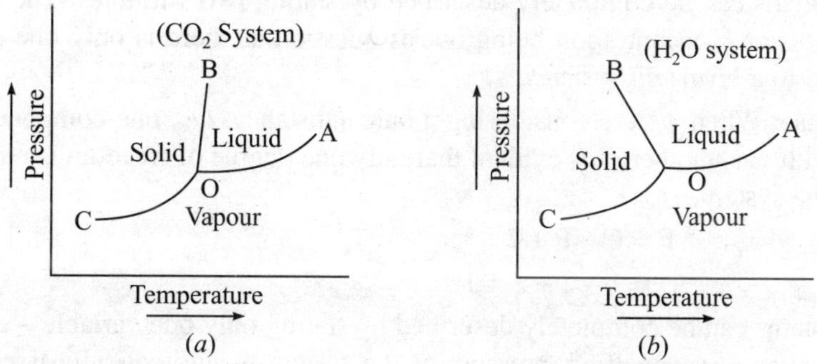

Fig. 2.1 Phase diagram of one component system

In phase diagrams of one component system, bivariant systems are represented by 'areas', univariant systems by 'lines' or 'curves' and non-variant systems by 'points'.

From the phase rule equation, it is quite clear that when there is one component existing in one phase, the value for the degree of freedom (F) is 2. Hence all systems of one component can be completely described graphically, by stating two variable factors; pressure and temperature on appropriate axis. From the statement of phase rule, in systems of one component it will be observed that:

(*i*) When the number of phases is reduced to one, the systems will have two degrees of freedom and are known as *Bivariant Systems.* A specified area represents a single phase and to locate any point in any area we have to specify both the variables factors.

(*ii*) When the number of phases is two, the systems will possess one degree of freedom an ure known as *Univariant systems.* A line or curve represents two phase in equilibrium and to locate a point on.

2.7 PHASE DIAGRAMS

Phase diagram in the complete description of behaviour of phases in equilibrium. The number of phases that exist together when the system is in equilibrium depends upon the conditionsof temperature and pressure or conditions of temerature and composition; pressure being kept constant. The conditions are determined experimentally and then substituted in temperature-pressure or temperature-composition diagram. The diagrams so obtained giving the conditions of equilibria between various phases of a substance are called **phase diagrams or equilibrium diagrams.**

Phase diagrams are thus maps which show the conditions under which the various phases are capable of stable existence and the condition under which one phase can be transformed into another. The diagram contain *a number of lines, areas and points of intersection,* the significance of each of which has been discussed below under the study of different systems. It is very easy to describe phase behaviour of a system by such diagrams and to know the conditions under which various phases will be present in the systems.

Let us consider one component systems with different number of phases.

(*a*) Single Phase: When a pure substance is in single phase and is under consideration, it is evident from the phase rule equations, that a maximum of two degree of freedom are required to be fixed in order to define the system.

$$F = C - P + 2 \qquad [\because \ C = 1, P = 1]$$
$$= 1 - 1 + 2 = 2$$

All such systems can be completely described by stating two variables. The two variables are temperature and pressure, concentration being out of question as there is only one component. Such systems are referred to a *bivariant systems.*

(*b*) Two Phase: When we are considering a pure substance, *i.e.,* one-component system in two phases in contact with one another, it is evident that only one degree of freedom is required to be fixed in order to define the system.

$$F = C - P + 2 \qquad [\because \ C = 1, P = 2]$$
$$= 1 - 2 + 2 = 1$$

All such systems can be completely described by stating only one variable—either temperature or pressure. If temperature is specified, pressure of the system is automatically fixed and vice versa. Such systems are referred to as *univariant systems.*

(*c*) Three Phases: When there are three phases in equilibrium in one component system, it is evident that the system has no degree of freedom.

$$F = C - P + 2 = 1 - 3 + 2 = 0$$

Such a system is refferred to as non-variant or invariant system.

2.8 THE WATER SYSTEM

This is most typical example of the one component system, in which the same chemical compound exists in the following three phases:

$$\underset{(s)}{\text{Ice}} \rightleftharpoons \underset{(l)}{\text{Water}} + \underset{(g)}{\text{Vapours}}$$

The conditions of temperature and pressure at which the various phases can exist have been determined experimentally and summed up in Fig. 2.2.

The phase diagram consists of:

(*i*) Three stable curves, OA, OB and OC and a metastable curve OA′,

(*ii*) Three areas, and

(*iii*) Definite point O.

A. Curves: (1) OA: This is the *vapour-pressure curve of water*, representing the equilibrium between *liquid water and vapour* at different temperatures. At any point on the curve the two phases are present. The curve starts from the point 0, the freezing point of water (0.0098°C at 4.578 mm) and ends at A, the critical temperature (374°C at 218 atms) beyond which the two phases merge into each other. Consider any point on the curve. For fixed value of pressure, temperature has also a fixed value. So by keeping the pressure constant, an attempt is made to raise the temperature then the liquid phase will disappear and if the temperature is lowered, then vapour phase will disappear. Therefore, to define the system completely at any time, it is necessary to mention temperature or pressure. In other words, pressure being fixed, it is not necessary to mention temperature; it is automatically fixed up. Thus water vapour system is univariant. It also follows from the phase rule equation.

$$H_2O_{(l)} \rightleftharpoons H_2O_{(g)}$$
$$F = C - P + 2$$
$$F = 1 - 2 + 2 = 1$$

At 100°C, the vapour pressure of water equals the atmospheric pressure (760 mm). This is, therefore, the boiling point of water. The curve OA extends upto the critical temperature of water (374°C), since above this temperature liquid water cannot exist.

2. OA′: The dotted curve OA′, continuation of OA curve represents the *vapour-pressure of super-cooled water*. It is a *metastable* curve. On slight disturbance, the super cooled phase at once changes to solid ice and the curve merges into OB.

$$H_2O_{(l)} \text{ (Supercooled)} \rightleftharpoons H_2O_{(s)}$$

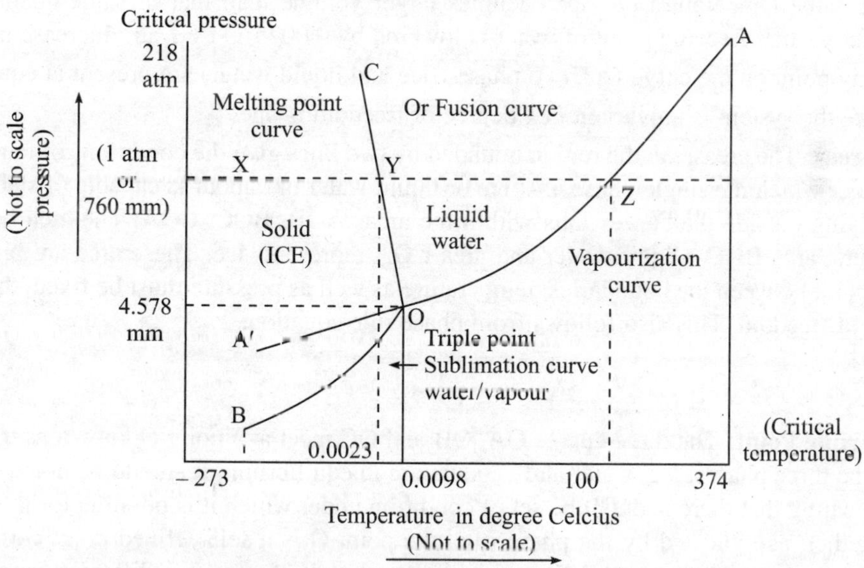

Fig. 2.2 Phase diagram of water system

The special features of vapour-pressure curve are:

(*a*) *There is no break in the vapour-pressure curve as long as the solid phase does not separate and*

(*b*) *The vapour pressure of the super-cooled water, which is an unstable phase is greater than of solid (ice) which is the stable phase at that temperature.*

3. OB: It is *the sublimation curve or the vapour pressure curve of ice*. It gives various values of temperature pressure at which ice and vapours can exist together.

$$H_2O_{(ice)} \rightleftharpoons H_2O_{(g)}$$

The curve starts from the point O, the freezing point of water and ends at B(– 273°C). As temperature decreases, the vapour-pressure of ice tends to become insignificant. Again we see that for a particular temperature there can be one and only one pressure value and *vice-versa.* In other words, ice vapour system is univariant. *i.e.,*

$$F = C - P + 2$$
$$F = 1 + 2 + 2 = 1$$

4. OC: It is the *melting point curve or fusion curve* and represents the equilibrium between ice and liquid water at various pressures.

In other words, it shows the effect of pressure on the melting point of ice. *It should be noted that the curve is inclind towards the pressure axis.*

$$H_2O_{(s)} \rightleftharpoons H_2O_{(l)}$$

The slope of the curve OC clearly indicates that expansion takes place on freezing of water and that melting point of ice is lowered by increase of pressure.

Normally melting point of solids is increased by increase of pressure because volume occupied by substance is the liquid form is usually greater than that is the solid form. But in case of water, the case is reversed. The volume occupied by ice is always greater than that of the same quantity of water. Thus when water changes to ice at the freezing point, the volume occupied by ice is larger than that of water at the same temperature *i.e.,* ice occupies larger volume than that of same quantity of water. The curve shows that freezing point of water is lowered by 0.0075°C by 1 atm increase in pressure.

At any point on the curve OC, two phases (ice and liquid water) are present at equilibrium.

Hence the system is univariant *i.e.,* degree of freedom is one.

B. Areas: The areas, *i.e.* the region bounded by two lines give the conditions of temperature and pressure under which the single phase, *i.e.,* ice or liquid water or vapour is capable of stable existence. Any point lying outside the curves falls within the areas AOB, BOC, COA. The area below AOB—water vapours, area BOC—liquid water and area COA represents ice. These area are bivariant, as to locate any point between the boundaries: temperature as well as pressure must be fixed, thus indicating two degree of freedom. This also follows from phase rule equation:

$$F = C - P + 2$$
$$F = 1 - 1 + 2 = 2$$

C. Triple Point: The three curves OA, OB and OC meet at a point O known as triple point. At this point, the three phases: ice, water and vapours are in equilibrium. It should be noted that it is only a point, indicating that there is only one set of condition under which it is possible for the three phases to exist together, as indicated by the phase rule, the point O is a self-defined point corresponding to definite temperature 0.0098°C and definite pressure, *viz.,* 4.58 mm. and is different from the ordinary freezing point *i.e.,* 0°C at 760 mm. pressure. If, either pressure or temperature be changed, then one of the three phases will disappear. Since on changing either pressure or temperature one of the phase disappears, therefore, the system at the point O has no degree of freedom, *i.e.,* it is invariant. This also follows from phase rule equation:

$$H_2(O)_{(s)} \rightleftharpoons H_2O_{(l)} \rightleftharpoons H_2O_{(g)}$$
$$F = 1 - 3 + 2 = 0$$

Table 2.1 The Water System

System as given by the	*Phases in Equilibrium*	*Degrees of Freedom*
Curves		
OA	Water ⇌ Vapour	1
OB	Ice ⇌ Vapour	1
OC	Ice ⇌ Water	1
OA′ (metastable)	Water ⇌ Vapour	1
Areas		
AOB	Vapour	2
COB	Ice	2
AOC	Liquid Water	2
Triple point		
O	Ice ⇌ Water ⇌ Vapour	0

2.9 SULPHUR SYSTEM

Sulphur shows two allotropic crystalline forms, rhombic and monoclinic, capable of existence under suitable conditions. These two forms can also exist in equilibrium with the liquid and vapour phase. The system, therefore, consists of **four phases**, *viz., (i) Rhombic sulphur, (ii) Monoclinic sulphur, (iii) Sulphur vapour and (iv) Sulphur liquid.* This is a typical example of one component system exhibiting polymorphism and solid transformation.

According to phase rule equation, all the four phases cannot co-exist at equilibrium. On substituting the values of component and phases in phase rule equation,

$$F = C - P + 2$$

$$F = 1 - 4 + 2 = -1$$

i.e., we get F = – 1. This is, of course, impossible and so far a system consisting of single component, it is not possible to have four phases in equilibrium together. However, it is possible to have three or lesser number of phases to exist together. The system will be invariant, univariant or bivariant and accordingly three, two or a single phases are present.

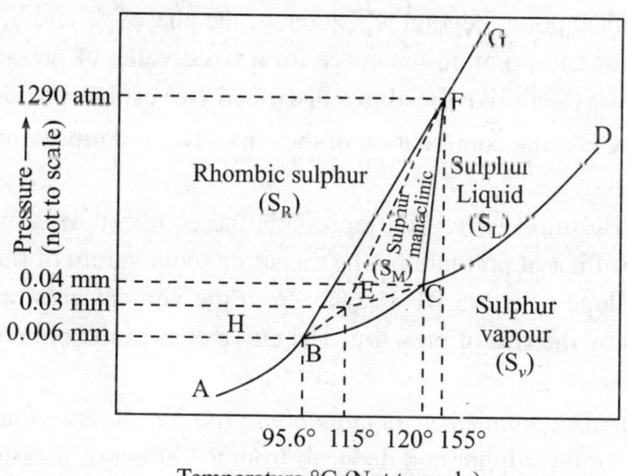

Fig. 2.4 Pressure-temperature phase diagram of the sulphur system

The pressure temperature diagram (Fig. 2.4) consists of *six stable curves* AB, BC, CD, BF, CE and FG and *four metastable curves.* The curves divide the diagram into *four areas* ABFG, ABCD, GFCD and BCF.

(A) Areas: The areas are bivariant to locate any point in the areas both temperature and pressure are to be fixed up. Each area represents the conditions for the stable existence of a single phase. In areas, thus, there are two degrees of freedom which also follows from the phase rule equation F = C – P + 2; putting the values C = 1, P = 1, we should get F = 1 – 1 + 2 = 2.

In the area ABFG only S_R (rhombic sulphur) exists while below in the area ABCD, only S_V, (Sulphur vapour) exists; to the right of GFCD, S_L (liquid sulphur) is present and in the region enclosed by BCF, S_M (monoclinic sulphur) exists.

(B) Curves: (1) AB: *It is the vapour pressure curve of rhombic sulphur.* The two phases present in equilibrium along this curve are S_R and S_V. The curve starts from the point B and ends at A(60°C). The point B corresponds to 95.6°C which is a transition temperature. Rhombic sulphur is stable below this temperature where as monoclinic sulphur is stable above this temperature. At this temperature both the forms exist in equilibrium.

If heating is done rapidly, then it is possible to heat rhombic sulphur above the transition point BE. BE *is the metastable curve* and on slight disturbance it merges into curve BC, the vapour pressure curve of monoclinic sulphur.

2. BC: *It is the vapour pressure curve of monoclinic sulphur.* The two phases present in equilibrium along this curve are S_M and S_V. The curve starts from the point B, the transition temperature of S_R and ends at C, the melting point of S_M (120°C).

Except at the point B, where three phases S_R, S_M and S_V and at any other point all along the curve only two phases S_V and S_M are in equilibrium. For a fixed value of temperature, pressure is automatically fixed. The degree of freedom, therefore is one.

If the temperature of monoclinic sulphur is lowered, then generally it would change into rhombic sulphur at 95.6°C, but it is possible to prevent this conversion by carefully cooling it. Then we will go along a curve BH which is a *metastable vapour pressure curve of monoclinic sulphur.*

3. CD: *It is the vapour pressure curve of liquid sulphur.* The two phases present in equilibrium along this curve are liquid sulphur, S_L and vapour sulphur, S_V. The curve starts at C, the melting point of S_M and ends at D, the critical point, beyond which out of the phases S_L and S_V, only S_V (vapour sulphur phase) can exist. Since for any point on the curve for a fixed value of pressure, the temperature has a fixed value, therefore, the system like the above discussed curves, has one degree of freedom.

The dotted curve CE is the continuation of the curve DC and represents *metastable vaporization curve of liquid sulphur.*

4. BF: *It is the transition curve* and represents the equilibrium between rhombic sulphur and monoclinic sulphur. The effect of pressure on the transition temperature of rhombic sulphur, is obtained from the is curve. The slope of curve BF is away from the vertical pressure axis, indicating that the transition point is raised by the rise of pressure. The curve ends at the point C, after which monoclinic sulphur phase disappears.

5. CF: *It is the melting point curve of monoclinic sulphur.* It represents the equilibrium between monoclinic sulphur and liquid sulphur and deduced from the effect of pressure on the melting point of monoclinic sulphur. The curve is inclined away from the pressure axis showing that the melting point is raised by the rise of pressure. The curve ends at F beyond which monoclinic sulphur ceases to exist in a stable condition. The system like the other curves, in univariant.

6. EF: This line represents the equilibrium between rhombic and liquid sulphur and is the *metastable melting point curve of rhombic sulphur.* Since the curve BF and CF meet at F, for which conditions are 155°C and 1290 atmospheres, it follows that if liquid sulphur were cooled at a pressure higher than 1290 atmospheres, then rhombic sulphur would crystallize at once, no monoclinic sulphur being formed. The curve passes through the point F and beyond this point it becomes a stable melting point curve FG of rhombic sulphur.

(C) Triple Points: There are three stable triple points, *viz.,* B, C and F in the phase diagram. At the **Point B,** *three phases in equilibrium are rhombic, monoclinic and vapour sulphur.* If one of the variables is changed, one of the phases disappears. Similarly, C is also a triple point between *monoclinic, vapour and liquid sulphur.* The point is again invariant. Further, lines BF and CF intersect at F to yield a three phase equilibrium between rhombic, *monoclinic and liquid sulphur.* Since on changing one of the variables, either pressure or temperature, one of the phases disappears, therefore the system at the triple point has no degree of freedom. This also follows from phase rule equation:

$$F = C - P + 2 = 1 - 3 + 2 = 0$$

Point B is obtained at 95.6°C and 0.006 nm pressure, while point C is obtained at (120°C and 0.04) and point F at (155°C and 1290 atm.) pressure. Three phases in equilibrium at these points are:

$$B(S_R - S_M - S_V);\ C(S_M - S_L - S_V) \text{ and } F(S_R - S_M - S_L)$$

Table 2.2: The Sulphur System

(System enclosed in brackets represents metastable curves)

Name of System	*Phases is Equilibrium*	*Degrees of freedom* $F = C - P + 2$
Areas		$F = 1 - 1 + 2 = 2$ (Bivariant)
(*i*) To the left of ABFG	S_R	Two
(*ii*) Enclosed by BCF	S_M	Two
(*iii*) To the right of GFCD	S_L	Two
(*iv*) Below ABCD	S_V	Two
Curves		$F = 1 - 2 + 2 - 1$ (Univariant)
(*i*) AB(BE)	$S_R \rightleftharpoons S_V$	One
(*ii*) BC(BH)	$S_M \rightleftharpoons S_V$	One
(*iii*) CD(CE)	$S_L \rightleftharpoons S_V$	Onc
(*iv*) FG(FE)	$S_R \rightleftharpoons S_V$	One
(*v*) BF	$S_R \rightleftharpoons S_M$	One
(*vi*) CF	$S_M \rightleftharpoons S_L$	One
Triple points		$F = 1 - 3 + 2 = 0$ (Invariant)
(*i*) B	$S_R \rightleftharpoons S_M \rightleftharpoons S_V$	Zero
(*ii*) C	$S_M \rightleftharpoons S_L \rightleftharpoons S_V$	Zero
(*iii*) F	$S_R \rightleftharpoons S_L \rightleftharpoons S_V$	Zero
(*iv*) E (Metastable)	$S_R \rightleftharpoons S_L \rightleftharpoons S_V$	Zero

2.10 APPLICATION OF PHASE RULE TO TWO COMPONENT SYSTEMS

A two component system can be defined as one in which it is possible to express the composition of all, the phases, in terms of two substances. A two component system may have as many as three degrees of freedom. This is because when P = 1, then F = C – P + 2 = 2 – 1 + 2 = 3

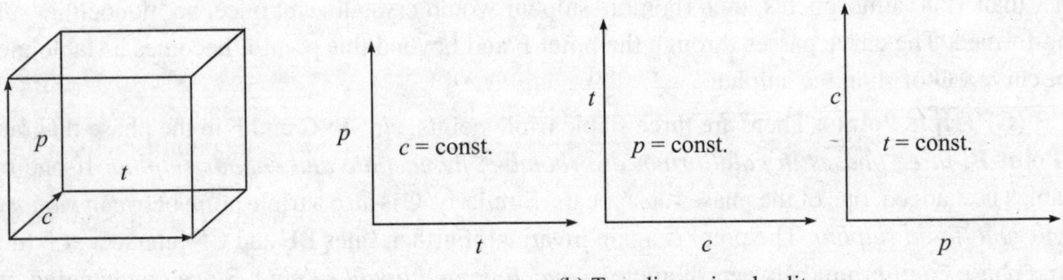

(*a*) Space model for a two component system. (b) Two dimensional split up.

Fig. 2.5 Depiction of phase diagram of 2-component systems.

Thus the system will have, besides temperature and pressure variables, one concentration variable for each phase present at equilibrium. Naturally, for the graphic representation of all possible conditions of equilibrium of such a system, a three-dimensional model is required in which pressure, temperature and concentration would be plotted along the three axes at right angles to one another (Fig. 2.5). However, in actual practice, a simple plain diagram with two variables only is considered; the third variables is considered as constant. In this way, use is made of the following three types of diagrams.

(*a*) *Pressure-temperature (p-t) diagrams.*

(*b*) *Temperature-composition (t-c) diagrams.*

(*c*) *Composition-pressure (c-p) diagrams.*

A two dimensional split up of such diagrams is shown in Fig. 2.5(*b*).

But in most of the cases, it is usually convenient to keep pressure constant at the atmosphere value and thus a *t-c* diagram is often met with. A diagram drawn for a constant value of temperature is called **isothermal**, while one obtained by keeping composition constant is called **isoplethal** diagram.

In case of only solid-liquid phases, small changes of pressures have very little effect on the system. In this equilibrium, pressure will not have considerable effect because only gaseous systems, which involve large changes of volume, are affected by pressure. Such systems are called **condensed systems.** The degree of freedom in such a case is reduced by one as the vapour pressures of solids are very small and experiments are performed at atmospheric pressure. Thus, we may use the **reduced phase rule equation** in such cases:

$$F' = C - P + 1$$

F′ in the above equation represents the number of degrees of freedom which the system can possess in addition to pressure, namely, temperature and concentration (or composition) of the constituent. Therefore, solid-liquid equilibria are represented on temperature composition diagram (*t-c* diagram) which are plotted by using thermal analysis procedure for different compositions of two component systems.

2.11 SIMPLE EUTECTIC SYSTEM

A binary system consisting of two substances, which are miscible in all proportions in the liquid phase, but do not react chemically, is known as the *'eutectic system'*. (The word eutectic means easy to melt). For example, a mixture of lead-silver, Bismuth-cadmium and of Potassium iodide-water are common eutectic systems.

Eutectic Point

Two solid substances capable of forming solid solution with each other have the property of lowering each other's freezing point; and the minimum freezing point attainable corresponding to the eutectic mixture is termed as the *eutectic* point. For example in Pb-Ag system, eutectic point is achieved when the composition is 2.6% Ag and 97.4% of Pb.

Systems giving rise to eutectic point are known as *eutectic systems*. The eutectic composition and temperature of two metals and salt-water system is given in Table 2.3.

Table 2.3 Some Eutectic Systems

S. No.	*Components are their melting point*	*Eutectic composition*	*Eutectic temperature*
1.	Ag(961°C); Cu(1083°C)	(71.8% Ag + 29.2% Cu)	778°C
2.	Ag(961°C); Pb(327°C)	(2.6% Ag + 97.4% Pb)	303°C
3.	Zn(419°C); Cd(323°C)	(67% Zn + 33% Cd)	270°C
4.	Zn(419°C); Al(658.7°C)	(95.64% Zn + 4.36% Al)	380.5°C
5.	Bi(273°C); Cd(323°C)	(60% Bi + 40% Cd)	140°C
6.	KCl(773°C); H_2O(0°C)	(20% KCl + 80% H_2O)	– 11°C
7.	KI(682°C); H_2O(0°C)	(52% KI + 48% H_2O)	– 22°C

Out of the above noted systems, only Pb-Ag systems have been discussed in this book.

Characteristics of Eutectic Point

1. Eutectic point represents the lowest or limiting temperature at which a liquid phase can exist in the system, or it is the maximum temperature upto which a solid phase can exist.
2. No other mixture containing the two components will have a melting point lower than the eutectic temperature.
3. Eutectic point has precise values of temperature and composition and it represents an invariant system.
4. If the liquid is cooled to just below the eutectic point, both the components of the eutectic simultaneously solidify without any change in the composition or temperature of the liquid phase.
5. An eutectic system can maintain its temperature constant for long periods.
6. When the liquid is cooled below the eutectic point, the components solidify in the form of small crystals intimately mixed with each other which fill in the spaces between the larger crystals of the pure components which are already separated out.
7. Eutectic mixtures appreciably contribute towards the strength of the solid structure in case of alloys.
8. Eutectic are mixtures of the components but not their compounds.

Applications of Eutectic Systems

1. Eutectic mixtures are used in preparing *Solders* which are employed for joining two pieces of metals. For example: Pb–Sn Soldiers.
2. Eutectic mixtures are used as *Safety fuses* employed in various equipment like automobiles, fire extinguishers, boilers, etc. For example: Wood metal (an alloy containing 50% Bi, 25% Pb, 12.5% Sn and 12.5% Cd) has melting point 65°C and is often used in devices such a pressure cookers and fire extinguishers in buildings as protection against accidental fire.

2.12 LEAD-SILVER SYSTEM

It is the simplest case of two component solid-liquid system involving eutectic formation.

The two components are lead and silver. The four phases involved in the system are: (*i*) Solid lead (*ii*) Solid silver (*iii*) Solution of lead and silver in the molten state and (*iv*) Vapour.

In this system, pressure will not have considerable effect because the vapour pressure of solids and molten metals are very small and can be neglected. The phase diagram, therefore, is drawn at atmospheric pressure so that no gaseous phase is present. The system can be studied by using reduced phase rule equation.

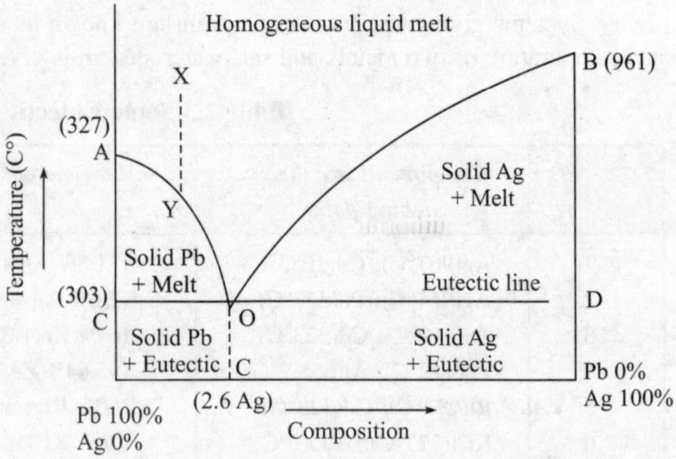

Fig. 2.6 Phase diagram for lead-silver system

According to the reduce phase rule equation when one phase is present, F = 2 – 1 + 2. Since F = 2, only two variables (excluding pressure which is considered constant) must be specified in order to describe the system and that the equilibrium can be represented in two dimensions by means of temperature-composition diagram (Fig. 2.6).

Curves and Points

(*i*) **Point A** represents the *melting or freezing point of pure lead (327°C)* and point *B* represents the *freezing point of pure silver* (961°C). Thus at *point A,* liquid lead is in equilibrium with solid lead and *B* liquid silver is in equilibrium with solid silver.

(*ii*) **Curve AO** represents the *freezing point of lead* to which successive small amounts of silver are added. It is seen that addition of silver to lead lowers the melting point of lead. All along the silver, which is added, goes into the solution while separation of solid lead takes place. *Along this curve, solid lead is in equilibrium with liquid melt* (solution of silver in liquid lead) *of varying composition.* The curve AO represents the freezing point of lead, separate from the melt varying containing varying amounts of silver. Along this curve two phases solid lead and liquid melt are present in equilibrium. Therefore, F = C – P + 1 = 2 – 2 + 1 = 1, *i.e.,* the system is univariant.

(*iii*) **Curve BO** represents the freezing point of silver to which successive small amounts of lead are added. It is observed that the melting point of silver is lowered, gradually by the addition of lead. All along this curve, the added lead goes into the solution and the separation of solid silver takes place. Thus, *along this curve, solid silver is in equilibrium with liquid melt* (solution of lead in liquid *varying composition*. The BO represents the freezing point of silver separated from silver containing varying amounts of lead. Along the curve BO two phases–solid silver and melt and the system is univariant as shown below:

$$F = C - P + 1 = 2 - 2 + 1 = 1$$

At the point O in the phase diagram lead and silver freeze out simultaneously. As the composition of the melt is the same as that of the solid mixture of lead and silver. The liquid melt out as if it was a pure compound. The whole of the melt freezes out at constant temperature, O is called *eutectic point* (Greek-low melting) and *the* solid mixture is called *eutectic* which has characteristic composition for each system.

It may be made clear here that the phase diagram will not be exactly similar at different pressures and the eutectic composition will also vary with pressure.

Below point O: In Fig. 2.6 only solid phase exists and the cooling of this solid eutectic is represented by a dotted line.

Eutectic point O: It is the point where two curves AO and BO meet. At this point solid lead and solid silver are in equilibrium with the melt, *i.e.,* three phases are present. Therefore, there is no degree of freedom $F = C - P + 1 = 2 - 3 + 1 = 0$.

This point O in Fig. 2.6 represents the lowest possible temperature (303°C) below which a liquid phase cannot exist and beyond which the liquid phase cannot be enriched in either composition by freezing out the other component. *Such a liquid mixture of two components Pb and Ag, which has the lowest freezing point compared to all other liquid mixtures is called the* **eutectic mixture.** The mixture freezes out completely at a constant temperature and on freezing gives the two components as solid phase in the same proportion in which they are present in the liquid phase. The temperature corresponding to the eutectic point O is called the eutectic **temperature.** In the above system, we see that the alloy silver (2.6%) with and lead (97.4%) melts at 303°C, a temperature lower than the melting point of silver (961°C) and of lead (327°C).

In the region or area above the curve AOB: Only the liquid alloy (melt of silver and lead) exists, *i.e.,* there is only one phase and it *has two degrees of freedom, i.e.,* temperature and concentration besides pressure. This system is, therefore, bivariant.

In the area AOC: Solid lead and liquid melt exists. The composition of the liquid melt, at any temperature can be obtained from the curve AO. At any point in this area, solid lead is in equilibrium with a liquid melt of definite composition.

In the area BOD: Solid silver and liquid melt exist. The composition of the liquid melt at any temperature can be obtained from the curve BO as in the case of area AOC.

In these areas AOC and BOD, there is one degree *of freedom temperature or composition besides pressure.*

Below line COD, there are two phases, solid lead and solid silver. In the region marked as solid Pb + eutectic, we find lead crystals and solid eutectic crystals whereas in the region marked solid Ag + eutectic, we find silver crystals and solid eutectic crystals. This can be examined under a microscope. Again this system has one degree of freedom besides pressure.

$$F = C - P + 1 = 2 - 2 + 1 = 1$$

The phase diagram provides very useful information about the system.

2.12.1 Pattinson's Process for Desilverisation of Lead

Galena (PbS, an ore of lead) is usually associated with silver. When galena is worked out for the extraction of lead, some traces of silver pass into it because silver is soluble in lead to some extent. Such a lead is known as *argentiferous* lead. The desilverisation of lead can be easily understood from the phase diagram of lead silver system shown in Fig. 2.7.

The agentiferous lead consisting of very small percentage of silver is first of all heated to a temperature well above its melting point, so that the system consists only of liquid phase represented by the point X in the figure. It is then allowed to cool, gradually along the line XY without any change in composition till the point Y is reached. As soon as this point is attained, lead starts separating out and the solution will contain respectively increasing amount of silver. Further cooling will shift the system till the eutectic point O is reached where the concentration of Ag increases up to 2.6 percent.

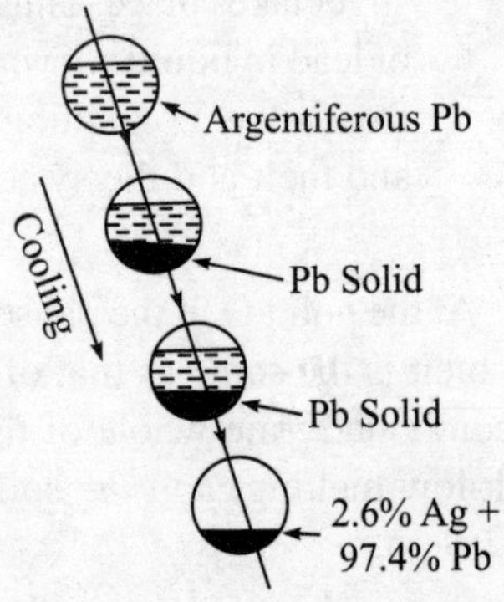

Fig. 2.7 Cooling of argentiferous lead in Pattinson process

The increase in the relative amount of Ag during the cooling process of molten argentiferous lead is shown in Fig. 2.7. The amount of Ag present in traces in the alloy increase upto 2.6 percent by gradually cooling the molten mass.

This alloy is then further refined by the process of cupellation. This above principle is used in Pattison's Process (*i.e.,* the process of increasing the relative proportions of silver in the alloy) for the desilverisation of lead.

PRACTICE PROBLEMS

1. Give an example of two component system involving eutectic formation.
2. What is the melting point of pure Pb and pure Ag?
3. What is the eutectic temperature in case of Pb–Ag system?
4. What is the composition of Pb and Ag at eutectic point?
5. Comment whether eutectic is a chemical compound or mixture?
6. Comment, whether the eutectic composition will change with change in pressure at which phase diagram is drawn.
7. Which is the physical state of the phases below eutectic point?
8. What is the degree of freedom at eutectic point?
9. What is the maximum percentage of Ag that can be obtained by Pattison's process?

Answers

1. Pb–Ag	**2.** Pb = 327°C, Ag = 961°C	**3.** 303°C
4. Ag = 2.6%, Pb = 97.4%	**5.** Mixture	**6.** Yes
7. Solid	**8.** Zero	**9.** 2.6%

2.13 SYSTEMS WITH CONGRUENT MELTING POINT

In this type of binary alloys, the components form one or more compounds having definite proportions. These compounds melt into the liquid state, having the same composition as the solid. Therefore, such compounds are called *compounds with congruent melting-points. Thus, a compound is said to possess congruent melting point, which melts sharply at a constant temperature into a liquid, having the same composition as that of the solid.*

Let us illustrate such a system by Zn–Mg binary alloys (Fig. 2.8). Zinc and magnesium melt at 420°C and 650°C (point A and C) respectively and they form a compound $MgZn_2$, *i.e.*, in which the ratio of Mg and Zn in molecular proportions is 1 : 2. This compound melts at *B* (590°C) which is congruent melting point.

2.13.1 Zinc-Magnesium Alloy System

This phase diagrams may be considered as equivalent to two phase diagrams of Pb–Ag type (as discussed earlier) placed side by side. The one *on the left consists of Zn and $MgZn_2$ system; while the other on the right is formed by the two component $MgZn_2$ and Mg. Zn–Mg system has four phases namely (i) Solid Magnesium (ii) Solid Zinc (iii) Solid $MgZn_2$ and (iv) liquid melt of Zn and Mg.*

Since the system is practically not effected by pressure, so it may be considered condensed system, so reduced phase rule equation, F′ = C – P + 1 is applicable. The phase diagram of the system comprised the following curves, points the areas.

Curves: The *curves AE_1* and *CE_2* represent the compositions of liquids in equilibrium with solid Zn and solid Mg respectively. The central portion of curve E_1BE_2 have a maxima at *B*. This gives the conditions of equilibrium of liquid with solid compound, $MgZn_2$. At *B*, the maxima the liquid has the same composition as the solid in equilibrium. Thus, *the temperature at B (590°C) represents the melting point of the compound $MgZn_2$*. As $MgZn_2$ can co-exist as solid and liquid of same composition, so $MgZn_2$ is said to possess *congruent melting point.*

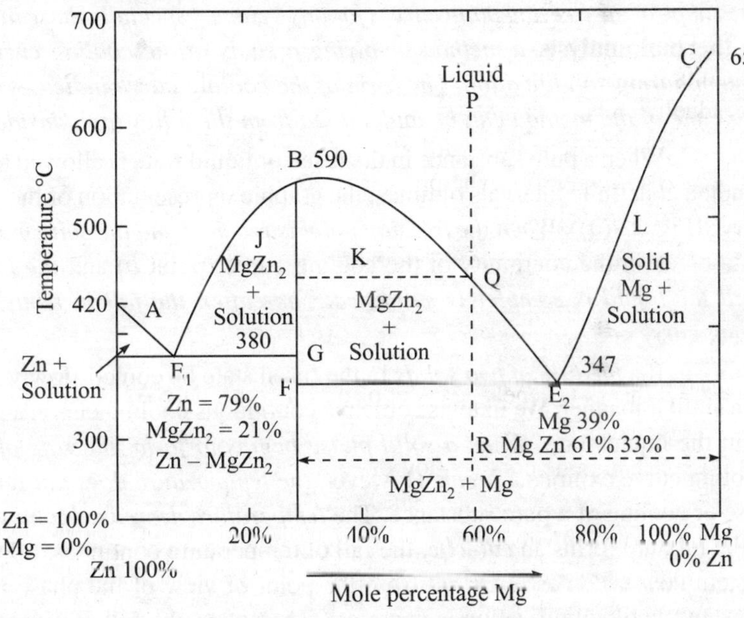

Fig. 2.8 Temperature composition diagram for Zn-Mg system.

The composition corresponding to maxima (congruent m. pt) is 33.3% Mg and 66.7% Zn. *i.e.*, in the ratio of 1 : 2 as $MgZn_2$. The sharpness of maxima gives the information about the stability of the compound formed. A sharp maxima indicate that the compound does not dissociate appreciably; whereas a *flat maxima indicates* the *appreciable dissociation,* as in the present case.

All along these three curves two phases are in equilibrium. Curve AE_1 is called *freezing point curve of zinc to which Mg is being added gradually*. The two phases in equilibrium are: Zn(*s*) and melt. Similarly curve CE_2 is called *the freezing point curve of magnesium to which zinc is being added gradually*. The two phases in equilibrium are: Mg(*s*) and melt. Curve E_1B E_2 is called the freezing melting point curve of the compound $MgZn_2$, where the two phases in equilibrium are solid $MgZn_2$ and melt. Solid MgZn $\rightleftharpoons$ Liquid $MgZn_2$. According to phase rule,

$$F = C - P + 1 = 2 - 2 + 1 = 1$$

The degree of freedom along these curves is one *i.e.,* the curves are univariant. It means that to locate any point on any of these curves, either temperature or composition is to be fixed.

Points E_1 and E_2 represent eutectic points. At these points, three phases are in equilibrium,

$$\text{Solid Zn} \rightleftharpoons \text{Solid } MgZn_2 \rightleftharpoons \text{Melt}$$

At E_1, the three phase are: Zn(*s*), melt and $MgZn_2(s)$

At E_2, the three phase are: Mg(*s*), melt and $MgZn_2(s)$

According to phase rule: $F = C - P + 1 = 2 - 3 + 1 = 0$

Molar composition at E_1 is 79% Zn and 21% $MgZn_2$: E_2 is 61%, $MgZn_2$ and 39% Mg.

2.14 THERMAL ANALYSIS (DETERMINATION OF EUTECTIC TEMPERATURE)

The shape of the freezing point curves for any system, especially those involving metals, can be determined by **thermal analysis**–*a method involving a study of the cooling curves of various compositions of a system during solidification. The form of the cooling curve indicates the composition of the solid. The principle of the method can be understood from the following consideration:*

1. When a pure substance in the fused or liquid state is allowed to cool slowly and the temperature is noted at definite intervals of times, the graphic representation of the rate of cooling will be continuous curve (Fig. 2.9(*a*)). *When the freezing point is reached and the solid makes its appearance,* it is indicated by a break in the continuity of the cooling curve (point *b*) and *the temperature will remain constant, until the liquid is completely solidified. Thereafter, the fall in temperature will again be continuous along curve cd.*

2. If a *mixture of two solids* in the fused state be cooled slowly and the cooling curve is obtained in a similar manner. We likewise obtain a continuous cooling curve, so long as the mixture *(or solution)* is in the liquid state. *When a solid phase begins to form, the rate of cooling abruptly alters* and the cooling curve exhibits a *break.* However, *the temperature does not remain constant,* as in the previous case of cooling of a pure substance. The *temperature decreases continuously,* but at a different rate and if the mixture forms an ***eutectic,*** the fall of temperature continues, till the eutectic point is reached. The system now becomes *invariant* from the point of view of the phase rule and the temperature remains constant; until solidification is complete. Thereafter, the fall of temperature becomes uniform, but the rate of fall is *quite different from the previous one.*

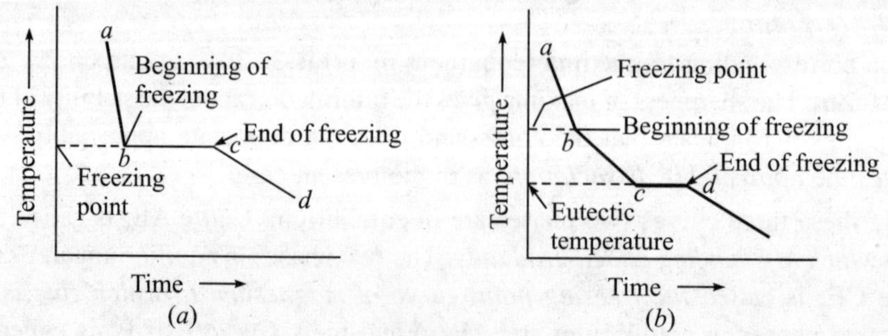

Fig. 2.9

From the cooling curve for any mixture of a definite composition, it is possible to obtain is: *(i) freezing point,* and *(ii)* the *eutectic temperature.*

(*a*) *The freezing point* **varies** *with the composition of the system, but the eutectic point remains* **constant** *for a given system.*

(*b*) The *nearer* the composition of the system to the eutectic, the *shorter* is the portion *bc* and the *more prolonged* is the half *cd.*

(*c*) If the mixture *coincides* with the eutectic composition, the curve shows no *break* corresponding to *bc, but* the break appears only at the **eutectic point,** (*c*).

(*d*) If the cooling curves of a series of alloys of known compositions are worked out and their freezing points are noted, by plotting freezing point against composition, *T-C curve* is obtained for the alloy system. However, in order to complete the diagram, it is necessary to know the freezing points of the pure components also.

(*e*) Now the cooling curve of an alloy of the same metals, but of unknown, composition is determined and its freezing point located in the *T-C diagram.* ***The composition corresponding to this freezing point yields the composition of the alloy.***

REVISION QUESTIONS

1. (*a*) State and explain what is meant by the terms phase, component and degree of freedom? How are they related?

(*b*) What is meant by (*i*) true equilibrium and (*iii*) metastable equilibrium. (*M.D.U., 2003*)

2. (*a*) How many number of phase and components are present in the following systems:

(*i*) A mixture of molten Lead, Tin and Bismuth.

(*ii*) Two ice cubes floating on water in a closed container in the presence of water vapour.

(*b*) Ammonium chloride is heated in a closed vessel. The system can be either a one component system or a two component system. Explain.

3. Draw a phase diagram for one component water system. Label it and discuss the importance of various points, lines and areas at equilibrium.

4. (*a*) What is meant by triple point of water? Why is it different from the normal melting point of ice?

(*b*) The melting point curve of water system has a negative slope whereas the melting point curve of sulphur has a positive slope. Explain.

(*c*) Justify the statement "The eutectic is a mixture not a compound" (*M.D.U. May 2008, 2009*)

5. (*a*) Explain the following terms used in phase rule.

(*i*) Tripple point

(*ii*) Critical point (*M.D.U., 2003*)

(*b*) What is eutectic? Is it is a compound or not?

(*c*) Draw and compare the cooling curve of a pure substance and a mixture of two solids. (*M.D.U., May, 2006*)

6. Explain, with suitable example, the following terms used in the phase rule study of heterogeneous equilibria:

(*i*) Triple point (*ii*) Transition point

(*iii*) Invariant system (*iv*) Eutectic point

(*v*) Eutectic (*vi*) Metastable equilibrium

7. (*a*) Draw and compare the cooling curve of a pure substance and a mixture of two solids. Mention the striking difference between the curves.

(*b*) Briefly describe a two component system which involves the formation of a compound with congruent melting point.

8. (*a*) Define and explain the term components of a system with suitable examples.

(*b*) Draw a labeled phase diagram of water system and the discuss the metastable curve and principles of freeze drying. (*P.T.U., June 2003*)

(*c*) What do you understand by Reduced Phase Rule? (*K.U., Dec. 2003*)

9. (*a*) Giving examples, explain the term: components of a system.

(*b*) Differentiate between True equilibrium and metastable equilibrium.

(*c*) Justify the statement, "The eutectic is a mixture and not a compound". (*K.U., June 2004*)

(*d*) Mention any two limitations of phase rule.

10. Calculate P, C and F in the following cases. (*K.U. 2007*)

(*i*) $NH_3(g)$ at 42°C

(*ii*) Water system at 4.578 mm of Hg and at 0.0098°C.

11. Define following terms used in phase equilibria with example: Phase, component, Degree of Freedom, Eutectic, Freezing point. (*K.U. June 2008*)

12. Draw a neat, cleaned and labelled sketch of the phase diagram of water system. Discuss all the equilibrium involved in it.

13. (*a*) The melting point curve in water system is inclined towards the pressure axis. Explain.

Hints: The slope of the curve indicates that melting point of ice is lowered by increase of pressure.

(*b*) Discuss the significance of triple point. (*P.T.U. May 2005*)

(*c*) A substance Z has its triple point at 18°C and at 0.5 atm its normal melting and boiling points are 20°C and 300°C respectively. Scketch the schematic phase diagram for Z. (*P.T.U. Dec. 2006, May 2007*)

14. Draw and discuss the phase diagram of lead-silver system. Explain the eutectic point and Pattison's process of desilverisation of lead. (*K.U. Jan. 2008*)

15. Draw and compare the cooling curve of a pure substance and mixture of two solids, mention the striking difference between the curves. (*M.D.U. May 2006*)

16. Explain the following:

(*i*) Reduced phase-rule equation (*ii*) Congruent melting point

(*iii*) Tripple point (*iv*) Peritectic point (*M.D.U., May 2007*)

17. What do you understand by congruent and incongruent melting point? Discuss Zn–Mg system in detail.

18. (*a*) Point out metastable curves in sulphur system diagram and tell the phases in equilibrium along each metastable curve.

(*b*) Point out the tripple points in the phase diagram of sulphur system and tell the phases in equilibrium at each tripple point.

19. With the help of neat clean and labelled diagram discuss the phase diagram of sulphur systems. (*K.U. June, 2006*)

❑❑❑

SECTION B

Chapter 3: Water Treatment

Chapter 4: Corrosion and its Control

3

Water Treatment

INSIDE THIS CHAPTER

3.1 INTRODUCTION

Water is surely God's greatest gift to mankind. The very existence of life cannot be imagined without water. One can live without food, shelter or any other thing but life without water is just not possible. Even our body contains around 70% water in it. From an engineer's point of view, water is no less important. It is required in boilers for production of steam, which acts as a source of energy, as a coolant in many power and chemical plants, in steel, paper, textile, ice and many other such industries. Despite its incredible importance, water is receiving least attention and is getting contaminated in ways beyond description.

Water is obtained from nature in various forms such as rain water, river water, sea water, spring water and well water. In addition to the inherent impurities water gets contaminated due to number of contaminants from sewage, industries etc. during usage. Water is not only essential for the lives of animals and plants but also occupies a unique position in industry. Surface and groundwater is normally used for industrial and domestic purposes. Each industry has its own specifications for water and so the treatment of water depends upon the purposes for which it is to be used. Some of the major parameters which are generally checked for assessing the water quality for a particular use are determination of *hardness, alkalinity, turbidity, total dissolved solids, suspended solids, dissolved oxygen (DO), Biological oxygen demand (BOD), chemical oxygen demand (COD) total residual chlorine, pH, temperature etc.* Hence a complete analysis with respect to these is desired for determining the suitability of water for a particular industrial use.

Depending upon the nature of impurities present in water, it has to be treated. In this chapter, we will study the determination of some of these parameters. We will also study hardness of water determination of hardness, BOD and COD, Boiler corrosion, sewage and its treatment as per syllabus of H.P. Technical university.

3.2 SOURCES OF WATER

Water present on earth passes through a remarkable cycle of changes. Thus from sea water it forms water vapours, which get condensed in colder regions and come down as rain water. This water flows on the earth surface as streams, rivers and comes back to sea water. The main sources of water are:

(i) Rain water *(ii) Surface water* *(iii) Groundwater*

***(i)* Rain water:** Since rain water is obtained by evaporation of surface water, it is considered as the purest form of natural water. But rain water during its downward path to the earth dissolves considerable amount of gases (*e.g.* CO_2, SO_2, NO, NO, etc.) and suspended solid particles present in the atmosphere. So it becomes polluted.

***(ii)* Surface water:** The water which comes to the surface through rain is further of three types:

(*a*) river water (*b*) lake water and (*c*) sea water.

(*a*) River water: River/stream/canal waters are inland waters fed by rains. This water is less pure than rain water. River water contains dissolved inorganic salts such as chlorides. In addition to the dissolved impurities obtained from the soil, river water also contains the organic products (from decomposition of plants and animals) and small particles of sand and rock as suspended impurities.

(*b*) Lake water: Lake water has nearly constant composition. In this type of water the main impurities are that of organic matter. The quality of water in some of the big lakes is so good that hardly any treatment is needed for its industrial uses. But the population growth and modern society is polluting even the lake water day-by-day.

(*c*) Sea water: This is the most impure form of natural water. River water join sea and thus give its impurities to sea. Secondly, continuous evaporation of water makes it continuously richer in dissolved impurities. Sea water is found to have about 3.5% of dissolved salts, out of which 2.6% is sodium chloride and hence sea water is always saltish in taste. Oceans cover about 70.8% of earths total surface and holds more than 97% of the water supply. This water, despite its abundance is not directly useful to man because it is not palatable as it contain about 2000 times more dissolved salts than fresh water.

***(iii)* Groundwater:** The rain water after reaching the ground is percolated in the soil. A large amount of rain water finds its way into the ground through permeable rocks, loose sand, gravel etc. When it reaches an impervious rock which prevents it from proceeding downwards, it collects in a pool and sometimes it rises to the surface in the form of spring water. During its passage downward into the ground, the suspended matter is left behind and organic matter is oxidised by bacteria. This water is extremely clear as a result of natural filteration through the sand bed. It is also free from NH_3 and organic impurities.

Water from wells and springs is extremely hard. It contains mainly bicarbonates, chloride and sulphates of Mg, Ca and other metals. Some spring waters possess minerals of medicinal value such as colloidal sulphur, hydrogen sulphide and borax etc.

3.3 IMPURITIES IN WATER

Water, from all the natural sources, has got impurities. The sources of these impurities are the following:

(*i*) Water takes up impurities from the ground, or soil with which it comes into contact.

(*ii*) Water becomes impure when it comes in contact with sewage or industrial wastes.

(*iii*) Organic impurities in water are generally introduced by the decomposition of plants and animals remains. Certain micro-organism and bacteria are also responsible for the organic impurities in water.

Nature of impurities in Water

The common impurities present in natural water may be grouped into the following groups.

(*i*) **Dissolved Gases:** Like oxygen, carbon dioxide, hydrogen sulphide etc. water may also contain some undesirable gases and odour due to the degradation of organic matter.

(*ii*) **Suspended Matter:** These are mostly insoluble minerals such as clay, sand etc. These causes turbidity to water.

(*iii*) **Microscopic Matter:** These consist mainly of bacteria and other micro-organisms, like algae and fungi.

(*iv*) **Dissolved Mineral Salts:** Dissolved salts are mainly the carbonates, bicarbonates, chlorides and sulphates of calcium, magnesium, iron sodium and potassium. The presence of these salts imparts a sort of hardness in water, *i.e.,* a hard water does not produce a lather with soap. Although hard water is not always harmful for drinking purposes, it causes a wastage of soap material in the laundry work.

(*v*) **Organic Matter:** It includes vegetable and animal matter, *i.e.,* decayed plants and animals, oil globules, carbohydrates etc.

(*iv*) **Colloidal Impurities:** They consist of proteins, amino acids, ferric hydroxide, clay and lumic acid etc.

3.4 HARD WATER AND HARDNESS

Water which does not produce lather with soap solution readily, but forms a white scum or precipitate is called hard water.

It is due to the presence of some dissolved salts in it. Soaps are sodium or potassium salts of higher fatty acids like stearic acid, oleic acid and palmitic acid. When hard water is treated with soap, it does not produce lather, rather soap gets precipitated in the form of insoluble salts of calcium and magnesium. No lather is formed until all these ions are completely removed and hence a large amount of soap gets wasted in the process.

$$\underset{\substack{\text{(Calcium chloride}\\\text{from hard water)}}}{CaCl_2} + \underset{\substack{\text{Sodium stearate}\\\text{(Soap)}}}{2C_{17}H_{35}COONa} \longrightarrow \underset{\substack{\text{Calcium stearate}\\\text{(Insoluble ppt. White scum)}}}{(C_{17}H_{35}COO)_2 \downarrow} + 2NaCl$$

Such precipitates are not formed with detergents, which are sulphonates of long chained alcohols.

Water, which easily forms lather with soap solution, is called soft water. Such water does not contain calcium and magnesium salts in it.

Difference between Hard Water and Soft Water

Hard Water	*Soft Water*
1. It does not produce lather readily on treatment with soap solution It form white scum or precipitates on treatment with soap solution.	1. It produces lather instantly on treatment with soap.
2. More time and fuel is consumed in cooling in hard water because boiling point is increased due to hardness.	2. Less time and fuel is consumed in cooling in soft water.
3. Detergent properties of soap are depressed in hard water so large amount of soap is consumed during its use.	3. Detergent properties of soap are not depressed in soft water so less amount of soap is consumed during its use.
4. Salts of calcium and magnesium are dissolved in it.	4. No salts of calcium and magnesium are dissolved in it.

3.4.1 Types of Hardness

Hardness of water is of two types:

(A) Temporary hardness *(B) Permanent hardness*

(A) Temporary hardness (Carbonate Hardness) or Alkaline Hardness

It is caused due to the presence of bicarbonates of calcium and magnesium. It is called temporary hardness because it can be removed easily by boiling.

The presence of bicarbonates in water is due to dissolution of insoluble carbonates in the presence of atmospheric CO_2.

$$CaCO_3 + CO_2 + H_2O \longrightarrow Ca(HCO_3)_2$$

On boiling, bicarbonates of calcium and magnesium get decomposed to give insoluble carbonates or hydroxides. These carbonates or hydroxides being insoluble are removed in form of precipitates and thus temporary hardness is removed.

$$Ca(HCO_3)_2 \xrightarrow{\Delta} CaCO_3 \downarrow + H_2O + CO_2 \uparrow$$

$$Mg(HCO_3)_2 \xrightarrow{\Delta} Mg(OH)_2 \downarrow + 2CO_2 \uparrow$$

Temporary hardness is also called *carbonate hardness or alkaline hardness.* It is due to the presence of bicarbonates of Ca and Mg, carbonates of Fe and other heavy metals to, small extent and hydroxides of the hardness-producing metal ions. It can be determined by titration with standard acid solution (say HCl) using methyl orange as an indicator.

(B) Permanent hardness (Non-carbonate hardness) or Non-alkaline Hardness

It is due to the presence of chlorides and sulphates of calcium and magnesium. It is called permanent hardness, because it cannot be removed by boiling.

However, it can be removed by using various chemical agents. Calcium and magnesium salts are not only salts responsible for hardness. They are mentioned because they are most commonly present. However, all soluble salts of heavy metals cause hardness.

It is also called *non-carbonate or non-alkaline hardness.* It can be determined by subtracting the alkaline hardness from the total hardness.

3.4.2 Units of Hardness

Hardness of water is due to the presence of number of dissolved salts in it but for comparing the hardness of different samples of water of varying composition, it is necessary to choose a reference standard. For this purpose, hardness is expressed in terms of *equivalents of calcium carbonate only.* The reason of choosing calcium carbonate as a reference standard is that it is the most insoluble salt that can be precipitated in water treatment. Moreover, its molecular mass being 100 equivalent weight being 50, the calculation become easy.

Hence whatever amount of dissolved salts are present in water, they are first converted into calcium carbonate equivalents. For example, if water contains 10 ppm of $MgCl_2$ and 8 ppm of $CaSO_4$, the calcium carbonate equivalents of these salts would be

$$\left.\begin{aligned} MgCl_2 &= \frac{10 \times 100}{95} = 10.53 \text{ ppm} \\ CaSO_4 &= \frac{8 \times 100}{136} = 5.88 \text{ ppm} \end{aligned}\right\} \text{where 95 and 136 molecular weights of } MgCl_2 \text{ and } CaSO_4 \text{ respectively}$$

Thus total hardness equal to 10.53 + 5.88 = 16.41 ppm expressed as Calcium Cardonate equivalents. In general, $CaCO_3$ equivalent of hardness

$$= \frac{\text{Mol. wt. of } CaCO_3 \text{ (100)}}{\text{Mol. wt. of the substance}} \times \text{Wt. of the substance producing hardness}$$

$$= \frac{\text{Eq. wt. of } CaCO_3 \text{ (50)}}{\text{Eq. wt. of substance}} \times \text{Wt. of hardness producing substance.}$$

Calculations of Equivalents of Calcium Carbonate

Dissolved salt	*Molar mass*	*Chemical equivalent*	*Conversion factor for calculation $CaCO_3$ equivalent*
$Ca(HCO_3)_2$	162	81	100/162
$Mg(HCO_3)_2$	146	73	100/146
$CaSO_4$	136	68	100/136
$CaCl_2$	111	55.5	100/111
$MgSO_4$	120	60	100/120
$MgCl_2$	95	47.5	100/95
$MgCO_3$	84	42	100/84
$CaCO_3$	100	50	100/100
$FaSO_4$, $7H_2O$	278	139	100/278
CO_2	44	22	100/44
$Mg(NO_3)_2$	148	74	100/148
HCO_3^-	61	61	100/122
OH^-	17	17	100/34
CO_3^{2-}	60	30	100/60
$NaAlO_2$	82	82	100/164
$Al_2(SO_4)_3$	342	57	100/114
H^+	1	1	100/2

Various units for expressing hardness of water are as under:

1. Parts of million (ppm)
2. Milligrams per liter (mg/l)
3. Degree French (°Fr)
4. Degree Clark (°Cl)

1. **Parts per million (ppm):** It is defined as number of parts of calcium carbonate equivalent hardness present per 10^6 parts of water. This is most common unit for expressing the hardness of water

$$1 \text{ ppm} = 1 \text{ mg of } CaCO_3 \text{ equivalent hardness in } 10^6 \text{ parts of } H_2O$$

2. **Milligrams per liter (mg/l):** It is defined as the number of milligrams of $CaCO_3$ equivalent hardness present per litre of water.

$$1 \text{ mg/l} = 1 \text{ part of } CaCO_3 \text{ equivalent per } 10^6 \text{ ml of water}$$
$$= 1 \text{ part of } CaCO_3 \text{ equivalent per } 10^6 \text{ parts of water}$$
$$= 1 \text{ ppm}$$

Thus, mathematically both the units are equal.

3. **Degree French (°Fr):** It is defined as the number of parts of $CaCO_3$ equivalent hardness present per 10^5 parts of water.

$$1\ °Fr = 1 \text{ part of } CaCO_3 \text{ equivalent hardness per } 10^5 \text{ parts of water.}$$

4. **Degree Clark (°Cl):** It is the number of grains (1/7000 lb) of $CaCO_3$ equivalent present per gallon (10 lbs or 70,000 grains)

Always remember that, 1 lb = 7000 grains and one gallon = 70,000 grains

It is defined as the number of parts of $CaCO_3$ equivalent hardness present per 70,000 parts of water.

$$1\ °Cl = 1 \text{ part of } CaCO_3 \text{ equivalent hardness per 70,000 parts of water}$$

Relationship among various units of hardness

$$1 \text{ ppm} = 1 \text{ mg/l} = 0.1\ °Fr = 0.07\ °Cl.$$

The above relation can be derived as follows: We know that,

$$1 \text{ ppm} = 1 \text{ part per 10,000 parts of water}$$
$$1\ °Cl = 1 \text{ part per 70,000 parts of water}$$
$$1\ °Fr = 1 \text{ part per 1,00,000 parts of water}$$

$\therefore$ 10,00,000 ppm = 70,000 °Cl = 1,00,000 °Fr

or 100 ppm = 7 °Cl = 10 °Fr

or 1 ppm = 0.07 °Cl = 0.1 °Fr

Degree of hardness:

It is the parts of calcium carbonate equivalent hardness in a definite number of parts of water, depending upon the units in which hardness is expressed.

NUMERICAL PROBLEMS BASED ON UNITS OF HARDNESS

Example 1: ***A water sample contains 204 mg of $CaSO_4$/L. Calculate the hardness in terms of $CaCO_3$ equivalent***

$$CaSO_4 \equiv CaCO_3$$

Solution: 136 g mol^{-1} 100 g mol^{-1}

$\therefore$ 136 mg/L of $CaSO_4 \equiv$ 100 mg/l of $CaCO_3$ equivalent

or $$204 \text{ mg/L of } CaSO_4 \equiv \frac{100 \times 204}{136} \text{ mg/L of } CaCO_3 \text{ equivalent} = 150 \text{ mg/L}$$

$\therefore$ Hardness = 150 mg/l of 1500 ppm

Example 2: ***A sample of groundwater has 150 ml/L of Ca^{2+} and 60 mg/L of Mg^{2+} ions. Find the total hardness expressed in milli-equivalent per litre and mg/L in terms of $CaCO_3$ equivalent.***

(UPT, May 2001)

Solution: Calculation of $CaCO_3$ equivalents.

Constituents	*Amount*	*Multiplication factor*	*$CaCO_3$ equivalent*
Ca^{2+}	150 ml/l	100/40	150 × 100/40 = 375 mg/l
Mg^{2+}	60 mg/l	100/24	60 × 100/24 = 250 mg/l

Total hardness = (375 + 250) = 625 mg/l = 625 ppm

$$= \frac{625}{50} \text{ m-eq./l} = 12.5 \text{ m eq./L}$$

Example 3: ***Calculate the amount of total, temporary and permanent hardness of water sample containing the following per litre: $Ca(HCO_3)_2$ = 162 mg, $Mg(HCO_3)_2$ = 73 mg, $MgCl_2$ = 95 mg, $CaSO_4$ = 136 mg.***

Solution:

$$162 \text{ mg of } Ca(HCO_3)_2 \equiv 162 \times \frac{100}{162} = 100 \text{ mg/l of } CaCO_3 \text{ equivalent}$$

$$73 \text{ mg of } Mg(HCO_3)_2 \equiv 73 \times \frac{100}{146} = 50 \text{ mg/l of } CaCO_3 \text{ equivalent}$$

$$95 \text{ mg of } MgCl_2 \equiv 95 \times \frac{100}{95} = 100 \text{ mg/l of } CaCO_3 \text{ equivalent}$$

$$136 \text{ mg of } CaSO_4 \equiv 136 \times \frac{100}{136} = 100 \text{ mg/l of } CaCO_3 \text{ equivalent}$$

Temporary hardness = Hardness due to $Ca(HCO_3)_2 + Mg(HCO_3)_2$
= 100 + 50 = 150 mg/L

Permanent hardness = Hardness due to $CaSO_4 + MgCl_2$
= 100 + 100 = 200 mg/L

∴ Total hardness = 150 + 200 = 350 mg/L

Example 4: ***A sample of water upon analysis gave the following data: $MgCl_2$ = 0.143 °Fr, $MgSO_4$ = 0.572 °Fr, $CaSO_4$ = 0.286 ° Fr, $Ca(HCO_3)_2$ = 2.316 °Fr. Calculate the hardness in ppm.***

Solution: Since 1 °Fr = 10 ppm.

$$0.143 \text{ °Fr of } MgCl_2 = 1.43 \text{ ppm} = 1.43 \times \frac{100}{95} \text{ ppm } CaCO_3 \text{ equivalent} = 1.51 \text{ ppm}$$

$$0.572 \text{ °Fr of } MgSO_4 = 5.72 \text{ ppm} = 5.72 \times \frac{100}{120} \text{ ppm } CaCO_3 \text{ equivalent} = 4.77 \text{ ppm}$$

$$0.286 \text{ °Fr of } CaSO_4 = 2.86 \text{ ppm} = 2.86 \times \frac{100}{136} \text{ ppm } CaCO_3 \text{ equivalent} = 2.11 \text{ ppm}$$

$$2.316 \text{ °Fr of } Ca(HCO_3)_2 = 23.16 \text{ ppm} - 23.16 \times \frac{100}{162} \text{ ppm } CaCO_3 \text{ equivalent} = 14.29 \text{ ppm}$$

∴ Total hardness in ppm scale = 22.68 ppm

Example 5: ***200 ml of water sample has a hardness equivalent to 25 ml of 0.08 N $MgSO_4$. Find the hardness in °Fr.***

Solution: 200 ml of water sample = 25 ml of 0.08 N $MgSO_4$
= 25 × 0.08 mL of 1 N $MgSO_4$
= 2 ml of 1 N $MgSO_4$ = 2 ml of 1 N $CaCO_3$ eq.
= 0.0021 of 1N $CaCO_3$ eq. = 0.002 × 50 g $CaCO_3$ eq.
= 0.1 g $CaCO_3$ eq. = 100 mg $CaCO_3$ eq.

$\therefore$ One liter (1000 ml) of water sample = $\dfrac{100 \text{ mg } CaCO_3 \text{ eq.} \times 1000 \text{ ml}}{200 \text{ ml}}$

= 500 mg $CaCO_3$ eq.

$\therefore$ Hardness of water = 500 mg L^{-1} = 500 ppm = 500 × 0.1 °Fr = 50 °Fr

Example 6: *Calculate the temporary hardness and permanent hardness of water in ppm and degree Clark from the following data:*

$Mg(HCO_3)_2$ = 16.8 mg/l; $MgCl_2$ = 19.0 mg/l; $Mg(NO_3)_2$ = 29.6 mg/l
$CaCO_3$ = 20.2 mg/l; $MgSO_4$ = 24.0 mg/l; KCl = 74.5 mg/l

Solution:

Compound	*Amount mg/L*	*Multiplication Factor*	*Equivalent of $CaCO_3$*
$Mg(HCO_3)_2$	16.8	$\frac{100}{146}$	$16.8 \times \frac{100}{146} = 11.5$
$MgSO_4$	24.0	$\frac{100}{120}$	$24.0 \times \frac{100}{120} = 20$
$MgCl_2$	19.0	$\frac{100}{95}$	$19.0 \times \frac{100}{95} = 20$
$Mg(NO_3)_2$	29.6	$\frac{100}{148}$	$29.6 \times \frac{100}{148} = 20$
$CaCO_3$	20.2		

KCl does not contribute toward hardness.

Temporary hardness = hardness due to magnesium bicarbonate and calcium carbonate

Temporary hardness = 11.5 + 22.2 = 31.7 ppm

= 31.7 × 0.07 °Cl = 2.219 °Cl

Permanent hardness = hardness due to $MgSO_4$ + $MgCl_2$ + $Mg(NO_3)_2$

Permanent hardness = 20 + 20 + 20 = 60 ppm = 4.20 °Cl

PRACTICE PROBLEMS

1. A sample of water is found to contain 40.5 mg/L $Ca(HCO_3)_2$, 46.5 mg/L $Mg(HCO_3)_2$, 27.6 mg/L $MgSO_4$, 32.1 mg/L $CaSO_4$ and 22.45 mg/L $CaCl_2$. Calculate the total hardness of water.
2. A water sample on analysis gave the following results.
   $Ca(HCO_3)_2$ = 105 ppm, $Mg(HCO_3)_2$ = 125 ppm, $CaSO_4$ = 75 ppm, $CaCl_2$ = 82 ppm, $MgSO_4$ = 26 ppm. Calculate the temporary and permanent hardness in °Fr.
3. A sample of water has a hardness of 5 10 ppm. Express the hardness in °Fr and °Cl.
4. How many grams of $MgCO_3$ dissolved per liter gives 84 ppm of hardness?
5. Calculate the temporary and permanent hardness of water sample with following analysis:
   $Mg(HCO_3)_2$ = 73 mg/l; $Ca(HCO_3)_2$ = 162 mg/; $CaSO_4$ = 136 mg/l; $MgCl_2$, = 95 mg/l; $CaCl_2$ = 111 mg/l and NaCl = 100 mg/l.

Answers

1. 123.7 ppm.
2. Temporary hardness = 15.04 °Fr. Permanent Hardness = 15.07 °Fr
3. 35.7 °Cl, 51 °Fr
4. 70.56×10^{-3} g
5. Temporary hardness = 25 ppm; Permanent hardness = 105 ppm (Total Hardness = 130 ppm)

3.5 HARDNESS (ANALYSIS OF HARD WATER)

As described earlier there are two types of hardness of water, viz. temporary and permanent. Temporary hardness is mainly due to the bicarbonates of calcium and magnesium, while the permanent hardness is due to chlorides and sulphates of calcium abd magnesium. Hardness of water can be estimated by a number of methods, discussed below.

EDTA Titration Method

Principle:

The EDTA titration method of determining hardness is based upon the fact that hardness producing ions like Ca^{+2} and Mg^{+2} can be estimated by titrating them with a standard solution of disodium salt of EDTA in the presence of indicator Eriochrome Black-T at a pH of 10.

The formula of EDTA is written as:

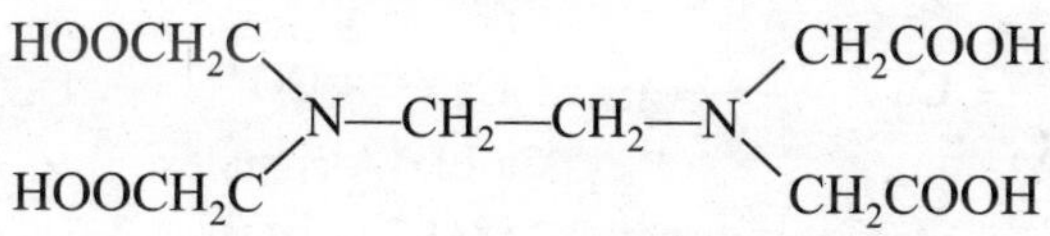

(Ethylene diamine tetraacetic acid)

Disodium salt of EDTA is:

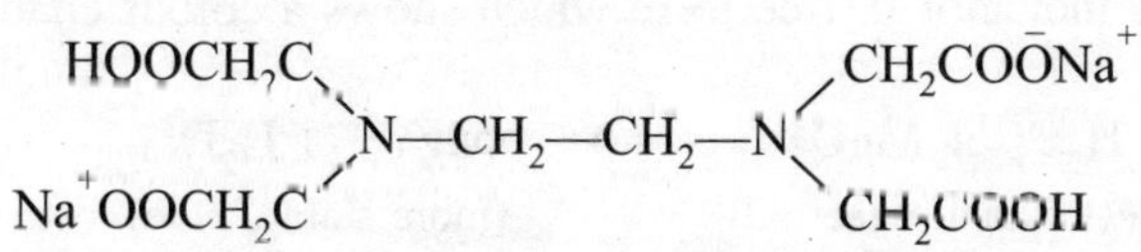

It is represented as Na_2H_2Y. It ionizes in aqueous solution to give $2Na^+$ and a strong chelating ion represented as H_2Y^{2-}.

$$Na_2H_2Y \longrightarrow 2Na^+ + H_2Y^{2-}$$

The formula of the complex formed by EDTA with Mg^{2+} or Ca^{2+} is shown below:

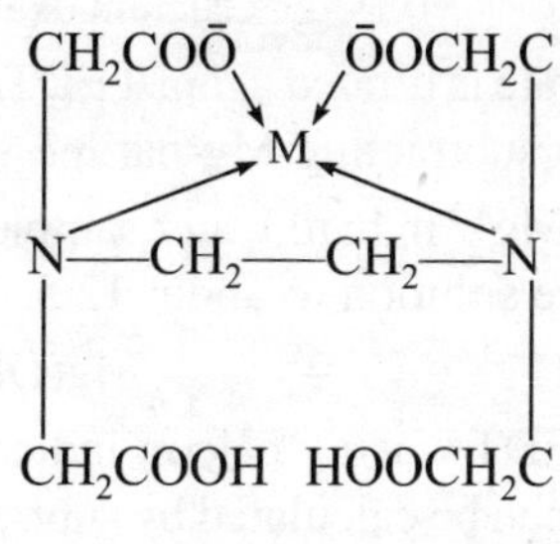

where $M = Mg^{2+}$ or Ca^{2+}

Indicator EBT give different colours at different pH values. Sodium salt of the indicator can be written as $Na^+ H_2 In^-$, which ionizes as:

$$\underset{\text{Red}}{H_2In^-} \underset{PH, 5.5}{\overset{PH, 7.0}{\rightleftharpoons}} \underset{\text{Blue}}{H_2In^{-2}} \underset{10.0}{\overset{12.5}{\rightleftharpoons}} \underset{\text{Orange yellow}}{In^{3-}}$$

During the titration with EDTA solution, EDTA combines with Mg^{2+} or Ca^{2+} ions present in hard water and form the stable complex. When all the Mg^{2+} or Ca^{2+} ions have formed the complex, then next drop of EDTA added displaces the EBT indicator from the unstable complex and the wine red colour changes to blue colour of EBT. Thus at end point,

$$\underset{\text{(Wine red)}}{[(Ca^{2+} \text{ or } Mg^{2+}) - \text{EBT}] \text{ complex}} + \text{EDTA} \longrightarrow [(Ca^{2+} \text{ or } Mg^{2+}) - \text{EDTA}] \text{ complex} + \underset{\text{(blue)}}{\text{EBT}}$$

Hardness of water is due to soluble salts of calcium and magnesium. So hardness can be estimated by knowing the amounts of Ca^{+2} and Mg^{+2} separately in water.

When Eriochrome Black-T indicator is added to hard water at a pH of 10, it reacts with Mg^{+2} of water to give a wine red coloured complex. Representing the EBT indicator at pH 10 as HD^{2-} (Blue), the complex formation may be represented as:

$$Mg^{+2} + \underset{\text{Blue}}{HD^{2-}} \longrightarrow H^+ + \underset{\text{wine red}}{MgD^-}$$

When EDTA solution is added from burette, free Ca^{+2} and Mg^{+2} of hard water react with it to form stable complexes:

$$\underset{\text{(EDTA)}}{H_2Y^{2-}} + Ca^{+2} \longrightarrow \underset{\text{Ca-EDTA complex}}{CaY^{2-}} + 2H^+$$

$$H_2Y^{2-} + Mg^{+2} \longrightarrow \underset{\text{Mg-EDTA complex}}{MgY^{2-}} + 2H^+$$

At the endpoint, EDTA abstracts Mg^{+2} from weak metal-indicator complex to form strong metal–EDTA complex liberating indicator in free form which shows a colour change from wine red to pure blue.

$$H_2Y^{2-} + \underset{\text{(less stable, wine red)}}{MgD^{2-}} \longrightarrow \underset{\text{(more stable, blue)}}{MgY^{2-}} + H_2D^{2-}$$

To get only Mg^{+2} hardness, Ca^{+2} is removed by precipitating it as calcium oxalate by adding calcium precipitating buffer $[NH_4Cl + NH_4OH + (NH_4)_2C_2O_4]$.

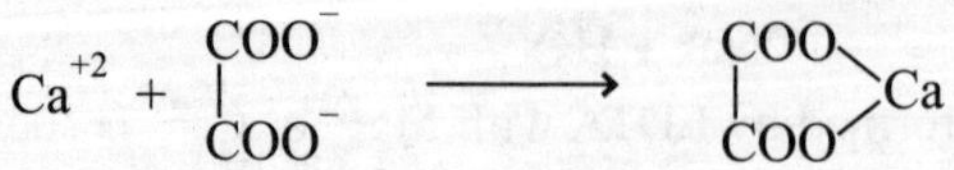

Calcium oxalate (White ppt.)

Precipitate is filtered and the filtrate is titrated against EDTA solution to give magnesium hardness. Calcium hardness can be calculated by subtracting Mg-hardness from total hardness.

Alternative Method for Ca^{2+}: Mg^{+2} in hard water sample is precipitated as $Mg(OH)_2$ by adding diethylamine which raises the pH of the solution to about 12.5

$$Mg^{+2} + 2OH^- \longrightarrow Mg(OH)_2$$

Ca^{+2} can then be titrated with EDTA using calcon indicator. Colour change from pink to pure blue marks the end point. Mg-hardness can be calculated by subtracting Ca-hardness from total hardness.

Indicators

(*i*) Eriochrome Black T-solution (EBT solution).

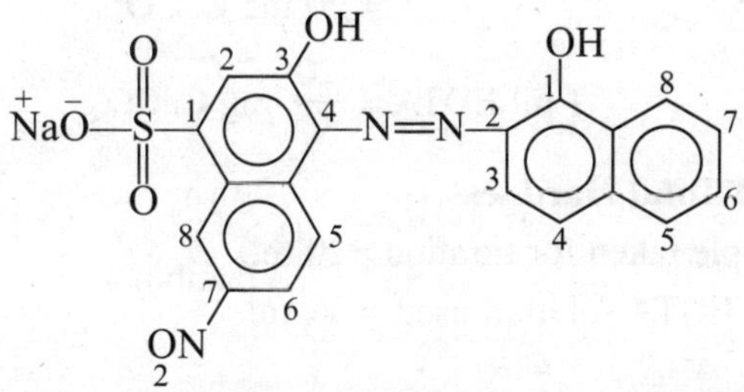

Sodium-1-hydroxy-2-naphthyloza-3-hydroxy-7-nitro naphthalene-1-sulphonate

End Point: Colour change from wine red to pure blue. (EDTA solution in burette).

Procedure:

(I) Preparation of standard hard water. Dissolve 1.0 gm of pure and dry calcium carbonate in the minimum amount of dil HCl. Evaporate the solution to dryness. Dissolve the residue in one litre of distilled water. The hardness of this solution is 1g/litre or 1000 mg/litre or 1000 ppm or 1 mg/ml.

(II) Standardisation of EDTA solution: Pipette out 20 ml of standard hard water and transfer it to the conical flask. Add 5 ml buffer solution and 5 drops of Eriochrome Black-T indicator solution. Colour of solution turns wine red. Titrate the flask solution against the EDTA solution from burette until the colour changes from wine red to clear blue at the end point. Note the burette readings. Repeat the procedure to get three concordant readings.

(III) Determination of total hardness: Pipette out 20 ml of given hard water sample in a conical flask. Add 5 ml of buffer solution and 5 drops of indicator solution. Titrate the wine red solution against EDTA solution in burette till the colour changes to pure blue. Repeat to get concordant readings.

(IV) Determination of permanent hardness: Take 100 ml of hard water sample in a 500 ml beaker and boil gently for half an hour. Cool, filter into a 100 ml measuring flask and make the volume upto the mark with distilled water. Stopper the flask and shake well. Take 20 ml of this solution and titrate with EDTA solution in the same manner as in step I and II. Take three concordant readings.

(V) Determination of magnesium hardness: Pipette out 100 ml of the hard water sample into a 500 ml beaker. Add 25 ml of calcium precipitating buffer solution while constantly stirring the mixture with a glass rod. Allow the precipitate to settle down for 30 – 40 minutes. Filter the solution through filter paper (whatman – 42) in a dry flask. Take 20 ml of filtrate and add 5 ml buffer solution and 5 drops of EBT indicator. Titrate against EDTA solution till wine red colour changes to pure blue.

(VI) Determination of calcium hardness: Pipette out 20 ml of hard water into the conical flask. Add 3 ml diethyl amine and 5 drops of calcon indicator. Shake the flask well and titrate it against standard EDTA solution until the colour changes from pink to pure blue. Repeat the procedure to get concordant readings.

Observations and Calculations

1 ml of standard hard water = 1 mg of $CaCO_3$

(I) Standardisation of EDTA solution:

Volume of S.H.W. taken for titration = 20 ml.

Let concordant volume of EDTA used = V_1 ml.

$$\text{Now, } V_1 \text{ ml of EDTA} = 20 \text{ ml S.H.W.} = 20 \times 1 \text{ mg of } CaCO_3 = 20 \text{ mg } CaCO_3$$

$$1 \text{ ml EDTA} = \frac{20}{V_1} \text{ mg } CaCO_3.$$

(II) Determination of Total Hardness:

Volume of water sample taken for titration =20 ml.

Let volume of EDTA solution used = V_2 ml.

$$20 \text{ ml of hard water} = V_2 \text{ ml EDTA} = V_2 \times \frac{20}{V_1} \text{ mg } CaCO_3.$$

$$1000 \text{ ml of hard water} = V_2 \times \frac{20}{V_1} \times \frac{1}{20} \times 1000 = \frac{V_2}{V_1} \times 1000 \text{ mg } CaCO_3.$$

or $$\text{Total hardness of water} = \frac{V_2}{V_1} \times 1000 \text{ ppm.}$$

(III) Determination of Permanent Hardness and Temporary Hardness:

Volume of hard water taken after boiling and filtering = 20 ml

Let volume of EDTA solution used = V_3 ml.

$$20 \text{ ml hard water} = V_3 \text{ ml EDTA} = V_3 \times \frac{20}{V_1} \text{ mg } CaCO_3.$$

$$1000 \text{ ml hard water} = V_3 \times \frac{20}{V_1} \times \frac{1000}{20} = \frac{V_3}{V_1} \times 1000 \text{ mg } CaCO_3$$

$$\text{Permanent hardness of water} = V_3 \times \frac{1000}{V_1} \text{ ppm.}$$

Temporary hardness of water = Total hardness-Permanent hardness

$$\left[\frac{V_2}{V_1} \times 1000\right] - \left[\frac{V_3}{V_1} \times 1000\right] \quad \frac{V_2 - V_3}{V_2} \times 1000 \text{ ppm.}$$

(IV) Determination of Mg-Hardness:

Total volume of solution = Volume of solution taken (100 ml) + Volume of buffer solution added (25 ml) = 125 ml.

Hardness of 125 ml prepared solution = Mg-Hardness of 100 ml. hard water sample

$$20 \text{ ml filtrate} = \frac{100}{125} \times 20 = 16 \text{ ml of hard water.}$$

Let volume of EDTA solution used = V_4 ml.

$$16 \text{ ml hard water} = V_4 \text{ ml EDTA} = V_4 \times \frac{20}{V_1} \text{ mg } CaCO_3.$$

$$1000 \text{ ml hard water} = V_4 \times \frac{20}{V_1} \times \frac{1000}{16} \text{ mg } CaCO_3.$$

Ca-hardness of water sample = Total hardness – Mg-hardness

(V) Determination of Ca-Hardness:

Volume of hard water taken for each titration = 20 ml

Let volume of EDTA solution used = V_5 ml

$$20 \text{ ml hard water} = V_5 \text{ ml EDTA}$$

$$= V_5 \times \frac{20}{V_1} \text{ mg } CaCO_3$$

$$1000 \text{ ml hard water} = V_5 \times \frac{20}{V_1} \times \frac{1000}{20} \text{ mg } CaCO_3$$

$$= V_5 \times \frac{1000}{V_1} \text{ mg } CaCO_3$$

Magnesium Hardness = Total Hardness – Calcium Hardness

Example 7: ***50 ml of a sample water consumed 15 ml of 0.01 M-EDTA before boiling and 5 ml of the same EDTA after boiling. Calculate the degree of total hardness, permanent hardness and temporary hardness.*** **(Anna, Nov. 2K, M.M.U., Mullana June 2008)**

Solution: 50 ml of water sample = 15 ml of 0.01 M-EDTA

$$1000 \text{ ml of water sample} = \frac{15 \times 1000}{50} \text{ ml of 0.01 M-EDTA}$$

= 300 ml of 0.01 M-EDTA

= 2 × 300 ml of 0.01 N-EDTA

(Molarity of EDTA = 2 × Normality of EDTA)

≡ 600 ml (or 0.6 L) of 0.01 eq $CaCO_3$

= 0.6 × 0.01 × 50 $CaCO_3$ eq

Hence, total hardness = 0.30 g or 300 mg of $CaCO_3$ eq

= **300 mg/L or 300 ppm**

Now 50 ml of boiled water = 5 ml of 0.01 M-EDTA

$$\therefore \quad 1{,}000 \text{ ml of boiled water} = \frac{15 \times 1000}{50} \text{ ml of 0.01 M-EDTA}$$

= 100 ml of 0.01 M-EDTA

= 200 ml (or 0.2 L) of 0.01 N-EDTA

= 0.2 × 0.1 × 50 g of $CaCO_3$ eq

= 0.1 g or 100 mg of $CaCO_3$ eq

Hence, permanent hardness = **100 mg/L or ppm**

∴ Temporary hardness = (300 – 100) ppm = **200 ppm**

Example 8: ***If 50 ml of a sample of hard water consumed 15 ml of 0.01 M-EDTA. What is the hardness of water?*** **(Anna, July 2K)**

Solution: Let us first calculate the molarity of hard water sample consuming 15 ml of 0.01 M-EDTA by applying molarity equation $M_1V_1 = M_2V_2$

$$50 \text{ ml} \times \text{Molarity of hard water} = 15 \text{ ml} \times 0.01 \text{ M-EDTA}$$

or
$$\text{molarity of hard water} = \frac{15 \times 0.01}{50} = 0.003 \text{ M} = 0.003 \times 2\text{N} = 0.006 \text{ N}$$

We know, strength in g/L = Molarity × Mol. Wt.

In order to express hardness in terms of $CaCO_3$ equivalents, we should take the molecular weight of $CaCO_3$ *i.e.,* 100.

∴ Strength of hard water = 0.003 × 100 g/L = 300 mg/L = 300 ppm.

Example 9: ***20 ml of standard hard water (containing 15 g $CaCO_3$ per liter) required 25 ml EDTA solution for end-point. 100 ml of water sample required 18 ml EDTA solution; while same volume of water after boiling required 12 ml EDTA solution. Calculate carbonate and non-carbonate hardness of water. Which buffer is used in this titration and what is its pH.*** **(K.U., Jan. 01)**

Solution: 1 L (or 1,000 ml) of SHW ≡ 15 g (or 15,000 mg) $CaCO_3$ eq

∴ 1 ml of SHW = 15 mg $CaCO_3$ eq

Now 25 ml EDTA solution = 20 ml SHW = (20 × 15) = 300 mg $CaCO_3$ eq

∴ 1 ml EDTA solution = 300/25 or 12 mg $CaCO_3$ eq

(*i*) Calculation of the total hardness of water

100 ml HW = 18 ml EDTA solution = 18 × 12 mg $CaCO_3$ eq

= 216 mg $CaCO_3$ eq

∴ 1000 ml of SHW = 2,160 mg $CaCO_3$ eq

∴ Total hardness = 2,160 mg/L (or **2,160 ppm**)

(*ii*) Calculation of non-carbonate hardness (Permanent hardness)

100 ml boiled hard water = 12 ml EDTA solution = 12 × 12 mg $CaCO_3$ eq

= 144 mg $CaCO_3$ eq

∴ 1,000 ml of boiled hard water = 1,440 mg $CaCO_3$ eq

∴ Non-carbonate hardness (Permanent hardness) = **1,440 mg/L (or 1,440 ppm)**

Hence, carbonate hardness (Temporary hardness) = (2,160 – 1,440) = **720 ppm**

The buffer used is $NH_4OH + NH_4Cl$ and its pH = 10.

Example 10: ***250 mg of $CaCO_3$ was dissolved in HCl and the solution was made 250 ml with distilled water. 50 ml of the above solution required 20 ml of EDTA. 50 ml of water sample consumed 25 ml of EDTA with EBT as indicator. Calculate the total hardness of water sample in ppm.***

(M.D.U. May 2007)

Solution: Normality of $CaCO_3$ solution $= \frac{0.25}{50} \times \frac{1000}{250} = 0.02$ N

(∴ Weight of $CaCO_3$ taken = 250 mg = 0.25g. Eq. weight of $CaCO_3$ = 50)

Suppose normality of EDTA solution is N_1

$$\underbrace{N_1V_1}_{\text{EDTA}} = \underbrace{N_2V_2}_{CaCO_3 \text{ solution}}$$

$$N_1 \times 20 = 0.02 \times 50 \text{ or } N_1 = \frac{0.02 \times 50}{20} = 0.05 \text{ N}$$

Gram equivalents of EDTA consumed by 50 ml of water sample $= \frac{20 \times 0.05}{1000} = 0.00125$

$\therefore$ Gram moles of EDTA consumed $= \frac{0.00125}{2} = 6.25 \times 10^{-4}$

$\therefore$ 1 gram mole of EDTA $= 10^5$ mg of $CaCO_3$

$\therefore$ 6.25×10^{-4} gram mole of EDTA $= 10^5 \times 6.25 \times 10^{-4} = 62.5$ mg of $CaCO_3$

Hence, total hardness present in 100 ml of water sample

$$= \frac{62.5}{50} \times 1000 = 1250 \text{ mg/L of } CaCO_3$$

$$= 1250 \text{ ppm} \quad \textbf{Ans.}$$

(*B*) O. Hehner's method

(*a*) Determination of temporary hardness:

It is based upon the fact that it can be removed by boiling. Thus, it can be determined by estimating the alkalinity of the water before and after boiling.

$$Ca(HCO_3)_2 \xrightarrow{\text{boil}} CaCO_3 \downarrow + H_2O + CO_2$$

$$Mg(HCO_3)_2 \xrightarrow{\text{boil}} Mg(OH)_2 + 2CO_2$$

Procedure: Take 50 ml of the sample water in a conical flask and add it 2–3 drops of methyl orange. Titrate it against N/50 HCl solution till the appearance of pink colour and note the volume of HCl consumed.

Now take 100 ml of the same sample water in a pyrex beaker and evaporate it to dryness. To the residue add 50 ml of distilled water, warm and then stir to dissolve the soluble portion. Filter off the insoluble material (carbonates), wash the residue with distilled water and then make up the filtrate to 100 ml in a measuring flask by adding more distilled water. Titrate 50 ml of this solution against N/50 using methyl orange as indicator and note the volume of acid used.

Calculation:

Let volume of N/50 acid used by 50 ml of water before boiling $= V_1$ ml

And volume of N/50 acid used by 50 ml of water after boiling $= V_2$ ml

$\therefore$ Volume of N/50 HCl used by temporary hardness present in 50 ml of water

$$= \mathbf{(V_1 - V_2) \text{ ml}}$$

$\therefore$ Alkalinity due to temporary hardness in 50 ml of water sample

$$= \mathbf{(V_1 - V_2) \text{ ml of N/50 HCl}}$$

$\therefore$ Alkalinity due to temporary hardness in terms of $CaCO_3$ eq

$$= \frac{(V_1 - V_2) \times 1 \times 50}{50 \times 50} \text{ g/L}$$

$$= \frac{(V_1 - V_2)}{50} \text{ g of } \mathbf{CaCO_3 \text{ eq } L^{-1}}$$

$$\text{Temporary hardness in ppm} = \frac{(V_1 - V_2)\times 10^3}{50} \text{ ppm}$$

$$= 20\,(V_1 - V_2) \text{ ppm}$$

$\therefore$ Temporary hardness $= \mathbf{20\,(V_1 - V_2)\ ppm}$

(b) Determination of permanent hardness:

It is based on the fact that it can be removed by boiling with sodium carbonate with which chlorides and sulphates of calcium and magnesium form insoluble carbonates.

$$CaCl_2 + Na_2CO_3 \longrightarrow CaCO_3 + 2NaCl$$
$$MgSO_4 + Na_2CO_3 \longrightarrow MgCO_3 + Na_2SO_4$$

In practice, an excess but known amount of standard sodium carbonate is used and the residual sodium carbonate is then determined by titrating against a standard acid solution. The amount of Na_2CO_3 used is equivalent to permanent hardness of water.

Procedure: Boil 50 ml of the water sample with 50 ml of N/50 sodium carbonate solution in a beaker. Cool the resulting solution, filter of the precipitate and the wash it with distilled water, collecting all the filtrate and washings in the conical flask. Titrate the contents of the conical flask against N/50 HCl using methyl orange as indicator.

Calculations

Let the volume of N/50 HCl used during titration = V ml

$\therefore$ Volume of N/50 Na_2CO_3 used for removing permanent hardness of the water sample

$$= (50 - V) \text{ ml}$$

$\therefore$ Strength of permanent hardness in terms of $CaCO_3$ eq

$$= \frac{(50 - V)\times 50}{50\times 50} \text{ g of } CaCO_3 \text{ eq}$$

$$= \frac{(50 - V)}{50} \text{ g of } CaCO_3 \text{ eq}$$

$$\therefore \text{ Permanent hardness} = \frac{(50 - V)\times 10^6}{50\times 10^3} \text{ ppm} = 20\,(50 - V) \text{ ppm}$$

Example 11: ***Calculate the temporary, permanent hardness and total hardness of a water sample which gave the following analysis:***

(i) 50 ml of water required 13 ml of N/50 HCl for neutralization to methyl orange end-point.

(ii) 50 ml of water after boiling required 8 ml of N/50 HCl for neutralization to methyl orange end point.

(iii) 50 ml of water sample was boiled with 55 ml of N/50 sodium carbonate solution and the resulting solution required 35 ml of N/50 HCl for neutralization to methly orange end-point.

Solution: (i) Temporary hardness: Volume of N/50 HCl required for removing temporary hardness of 50 ml of water sample = 13 – 8 = 5 ml

Alkalinity due to temporary hardness in 50 ml water sample

$$= (13 - 8) = 5 \text{ ml of N/50 HCl}$$

Alkalinity due to temporary hardness in 50 ml of water sample in terms of $CaCO_3$ eq

$$= \frac{5\times1\times50}{50\times50} \qquad \text{(Eq. wt. of } CaCO_3 = 50)$$

$= 0.1$ g of $CaCO_3$ eq. per litre

$$\therefore \quad \text{Temporary Hardness} = \frac{0.1\times10^6}{10^3} = 100 \text{ ppm}$$

(ii) **Permanent hardness:** Volume of N/50 Na_2CO_3 required for removing permanent hardness of 50 ml of water = 55 – 35 = 20 ml

Permanent hardness in grams per litre in terms of $CaCO_3$ eq

$$= \frac{20\times1\times50}{50\times50} = \frac{20}{50} \text{ g of } CaCO_3 \text{ eq}$$

$$\text{Permanent hardness} = \frac{20\times10^6}{50\times10^3} = 400 \text{ ppm}$$

Total hardness of water sample = 100 ppm + 400 ppm = 500 ppm.

Example 12: ***In an experiment to determine the hardness of a sample of water, 25 ml of N/50 Na_2CO_3 solution was added to 100 ml of water sample. After completion of precipitation of insoluble carbonate, the unreacted Na_2CO_3 was titrated against N/50 H_2SO_4 solution, when 10 ml of the acid was required. Calculate the hardness and comment on the nature of hardness so determined.***

Solution: 100 ml of water = 25 ml of N/50 Na_2CO_3 + 10 ml of N/50 H_2SO_4

Now 10 ml of N/50 H_2SO_4 = 10 ml of N/50 Na_2CO_3

$\therefore$ Amount of Na_2CO_3 consumed for removing permanent hardness

= (25 – 10) ml = 15 ml of N/50 Na_2CO_3

= (15/50) ml of Na_2CO_3

100 ml of water requires 15/50 ml of Na_2CO_3 for hardness removal

$$1000 \text{ ml of water requres } \frac{15}{50} \times 10 \text{ ml of } Na_2CO_3 = 3 \text{ ml of } Na_2CO_3$$

$$3 \text{ ml of } Na_2CO_3 - 3 \text{ ml of } Na_2CO_3 \text{ hardness} = \frac{3\times50}{1{,}000} \text{ g}$$

$$3 \text{ ml of N–}Na_2CO_3 = \frac{3\times50}{1{,}000} \text{ g of } CaCO_3 \text{ eq} = 0.150 \text{ g of } CaCO_3 \text{ eq.}$$

$\therefore$ Hardness of water = 0.150 g/L or 150 mg/L or **150 ppm**

Since sodium carbonate removes permanent hardness causing ions, so the above result represents **permanent hardness of water.**

(*C*) Soap titration method

Principal: This method is based on the fact that soluble soap (sodium or potassium salt of a higher fatty acid) when added to hard water first forms insoluble compounds with the metal ions producing hardness and thus does not produce any lather. But as soon as such metal ions are precipitated out, addition of further drop of soap solution will produce lather.

$$\underset{\text{Sod stearate (soap)}}{2C_{17}H_{35}COONa} + Ca(HCO_3) \longrightarrow \underset{\text{Cal. Stearate}}{(C_{17}H_{35}COO)_2Ca} \downarrow + 2NaHCO_3$$

$$2C_{17}H_{35}COONa + MgCl_2 \longrightarrow (C_{17}H_{35}COO)_2Mg \downarrow + 2NaCl$$

$$2C_{17}H_{35}COONa + CaSO_4 \longrightarrow (C_{17}H_{35}COO)_2Ca \downarrow + Na_2SO_4$$

Thus the total hardness can be determined by titrating the water sample against a standard soap solution. On the other hand permanent hardness can be determined by first precipitating the temporary hardness by boiling and then titrating the filtrate against standard soap solution. The end point is the appearance of a stable lather at least for two minutes.

Procedure:

1. **Preparation of standard hard water.** Dissolve 1.0 gm of pure and dry calcium carbonate in the minimum amount of dil HCl. Evaporate the solution to dryness. Dissolve the residue in one litre of distilled water. The hardness of this solution is 1g/litre or 1000 mg/litre or 1000 ppm or 1 mg/ml.
2. **Preparation and standardization of soap solution:** Dissolve 10 gm of pure and dry soap in one litre of 80% alcohol. Take this solution in the burette. Take 50 ml of standard hard water in a stoppered bottle and add to it soap solution from the burette drop wise and with constant shaking till a permanent (at least for a minute) lather is formed. Let the volume of soap used be V_1 ml.
3. **Lather factor of distilled water:** Titrate 50 ml. of distilled water against the above soap solution till permanent lather is formed. Let the volume of soap used is V ml. This is lather factor and must be subtracted from all the titre values.
4. **Total hardness:** Titrate 50 ml of hard water against the soap solution as above. Let the volume of soap used is V_2 ml.
5. **Permanent hardness:** Boil 250 ml of the water sample in a pyrex beaker till the volume is reduced to 50 ml to convert all bicarbonates to carbonates. Filter off the precipitate, wash the ppt. with the distilled water and collect all the filtrate and washings in a 250 ml. measuring flask. Make up the solution with distilled water. Titrate 50 ml. of this solution against soap solution and let V_3 ml of soap solution is used.

Calculations:

***(i)* Strength of soap solution.** Since 1 ml of standard hard water contains 1 mg of hardness causing substances in terms of $CaCO_3$, equivalent

50 ml of standard hard water = 50 mg of $CaCO_3$

Also 50 ml of standard hard water = $(V_1 - V)$ ml. of soap solution

$(V_1 - V)$ ml of soap solution = 50 mg of $CaCO_3$

or $$1 \text{ ml of soap solution} = \frac{50}{(V_1 - V)} \text{ mg of } CaCO_3$$

***(ii)* Calculation of total hardness**

Since $$1 \text{ ml of soap solution} = \frac{50}{(V_1 - V)} \text{ mg of } CaCO_3$$

$$(V_2 - V) \text{ ml of soap solution} = \frac{(V_2 - V) \times 50}{(V_1 - V)} \text{ mg of } CaCO_3$$

$$\text{Thus 50 ml of the sample water contains} = \frac{(V_2 - V) \times 50}{(V_1 - V)} \text{ mg of } CaCO_3$$

$$\therefore \quad 1000 \text{ ml (1 litre of water sample)} = \frac{(V_2 - V) \times 1000 \times 50}{(V_1 - V) \times 50} \text{ mg of } CaCO_3$$

$$\therefore \quad \text{Total hardness of water sample} = \frac{(V_2 - V) \times 1000}{(V_1 - V)} \text{ mg/litre} = \frac{(V_2 - V) \times 1000}{(V_1 - V)} \text{ ppm}$$

(*iii*) Calculation of permanent hardness

$$\text{Similarly,} \quad \text{Permanent hardness} = \frac{(V_3 - V) \times 1000}{(V_1 - V)} \text{ mg/litre} = \frac{(V_3 - V) \times 1000}{(V_1 - V)} \text{ ppm}$$

(*iv*) Temporary hardness = Total hardness – Permanent hardness

$$= \frac{(V_2 - V) \times 1000}{(V_1 - V)} - \frac{(V_3 - V) \times 1000}{(V_1 - V)}$$

$$= \frac{100}{V_1 - V} [(V_2 - V) - (V_3 - V)] = \frac{1000 \times (V_2 - V_3)}{V_1 - V} \text{ ppm}$$

Example 13: *50 ml of water sample during each time of titration against soap solution gave the following results:*

(i) Lathter factor = 0.5 ml

(ii) Total hardness volume = 15.5 ml

(iii) Permanent hardness volume = 12.5 ml

(iv) Standard hard water volume = 40.5 ml

Standard hard water is prepared by dissolving 0.2 g of $CaCO_3$ per litre. Calculate total, temporary and permanent hardness of water sample in ppm. **(Amaravati, Dec. 2 K)**

Solution: 1 L of SHW = 20 mg $CaCO_3$ eq.

$$\therefore \quad 50 \text{ ml of SHW} = \frac{200 \times 50}{1000} \text{ mg } CaCO_3 \text{ eq}$$

$= 10$ mg $CaCO_3$ eq.

Now 50 ml SHW = 10 mg $CaCO_3$ eq. = 40.5 – 0.5 = 40 ml soap solution.

∴ 40 ml soap solution = 10 mg $CaCO_3$ eq

1 ml soap solution = 10/40 or 0.25 mg $CaCO_3$ eq

(*i*) Calculation of total hardness of water:

50 ml Hard water = 15.5 – 0.5 or 15 ml of soap solution

= 15 × 0.25 mg $CaCO_3$ eq = 3.75 mg $CaCO_3$ eq.

∴ 1 L Hard water = 3.75 × 1000/ 50 or 75 mg $CaCO_3$ eq.

Hence, total hardness = **75 mg/L or 75 ppm**

(*ii*) Calculation of permanent hardness of water:

50 ml boiled hard water = (12.5 – 0.5) or 12 ml of soap solution

= 12 × 0.25 mg $CaCO_3$ eq.

∴ 1 L boiled hard water = 3 × 1000/50 or 60 mg $CaCO_3$ eq.

∴ Permanent hardness = **60 mg/L or 60 ppm**

Hence, temporary hardness = (75 – 60) = **15 ppm.**

Example 14: ***Titration of 50 ml of water at each time with standard soap solution gave the following results:***

(i) Lather factor = 0.4 ml

(ii) Total hardness = 8.2 ml

(iii) Permanent hardness = 2.5 ml

(iv) Standard hard water = 19.9 ml

Calculate each type of hardness present in water.

Solution: V = 0.4 ml V_1 = 19.9 ml V_2 = 8.2 ml V_3 = 2.5 ml

$$\therefore \quad \text{Total hardness} = \frac{V_2 - V}{V_1 - V} \times 1000 \text{ ppm} = \frac{7.8 \times 1000}{19.5} = 80 \text{ ppm}$$

$$\text{Similarly,} \quad \text{Permanent hardness} = \frac{V_3 - V}{V_1 - V} \times 1000 \text{ ppm}$$

∴ Permanent hardness in sample water

$$= \frac{2.1 \times 1000}{19.6} \text{ ppm} = \mathbf{21.5 \text{ ppm}}$$

Hence, Temporary hardness = Total hardness – Permanent hardness

= (80 – 21.5) ppm = **58.5 ppm.**

3.6 AEROBIC AND ANAEROBIC OXIDATION

Domestic effluents, Industrial effluents and *agriculture waste* pollute the water by adding various pollutants. When this polluted water is consumed by human beings, it causes various short-term or long-term diseases. 5 to 7 ppm of dissolved oxygen is present in unpolluted water and is a must for supporting aquatic lives. In presence of a good amount of dissolved oxygen (> 8 ppm), aerobic bacterias lead to oxidation of organic compounds present in water. This kind of oxidation is called '*aerobic oxidation.*' But if the dissolved oxygen is (< 5 *ppm*), *anaerobic oxidation* of organic compounds present in water takes place by anaerobic bacteria.

Excessive nutrients, such as nitrates and phosphates, commonly originate in domestic sewage, run off from domestic fertilizers, waste material from animal feed, packing plants etc. The nutrients are responsible for water pollution primarily because they stimulate the growth of micro-organisms which increases the Biochemical Oxygen Demand (BOD).

Micro-organisms can utilize carbohydrates, proteins and oils and fats as feed and oxidize them to obtain energy for their life processes. Some bacteria can also utilize reduced inorganic materials like Fe^{2+}, S^{2-} and ammonia as source of energy. In the biological degradation of sewage and other waste material, which is brought about by a diverse group of living organisms or bacteria, organic matter is first converted into fatty acids which by successive oxidation at β-carbon atom, are broken down into fragments consisting of acetic acid. If sufficient oxygen is present, that is in aerobic systems, oxygen is the ultimate electron acceptor which is reduced while organic matter is being oxidised to CO_2 and H_2O.

$$CH_3COO^- + 2O_2 \xrightarrow{\text{bacteria}} CO_2 + HCO_3^- + H_2O$$

Inorganic materials like S^{2-} and reduced nitrogen (NH_3 or NH_4^+), if present are oxidised to sulphate and nitrate.

$$S^{2-} + 2O_2 \xrightarrow{\text{bacteria}} SO_4^{2-}$$

$$NH_3 + 2O_2 \xrightarrow[\text{bacteria}]{\text{Nitrifying}} NO_3^- + H^+ + H_2O$$

Under *anaerobic* condition, *i.e.* in the absence of oxygen, SO_4^{2-}, PO_4^{2-} and CO_2, whatever available, can act as electron acceptor and reduced to H_2S.

Sewage and other oxygen decaying wastes are termed as water pollutants. To avoid development of nuisance or septic conditions, sufficient oxygen must be present or supplied externally to maintain aerobic conditions throughout.

Products of aerobic and anerobic oxidation are summarized below:

		Compounds of			
		C	*N*	*S*	*P*
Products formed under	*Aerobic oxidation*	CO_2	NH_3, HNO_3	H_2SO_4	H_3PO_4
	Anaerobic oxidation	CH_4	NH_3, RNH_2	H_2S	PH_3

If the water is polluted with large amount of organic matter, a large amount of dissolved oxygen is rapidly consumed in the biological aerobic oxidation. The decrease in the dissolved oxygen in turn decreases the population of aquatic life. This *biological oxygen demand* (*BOD*) can thus be an index of water pollution.

3.7 DISSOLVED OXYGEN (DO)

Determination of DO is important for boiler feed water and also in sanitary engineering practice. DO is needed for living organisms to maintain biological processes. DO test is the basis of BOD test which is an important parameter in evaluating the pollution potential of domestic and industrial wastes.

The principle involved in this method for determining DO is to bring about the oxidation of KI to I_2 with the DO present in the water sample after adding $MnSO_4$, Na OH and KI. The liberated iodine is titrated against standard hypo solution using starch as indicator.

$$MnSO_4 + 2KOH \longrightarrow Mn(OH)_2 + K_2SO_4$$

$$2Mn(OH)_2 + O_2 \longrightarrow \underset{\text{Basic Manganic Oxide}}{2MnO(OH)_2}$$

$$MnO(OH)_2 + H_2SO_4 \longrightarrow MnSO_4 + 2H_2O + O$$

$$2KI + H_2SO_4 + O \longrightarrow K_2SO_4 + H_2O + I_2$$

$$\underset{\text{hypo}}{I_2 + Na_2S_2O_3} = Na_2S_4O_6 + 2\,NaI$$

Iron and nitrite interfere with this procedure. The interference of nitrite can be eliminated by adding sodium azide (NaN_3). If ferric iron is present, phosphoric acid is used instead of H_2SO_4.

Procedure: The water sample is taken in a 300 ml BOD bottle. Then, 2 ml of $MnSO_4$ and 2 ml of alkaline iodide-azide reagent are pipetted out into it in such a way that the tip of the pipette dips below the liquid surface while adding the reagents. The bottle is stoppered immediately and the solution is mixed well by inverting the bottle gently 3 to 4 times. After 2 minutes, 1 ml of con. H_2SO_4 is added, the stopper is inserted and the solution is mixed well until the precipitate goes into solution. After allowing to stand for 5 minutes, 203 ml of the clear solution are transferred into a conical flask and titrated against standard N/20 hypo solution (which is earlier standardised using N/20 $K_2Cr_2O_7$ solution). The titre value, V is noted.

The Dissolved Oxygen content of the water sample can be calculated as follows:

$$DO = V \times N \times \frac{1}{200} \times 8 \times 100 \text{ mg/l or ppm.}$$

where V is the titre value and N is normality of hypo.

3.8 BIO-CHEMICAL OXYGEN DEMAND (BOD)

The amount of oxygen required by a mixed population of micro-organism in oxidising organic matter present in a sample under strictly aerobic conditions, is generally known as B.O.D. and is directly related to the extent pollution (by sewage or other oxygen-demanding wastes).

Biological Oxygen Demand (BOD) is defined as the quantity of dissolved oxygen required by aerobic bacteria for the oxidation of organic matter under aerobic conditions.

The demand for oxygen is proportional to the amount of organic waste to be degraded aerobically. Hence, BOD approximates the amount of oxidizable organic matter present in the solution. When the BOD is high, the dissolved oxygen becomes low. Hence, greater the BOD, greater is the pollution. BOD is thus an indication of extent of pollution, this would be more clear from the BOD values of following effluents:

The results of B.O.D. is expressed in milligrams oxygen per litre of water (mg/litre) or in ppm. The pure drinking water, on the monthly average, should have 5 day BOD in the range of 0.75–1.5 mg/litre. BOD value of sewage runs from 200–400 mg of oxygen per litre of water. In fact, drinking water should have BOD less than 1 ppm.

BOD of say, 90 mg/litre of polluted water indicates that the biodegradation of organic matter in one litre of sample consumes 90 mg of oxygen. BOD is an indicator and not an pollutant. BOD values are very important, when we know that oxygen supply dissolved in water is so reduced that fish no longer survive in water and when dangerous bacteria exist in water.

Method For Determination of BOD

The sample of water is incubated for 5 days at 20°C (roughly a mean value of temperature of natural bodies of water) and the BOD values are reported as BOD_5. A polluted sample may consume more oxygen in 5 days than present in water (nearly 9 mg/litre at 20°C) and so, before analysis, is diluted with a specially prepared Dilution water. The diluted sample is taken in two bottles. The D.O. (dissolved oxygen) is one bottle is determined immediately and in the other after 5 days of incubation. The BOD of the sample in then given by

$$\text{BOD} = \frac{D_1 - D_2}{A} \times B \text{ mg/L}$$

where D_1 = D.O. of the sample in mg/L at the start of the experiment

D_2 = D.O. of the sample in mg/L after 5 days

A = ml of the sample before dilution

B = ml of the sample after dilution

Importance of BOD

(*i*) BOD is most important in sewage treatment, as it indicates the amount of decomposable organic matter in the sewage.

(*ii*) Greater the concentration of organic matter in the sewage, greater the BOD.

(*iii*) It has special signification in pollution control, because it tells us the degree of pollution at any time in sewage waste.

(*iv*) It also tells us about the amount of dissolved oxygen in water because it will be so greatly reduced that fish no longer survive in water.

3.9 CHEMICAL OXYGEN DEMAND

It is an index of the total organic content of water oxygen demanding substances in water. It is a satisfactory method for determining the organic load of a water body, which is preferable to the bio-chemical oxygen demand. COD is readily measurable parameter for streams and industrial waste water studies and control of water treatment plants. The method is based on the chemical oxidation of material in the presence of a catalyst by $Cr_2O_7^{2-}$ in 50% H_2SO_4.

COD may, therefore, be defined as the amount of oxygen required by organic matter in a sample of water for its oxidation by a strong chemical oxidising agent such as $K_2Cr_2O_7$. It is expressed as ppm of oxygen taken form a $K_2Cr_2O_7$ solution in two hours.

Method for Determination of COD

A known amount of $K_2Cr_2O_7$ is added to a measured amount of the sample and the mixture is boiled with concentrated sulphuric acid. After refluxing for two hours, the unreacted dichromate ($Cr_2O_7^{2-}$) is determined by titration against a standard ferrous ammonium sulphate (Mohr's salt) solution using ferroin as the indicator.

Ag_2SO_4 catalyses the oxidation of straight chain aliphatic compounds, aromatic hydrocarbons and pyridine. $HgSO_4$ ties up Cl^- ions as soluble complex and prevents its interference. The difference between the dichromate originally present (added) and the dichromate remaining unreacted gives the amount of dichromate used for the oxidation of organic matter.

Therefore $$\text{C.O.D.} = \frac{(B - A)}{4 \times 50} \times 9 \text{ mg/L}$$

$$= \frac{(B - A) \times 8 \times 1000}{4 \times 50} \text{ mg/L}$$

$$= (B - A) \times 40 \text{ mg/L}$$

where B – is the volume of dichromate originally present.

A – is the volume of dichromate remaining unreacted.

Importance of COD

1. The COD test is, therefore, widely used for measuring the pollutional strength of domestic and industrial wastes.
2. The major advantages of COD test is that the determination is completed in 3 hours, as compared to the 5 days required for the BOD determination.
3. COD is important in management and design of treatment plant.
4. It is used in calculating the efficiency of treatment plants and proposing standards for discharging domestic effluents in various types of water streams.

3.9.1 Distinction Between BOD and COD

The BOD and COD tests employed for the measurement of pollutional strength of waste water differ in the following respects:

BOD	*COD*
(*a*) BOD is the amount of dissolved oxygen required by bacteria while oxidizing organic matter under aerobic conditions in the waste water.	(*a*) COD is a measure of both biologically oxidisable (like glucose) as well as biologically inert (like cellulose) organic matter present in the waste water.
(*b*) The BOD determination takes a long period of 5 days.	(*b*) It can be determined in short period of time around 2-3 hours.
(*c*) BOD values are useful generally in process design and loading calculations, measurement of treatment efficiency and operation stream pollution control,and in determining the self purifying capacity of streams.	(*c*) COD is important in management and design of treatment plants because of its rapid determination. It is used in calculating the efficiency of treatment plants and proposing standards for discharging domestic effluents in various types of water streams.

NUMERICAL PROBLEMS ON BOD AND COD

Example 15: ***10 ml of a polluted water sample was diluted to 600 ml and equal volumes were poured into two B.O.D. bottles. D.O. in one bottle was determined immediately (Blank) by Winkler's method and 204.4 ml solution required 3.9 ml of N/40 $Na_2S_2O_3$. The second sample was incubated for 5 days and in the D.O. determination, 204.4 ml solution requirted 2.1 ml of N/40 $Na_2S_2O_3$. Calculate B.O.D. of the polluted water.***

Solution:

$$D_1 = 3.9 \text{ ml}$$

$$D_2 = 2.1 \text{ ml}$$

$$A = 10 \text{ ml}$$

$$B = 600 \text{ ml}$$

$$\text{B.O.D.} = \frac{D_1 - D_2}{A} \times B \text{ mg/L}$$

$$= \frac{(3.9 - 2.1)}{10} \times 600 = \frac{1.8 \times 600}{10}$$

$$= 108 \text{ mg/L}$$

Example 16: ***100 ml of waste water (containing 820 ppm dissolved oxygen) was diluted with 100 ml of distilled water and kept in a bottle at 20°C for 5 days. The oxygen content of the resulting water was then found to be 260 ppm. Calculate the BOD of the sample.***

Solution: BOD = [orginial D.O. – Final D.O.] × Dilution factor

= (820 – 260) ppm × (200 ml/100 ml)

= 560 × 200 = 1120 ppm.

Example 17: ***Calculate the COD of the effluent sample when 25 cm³ of an effluent requires 9.3 cm³ of 0.001 M $K_2Cr_2O_7$ for oxidation [Molecular mass of $K_2Cr_2O_7$ = 294].***

Solution: 1,000 cm³ (or ml) of 0.001 M $K_2Cr_2O_7$ = 6 × 8 × 0.001 g of oxygen

∴ 9.3 cm³ (or ml) of 0.001 M $K_2Cr_2O_7 = \dfrac{6 \times 8 \times 0.001 \times 9.3}{1,000}$ g of oxygen

∴ 25 cm³ of effluent contains $= \dfrac{6 \times 8 \times 0.001 \times 9.3}{1,000}$ g of oxygen

or 1,000 cm³ of effluent contains $= \dfrac{6 \times 8 \times 0.001 \times 9.3 \times 1,000}{1,000 \times 25}$ g O_2

$$= \frac{6 \times 8 \times 0.001 \times 9.3}{25} \text{ g } O_2$$

$$= \frac{6 \times 8 \times 0.001 \times 9.3 \times 1,000}{25} = \mathbf{17.85 \text{ mg } O_2}$$

Example 18: ***50 ml of sample for COD analysis was reacted with 15 ml of 0.25 N potassium dichromate solution and after the reaction, the unreacted dichromate required 20 ml 0.1 N FAS for reaction. Under identical condition, 15 ml of dichromate solution mixed with 25 ml of distilled water required 35 ml of 0.1 N FAS. What is the COD of the sample?***

Solution: Difference in volume of FAS required in the blank (distilled water) and the sample

= (35 – 20) ml = 15 ml

Now, 1000 ml of 1 N FAS contains = 8 oxygen

or 15 ml of 0.1 N FAS contains $= \dfrac{8 \times 15 \times 0.1}{1,000}$ g oxygen

∴ 25 ml of water sample contains $= \dfrac{8 \times 15 \times 0.1}{1,000}$ g O_2

or 1.000 ml of water sample contains $= \dfrac{8 \times 15 \times 0.1 \times 1,000}{1,000 \times 50}$ g O_2

= 0.240 g or 240 mg oxygen

Hence, COD of sample = 240 mg/L

Example 19: ***5 ml of a sewage water sample was refluxed with 10 ml of 0.25 N $K_2Cr_2O_7$ solution in presence of dil. H_2SO_4. The unreacted dichromate required 6.5 ml of 0.1 N ferrous ammonium sulphate solution. 10 ml of the same $K_2Cr_2O_7$ solution and 25 ml of distilled water under the same conditions as the sample required 27.0 ml of FAS of 0.1 N. Calculate the COD of the sample.***

Solution: Difference in volume of FAS required in the blank and sample.

$$= 27.0 - 6.5 = 20.5 \text{ ml}$$

∴ Now 1,000 ml of 1 N FAS = 8 g of oxygen

or 20.5 ml of 0.1 N FAS = $\frac{8 \times 20.5 \times 0.1}{1,000}$ g of O_2

∴ 25 ml of sample contains = $\frac{8 \times 20.5 \times 0.1}{1,000}$ g of O_2

or 1,000 ml sample contains = $\frac{8 \times 20.5 \times 0.1 \times 1,000}{1,000 \times 25}$ g of O_2

= 0.655 g of O_2

= 655 mg of O_2

Hence, COD = 655 mg/litre or ppm.

PRACTICE PROBLEMS

1. If 4.0 ml of raw sewage has been diluted to 400 ml, and the DO concentration of the diluted simple at the beginning of the BOD test was 9 mg/L, and 5 mg/l after 5 days incubation at 20°C. Find the BOD of raw sewage.
2. A 2.5% solution of a sewage is incubated for 5 days at 20°C. Find the BOD of the sewage, if the depletion of oxygen was found to be 5 mg/L.
3. If 3.0 ml of raw sewage has been diluted to 300 ml, and the DO concentration of the diluted simple at the beginning of the BOD test was 8 mg/L, and 5 mg/L after 5 days incubation at 20°C. Find the BOD of raw sewage.
4. A 1.5% solution of a sewage sample is incubated for 5 days at 20C. Find the BOD of the sewage, if the depletion of oxygen was found to be 3 mg/L.

Answers

1. 400 mg/L.
2. 200 mg/L.
3. 300 mg/L.
4. 199.8 mg/L.

3.10 BOILER CORROSION

It is defined as the decay of boiler material by attack of some chemicals. It is one of the most serious problems caused by hard water. It affects the boiler tubes, drums, super heaters and condensers of the boilers. It can be due to the following factors.

1. Dissolved oxygen,
2. Mineral acids,
3. Dissolved carbon dioxide, and
4. Galvanic cell formation.

1. ***Dissolved oxygen:*** Solubility of oxygen decreases with increase in the temperature. Therefore, at high temperature, the entire oxygen present in water is liberated, which attacks the iron parts of the boilers.

$$2Fe + 2H_2O + O_2 \longrightarrow 2Fe(OH)_2$$

$$4Fe(OH)_2 + O_2 \longrightarrow \underbrace{2Fe_2O_3 \cdot 2H_2O}_{\text{Rust}}$$

Dissolved oxygen can be removed by following methods.

(*a*) Addition of chemicals

$$\underset{\text{(Sod. Sulphide)}}{2Na_2SO_3} + O_2 \longrightarrow 2Na_2SO_4$$

$$\underset{\text{(Sod. Sulphide)}}{Na_2S} + 2O_2 \longrightarrow Na_2SO_4$$

$$\underset{\text{(Hydrazine)}}{N_2H_4} + O_2 \longrightarrow N_2\uparrow + 2H_2O$$

Out of these chemicals hydrazine is preferred for removal of oxygen, as it leaves no salts behind. Moreover, the nitrogen gas produced in the reaction is harmless. However, excess of hydrazine should be avoided as it gives ammonia on decomposition, which attacks copper parts of the boiler.

$$3N_2H_4 \longrightarrow 4NH_3 + N_2$$

Na_2SO_4 obtained from Na_2S and Na_2SO_3 decomposes to give sulphur dioxide which combines with water to give H_2SO_3 (sulphurous acid) which is again corrosive.

(*b*) Mechanical deaeration: Solubility of gases decreases with increase in temperature and decrease in pressure. Therefore, for deareation hot feed water is injected as a fine spray into a vacuum chamber through perforated plates. Due to perforated plates more surface area is exposed. The chamber is heated from outside, so the dissolved gases get liberated and are sucked out by a vacuum line.

2. ***Mineral acids:*** Mineral acids may be present in feed water as such or are produced by hydrolysis of magnesium salts.

$$MgCl_2 + H_2O \longrightarrow Mg(OH)_2 + 2HCl$$

The acid produced in the above reaction attacks the iron parts of the boiler.

$$Fe + 2HCl \longrightarrow FeCl_2 + H_2$$

$$FeCl_2 + 2H_2O \longrightarrow Fe(OH)_2 + 2HCl$$

So HCl is again generated. Therefore, even a small amount of magnesium chloride will cause corrosion of large amount of iron.

Removal of mineral acids: The mineral acids present in feed water can be removed by adding calculated amount of alkali. It is suggested that except in high pressure boilers, the boiler feed water should have a pH value of about 10.5 so that the acid may not attack iron.

3. ***Dissolved carbon dioxide:*** Carbon dioxide may be present in the hard water in dissolved form or is obtained by decomposition of bicarbonates.

$$Ca(HCO_3)_2 \xrightarrow{\Delta} CaCO_3\downarrow + H_2O + CO_2\uparrow$$

The CO_2 so obtained combines with water to give carbonic acid (H_2CO_3). Which has a corrosive effect on the equipment.

Removal of CO_2

(*a*) *By addition of lime* $\quad Ca(OH)_2 + CO_2 \longrightarrow CaCO_3\downarrow + H_2O$

(*b*) *Addition of ammonia* $2NH_3 + H_2O + CO_2 \longrightarrow (NH_4)_2\ CO_3$

However, excess of ammonia should be avoided as it attacks copper material.

(*c*) *Mechanical Deaeration:* Same as for removal of oxygen.

4. *Galvanic cells:* Corrosion can also be due to simple galvanic cells which iron forms with some of the boiler fittings made of other materials or with impurities. It can be avoided by suspending zinc plates. Since zinc is more electro positive than iron, it gets corroded and iron is saved.

3.11 SEWAGE AND ITS TREATMENT

Composition of Sewage: Sewage is the liquid waste, which consist of about 99.9 percent water and 0.1% of organic and inorganic solids in dissolved, suspension and colloidal states. The organic matter consists of carbohydrades, fats, proteins, urea, amines, amino acids and products of their decomposition. The inorganic solid consists of ash, cinder, grit, sand, mineral salts and mud etc. In addition to these solids, sewage contains living organisms such as bacteria, fungi, algae, virus, protoza etc. It may contain toxic substances like Pb, As, Hg, cynides etc., if industrial effluent forms a part of sewage. In fact the composition of sewage depends upon the sources from where it is formed.

The sewage includes:

(*i*) domestic sewage, which contains human excreta as well as discharges from kitchens, baths, lavatories etc.

(*ii*) industrial wastes from manufacturing processes such as tanneries, slaughter-houses, distilleries, textile mills, laundries, chemical plants etc.

(*iii*) storm water, which is rain water from houses, roads etc.

3.11.1 Decomposition of Sewage

Most of the organic matter present in sewage is decomposed readily through biological actions, and is called *biodegradable organic matter.* Biological decomposition can be of the following two types:

(*i*) *Aerobic decomposition or oxidation and*

(*ii*) *Anaerobic decomposition or oxidation*

Aerobic decomposition: The bacteria which carry out the decomposition of organic matter is the presence of free dissolved oxygen in sewage and consume organic matter as their food are called *aerobic bacteria.* The decomposition of organic matter brought about by aerobic bacteria giving inoffensive (not foul smelling) end products is called aerobic decomposition. Aerobic decomposition takes place in two stages. In the first stage carbonaceous matter is oxidised to CO_2 and H_2O.

In the second stage nitrogeneous organic matter is oxidised to nitrates (NO_3^- ions), write it in followed by sulphourous organic matter to sulphates (SO_4^{2-} ions) and phosphorus organic matter to phosphates (PO_4^{3-} ions). The end products formed are CO_2, nitrates, sulphates are phosphates, which are stable and inoffensive. Water heat energy and additional bacteria are also produced in aerobic decomposition. The treatment unit which work on aerobic decomposition are *aeration tanks, trickling filters, oxidation ponds, oxidation ditches, contact bed etc.*

Anaerobic decomposition: When the free dissolved oxygen is exhausted, aerobic decomposition subsides and anaerobic decomposition begins. It is carried out by anaerobic bacteria the absence of tree dissolved oxygen and light. When the dissolved or free oxygen supply is below a certain limit, the sewage is called *stale* and anaerobic bacteria bring about purification producing CH_4, H_2S, PH_3 which give offensive odour. In anaerobic decomposition, hydrolysis takes place, *i.e.*, hydrogen and oxygen in the water molecule are separated and combined with carbon, nitrogen and sulphur. The end products formed are methane, hydrogen, ammonia, carbon dioxide, hydrogen sulphide, organic acids etc.

In the first phase of decomposition, large complex organics are broken down into simple amino acids, fatty acids, sugar and glycerol. In the next phase, the simple organic substances are absorbed by the bacteria and new bacterial cells are produced. The final end products have offensive odour and are toxic to the organisms themselves. It may be noted that in anaerobic oxidation, the micro-organism extract the required amount of oxygen from the organic matter, nitrites and sulphates present in the sewage. When the anaerobic decomposition is in progress, the sewage is called ***septic sewage.*** The various treatments that work on anaerobic decomposition or putrefication alone are *septic tanks, imhoff tanks, and sludge digestion tanks.*

Anaerobic decomposition is generally avoided because during this, the sewage becomes septic and foul gases are evolved, and also anaerobic decomposition is slower than aerobic decomposition.

3.11.2 Sewage Treatment

The sewage should be suitably treated before releasing it into water bodies (*i.e.* streams, rivers, lakes and oceans) or on land. The main objective of sewage treatment are:

(*i*) To eliminate the chances of contamination of water bodies from where water is supplied to public in towns and cities.

(*ii*) To remove the foul smelling and undesirable elements in the sewage, which may cause nuisance otherwise.

(*iii*) To prevent the destruction of fish and other aquatic life in rivers, canals etc.

The sewage treatment processes are generally classified as:

(*a*) *Preliminary and primary treatment*

(*b*) *Secondary or biological treatment*

(*c*) *Tertiary or final treatment*

(*a*) Preliminary and Primary Treatment: Preliminary treatment is carried out for the removal of the floating materials (such as leaves, rags, paper, wood, dead animals etc.), the heavy settleable inorganic solids (such as sand grit etc.) and the organic materials (such as fats, oils and greases) from the sewage. This involves in passing the sewage through *bar screens* (to remove large suspended floating matter) and *mesh screens* (to remove coarse solids, gravels, silt etc.)

The processes generally used in preliminary treatment are screening, grit chamber and skimming tanks. Primary treatment is carried out for the removal of large suspended organic solids present in the sewage. It is usually accompanied by sedimentation.

For this, continuous flow type sedimentation tanks are employed. Chemical treatment is sometimes given to sewage, just prior to sedimentation. This results in more rapid and complete removal of suspended matter. The chemicals employed are alum, ferrous sulphate etc. These produce large gelatinous flocs, which entrap finely divided organic matter and settle rapidly.

$$Al_2(SO_4)_3 + 6H_2O \longrightarrow 2Al(OH)_3 \downarrow + 3H_2SO_4$$

The main purpose of adding chemical coagulant is its ability to remove colloidal materials. The primary process reduces the BOD about 30 to 35 percent.

(*b*) Secondary or Biological treatment: It is carried out for the removal of fine suspended or dissolved organic material present in the sewage obtained from primary treatment unit. It is essentially an *aerobic chemical oxidation.* In this process, sewage is filtered through specially designed *sprinkling filters,* so that *aerobic conditions* are maintained all the times. During this aerobic oxidation process, the carbon of the organic matter is converted into CO_2; the nitrogen into NH_3 and finally into nitrates and nitrates. The dissolved bases, present in the sewage water, then form salts such as ammonium nitrite, ammonium nitrate, calcium nitrate, etc. Thus:

$$\underset{\text{Urea}}{NH_2CONH_2} \xrightarrow[\text{Oxidation}]{\text{aerobic}} CO_2 + 2NH_3$$

$$N_3 \xrightarrow{\text{Oxidation}} HNO_3, HNO_2 \text{ etc.}$$

$$HNO_3, NNO_2 \xrightarrow[\text{of soil}]{+\text{ Bases}} NH_4NO_2, NH_4NO_3, Ca(NO)_2, Ca(NO_3)_2, KNO_2, \text{etc.}$$

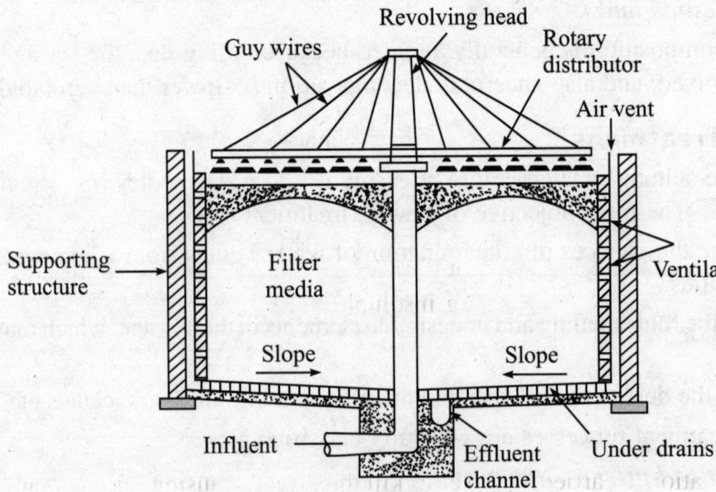

Fig. 3.1 Cross-section of Trickling Filter

Trickling filters are mostly employed for this purpose. These are either rectangular or circular in shape and about 2 m deep (Fig. 3.1). They are filled with bricks. The *underdrain system* is provided at the bottom of bed to collect effluent. Sewage is delivered to the filters by means of a rotary distributor.

The sewage is allowed to trickle over the filtering medium by means of rotating arm distributor. As the trickled sewage starts percolating downwards, through the filtering media, microorganism present in sewage grow on the surface of aggregates using organic material of the sewage as food. An aerobic condition always prevail the purified sewage is removed from the bottom through the effluent channel trickling filter remove about 90% BOD.

Activated Sludge Process

In this process required amount of oxygen of air is passed the sewage containing bacteria. The aeration is carried out in the presence of sludge obtained from the previous operation to quicken aerobic oxidation process. The sludge so added from the previous operation is known as *activated sludge,* since it contains a lot of bacteria and organic matter. After mixing the sedimented sewage with activated sludge, the mixture is taken in aeration tanks and the mixed liquor is simultaneously aerated and agitated for 4 – 6 hours. First carbon from the organic matter is oxidised to CO_2 and H_2O, followed by the oxidation of nitrogen to nitrates and nitrates. After aeration, the effluent is sent to setting or sedimentation tanks, where sludge is deposited and clean liquid, free from the bacteria is drawn off. A part of the settled sludge is sent back for seeding fresh, sludge is subjected to dumping in sea or spreading uniformly over soil, followed by ploughing. Sometimes, the digestion process *i.e.,* sludge is kept in a closed tank in the absence of air for a prolonged period of month or so, when it suffers anaerobic decomposition, yielding gases like methane, hydrogen sulphide, ammonium sulphate and phosphine. The gas can be used as *fuel* either for city supply or power generation. The secondary treatment redices the BOD about 70 to 95 percent. A flow diagram of the secondary treatment is given below in Fig. 3.2.

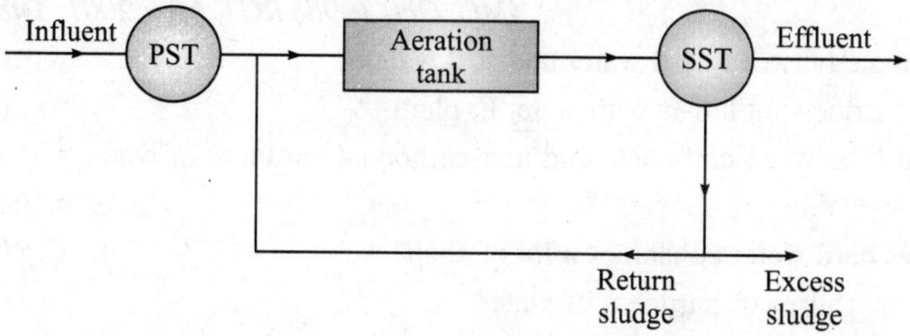

Fig. 3.2 Flow Diagram of a conventional Activated Sludge Process

***(c)* Tertiary or Final treatment:** Tertiary treatment is also called advanced treatment and its objectives are to decrease the load of nitrogen and phosphorus compounds present in the effluents.

It involves treatment of effluent from secondary treatment with calcium oxide (lime), which reacts with phosphours compounds forming insoluble calcium phosphate, which settles down at the bottom of the tank, from where it is filtered off.

The nitrogen load is reduced by removing dissolved NH_2 gas. This is carried out by transferring the filtered water into a metal tower, containing a series of small baffle plates. Air is forced upwards through the trickling effluent, where by ammonia gas is removed.

Finally chlorination is carried out so as to kill the disease causing micro organisms that might be present in sewage waste water. The so-treated water is then finally discharged into rivers/lakes, since it can be used after normal water treatment processes.

The sewage treatment is usually confined upto the secondary treatment. The effluent obtained from secondary unit is disposed off into the water bodies (rives/streams) Tertiary treatment has to adopted when the ouftall of sewage into the water body is very near to the water intake of some near by town on the down stream side of water body. The primary and secondary sewage treatment process is illustrated schematically in Fig. 3.3.

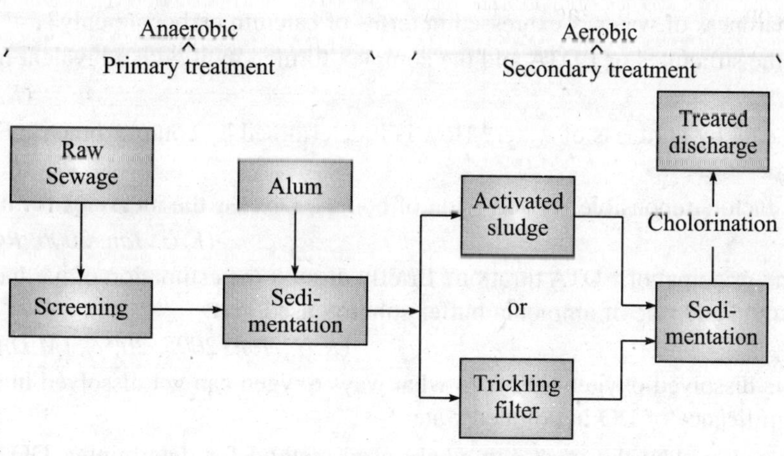

Fig. 3.3 Flow Diagram of Sewage treatment

REVISION QUESTIONS

1. What is meant by hardness of water? What are the common units of measurement of hardness of water? Give relationship between various units. *(MG, Oct. 1998; RGT, May 2001; Dibrugarh Dec. 2K)*
2. (*a*) Defferentiate between hard water and soft water. *(K.U., May 1999)*
   (*b*) Hard water does not lather with soap. Explain. *(K.U., May 1999)*
   (*c*) Distinguish between carbonate and non-carbonate hardness of water with examples. *(Anna, July 2K, Jan. 2001)*
   (*d*) Why does hard water consumes a lot of soap? *(U.P.T., Jan. 2001)*
   (*e*) Explain the causes of hardness of water. *(R.G.T., July 2001)*
   (*f*) What are the disadvantage of using hard water in boilers? *(P.T.U., Dec. 2K)*
3. (*a*) Distinguish between carbonate and non-carbonate hardness of water with example.
   (*b*) What is demineralised water? How is it different from soft water? *(Anna July 2000, Jan. 2001)*
4. (*a*) Name various units of hardness and the relationship among them *(M.U.U. May, 2008)*
   (*b*) Name the three gases dissolved in water responsible for causing corrosion *(M.D.U., May 2008)*
   (*c*) Define hardness of water. What are the units of hardness? Why hardness is measured in terms of $CaCO_3$ *(M.D.U. Dec. 2008)*
5. (*a*) 50 ml of a given water sample consumed 15 ml of 0.01 M EDTA solution before boiling and 5 ml of the same EDTA solution after boiling. Calculate the various types of hardness in ppm as $CaCO_3$ equivalents. *(K.U., Jan. 2005)*
   (*b*) Write the structural formulae of EDTA and the indicator used in EDTA titrations. *(K.U., Jan 2005)*
6. Define the term hardness. From where is it introduced in the water? Describe a suitable method for the determination of hardness.
7. (*a*) What are the type of hardness? Explain its causes. Write the units of hardness. How are they related each other. *(K.U., Jan. 2005, 2006, June 2006)*
   (*b*) Why hardness of water is expressed in terms of calcium carbonate only?
   (*c*) Write the structures of EDTA and the complex formed by it with a bivalent metal ion. *(K.U., Jan. 2006)*
8. What is meant by hardness of water? How is it determined by complexometric titration (EDTA method) method? *(K.U., Jan. 2006)*
9. Name the factors reponsible for corrosion of boilers. Discuss the measures for its prevention. *(K.U. Jan. 2006, RGT, May 2001)*
10. What is the principal of EDTA titration? Briefly discuss the estimation of hardness of water by EDTA method and role of ammonia buffer solution used in it. *(K.U., Jan. 2005, 2007), (M.D.U., May, 2004)*
11. (*a*) What is dissolved oxygen (DO)? In what ways oxygen can get dissolved in water? What is the significance of DO in polluted water?
    (*b*) Name and explain the most commonly used method for determining DO in water/waste water.

12. What do you mean by the term BOD? How can it be experimentally determined?
13. What is COD? How can it be expermentally determined? Give the principle, procedure and precautions of the method.
14. Distinguish between the following
    (*a*) First stage and second stage BOD.
    (*b*) BOD and COD
    (*c*) Aerobic and Anaerobic decompositions.
15. Differentiate between BOD and COD. Discuss the importance of these tests in the analysis of domestic sewage/waste water. What are the merits and demerits of these tests?
16. What is the composition of sewage? How does the decomposition of sewage takes place? What are the main objectives of sewage treatment?
17. Classify the sewage treatment processes and briefly explain each class.
18. Explain in detail the secondary and tertiary stage of sewage treatment.
19. Write the structure of metal EDTA complex and EBT indicator. Describe the process of estimation of Ca and Mg with EDTA. (*M.D.U., Dec., 2006*)
20. (*a*) 250 mg of $CaCO_3$ was dissolved in HCl and the solution was made 250 ml with distilled water 50 ml of above solution required 20 ml of EDTA 50 ml of water sample consumed 25 ml of EDTA with EBT indicator. Calculate the total hardness of water sample in ppm. (*M.D.U., May, 2007*)
    (*b*) Why hardness of water is expressed in terms of $CaCO_3$ only? (*M.D.U., Dec., 2007*)
21. (*a*) Why $NH_3 - NH_4Cl$ buffer solution is added in determination of hardness of water by EDTA method?
    (*b*) What type of ligand is EDTA? Why the complex of Ca^{2+} or Mg^{2+} with EDTA is more stable than the complex of Ca^{2+} or Mg^{2+} with EBT?
    [**Ans.** EDTA is a hexadentate ligand. It form 4 or 6 bonds with the divalent cations (Ca^{2+} or Mg^{2+}) thus entrapping it strongly. Such type of compounds are known as chelate compounds which are very stable.]
22. (*a*) What are the disadvantages of using hard water in boiler operations?
    (*b*) What are the reasons of boiler corrosion?

4

Corrosion and its Control

INSIDE THIS CHAPTER

4.1 INTRODUCTION

As soon as the metals are extracted from their ores, the reverse process begins *i.e.,* nature tries to convert them back into the form in which they occur. This is due to the chemical attack of the gases present in the atmosphere on the surface of the metal converting it into compounds such as oxides, sulphides, carbonates, sulphates etc. It can also take place due to electrochemical interaction of metal with the environment.

The process of slow destruction of metal due to chemical or electrochemical attack of the environment at their surfaces resulting into the formation of compounds such as oxides, carbonates, sulphides, sulphates etc. is called ***corrosion.***

The most common example of corrosion is the rusting of iron. *Rust is hydrated ferric oxide,* $Fe_2O_3\ xH_2O$. Some other examples include tarnishing of silver, development of green coating of basic carbonate $[CuCO_3 + Cu(OH)_2]$ on copper and bronze etc.

Corrosion occurs at the surface of metal and is a slow process, but its ill effects are numerous. It destroys metallic equipments, machines and instruments and reduces their efficiency. It is, therefore, important to study its mechanism and measures to control it to minimize the loss due to corrosion.

In this chapter we would learn types of corrosion, their mechanism of occurrence, factors effecting corrosion and measures which are usually adopted for their prevention.

4.2 TYPES OF CORROSION

Various types of corrosion processes are given below in the form of a table. The shaded areas represent the main types, which have been sub-divides into the types related to each main type.

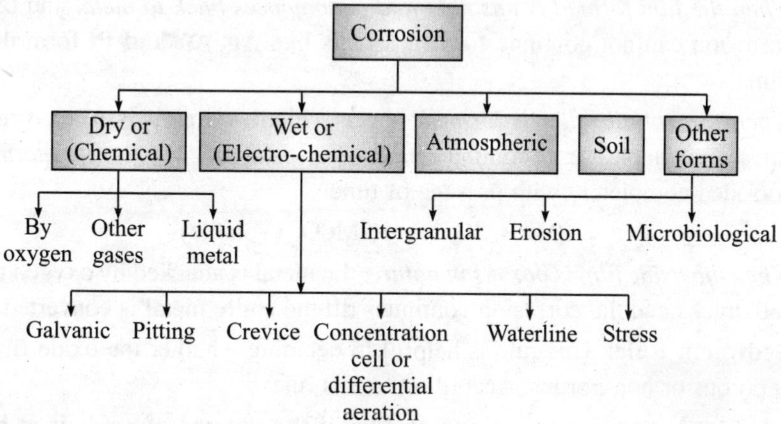

The details of only those types which are in the syllabus of H.P. Tech. university discussed here.

4.3 DRY OR CHEMICAL CORROSION

The *direct chemical action of atmospheric gases such* as oxygen, *halogens, sulphur dioxide* and *hydrogen sulphide with metals* resulting into the formation of compounds such as oxides, *halides, sulphates* and sulphides is referred to as chemical corrosion.

It is mainly of the following three types.

(*a*) Corrosion by oxygen (oxidation corrosion): Atmospheric oxygen reacts with metals in the absence of moisture forming their respective oxides.

Alkali metals (Li, Na, K, etc.) and alkaline earth metals (Be, Ca, Sr, etc.) are oxidized at low temperatures, whereas all other metals (except, Ag, Au and Pt) are oxidized at high temperatures.

Mechanism: The process of oxidation starts at the exposed surface of metal. The metal loses electrons and changes to metal ions, whereas oxygen accepts the electrons and changes to oxide ion.

$$2M \longrightarrow 2M^{n+} \text{ (Metals ions)} + 2ne^- \text{(loss of electrons, oxidation)}$$

$$\frac{n}{2} O_2 + 2ne^- \longrightarrow nO^{2-} \text{ (Oxide ions) (Gain of electrons, reduction)}$$

overall reaction

$$2M + \frac{n}{2} O_2 \longrightarrow \underset{\text{Metal ion}}{M^{n+}} + \underset{\text{Oxide ion}}{2nO^{2-}} \longrightarrow \underset{\text{(Metal oxide)}}{M_2O_n}$$

This oxide layer formed acts as a barrier towards further oxidation. The continuation of the oxidation process depends upon two factors (*i*) the nature of the oxide film formed (*ii*) the rate of diffusion of the metal ion and the oxide ion through the layer formed.

(i) Nature of oxide film formed on the surface

(a) *When the oxide film is stable,* tightly adhering and impervious in nature, it acts as a protective coating and further action of atmospheric oxygen on the metal is prevented. For example, oxide films formed on **Al, Cu, Sn and Pb** are of this type.

(b) *When the film formed is unstable and decomposes back to metal* and oxygen, oxidation corrosion can not continue further. Metals like **Ag, Au and Pt** form this type of oxide films.

(c) *When the volatile film is formed,* it gets volatilized as it is formed making the metal surface available for continuous attack. For example, *Mo forms volatile oxide* and it is corroded completely with passage of time.

$$2Mo + 3O_2 \longrightarrow 2MoO_3 \text{ (Volatile)}$$

(d) *When the oxide film is porous in nature,* the metal is attacked by oxygen through the pores and cracks and the corrosion continues till the entire metal is converted into its oxide.

Piling-Bedworth Rule: This rule is helpful in deciding whether the oxide film formed on the metal surface is porous or non-porous. According to this rule:

"An oxide film is non-porous *i.e.,* protective, if the volume of oxide is at least equal to or greater than the volume of metal from which it is formed. However, if the volume of the oxide is less than the volume of the metal, the oxide film is porous *i.e.,* non-protective."

Metals like Al form oxide having greater volume as compared to the metal, so they form non-porous oxide layer and further oxidation of the metal is prevented. On the other hand, alkali and alkaline earth metals form oxides having lesser volume as compared to the metal, they form porous oxide films which allow further attack to atmospheric oxygen through the pores and the metals are completely corroded.

(ii) Rate of diffusion of metal ions: The metal oxide film formed at the surface acts as an interface. For the oxidation to continue further, the metal ions (M^{n+}) must diffuse outwards and the oxide ion (O^{2-}) inward through the interface for further action between them. This happens through the pores of the porous film only and not through the non-porous film. The metal ions being smaller in size as compared to the oxide ions diffuse outward at a faster rate than the oxide ions diffusing inward.

(b) Corrosion by other gases: Besides oxygen, gases like Cl_2, SO_2, H_2S, CO_2 and F_2 also react with metals. The corrosive effect of these gases depend upon their chemical affinity for the metal and also upon the nature of the compound formed by them *i.e.,* whether the film formed is protective or non-protective. For example, Cl_2 reacts with silver forming a protective (non-porous) film of AgCl which prevents further attack of Cl_2 on Ag. But Cl_2 forms a volatile chloride of tin. Similarly, H_2S gas at high temperature reacts with iron forming a porous FeS film and corrodes iron.

(c) Liquid metal corrosion: When liquid metal flows over solid metal at high temperature, corrosion takes place which is called *liquid corrosion.* It weakens the solid metal due to (*i*) penetration of the liquid metal, (*ii*) dissolution of the solid metal in liquid metal.

A common example of this type of corrosion is the corrosion of cadmium rods taking place due to sodium metal acting as coolant in nuclear reactors.

4.4 WET OR ELECTROCHEMICAL CORROSION (THEORY)

Wet corrosion is more common than dry corrosion. It occurs mostly under wet or moist conditions through the formation of electrochemical cells and is therefore, referred to as electrochemical corrosion. The **mechanism/theory** of electrochemical corrosion involves:

(i) The existence of separate anodic and cathodic areas between which current flows through the conducting solution.

(ii) Oxidation (loss of electrons) take place at the anodic area and the metal is destroyed by either dissolution or combination with oxygen. Hence corrosion always takes place at the anode.

$$M \longrightarrow M^{n+} + ne^- \text{ (oxidation)}$$

M^{n+} forms compounds such as oxides or dissolves in solution

(iii) Reduction (gain of electrons) takes place at the cathode. The electrons from the anode are accepted by the dissolved oxygen forming ion such as OH^- or O^{2-} ions.

$$\frac{1}{2}O_2 + H_2O + 2e^- \longrightarrow 2OH^- \text{ (Reduction)}$$

or

$$O_2 + 2e^- \longrightarrow 2O^{2-} \text{ (Reduction)}$$

(i) The metallic ions formed at the anode and non-metallic ions formed at the cathode diffuse through the conducting medium and combine to form the corrosion product some where between anodic and cathodic area.

Wet corrosion involves flow of electron current, between the anodic and cathodic areas. The anodic reaction involves dissolution of metal as corresponding metallic ions with the liberation of free electrons. Whereas, the cathodic reactions accept electrons by two ways: (*i*) *evolution of hydrogen* or (*ii*) *by absorption of oxygen.* The evolution of hydrogen and absorption of oxygen in cathodic reaction depends on the nature of corrosive environment.

(*i*) **Evolution of Hydrogen:** The process of corrosion in which H_2 is liberated is called *evolution of hydrogen type corrosion.* This type of corrosion usually takes place in acidic environment. Thus, acidic medium (absence of oxygen) hydrogen ions acquire electrons with the liberation of H_2 gas cathodic reaction, whereas in the anodic reaction metal atoms lose their electrons to the environment and pass into solution in the form of +ve ions.

$$2H_3O^+ + 2e^- \longrightarrow 2H_2O + H_2\uparrow$$

or simply $2H^+ + 2e^- \longrightarrow H_2\uparrow$ (Cathodic reaction)

In neutral and alkaline medium (in the absence of oxygen) the reaction taking place will be

$$2H_2O + 2e^- \longrightarrow 2OH^- + H_2\uparrow \text{ (cathodic reaction)}$$

whereas $Fe \longrightarrow Fe^{2+} + 2e^-$ (anodic reaction)

e.g., if iron metal is used, the dissolution of iron as Fe^{2+} ions with the evolution of electrons in anodic reaction.

$$Fe \longrightarrow Fe^{2+} + 2e^- \text{ (oxidation)}$$

These electrons flow through the metal, from anode to cathode, where H^+ ions of acidic solution are eliminated as H_2.

$$2H^+ + 2e^- \longrightarrow H_2\uparrow \text{ (Reduction)}$$

Overall reaction: $Fe + 2H^+ \longrightarrow Fe^{2+} + H_2$

Therefore, this type of corrosion cause displacement of hydrogen ion from the acidic solution by metal ions.

The above statements can be easily understood from Fig. 4.1.

Thus, from the diagrams, it is clear that anodic areas are very large in comparison to cathodic areas, All the metals, above hydrogen in the electrochemical series have a tendency to get dissolved in acidic solution with simultaneous liberation of hydrogen.

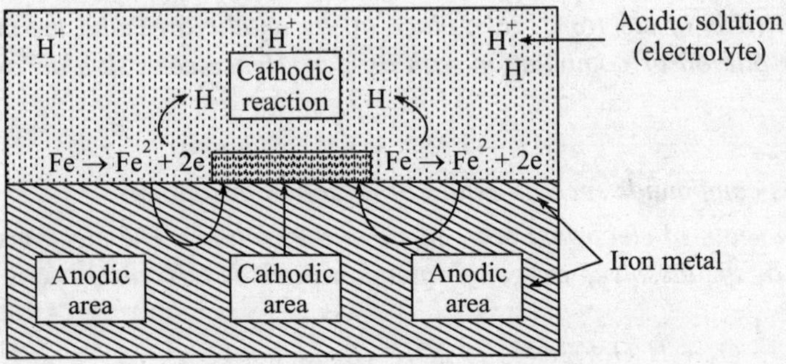

Fig. 4.1 Mechanism of wet corrosion by hydrogen evolution

***(ii)* Absorption of Oxygen:** This type of corrosion occurs in neutral environment (such as NaCl solution used as electrolyte). The popular examples of absorption of oxygen type corrosion is rusting of iron in neutral aqueous solution of NaCl in the presence of atmospheric oxygen.

The surface of iron is usually covered with a film of iron oxide. If there are some cracks in iron oxide film, an anodic area is developed on the surface, whereas the wet metal part act as cathodic area. In this corrosion reaction cathodic area is much more large in comparison to anodic area.

The above overall reaction mechanism is represented diagrammatically in Fig. 4.2.

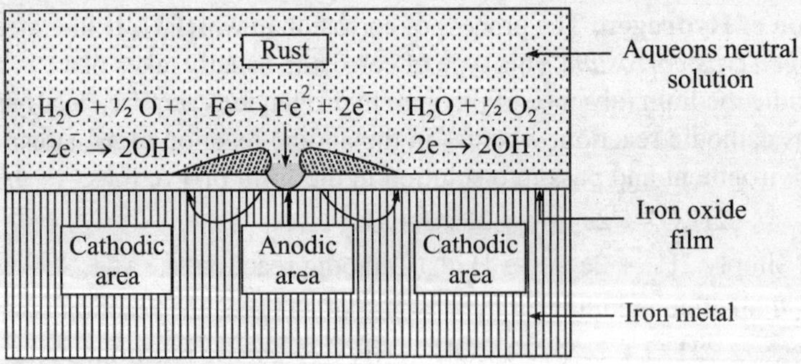

Fig. 4.2 Mechanism of wet corrosion by absorption of oxygen type corrosion

Thus, from the above diagram it is clear that at the cathodic area, iron metal dissolves with the liberation of electrons as

$$Fe \xrightarrow{\text{Oxidation}} Fe^{2+} + 2e^-$$

The above evolved electrons flow from anodic areas through iron metal where electrons are taken up by the dissolved oxygen as

$$\frac{1}{2}O_2 + H_2O + 2e^- \xrightarrow{\text{Reduction}} 2OH^-$$

The OH^- ions formed at cathode as well as Fe^{2+} ions formed at anode combine and from ferrous hydroxide $Fe(OH)_2$ as precipitate

$$Fe^{2+} + 2OH^- \longrightarrow Fe(OH)_2 \uparrow$$

In the presence of sufficient oxygen, $Fe(OH)_2$ can be easily oxidized into ferric hydroxide $Fe(OH)_3$

$$4Fe(OH)_2 + 2H_2O + O_2 \longrightarrow \underset{\text{Yellow rust}}{4Fe(OH)_3} \downarrow$$

actually $2Fe(OH)_3$ corresponds to $Fe_2O_3.\ H_2O$.

If the supply of oxygen is limited then black anhydrous magnetite *i.e.*, ferroferri oxide is formed as

$$3Fe(OH)_3 \longrightarrow Fe_3O_4.6H_2O \text{ or } Fe_2O_3.FeO.6H_2O$$

As we know that Fe^{2+} ion is smaller in size than OH^- ion, the Fe^{2+} ion will diffuse more rapidly, so corrosion occurs at the anode, but rust is deposited at the cathode or near the cathode.

Thus, electrochemical Corrosion is due to the flow of electrons from anodic area towards cathodic area through a conducting solution. It is very common type of corrosion. Such type of corrosion take place in following conditions.

(*i*) When two dissimilar metals or alloys are in contact with each other in the presence of a conducting medium (aqueous solution, moisture, etc.).

(*ii*) Separate anodic and cathodic areas between which the current flows through the conducting medium.

(*iii*) Oxidation takes place at anode and reduction takes place at cathode.

A common example of electrochemical is rusting of iron, which is discussed below:

Rusting of Iron

When an iron piece is exposed to moist air or moisture, a reddish brown layer is formed on the iron surface. This reddish brown layer is called rust and the phenomenon is known as rusting. Actually rust is mainly composed of hydrates ferric oxide ($Fe_2O_3.xH_2O$) or iron (II, III) oxide ($Fe_3O)_4$. The exact composition of rust depends upon the atmospheric conditions which are responsible for rusting. The layer of rust is porous so the inside layer of iron metal further undergoes corrosion. In this way the process of corrosion or iron metal continues till it is completely oxidized.

Factors Involving in Rusting

Following are the factors for favouring the rusting the iron:

(*i*) Presence of air,

(*ii*) Presence of moisture,

(*iii*) Presence of electrolytic impurities in water,

(*iv*) Presence of impurities in metals,

(*v*) Nature of metal.

4.4.1 Comparison of Chemical with Electrochemical Corrosion

The comparison of these two main types of corrosion of metals has been summed up in the following table 4.1

Table 4.1 Comparison of Chemical With Electrochemical Corrosion

Chemical corrosion	*Electrochemical corrosion*
1. *It occurs in dry condition.*	1. *It occurs in wet condition* in the presence of moisture and electrolyte.
2. It involves the chemical attack of oxygen or other gases on the surface of metal.	2. It involves electrochemical attack of corrosive environment on the surface of the metal.
3. It can occur at homogeneous as *well as* heterogeneous metal surfaces.	3. It occurs only on heterogeneous metal surface.
4. It is uniform in nature throughout the surface of the metal.	4. It is not uniform. If the area of anode is small, pitting corrosion takes place.
5. It is a slow process taking place by chemical reaction of atmospheric gases.	5. It is a fast electrochemical process. It proceed through the formation of electrochemical cells.
6. Corrosion products accumulate at the place where corrosion occurs *i.e.,* at the anode.	6. Corrosion products accumulate some where between the area of anode and cathode.
7. Its mechanism is explained on the basis of absorption.	7. Its mechanism is explained on the basis of electrochemical reactions.

4.5 TYPES OF ELECTROCHEMICAL CORROSION

Electrochemical corrosion is further Classified as under:

1. Concentration cell corrosion or Different aeration corrosion: This type of corrosion occurs when the *metal surface is exposed* to an electrolyte of varying concentration and varying aeration. It also occurs when one part of the metal is exposed to a different air concentration from the other part. It has been observed that poorly oxygenated parts become anodic and get corroded, whereas more oxygenated part acts as cathode.

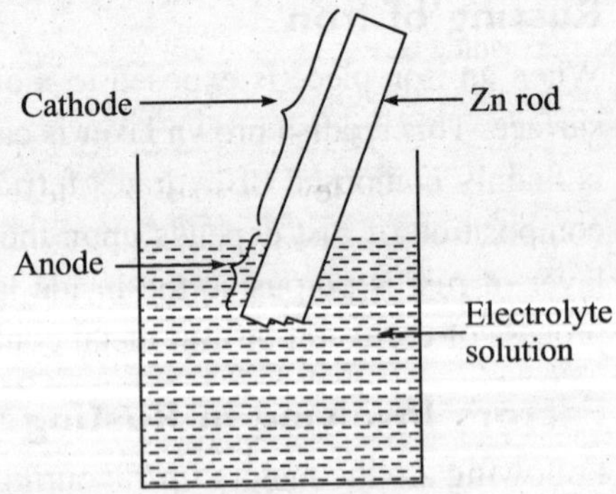

Fig. 4.3 Differential aeration corrosion

It occurs when a metallic surface is partially immersed in a solution and partially exposed to the atmosphere. Consider a metal (say Zn) is partially immersed in a dilute solution of NaCl and solution is not agitated for long. The part of the metal immersed in solution has less access to oxygen and act as anode, whereas the part above and closely adjacent to the waterline is more strongly aerated and act as cathode (Fig. 4.3). So, a difference of potential is created which causes flow of electrons between two differentially aerated areas of the same metal. Zinc change to Zn^{2+} ions at the anodic area and oxygen takes up electrons at the cathodic area of the metal to form hydroxial ions. The Zn^{2+} and OH^- ions interact to form $Zn(OH)_2$.

At anode Zn $\longrightarrow$ $Zn^{2+} + 2e^-$ (oxidation)

At anode $\frac{1}{2}O_2 + H_2O + 2e^-$ $\longrightarrow$ $2OH^-$ (reduction)

$Zn^{2+} + 2OH^-$ $\longrightarrow$ $Zn(OH)_2$

2. Pitting corrosion: A small part of the metal surface covered by an impurity or water has a low oxygen concentration and it acts as an anode, whereas the uncovered metal surface acts as a cathode due to high oxygen concentration. The small covered area acting as anode gets corroded giving rise to localized spots (pits), pinholes and cavities in the metal. Once a small pit is formed, the rate of corrosion increases. The electrochemical reaction may be represented as:

At anode: Fe $\longrightarrow$ $Fe^{2+} + 2e^-$

At cathode: $\frac{1}{2}O_2 + H_2O + 2e^-$ $\longrightarrow$ $2OH^-$

$Fe^{2+} + 2OH^-$ $\longrightarrow$ $Fe(OH)_2 \xrightarrow{\text{Oxidation}} Fe(OH)_3$

Pitting corrosion is usually caused due to *cracking of the protective film on a metal at specific points.* The cracks serve as initiators of corrosion by creating two areas, one fairly accessible to oxygen and the remaining poorly accessible. The presence of impurities such as dust, sand, scale etc. also cause pitting corrosion.

Pitting corrosion generally occurs due to the following reasons:

1. Local strains on metal.
2. Non-uniformly polishing of metals etc.
3. Rough surface of metal.
4. Scratching of metals.
5. Designing of metal (mainly sharp corners or bent of metals).

It may be prevented by:

1. Proper polishing of metal.
2. Proper designing of metal.
3. Purifying the metal.

3. Water line corrosion: When water is stored in a metallic tank, it is observed that corrosion takes place along a line just below the level of water in the container. The area below the level of water is poorly oxygenated and acts as anode, whereas area above the water level is highly oxygenated and acts as a cathode. The part of the steel tank just below the water level is corroded, whereas the area above the water level remains protected from corrosion (Fig. 4.4).

At anode: Fe $\longrightarrow$ $Fe^{2+} + 2e$ (oxidation)

At Cathode: $2H^+ + \frac{1}{2}O_2 + 2e^-$ $\longrightarrow$ $2OH^-$ (reduction)

$Fe^{2+} + 2OH^-$ $\longrightarrow$ $Fe(OH)_2 \xrightarrow{\text{Oxidation}} Fe(OH)_3$

Water line corrosion is a problem for the marine engineers when marine plants are attached to the ships in ocean, waterline corrosion comes into action and it can be checked by anti fouling paints to some extent.

Important Features:

(*i*) It is increased by the presence of salt like chlorides, bromides etc. present in water.

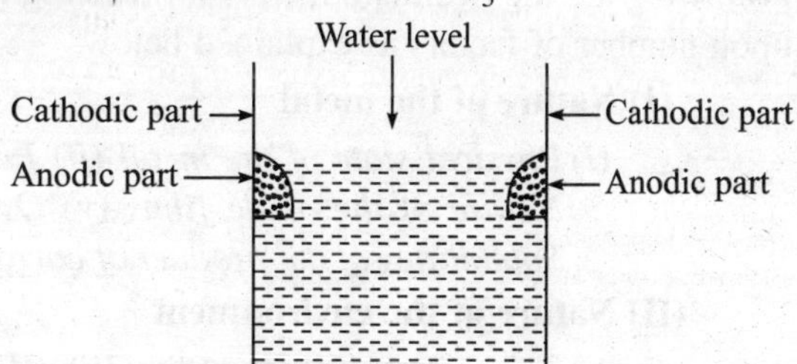

Fig. 4.4 Water line corrosion

(*ii*) It gets retarded by the presence of anodic inhibitors like carbonates, phosphate and silicate, hydroxide etc.

(*iii*) It is reduced if the water is relatively free from acidic impurities.

(*iv*) It is generally accelerated due to attachment of marine plants on the sides of ships. Antifouling paints are used to minimize this type of corrosion.

4. Stress Corrosion: During the manufacture of metals and different metallic equipment, metals undergo certain processes such as bending, pressing, rolling, quenching and annealing etc. which cause stress on certain areas of the metal. Such areas under stress act as anode, whereas other areas act as cathode. The potential difference is thus developed between the strained (under stress) and unstrained part of the metal. This accelerates corrosion of the strained portion of the metal in corrosive environment.

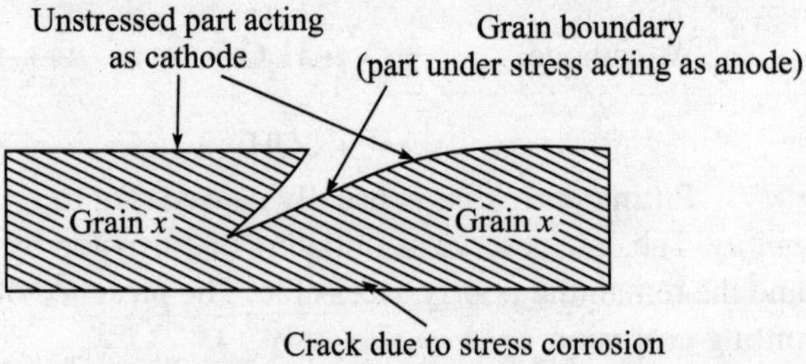

Fig. 4.5 Stress Corrosion

Favourable Conditions: It is represented in Fig. 4.5. For stress corrosion to take place, the basic requirements are (*i*) stressed metallic areas which is developed due to processes such as quenching, annealing, bending etc. and is referred to as tensile stress, (*ii*) corrosive environment, which is specific for particular metal alloys. For instance, caustic alkali is specific corrosive environment for mild steel, acid chloride solution for stainless steel and traces of ammonia for brass. The stressed part of the metal acquires a higher electric potential and becomes chemically active towards the corrosive environment and gets corroded. The following are some examples of stress corrosion.

(*i*) Stress corrosion of brass: The main components of brass are zinc and copper. Both the components are prone to stress corrosion in ammonical environment. The metal (Zn as well as Cu) lose electron change to their respective ions at the stressed portion of the equipment. The ions so formed interact NH_3 or amines to form their respective complexes as shown below:

$$Zn \longrightarrow Zn^{2+} + 2e^- \text{ and } Cu \longrightarrow Cu^{2+} + 2e^-$$

$$Zn^{2+} + 4NH_3 \longrightarrow [Zn(NH_3)_4]^{2+} \text{ and } Cu^{2+} + 4NH_3 \longrightarrow [Cu(NH_3)_4]^{2+}$$

This corrosion in generally referred to a **season cracking**.

(*ii*) Stress corrosion of mild steel: Mild steel undergoes stress corrosion in caustic alkalies at high temperature and pressure. This process of corrosion is referred to as ***caustic embrittlement*** of the steel boilers, which has been discussed in detail in the chapter on water.

4.6 FACTORS INFLUENCING CORROSION

Following are the two main factors influencing the rate and extent of corrosion; which in turn depends upon number of factors as explained below.

(I) Nature of the metal

(*i*) *Physical state of the metal* (*ii*) *Purity of metal* (*iii*) *Position in the galvanic series* (*iv*) *Nature of the oxide films* (*v*) *Over voltage* (*vi*) *Volatility of corrosion product* (*vii*) *Solubilities of the product of corrosion* (*viii*) *Relative areas of the anode and cathode*

(II) Nature of the environment

(*i*) *Temperature* (*ii*) *Humidity* (*iii*) *pH* (*iv*) *Nature of electrolyte* (*v*) *Conductance of corroding medium* (*vi*) *Presence of impurities in atmosphere* (*vii*) *Formation of oxygen concentration cell.*

(a) Nature of the Metal

(*i*) **Physical State of the Metal:** The rate of corrosion is influenced by the physical state of the metal such as grain size, orientation of crystal, stress, etc. The smaller the grain size of the metal or alloy, the greater will be its solubility in corroding medium and hence greater will be its corrosion. Moreover, areas under stress, even in a pure metal, tend to be anodic and corrosion takes place at these areas.

(*ii*) **Purity of Metal:** Pure metals are generally corrosion resistant. The rate and extent of corrosion increase with the increase of the impurities, because the impurities present in the metal form minute galvanic cells and anodic part gets corroded, while in an alloy, the solid solutions are homogeneous and do not form galvanic cells.

(*iii*) **Position in the Electrochemical/Galvanic Series:** The extent of corrosion depends upon the position of the metal in the electrochemical series and galvanic series. When two metals are in electrical contact in the presence of an electrolyte, the more active metal (or higher up in the series) suffers corrosion. The rate and severity of corrosion depends upon the difference in their positions. Greater is the difference, the faster is the corrosion of the anodic metal or alloy. A metal high in the electrochemical series is more anodic and undergoes corrosion faster than the metal below it. For example, Li corrodes faster than Mg, Zn corrodes faster than Fe, Cu corrodes faster than Ag and so on.

Galvanic Series

However, some exceptions to this generalisation have been noticed. For instance, Zn which is below Ag in the electrochemical series is corroded faster than Ag. Similarly Ag which is below Ti in the series is corroded faster than Ti. These observations contrary to that predicted by electrochemical series have been explained on the basis that metals like Al and Ti form strongly adhering oxide layers on their surface, which resist corrosion (Passivity). Due to formation of oxide layer, their effective electrode potentials become positive or less negative. As a result, the electrochemical series do not account for the corrosion behaviour of all metals. Hence, another series called galvanic series (Table 4.2).

Which provides a more reliable information about the corrosion of not only metals but alloys also, has been framed. These series have been prepared by studying the corrosion behaviour of metals and alloys in a given environment like sea water. The difference between these two series has been summarized below:

Table 4.2 Galvanic series (on the basis of relative oxidation potential in sea water)

Metal / Alloy
Magnesium
Magnesium alloys
Zinc
Aluminium
Aluminium alloys
Low carbon steel
Cast Iron
Stainless steel (active)
Lead-tin alloys
Lead
Tin
Brass
Copper
Bronze
Copper-nickel alloys
Inconel
Silver
Stainless steel (passive)
Monel
Graphite
Titanium
Gold
Platinum

↑ Active (Anodic)　↓ Noble (Cathodeic)

Electrochemical Series	*Galvanic Series*
1. This series predicts the relative displacement tendencies of metals and non-metals in electrolytes containing particular concentration of salts of the same metal.	1. This series predict the corrosion tendencies of metals and non-metals in actual environment.
2. It comprises metals and non-metals.	2. It comprises metals and alloys.
3. The position of a metal in this series is permanently fixed the form of alloy.	3. The position of a given metal may shift in this series. The position of a metal, when present in is different from that of pure metal.
4. The electrode potentials are measured by using standard hydrogen electrode (SHE) as a reference electrode.	4. The electrode potentials are measured by using calomel electrode as a reference electrode.
5. It gives no information regarding position of alloys.	5. The corrosion behaviour of alloys can also be studied from their position in this series.

***(iv)* Nature of the Oxide Films:** In aerated atmosphere practically all metals get covered with a thin surface film of metal oxide having a thickness of a few angstroms (1 Å = 10^{-10} m). The ratio of the volume of the metal oxide to the metal is known as specific volume ratio. *Greater the specific volume ratio, lesser is the oxidation corrosion rate e.g.,* the specific volume ratio of Ni, Cr and W are 1.6, 2.0 and 3.6 respectively, which indicates that the rate of oxidation at elevated temperature is least for tungsten (W).

***(v)* Over voltage:** In electrolysis, the potential at which a metal ion starts depositing on the cathode (deposition potential) is very close to the reversible electrode potential of that metal.

For instance, the reversible electrode potential of Zn is – 0.76 V. It has been observed that potential at which Zn^{2+} ion deposit on Zn rod used as cathode is also close to or only slightly higher *i.e.* more negative than – 0.76 V. However, in case of evolution of gases, say hydrogen at the cathode or oxygen at the anode, the potential required is much higher that of the reversible potential of hydrogen or oxygen electrode in the same solution.

The difference between the potential of the electrode when gas evolution is actually observed and theoretical value for the same evolution is called the overvoltage.

Now, let us apply this concept to the corrosion of metals, consider the dissolution of Zn in hydrochloric acid. The standard electrode potential of Zn(E°) is – 0.76 V. It should, therefore, dissolve, readily in dilute HCl evolving hydrogen gas. However, in actual practice, pure Zinc does not dissolve is dil HCI and no hydrogen gas is evolved. This is because of hydrogen overvoltage, the overvoltage of Zn being 0.070 V. But when pure Zn is connected to Cu and both are dipped in dil HCI or H_2SO_4, the dissolution of Zn takes place and hydrogen is evolved at the copper electrode readily. The evolution of H_2 gas takes place at Cu because its hydrogen overvoltage is only 0.20 V. Similar effect is observed if Zinc contains copper or any other metal having low value of hydrogen overvoltage as an impurity. In such a case zinc dissolves producing Zn^{2+} ions at the anodic area and hydrogen is liberated at the copper part *i.e.,* cathodic area. This results in corrosion or dissolution of Zn. The same situation can arise with any other base metal (iron, lead etc.), if it contains a metal with low hydrogen voltage as an impurity.

However, in the presence of oxygen in air or an oxiding agent which acts as depolarizer, there may be no evolution of hydrogen gas as it gets oxidized as soon as it is formed. Therefore, in the presence of air or oxygen, base metals such as zinc, lead, iron etc. containing even a high overvoltage metal as impurity are corroded in acid solution. For example, iron present in tin does not corrode at all in the absence of oxygen or air due to high overvoltage of iron over tin, but in the presence of oxygen of air, it dissolves in acid medium. It is also well known that pure zinc and pure lead does not dissolve in acid solution ordinarily, but they dissolve when air is bubbled through the solution.

***(vi)* Volatility of Corrosion Product:** If the corrosion product is more volatile, then metal surface is exposed for further attack *i.e.,* corrosion rate will be more *e.g.,* Molybdenum oxide (MoO_3), the oxidation corrosion product of molybdenum is volatile.

***(vii)* Solubilities of the Product of Corrosion:** If the corrosion product is soluble, corrosion of the metal will proceed faster. On the other hand, if corrosion product formed is insoluble, then the protective film formed tends to suppress corrosion.

***(viii)* Relative Areas of the Anode and Cathode:** When two steel plates of the same area are separately connected one to a copper plate (cathode) having the same area and the other to a copper plate having a much larger areas, it can be observed that the latter couple produces greater amount of current and the corrosion of the steel anode in this case will be much greater than the first one. It may be broadly concluded that when anodic polarization is negligible and conductance remaining more or less same, *corrosion of the anode is directly proportional to the area of the cathode.*

***(b)* Nature of the Environment:**

***(i)* Temperature:** With increase of temperature of environment, the reaction as well as diffusion rate increase, thereby corrosion rate is generally increased.

***(ii)* Humidity:** The higher the moisture content in the atmosphere, the greater is the rate and extent of the corrosion because moisture acts as medium for oxygen in air and behaves as an electrolyte in air *e.g.,* atmospheric corrosion of iron is slow in dry air compared to moist air.

***(iii)* pH:** Acidic media are generally more corrosive than alkaline and neutral media. The amphoteric metals such as Al, Zn, Pb, etc. dissolve, corroded in alkaline solution. The corrosion rate of iron in oxygen free water is slow, until pH is below 5. The corresponding corrosion rate in presence of oxygen is much higher. Zinc is rapidly corroded, even in weakly acidic solutions such as carbonic acid suffers minimum corrosion at pH = 11.

***(iv)* Nature of Electrolyte:** Chloride ions present in the medium destroy the passive film and corrode many metals and alloys. On the contrary, some anions like silicate may from an insoluble product (silica gel) which inhibit corrosion.

***(v)* Conductance of Corrosion Medium:** The corrosion current depends upon the internal resistance of the short circuited cells which in turn depends on the conductance of the electrolytic medium. If the conductance of the soil is more, the corrosion of underground structures will also be more, consequently there will be more severe damage to the metallic structures in the presence of stray currents from power leakage. Conductance of clay and mineral soil is more than that of dry sandy soil, hence corrosion of underground metallic structures is more in clay and mineralized soil.

***(vi)* Presence of Impurities in Atmosphere:** Atmosphere, in the vicinity of industrial areas, contains corrosive gases like SO_2, H_2O and CO_2 and fumes of HCl and H_2SO_4 etc. In the presence of these gases, the acidity of the liquid, adjacent to the metal surfaces increases and its electrical conductivity also increases. This consequent results in an increase of corrosion on the exposed metal surfaces. Similarly, in marine atmosphere, presence of Na^+ and Cl^- leads to increase the conductivities of liquid layer in contact with the metal surfaces thereby increase the corrosion.

(*vii*) **Presence of Suspended Particles in Atmosphere:** Two types of suspended particles are present in atmosphere *viz.,* chemically active and chemically inactive. The chemically suspended particles like NaCl, $(NH_4)_2$ SO_4 absorb moisture and thus act as strong electrolytes, thereby enhance corrosion, whereas chemically inactive suspended particles like charcoal, absorb both moisture as well as sulphur gases and thus slowly enhances corrosion rate.

(*viii*) **Formation of Oxygen Concentration Cell:** The rate of corrosion increases with increasing supply of oxygen. Further differential aeration will also promote corrosion by the formation of concentration cells. The region where the oxygen concentration is less becomes anodic and suffers corrosion. The mechanism of corrosion by differential aeration has already been discussed under the electrochemical theory of corrosion.

4.7 PREVENTIVE MEASURES

Protection against corrosion means not allowing corrosion reactions to take place. Noble metals do not corrode but they cannot be used for common purposes, because of their high cost. We have to use other metals and alloys in the fabrication of many kinds of machinery and equipments and adopt measures to protect these from corrosion.

Pre-treatment of Metals: The metal should be cleaned or descaled before protective measures are adopted. This pretreatment usually consists of *degreasing and descaling.* Oily and greasy surface films are dissolved out by exposure to volatile organic solvents such as trichloroethylene ($CHCl = CCl_2$), tetrachloroethylene ($Cl_2C = CCl_2$) etc. Descaling can be affected either mechanically or chemically. *Acid pickling* is more convenient method of removing scale than mechanical cleaning. Metals like iron and steel are cleaned by immersing in warm dil H_2SO_4 or in cold HCl solution, whereas copper, brass or nickel are dipped in dil HNO_3 or a mixture of dil HNO_3 and dil H_2SO_4. However, Al is not pickled in acid solution, but in alkaline solution.

Before any coating can be applied it is necessary to properly prepare the metal surface. If the metal surface has not been properly cleaned, the protective coating will not stick to the surface and more severe corrosion may result at points exposed to the corrosive environment.

Oil and grease are removed from the surface by washing with soap or alkali solution in water. Several patented compositions containing alkalies, phosphates and silicates are marked for removing oil also. Rust or scale is mostly removed by dipping the material in dil. (5 to 10%) H_2SO_4 and by mechanical rubbing with a wire brush. Large equipments or bridges are cleaned by the use of wire brush or by sand blasting.

The process of *sand blasting* consists in throwing a blast of sand mixed in air under a pressure of 25 to 100 atmospheres on the metal surface to be cleaned. The sand blast removes any scale (especially oxide scales) present and also hardens the metal surface. The process is especially suitable for large steel surfaces. Paint coatings on sand blasted surfaces are more durable than those on surfaces cleaned by other methods, because beside cleaning it make the surface rough. However, the process requires expensive equipment and is injurious to the health of the workers.

Another method of cleaning is by applying flame from a kerosene burner. The old paint, oil or grease burns and the scale is removed because of sudden expansion of the metal.

Corrosion can be controlled by a number of ways. Some of these protective measures are explained below:

4.7.1 Proper Design

The design of the material should be such that corrosion should be minimum. Even if it occurs, it should not result in intense and localized corrosion. Important design principles are:

(*i*) Avoid the contact of dissimilar metals in the presence of a corroding environment. If this principle is not followed, then corrosion is localized on the more active metal; while the less active metal remains protected.

(*ii*) When two dissimilar metals are to be in contact, the anodic material should have as large area as possible; whereas the cathodic metal should have as much small area as possible.

(*iii*) If two dissimilar metals in contact have to be used, they should be as close to each other as possible in the electrochemical series.

(*iv*) Whenever the direct joining of dissimilar metals, is unavoidable, as insulating fitting (Nylon washer etc.) may be applied in-between them to avoid direct metal-metal electrical contact.

(*v*) The anodic metal should not be painted or coated, when in contact with a dissimilar cathodic metal, because any break in coating would lead to rapid localized corrosion.

(*vi*) Angles, corners, edges etc. should be avoided in construction, because they increase the chances of corrosion. For this reason L, T and U shaped structures should be avoided as far as possible Fig. 4.6(*a*).

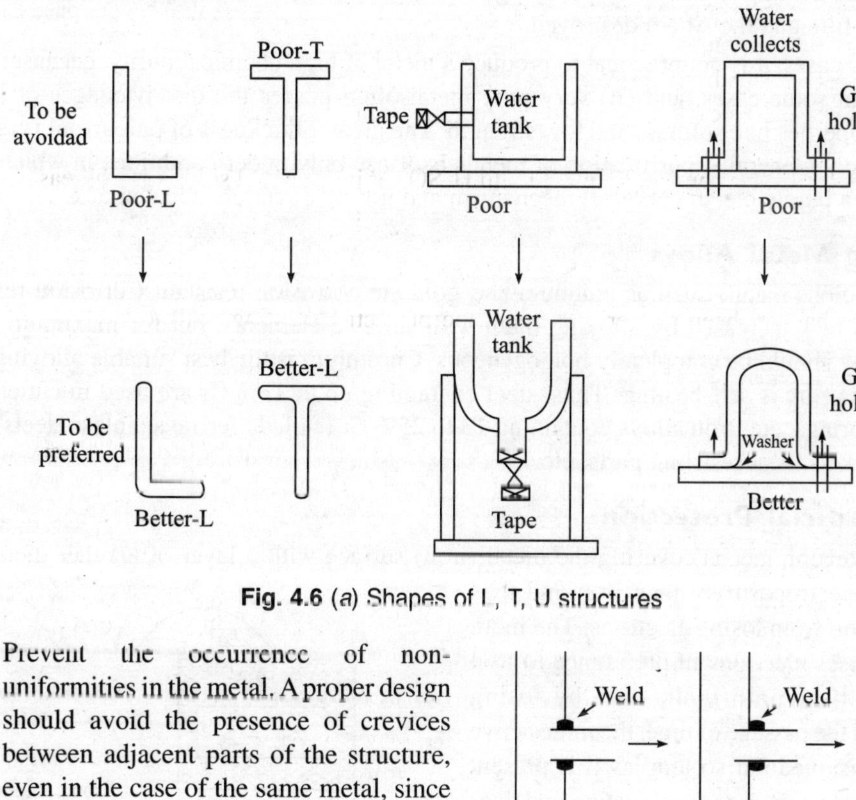

Fig. 4.6 (*a*) Shapes of L, T, U structures

(*vii*) Prevent the occurrence of non-uniformities in the metal. A proper design should avoid the presence of crevices between adjacent parts of the structure, even in the case of the same metal, since crevices permit concentration differences. Bolts and rivets are, therefore, undesirable and these should preferably be replaced by a weld Fig. 4.6(*b*).

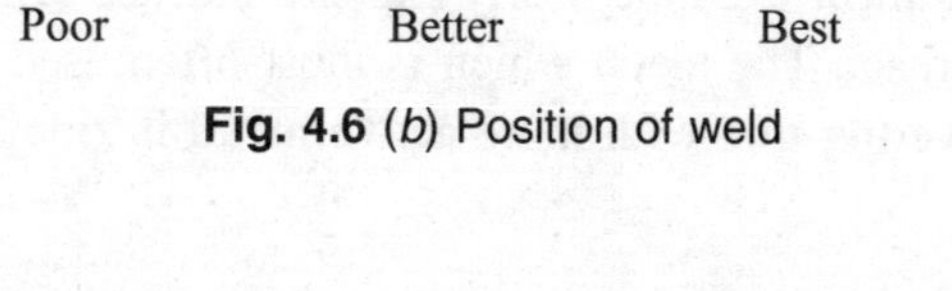

Fig. 4.6 (*b*) Position of weld

(*viii*) It is desirable that the design allows for adequate cleaning of the critical parts of the equipment so that deposition of dirt etc. is avoided. Sharp corners and recesses should be avoided, because they favour the formation of stagnant areas and accumulation of solids, etc.

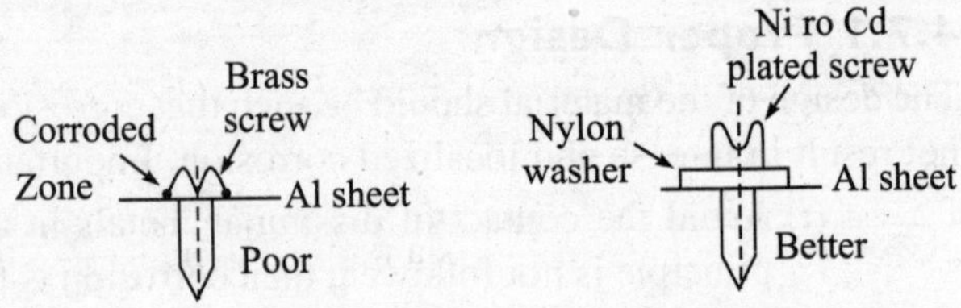

Fig. 4.6 (*c*) Use of washer to prevent corrosion

(*ix*) Whenever possible, the equipment should be supported on legs to allow free circulation of air and prevent the formation of stagnant pools or damp areas Fig. 4.6(*a, b, c*).

4.7.2 Using Pure Metal

Impurities in a metal cause heterogeneity which decreases corrosion-resistance of the metal. Thus, the corrosion-resistance of a given may be improved by increasing its purity. Purification of metals like Al, Mg, etc. provides a coherent and impervious protective oxide film on their surfaces, when exposed to environment. However, corrosion-resistance of a purified metal depends on the nature of corrosive environment. For example, the corrosion resistance of Al depends on its oxide film formation, which is highly protective only on the pure metal.

However, nothing is gained by purifying Al, if it is to be used in alkaline environment, in which both the oxide film and the Al are destroyed.

In many cases, it is not practical to produce a metal of high chemical purity, because: (*i*) of cost consideration in some cases, and (*ii*) very pure metal often posses the disadvantages of inadequate mechanical properties like softness and low strength. The greater thickness of pure metal is required for a proper strength. Generally, purification of metals is of use only under conditions in which corrosion takes place by a purely electrochemical mechanism and not by direct chemical attack.

4.7.3 Using Metal Alloys

Noble, but precious metals such as platinum and gold are corrosion-resistant Corrosion-resistance of most metals is best increased by alloying them with suitable elements, but for maximum corrosion-resistance, alloy should be completely homogeneous. Chromium is the best suitable alloying metal for iron or steel. Its film is self-healing. Thus, steel containing up to 13% Cr are used in cutlery, surgical instruments, springs, etc. Iron alloys containing 13 to 25% Cr (called "ferrite stainless steels") are used in turbine brackets, heat resisting parts, etc.

4.7.4 Sacrificial Protection

Sacrificial protection means covering the metal (iron) surface with a layer of another metal which is more active (electropositive) than iron and thus prevents the iron from losing electrons. The more active metal loses electrons in preference to iron and converts itself into ionic state by losing electrons. With the passage of time, the more active metal gets consumed but so long as it is present there, it will protect the iron from rusting and does not allow even the nearly exposed surface of iron to react. The metal which is most often used for covering iron with more active metal is zinc and

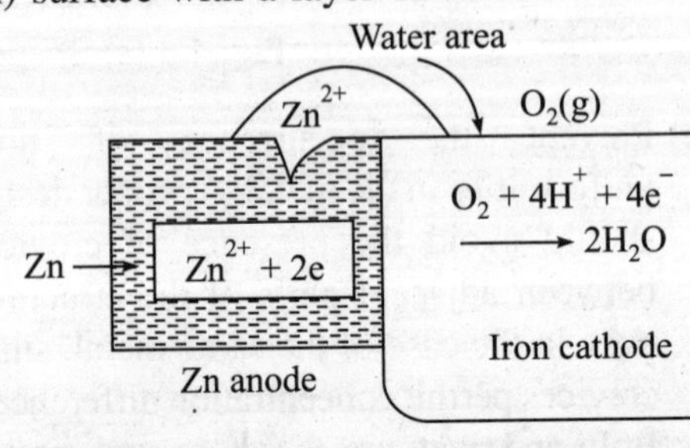

Fig. 4.7 Galvanised iron cathodic protection of iron in contact with zinc provided by galvanization

the process is called *galvanization* (Fig. 4.7). Zn being more electropositive than Fe acts as anode and it gets corroded, whereas iron remains protected till Zn is present.

The layer of zinc on the surface of iron, when comes in contact with moisture, oxygen and carbon dioxide in air it forms a protective invisible thin layer of basic zinc carbonate $ZnCO_3.Zn(OH)_2$. When such a layer is formed on the galvanised iron sheets, the sheets lose their luster, but such a layer tends to protect it from further corrosion.

Iron can be coated with copper by electro-deposition from a solution of copper sulphate or with tin by dipping into molten metal. Now if the coating is broken, iron is exposed and iron being more active than both copper and tin, is corroded. *Deposition of tin or copper on iron is not very useful because the corrosion of iron is increased in case of any scratch or crack on the coating.*

$$(E^\circ_{Zn^{2+},Zn} = -0.76 \text{ V and } E^\circ_{Fe^{2+},Fe} = -0.44 \text{ V})$$

$$(E^\circ_{Sn^{2+},Sn} = -0.14 \text{ V and } E^\circ_{Cu^{2+},Cu} = 0.34 \text{ V})$$

Here iron corrodes more rapidly than it does in the absence of tin. At times zinc, magnesium and aluminium powders mixed with paints provide decorative protective coating also.

4.7.5 Electrical Protection

(*i*) Cathodic Protection: The iron object to be protected from corrosion is connected to a more active metal either directly or through a wire. *The iron object acts as cathode and the protecting metal acts as anode.* The anode is gradually used up due to the oxidation of the metal to its ions due to loss of electrons. Hydrogen ions collect at the iron cathode and prevent the rust formation. The iron object gets protection from rusting as long as some of the active metal is present. *Metals widely used for protecting iron objects from rusting are magnesium, zinc and aluminium which are called sacrificial anodes.* Magnesium is oftenly employed in the cathodic protection of iron pipes buried in the moist soil, canals, storage tanks etc. Pieces of magnesium are buried along the pipeline and connected to it by the wire as shown in Fig. 4.8.

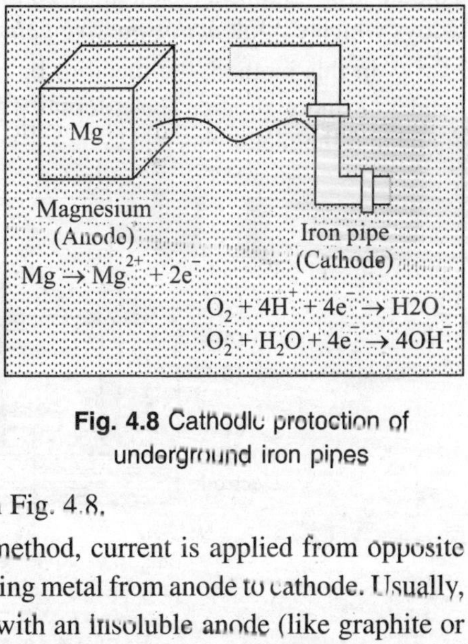

Fig. 4.8 Cathodic protection of underground iron pipes

(*ii*) Impressed current cathodic protection: In this method, current is applied from opposite direction to nullify the corrosion current, and convert the corroding metal from anode to cathode. Usually, the impressed current is derived from a direct current source with an insoluble anode (like graphite or platinum). Usually, a sufficient d.c. current is applied to an insoluble anode, buried in the soil, and connected to the metallic structure to be protected (Fig. 4.9). To increase the electrical contact between the anode and the soil, the anode is buried in a back fill such as gypsum under the earth. Thus electrons are *supplied from an external cell so that the object itself becomes cathodic and is not oxidized.* This type of cathodic protection has been applied to open water box coolers, water-tanks, ships, etc. This kind of protection technique is, particularly, useful for large structures for long term operations.

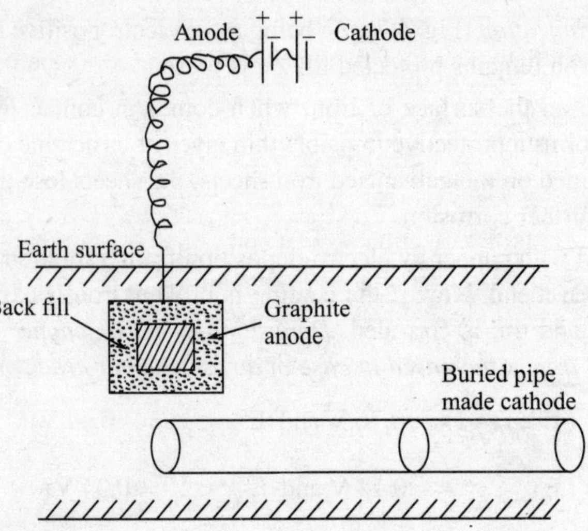

Fig. 4.9 Impressed current cathodic protection

Anodic Protection

Anodic protection is a technique used to prevent corrosion, but it is applicable to only those metals and alloys which exhibit ***active passive behaviour***. For example, Fe, Ti, Al and Cr and their alloys.

In this method, the metal to be protected is passivated by applying current in a direct ion that renders it more anodic. The externally applied current deposits a protective film on the surface of the metal and the metal gets passivated.

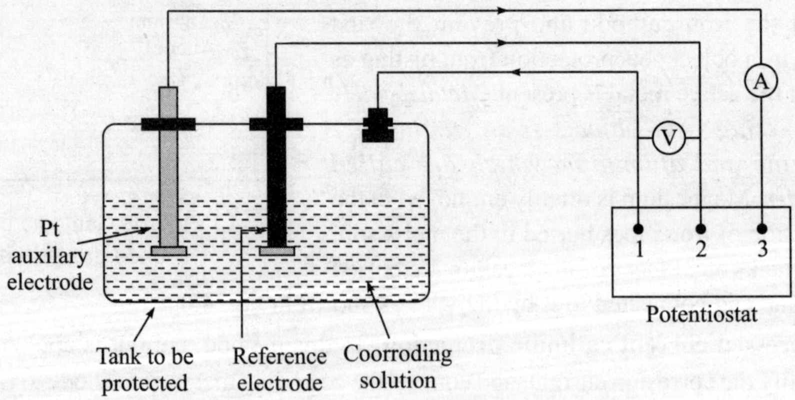

Fig. 4.10 Anodic protection system for a tank.

The device used in this method is called a ***potentiostat.*** It is an electronic device which maintains the metal at a constant potential with respect to a reference electrode. A potentiostat used for the protection of a tank is shown in Fig. 4.10. It consists of three terminals. The terminal 1 is connected to the tank to be protected. The terminal 2 is connected to a platinum auxilary electrode immersed in the tank, whereas terminal 3 is connected to a reference electrode (*e.g.*, calomel electrode) also immersed in the tank. The role of potentiostat is to maintain a constant potential between the tank and the reference electrode. The optimum potential needed for the protection of the tank is determined by electrochemical measurements.

Anodic protection is a versatile technique and can protect a metal against very aggressive corrodents. Although the installation cost of the technique is high, but the operating costs are much lower the only limitation is that the technique is applicable to only a few metals which are able to exhibit active-passive behaviour.

4.7.6 Protective Coating or Barrier Pretection

The metal surface may be coated for protection against corrosion. Most of these coatings act as a barrier between the metallic surface and the corrosive environment and therefore this method is, also called barrier protection.

Following are the main coatings, generally used for corrosion protection:

(A) Metallic coating (B) Inorganic non-metallic coating (C) Organic coating

(A) Metallic coating: The metals used for coating may be put under two categories, so metallic coating is of the following two types:

(i) Anodic Coating: Metals such as Zn, Mg, Cd, Al etc. which are anodic as compared to the metal to be protected, they stand above Fe in the galvanic series, the metals under this category being more anodic get themselves corroded, saving the iron which is to be protected. *Anodic coating protect the under lying base metal sacrificially.* In such cases even if the surface is scratched the metal to be protected does not get corroded. The functioning of anodic category has been represented by taking Zn as an example in Fig. 4.11. These metals may be applied on the surface by hot dipping or spraying.

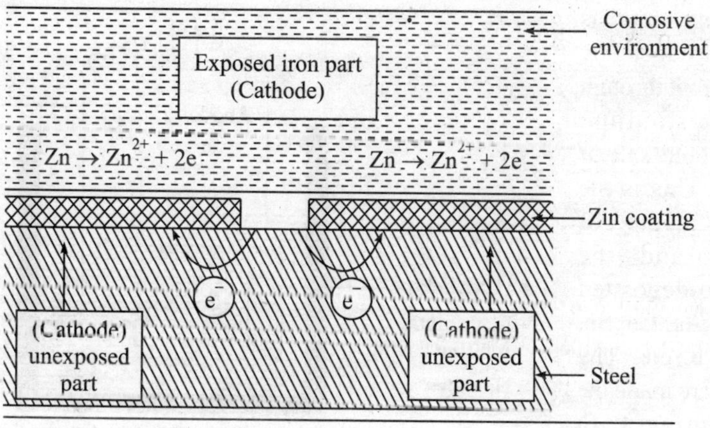

Fig. 4.11 Functioning of anodic coating

(ii) Cathodic Coating: Metals such as Sn, Ni, Cu etc., which are cathodic as compared to the metal to be protected, they stand below Fe in the galvanic series. The metals which fall in this category being cathodic to iron are more resistant to corrosion and offer protection so long as the surface is completely covered. *If the coating breaks or is applied in a manner that pinholes are left, these points at which the iron surface is exposed get very severely corroded because they are anodic to the large cathodic surface. Hence great care has to be exercised while applying these coating.* The functioning of cathodic metals has been represented by taking Sn as an example and is explained in Fig. 4.12.

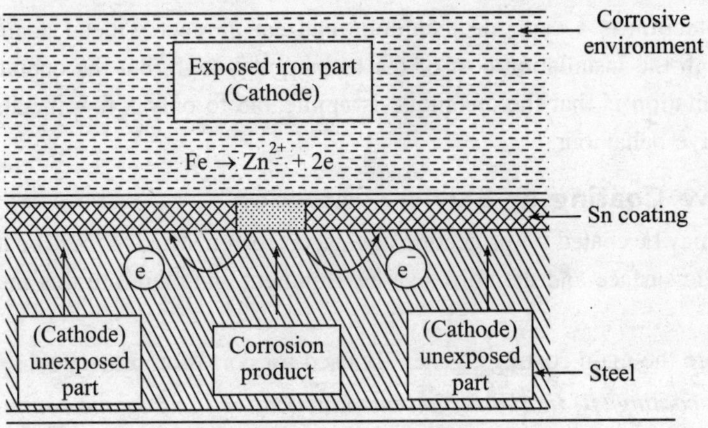

Fig. 4.12 Functioning of cathodic coating

Metallic coating may be applied by any one of the following ways:

(*i*) *Hot dipping* (*ii*) *Electroplating*

(*iii*) *Metal spraying* (*iv*) *Metal cladding*

(*v*) *Cementation* (*vi*) *Tinning*

Generally, the cathodic metal coating is applied by electroplating because by this method a more uniform coating which completely covers the surface is produced.

Electroplating: In this process, the coating metal is deposited on the base metal by passing a direct current through an electrolytic solution, containing the soluble salt of the coating metal. This is the most widely used method of coating metals; and the commonly electro-deposited metals include Au, Ag, Cu, Sn, Ni, Cd, Cr, Zn, Pb, etc. The metals to be plated are made the cathode in a plating bath containing the ions of the metal to be deposited. Anodes are usually, either the metals to be deposited, or an inert material of good electrical conductivity (graphite). Usually, conditions are so adjusted that the rate of dissolution at the anode and the rate of deposition at the cathode are almost equal, so that the composition of the electrolyte remains unchanged. Electroplating is carried out not only to save the base metal from corrosion, but to enhance its physical appearance *i.e.,* to decorate it also.

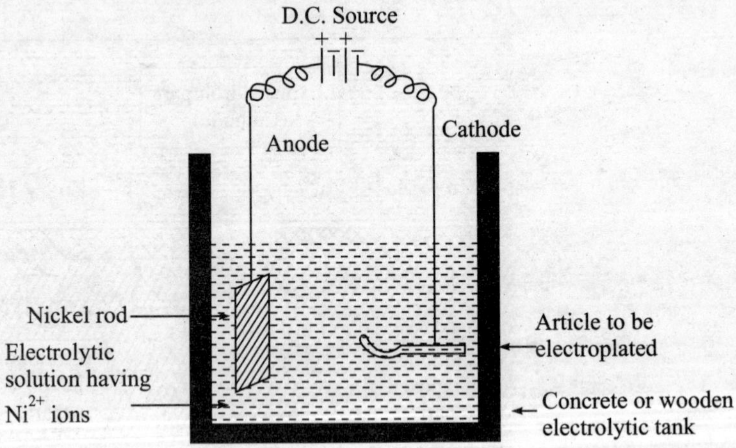

Fig. 4.13 Electroplating

To illustrate the process of electroplating let us take one example (say Nickel plating). The article to be electroplated is first cleaned *i.e.,* subjected to degreasing and descaling. Oils, greases etc. are removed by treatment with organic solvents like tetrachloroethylene. Surface scales, oxides etc. are removed by treatment with dilute acids (HCl or H_2SO_4). The cleaned metal in then made cathode in a electrolytic bath usually made of ***concrete*** or wood, Fig. 4.13, the electrolyte is solution of salt containing Ni^{2+} ions having good solubility and electric conductivity, usually a double salt $[NiSO_4(NH_4)_2SO_4]$ is taken in this case. The anode is the pure metal itself *i.e.,* Nickel rod or graphite. On passing electric current from a d.c. source (bettery), Ni^{2+} ion present in the electrolyte solution move towards the cathode and get deposited on it in the form of a uniform metalic layer. An equivalent amount of nickel is dissolved from the nickel anode into the solution. In case of graphite as anode, the concentration of electrolytic solution has to be maintained by adding the requisite amount of the double salt of nickel.

Protection of Iron by Tin (Tinning): Iron can also be protected by depositing a thin layer of tin on it. The process is called tinning. Tinning can protect iron as long as the coating is intact. If the coating develops scratches or cracks, iron is not protected any more. The protective action of tin is different from that of zinc which protects iron even when the coating develops cracks. This is because the standard reduction potential of iron is less than that of tin.

$$E^{\circ}_{Sn^{2+}/Sn} = -0.44 \text{ V}$$

$$E^{\circ}_{Sn^{2+}/Sn} = -0.14 \text{ V}$$

This shows that iron possesses a greater tendency to get oxidised as compared to tin. Therefore, if tin coating develops scratches or cracks, iron present in the exposed part will undergo oxidation *i.e.,* it will act as anode in preference to tin, which will now act as cathode and will get rusted. *In such a case the rusting is much more rapid as compared to that in case of an unprotected iron piece.*

Protection of copper by tin: Tin can protect copper as efficiently as zinc protects iron. This is due to the fact that the standard reduction potential of tin is less than that of copper.

$$E^{\circ}_{Sn^{2+}/Sn} = -0.14 \text{ V}$$

$$E^{\circ}Cu^{2+}/Cu = +0.34 \text{ V}$$

Thus, tin has a much greater tendency to get oxidised as compared to copper. Therefore, if tin layer develops a crack, tin itself undergoes oxidation in preference to copper and copper remains protected.

From the above discussion, it may be concluded that *a more electropositive metal (having a lower value of standard reduction potential) can sacrificially protect a lesser electropositive metal when used in the form of a thin layer on the metal to be protected.* This is why highly electropositive metals like Al, Zn, Mg etc. (in powdered form) are mixed with paints used to protect inferior metals like iron.

(*i*) **Surface conversion or Chemical conversion coating or Ceramic coatings:** These coatings are produced by covering the surface of a metal or alloy, by chemical or electro-chemical methods. These coatings afford good protection of the base metal from corrosion in some environments and sometimes are of decorative value. Further, many of these coatings are particularly useful to serve as excellent basis for the application of paints, enamels and other organic protective coatings.

The most commonly used surface conversion coatings are chromate coatings, phosphate coatings and oxide coatings. Chromate coatings are applied on Al, Zn, Mg, etc. by immersing the metal first in acid chromate bath followed by that of neutral chromate. Oxide coatings can be obtained by treating the metal with alkaline oxidizing agents. Protective films of magnetic oxide on iron can be obtained by passing steam over the heated metal.

(*ii*) **Anodized oxide coatings:** Anodized oxide coatings are, generally, produced on non-ferrous metals like Al, Zn, Mg and their alloys by anodic oxidation process, in which the base metal is made as anode. It is carried out by passing a moderate direct electric current through a bath in which the metal or alloy is suspended from anode. The bath usually contains sulphuric, phosphoric, chromic, oxalic or boric acid. After anodizing, the oxide coating is 'sealed' by immersing in boiling water. As the anodized coatings are somewhat thicker than the natural oxide films, so they posses improved resistance to corrosion as well as to mechanical injury.

(*iii*) **Vitreous or Enamel Coating:** Vitreous or porcelain enamel coatings are modified glass-like materials having different compositions which are usually applied on steel and cast iron equipment. The metal part to be enameled is first cleaned carefully. *"Frit"* is prepared by fusing together refractory acidic substance with basic fluxes (*e.g.*, borax, cryolite, soda ash, feldspar, etc.). The 'frit' is applied to the metal either by wet or by dry process.

High proportion of silica and feldspar and lower proportions of fluxes give acid-resistant enamel. Ceramic protective coatings made from high reaeractory oxides such as Cr_2O_3 afford protection of the metal from corrosion even at high temperatures.

(C) Organic Protective Coating: Organic coatings are inert organic barriers applied on metallic surfaces and other constructional materials for both corrosion protection as well as decoration. They include (*i*) Paints (*ii*) Varnishes (*iii*) Enamels (*iv*) Lacquers (*v*) Emulsions (*vi*) Plastic coating. The protective value of such coatings depends on:

(*i*) its chemical inertness to the corrosive environment

(*ii*) its good surface adhesion

(*iii*) its impermeability to water, salts and gases and

(*iv*) its proper application method

(i) Paints: Paints are a viscous suspension of finely divided solids called pigments in a fluid medium which on drying, yields an impermeable film having considerable covering power. The medium, consists of non-volatile film forming materials like drying oils and resins in suitable volatile solvent (thinners). When a properly treated metal surface is painted, the thinner evaporates leaving behind the film and the pigment on the surface of metal. The mixture gradually dries into a solid film. The film formation can be accelerated by the addition of driers in case of drying oil, which are metallic soaps. The important constituents of a paint are:

(*a*) Pigment — (*b*) Vehicle or medium (drying oil)

(*c*) Thinner — (*d*) Drier

(*e*) Fillers or extenders — (*f*) Plasticizers

(ii) Varnishes: Varnishes are solutions of natural or synthetic resins in a vehicle consisting of a drying oil, a thinner and a drier. A varnish does not contain any pigment. It is used as a protective and or decorative coating of suitable surfaces and dried by evaporation, oxidation and polymerization of its constituents, leaving behind transparent, glossy, lustrous and durable film. Varnishes are of two types:

(*i*) Oil varnish — (*ii*) Spirit varnish

The properties of a varnish depend upon the nature of the resin and the oil used. Varnishes are used for protection from corrosion. They are mainly used for wooden furniture to improve their brightness.

(iii) Enamels: Enamels are the pigmented varnished or lacquers. They contain pigments, oil or resin vehicle, driers and thinners. Enamels give the combined advantages of a paint and varnish and produce harder, smoother, fore lustrous and more glossy finish than ordinary paints.

(iv) Lacquers: Lacquers are solutions of film forming materials like cellulose derivatives, resin and plasticizer in a suitable solvent, *e.g.*, spirit varnishes, solution of shellac in methanol etc. in which the film formation takes place entirely by evaporation of the solvent.

Lacquers are used for protection from corrosion and for interior decoration. They give tough, durable and glossy coating. They are widely used in automobiles.

(*v*) ***Emulsion Paints:*** These are dispersions of synthetic resin latex in water containing oil, resin, pigment and other constituents. The latex (rubber) used are styrene butadiene copolymers. PVC, polytyrene etc.

(*vi*) ***Plastic coating:*** These days costly metallic components like micrometers, gauges, verniers and surgical tools are coated with plastics such as PVC, PVA, PMMA, Teflon etc. These polymers prevent corrosion of the metal because they have chemical resistance better than any other coating. But a crack in the coating will lead to corrosion which should be avoided. The material to be coated is thoroughly washed and dipped in liquid plastic or solution of plastic in a volatile solvent. When the plastic layer adheres to the metallic surface, the material is removed, cooled and dried.

REVISION QUESTIONS

1. Define corrosion of metals. What are different types of corrosion? Explain the electrochemical theory of wet corrosion, giving its mechanism.
   (*Tripura, 2K; SV, Aug.2001; Anna, May 2001; UPT, Jan 2001; PT, Dec. 2K; SK, June 2001; Surat, Nov 2K; Calicut, June 2001*)
2. (*a*) Explain rusting of iron with the help of electrochemical theory of corrosion.
   (*K.U., May 1999; VT March 1999; Raj, May 2001*)
   (*b*) Differentiate between chemical corrosion and electrochemical corrosion.
   (*c*) What will happen when stainless is immersed in aerated and deaerated sodium chloride solution? (*Surat, June 01*)
3. (*a*) Write chemical reactions taking place during stress corrosion of brass. (*V.T., Aug. 2K*)
   (*b*) Illustrate the reactions involved in differential aeration corrosion will reference to the iron materials.
   (*c*) Define corrosion. Explain dry corrosion and its mechanism.
4. Explain the mechanism of following types of corrosions.
   (*a*) Chemical corrosion (*Andhra, June 2001*)
   (*b*) Stress corrosion
   (*c*) Water-line, corrosion (*K.U.; Dec. 2K; Amravati, Dec. 2K; Sindri, May 2001*)
   (*d*) Pitting corrosion (*Anna, Nov. 2K, May 2001; Amravati, Dec 2K*)
   (*e*) Bimetallic or galvanic corrosion (*Surat, Nov. 2K; Anna, Nov. 2K*)
5. Mention different methods used for prevention of corrosion of metals and discuss any one method (*Tripura, 2K*)
   (*b*) How is steel protected cathodically? Discuss with an example.
   (*c*) Write short note on cathodic protection. (*K.U., Jan. 2003*)
6. (*a*) How material selection and design can prevent corrosion? What protective coating can be used for corrosion prevention? (*K.U., Jan. 2001*)
   (*b*) Explain sacrifical anodic protection of controlling corrosion.
   (*Amravati, Dec. 2K; Surat, Nov. 2K, Anna, May*)
7. (*a*) Explain the corrosion phenomenon including oxide film growth law.
   (*b*) What do you mean by stress corrosion, explain.
   (*c*) Write in brief about the pitting and soil corrosion.

8. (*a*) Why impure metals are more susceptible to corrosion than pure metals.

(*b*) What are the factors which affect corrosion. (*U.P. Tech., Jan. 2001*)

9. (*a*) Comment on the use of Al in place of Zn for the cathodic protection of iron from rusting. (*U.P. Tech. Jan. 2001*)

(*b*) Write a short note on corrosion by Differential Aeration.

(*c*) Discuss electroplating for corrosion protection.

10. (*a*) Explain hydrogen evolution type corrosion.

(*b*) Explain the difference in the use of Anodic and Cathodic coatings for corrosion prevention.

(*c*) State the law of oxide film growth. Give one example of each. (*P.T.U., July 2001*)

11. (*a*) Giving two examples of each, explain the difference between Galvanic corrosion and concentration cell corrosion.

(*b*) Discuss the factors influencing corrosion. (*P.T.U., Dec. 2003*)

(*c*) Explain how corrosion can be controlled by proper designing. (*K.U., Dec. 2003*)

12. (*a*) Discuss the role and nature of oxide formed in oxidation corrosion. State and explain Pilling-Bedworth rule.

(*b*) Describe the mechanism of electrochemical corrosion.

(*c*) Write a short notes on:

(*i*) Pitting corrosion

(*ii*) Role of sacrificial anode in corrosion control (*K.U. June 2004*)

13. (*a*) Explain the Electrochemical theory of corrosion. (*K.U. Jan. 2005, June 2005*)

(*b*) Discuss the various factors affecting corrosion.

(*c*) Iron corrodes faster than aluminium, even though iron is placed below aluminium in the electrochemical series. Why?

14. Explain the mechanism of hydrogen evolution and oxygen absorption in electrochemical corrosion. Illustrate your answer with figures. (*K.U. Jan. 2005, 2006, June 2006*)

15. (*a*) Write short note on:

(*i*) Stress corrosion

(*ii*) Water line corrosion (*K.U. Jan. 2006, 2007*)

(*b*) How are the metals protected against corrosion by modifying the environment? (*K.U. Jan. 2006*)

16. Explain rusting of iron with the help of electrochemical theory of corrosion. (*M.M.U. Mullana, June 2008*)

17. What type of corrosion is involved in rusting of iron? Give its mechanism. (*M.D.U. May 2004*)

18. Why does corrosion generally occurs of anode? Describe the anodic protection of metals against corrosion. (*M.D.U. Dec. 2004*)

19. Explain the mechanism of Galvanic corrosion and differential aeration corrosion. (*M.D.U. Dec. 2005*)

(*i*) Water line corrosion

(*ii*) Pitting corrosion

(*iii*) Stress corrosion

(*iv*) Bimetallic corrosion

20. Explain the mechanism of hydrogen evolution and oxygen absorption in electrochemical corrosion. *(M.D.U. May 2007)*

21. Describe the cathodic protection method for corrosion. Discuss its merits and demerits. *(M.D.U. May, 2007)*

22. (*a*) Explain "Rusting of iron" with the help of electrochemical theory of corrosion.
(*b*) Distinguish between Galvanising and tinning process.
(*c*) What is pilling-Bedworth rule? Explain it *(M.D.U., May, 2008)*

23. (*a*) Describe the mechanism of electrochemical corrosion.
(*b*) Discuss the following:
(*i*) Sacrifical Anodic Protection.
(*ii*) Galvanic corrosion. *(M.D.U., May, 2008)*

24. (*a*) Why does corrosion generally occurs at anode? Describe the anodic protection of metals against corrosion. *(M.D.U., Dec., 2004)*
(*b*) Describe the cathodic protection method for corrosion. Describes its merits and demerits. *(M.D.U., May, 2007)*

25. (*a*) Distinguish between Galvanizing and Tinning process.
(*b*) Explain the mechanism of chemical corrosion. *(M.D.U., May, 2008)*

26. (*a*) Comment on the use of Al in place of Zn for the cathodic protection of iron from rusting *(UP Tech., Jan. 2001)*
(*b*) Explain hydrogen evolution type corrosion. *(P.T.U., Jan. 2001)*
(*c*) Give reasons for the following:
(*i*) Corrosion of water filled steel tanks occurs below the water line.
(*ii*) A copper equipment should not possess a small steel bolt. *(P.T.U., Dec. 2002)*
(*d*) Write a short note on metallic coatings for corrosion prevention *(P.T.U., June. 2003)*

27. (*a*) How is metal structure protected by cathodic protection method?
(*b*) Explain the effect of the following factors on the rate of corrosion
(*i*) Nature of corrosion product (*ii*) Anodic and cathodic areas,
(*iii*) pH *(V.T.U., B.E., April 2000)*

28. (*a*) Discuss the influences of following factors on the rate of metallic corrosion.
(*i*) Nature of corrosion product
(*ii*) Polarization of the electrode surface.
(*b*) Explain the cathodic protection given to metals exposed to corrosion environment. *(V.T.U., B.E., August 2001)*

29. Define corrosion. How electrochemical theory of corrosion accounts for rusting of iron? Give the reaction include. *(V.T.U., B.E., August 2002)*

30. (*a*) Describe the electrochemical theory of corrosion.
(*b*) Discuss the following factors influencing the rate of corrosion.
(*i*) Nature of corrosion product (*ii*) Temperature (*iii*) pH.
(*c*) Explain the differential metal, water line and pitting corrosion with examples. *(V.T.U., B.E., Jan./Feb. 2003)*

31. (*a*) What is corrosion? Explain the rusting of iron based on electrochemical theory.
(*b*) Explain the effect of the following factors on the rate of corrosion:
(*i*) Nature of the metal.
(*ii*) Relative areas of the anode and cathode.
(*iii*) Nature of the corrosion product. (*V.T.U., B.E., July/Aug. 2004*)

32. (*a*) What in metallic corrosion? Explain the electrochemical theory of corrosion taking iron as an example.
(*b*) Explain what type of corrosion occurs when
(*i*) Screw and washer are made of different metals.
(*ii*) Presence of NaOH in mild steel boiler under stress. (*V.T.U., B.E., Feb. 2005*)

❑❑❑

SECTION C

Chapter 5: Instrumental Methods of Analysis

Chapter 6: Fuels and Combustion

5

Instrumental Methods of Analysis

INSIDE THIS CHAPTER

5.1 INTRODUCTION

Spectroscopy is the branch of science which deals with transitions occurring in a molecule when it interacts with electromagnetic radiations. Our knowledge about molecular structure is derived indirectly from different spectroscopic techniques. It is not possible to see atoms and molecules even with the most powerful microscope because of their infinitesimal small dimensions. As atoms give line spectra and each atom gives its characteristic spectrum, so the molecular spectra depend upon the nature of atoms involved and also upon the arrangement of the atoms in the molecules. Atomic spectra is due to electronic transitions only *i.e.*, the electrons shift from one energy level to another. But in molecular spectra, rotational and vibrational transitions also take place in addition to electronic transitions and this spectra is, therefore, much complicated than atomic spectra, even then, the molecular spectra is of much help to a chemist in elucidating the structure of molecules. But the structures of molecules have been investigated indirectly by molecular spectroscopic techniques. Besides molecular structure, spectroscopy can also be employed to study the followings:

(*i*) rates of reaction *i.e.*, kinetic studies,

(*ii*) detection of impurities in a sample,

(*iii*) identification of unknown compounds,

(*iv*) determination of molecular weights,

(*v*) determination of dissociation constants of acid and bases,

(*vi*) detection of steric hindrance,

(*vii*) determination of intra and inter molecular hydrogen bonding,

(*viii*) distinction between cis-trans isomers and detection of aromaticity etc.

Some general features of spectroscopy common to all type of spectra and a brief study of some spectroscopic techniques have been undertaken in this chapter.

5.2 ELECTROMAGNETIC SPECTRUM

The visible light represents only a small part of the entire electromagnetic spectrum, which extend from high energy cosmic rays to low energy radio or radar waves. The complete electromagnetic spectrum gives the arrangement of various types of electromagnetic radiations in order of their increasing (or decreasing) wave lengths or frequencies (Fig. 5.1)

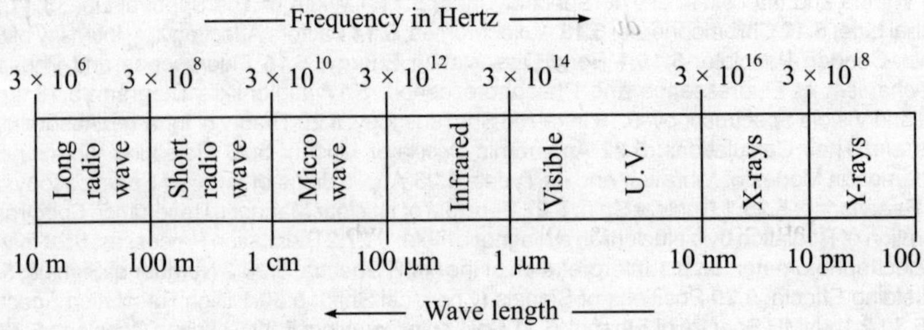

Fig. 5.1 Electromagnetic spectrum

Table 5.1 illustrates the various regions into which electromagnetic radiation has been divided. It also includes the types of spectra that can be studied in each region of electromagnetic radiation.

Table 5.1: The Region of Electromagnetic Radiation

Electromagnetic Radiation	*Frequency* (v)	*Wavelength* (λ)	*Spectrum*
Radio frequency region	$3 \times 10^6 - 3 \times 10^{10}$ Hz	10 m – 100 m 100 m – 1 cm	Nuclear magnetic resonance Electrons spin resonance
Microwave region	$3 \times 10^{10} - 3 \times 10^{12}$ Hz	1 cm – 100 μm	Microwave (rotational) spectroscopy
Infra-red region	$3 \times 10^{12} - 3 \times 10^{14}$ Hz	100 mm – 1 μm	Infra-red (vibrations) spectroscopy
Visible and ultraviolet	$3 \times 10^{14} - 3 \times 10^{16}$ Hz	1μm – 10 nm	*u.v.* – visible (electronic) spectroscopy
X-ray region	$3 \times 10^{16} - 3 \times 10^{18}$ Hz	10 nm – 100 pm	Photoelectron spectroscopy
γ-ray region	$3 \times 10^{18} - 3 \times 10^{20}$ Hz	100 pm – 1 pm	Energy changes involves the rearrangement of nuclear particles

$1\mu = 10^{-4}$ cm; $1\ m\mu = 1$ nm $= 10^{-7}$ m $= 10$ Å

5.3 ATOMIC AND MOLECULAR SPECTROSCOPY

Atomic Spectroscopy

It is obtained by the interaction of electromagnetic spectrum with atoms. Each atom gives its own characteristic spectrum in the form of lines. It is, therefore referred to as **line spectrum** also. It can be obtained either by absorption or by emission of radiation. Depending upon the mode of formation, atomic spectrum is called absorption or emission spectroscopy. The characteristic line spectrum obtained in case of an atom is due to the electronic transitions between the energy levels.

Molecular Spectroscopy

A molecule also can absorb or emit energy as a result of transitions between different electronic energy levels. From Fig. 5.2 it is evident that each **electronic level** consists of a number of **vibrational sub-levels** represented by v = 0, 1, 2, 3 etc. and further each vibrational level consists of a number of **rotational sub-levels** represented by j = 0, 1, 2, 3 etc. v and j are called *vibrational and rotational* **quantum numbers** respectively. However, a molecule can absorb a quantum of energy and increase its vibrational energy and rotational energy also. The quanta associated with these three different types of energy levels namely *electronic, vibrational and rotational* are of very different energy. The energies of the quanta that are absorbed or emitted by a molecule can be measured by spectroscopic means.

The molecular spectra arise due to changes in rotational, vibrational and electronic energies possessed by molecules. The total energy of a molecule according to **Born-Oppenheimer approximation**, *is given by:*

$$E = E_{tr} + E_{rot} + E_{vib} + E_{el} \quad \text{where} \quad E_{el} >>> E_{vib} > E_{rot} > E_{tr}$$

where E_{tr}, E_{rot}, E_{vib} and E_{el} represents translational, rotational, vibrational and electronic energies respectively. The translational energy is not quantised, whereas all the other energies are quantised. Since the translational energy is negligibly small, we can write the above approximation as:

$$E = E_{rot} + E_{vib} + E_{el}$$

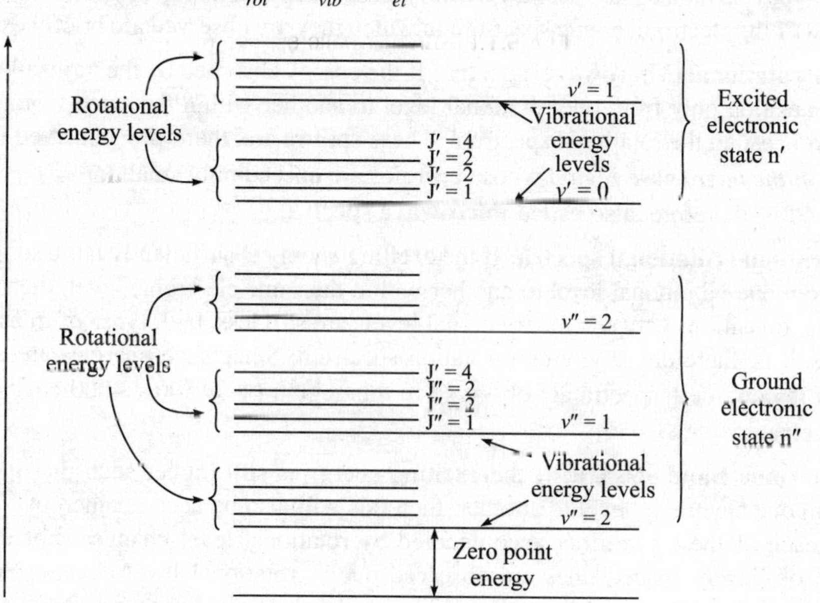

Fig. 5.2 Energy level diagram showing the various molecular energies

Responding to the three types of energies possessed by molecules three quantum numbers namely rotational quantum number (j), vibrational quantum number (v) and electronic quantum number (x) have been assigned.

A diagram illustrating the distribution of different energy levels is shown in Fig. 5.2. Two electronic levels n'' and n' are shown. The lower level n'' has a number of vibrational energy level denoted by the vibrational quantum number $v'' = 0, 1, 2...$ and each vibrational level has its own rotational energy levels represented by the rotational quantum number $J'' = 1, 2, ...$ In the same way, the upper electronic level n' has vibrational levels $v' = 0, 1, 2, ...,$ and each vibrational level has rotational energy level $J' = 1, 2 ...$

Transitions between vibrational levels within the same electronic level give rise to spectra which are called **vibrational spectra**. In the near infra-red, transitions between different rotational levels with in the same vibrational level are responsible for **rotational spectra** in the far infra-red or microwave region. Infact a transition between two electronic levels is usually accompanied by changes in the vibrational and rotational quantum numbers, so that electronic spectra are really electronic vibration-rotational spectra. In the same way, vibrational spectra are really **vibrational-rotational spectra** and are normally given this name. However, very small changes are required for transitions between rotational levels alone. Hence the spectra corresponding to the smallest changes in energy are thus called **pure rotational spectra**.

Because of a large number of energy levels involved for transition, therefore molecular spectra are much more complex than atomic spectra. However, just as atomic spectra help in the study of the structure of the atoms, the study of molecular spectra gives even more information about the structure and properties of molecules such as bond lengths, bond angles, bond strength, shapes, dipole moments etc.

5.4 TYPES OF MOLECULAR SPECTRA

The different types of molecular spectra generally observed, depending upon the energies absorbed, and the regions of the electromagnetic spectrum in which they are observed are briefly explained below:

1. Pure rotational (Microwave) spectra: If the energy absorbed by the molecules is so low that it can cause transition only from one rotational level to another within the same vibrational level, the result obtained is called the rotational spectrum. These spectra are, therefore, observed in the *far infra-red region or in the microwave region* whose energies are exceedingly small ($\bar{v} = 1 - 100\ cm^{-1}$). The spectra obtained is, therefore, also called **microwave spectra.**

2. Vibrational rotational spectra: If the exciting energy is sufficiently large so that it can cause a transition from one vibrational level to another within the same electronic level, then as the energies required for the transitions between the rotational levels are still less, both types of transitions will take place. The result is, therefore, a vibration-rotation spectrum. Since such energies are available in the near infra-red region, these spectra are observed in this region ($\bar{v} = 500 - 4000\ cm^{-1}$) and are called **infra-red spectra.**

3. Electronic Band spectra: If the exciting energy is still higher such that it can result in a transition from one electronic level to another, then this will also be accompanied by vibrational level changes and each of these is further accompanied by rotational level changes. For each vibrational changes a set of closely spaced lines are observed due to rotational level change. Such a group of closely spaced rotational lines is called a band. Thus for a given electronic transition, a set of bands is observed. This set of bands is called *a band group or a band system*. Each electronic transition gives a band system. The complete set of band systems obtained due to different electronic transitions gives the electronic band spectrum of the gaseous molecule.

Thus, whereas atoms give the line spectra, molecules give band spectra. In electronic band spectra high excitation energies are available in the visible and ultraviolet regions, these spectra are observed in the visible region (12,500 – 25,00 cm^{-1}) and ultraviolet region (25,000 – 70,000 cm^{-1}).

4. Raman spectra: This is also a type of vibrational rotation spectrum but is based on scattering of radiation and not on the absorption of radiation by the sample. It is based upon the principle that when a sample is hit by monochromatic radiation of the visible region and *scattering is observed at right angles to the direction of the incident beam,* the scattered radiation have frequency equal to that of the incident beam (called *Rayleigh scattering*) as well as frequencies different (higher as well as lower) than that of the incident beam (called *Raman scattering*). The difference in the frequencies of the incident beam and that of the scattered beam (called *Raman frequencies*) are similar to those needed for the vibrational and rotational transitions. However, by suitably adjusting the frequency of the incident radiation, Raman spectra are observed in the visible region (12,500 – 25,000 cm^{-1}).

5. Nuclear Magnetic Resonance (NMR) spectra: This type of spectrum arises from the transitions between the nuclear spin energy levels of the molecule (*involving reversal of nuclear spin*) when an external magnetic field is applied on it. This energies involved in these transitions are very high which lie in *the radio frequency region* (5 – 100 MHz). The method is based upon applying such frequencies on the sample so that it is in resonance with the applied frequency of proton.

6. Electron Spin Resonance (ESR) spectra: This type of spectrum arises from the transitions between the electron spin energy levels of the molecule (*involving reversal of electron spin*) when an external magnetic field is applied on it. These involve frequencies corresponding to microwave region (2000 – 9600 MHz). Frequencies of this range are applied on the sample to bring the sample in resonance condition.

5.5 ULTRAVIOLET AND VISIBLE SPECTROSCOPY

To understand why some compounds are coloured and others are not, and to determine the relationship of conjugation to colour, we must make accurate measurements of light absorption at different wavelengths in and near the visible part of the spectrum. Commercial optical spectrometers enable such experiments to be conducted with ease, and usually survey both the near ultraviolet and visible portions of the spectrum.

The visible region of the spectrum comprises photon energies of 36 to 72 kcal/mole, and the near ultraviolet region, out to 200 nm, extends this energy range to 143 kcal/mole. Ultraviolet radiation having wavelengths less than 200 nm is difficult to handle, and is seldom used as a routine tool for structural analysis.

The energies noted above are sufficient to promote or excite a molecular electron to a higher energy orbital. Consequently, absorption spectroscopy carried out in this region is sometimes called **"electronic spectroscopy"**.

5.6 LAWS GOVERNING LIGHT ABSORPTION-LAMBERT LAW AND BEER'S LAW

Lambert law

Light incident upon a system can be transmitted, refracted, scattered, or absorbed. The fraction of incident light absorbed by a system depends on the thickness of the medium that is traversed. The intensity of absorption at a particular wavelength can be determined by passing a monochromatic beam of light through a sample of known thickness and measuring the intensity of transmitted light which is found to decrease. The decrease of intensity of incident light at any given wavelength on passing through a transparent absorbing substance is given by **Lambert Law**, according to which,

"The rate at which the intensity of radiation decreases with thickness of the medium is proportional to the intensity of the incident light."

Mathematically, it may be expressed as

$$-\frac{dI}{dx} \propto I \quad \text{or} \quad -\frac{dI}{I} = \alpha\, dx \qquad \ldots(5.1)$$

Where I is the intensity of light of a particular wavelength (energy per unit area per unit time); dI is the change in light intensity produced by absorption radiation on passing through of a thin layer of thickness dx (Fig. 5.3); α is called absorption coefficient or extinction coefficient. The minus sign is introduced because there is reduction in intensity.

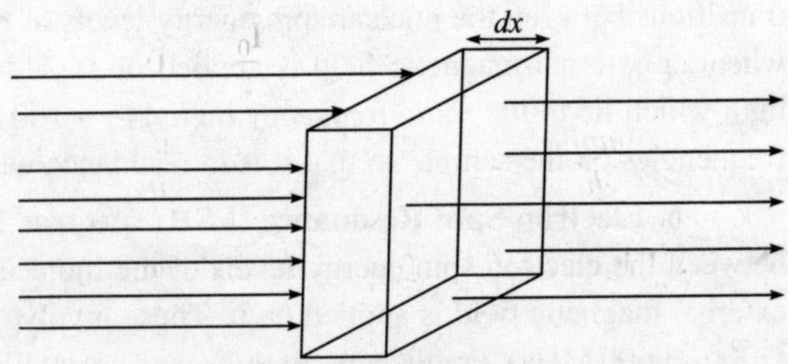

Fig. 5.3 Lambert Law

Equation (5.1) can be also written as:

$$+\frac{dI}{I} = -\alpha\, dx \qquad \ldots(5.2)$$

Integrating equation (5.2) and applying boundary conditions,

when $x = 0$, $I = I_0$ and when $x = x$, $I = I$

$$\int_{I_0}^{I} \frac{dI}{I} = \int_{x=0}^{x=x} -\alpha\, dx$$

or

$$\ln \frac{I}{I_0} = -\alpha x \quad \text{or} \quad \frac{I}{I_0} = e^{-\alpha x} \qquad \ldots(5.3)$$

or

$$I = I_0\, e^{-\alpha x} \qquad \ldots(5.4)$$

Equation (5.3) can also be rewritten as

$$2.303 \log \frac{I}{I_0} = -\alpha x$$

$$\log \frac{I}{I_0} = -\frac{\alpha}{2.303} = -\alpha' x \qquad ...(5.5)$$

or

$$\frac{I}{I_0} = 10^{-\alpha' x} \qquad ...(5.6)$$

or

$$I = I_0\, 10^{-\alpha' x} \qquad ...(5.7)$$

The equantity $\alpha' = \frac{\alpha}{2.303}$ is called **absorption coefficient or absorptivity** of the substance.

Equation 5.5 can be re-written as

$$\log \frac{I_0}{I} = \alpha' x \quad \text{or} \quad \alpha' = \frac{1}{x} \log \frac{I_0}{I} \qquad ...(5.8)$$

The factor $\log \frac{I_0}{I}$ is called optical density or **absorbance (A)** and $\frac{I}{I_0}$ is termed as **transmittance (T).**

when

$$\frac{I_0}{I} = 10 \log I_0/I \Rightarrow \alpha' = \frac{1}{x}$$

This implies that absorptivity is equal to the reciprocal of thickness of that layer in which the intensity of light falls to one-tenth of its original value.

Note: If I_t is the intensity of transmitted light and I_0 is the intensity of incident beam of light then intensity of absorbed light I_a, is given by

$$I_a = I_o - I_t \qquad ...(5.9)$$

Lambert Beer's Law: Later Beer found that the absorption coefficient (α) was proportional to the concentration '*c*' of solution and combined it with Lambert law, this is known as Lambert Beer law and is stated as:

"On passing a beam of monochromatic radiation through a solution, the rate of decrease of intensity with thickness of the solution is proportional to the intensity of incident radiation as well as concentration of the solution".

Mathematically,

$$-\frac{dI}{dx} \propto cI \quad \text{or} \quad -\frac{dI}{I} = \epsilon' c dx \qquad ...(5.10)$$

where ϵ is the absorption or extinction coefficient, Integration of Equation (5.10) gives

$$\int -\frac{dI}{I} = \int \epsilon' c dx \qquad ...(5.11)$$

or $\ln I = \epsilon' cx$ + constant of integration ...(5.12)

At $x = 0, I = I_o$ and at $x = x, I = I$

Substituting these boundary conditions, we get

$$-\ln \frac{I}{I_0} = \epsilon' cx \qquad ...(5.13)$$

or $$I = I_0 e^{-\epsilon' cx} \qquad ...(5.14)$$

Equation (5.14) is one of the well known form of Lambert-Beer's law, Changing natural logarithm in Equation (5.13) to the logarithm to the base 10 we get,

$$-2.303 \log \frac{I}{I_0} = \epsilon' cx \qquad ...(5.15)$$

or $$-\log \frac{I}{I_0} = \frac{\epsilon'}{2.303} cx \qquad ...(5.16)$$

or $$\log \frac{I_0}{I} = \frac{\epsilon'}{2.303} cx = \epsilon\ cx = A,$$

where $$\epsilon = \frac{\epsilon}{2.303} \qquad ...(5.17)$$

The quantity $\log \frac{I_0}{I}$ is known as the absorbance (A) and ϵ is called the **molar absorption coefficient**.

The equation (5.17) tells that *the absorbance A is directly proportional to the concentration of the solution and the length of the path.*

The length of the path can be kept constant and hence absorbance measurement can be used for measuring unknown concentration of the coloured solutions. The molar absorption coefficient depends on temperature T at which the solution is kept and wavelength λ of the incident light.

From equation 5.17, $$\log \frac{I_0}{I} = \epsilon\ cx \qquad ...(5.18)$$

If $$C = 1 \text{ and } \log \frac{I_0}{I} = 1 \quad \text{then} \quad \epsilon\ x = 1$$

or $$\epsilon = \frac{1}{x}\left(\log \frac{I_0}{I} = 1, \text{show } \frac{I_0}{I} = 10\ \textit{i.e.}, I = \frac{1}{10} I_0\right)$$

This gives us the definition of ***Molar Absorption Coefficient****. Thus Molar Absorption Coefficient is the reciprocal of that thickness of the solution layer of one molar solution which decreases the intensity of light to one tenth of its original value.*

Relation between transmittance and absorbance of solution

The absorbance of solution is denoted by A and is given as:

$$A = \log_{10} \frac{I_0}{I} \quad \text{or} \quad A = -\log_{10} \frac{I}{I_0}$$

or $$A = -\log_{10} T \qquad ...(5.19)$$

A plot of absorbance versus concentration of the solution and that of transmittance versus concentration has been shown by Fig. 5.4 (*a*) and (*b*) respectively.

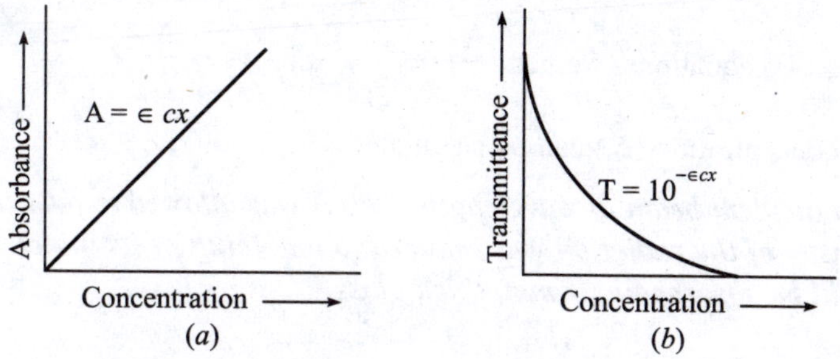

Fig. 5.4 Plots of Absorbance and transmittance versus concentration

SI Units of molar extinction coefficient

$$\log \frac{I}{I_0} = -\epsilon' cx \quad \text{or} \quad \epsilon' = -\frac{1}{cx} \log \frac{I}{I_0}$$

Here, x, thickness of solution in meters, c the concentration of solution in mol m^{-3} and $\frac{I_0}{I}$ is a dimensionless quantity.

Putting the units of the terms in the above expression, we get

$$\epsilon' = \frac{1}{\text{mol } m^{-3}} \frac{1}{m} = \frac{1}{m^{-2} \text{ mol}}$$

So, the units of molar extinction coefficient ϵ' are m^2 mol^{-1}.

Limitations of Lambert-Beer Law

(*i*) The law is not obeyed if radiation used is not monochromatic.

(*ii*) The temperature of the solution should not be allowed to increase as it will show bathochromic effect on ions in solution *i.e.*, the absorption band in u.v.–visible spectra shifts towards longer wavelength.

(*iii*) The law is applicable for dilute solutions only as in concentrated solution the refractive index of the solution changes and high concentration of solute shows strong inter ionic interactions which can disturb the ability of a solute to absorb a given wavelength of the incident radiation.

Determination of concentration of an unknown solution by Lambert Beer's Law:

According to Lambert-Beer's Law: $\log \frac{I}{I_0} = -\epsilon' cx$

or
$$\log \frac{I_0}{I} = -\epsilon' cx = A$$

where A is the absorbance and the other terms have their usual significance.

For determining the concentration of an unknown solution, Absorbance of a solution of known concentration (c_1) is measured and given as:

$$A_1 = \epsilon' c_1 x$$

The absorbance of solution of unknown concentration (c_2) is then measured and given as:

$$A_2 = \epsilon' c_2 x$$

From the above two equations, we get: $\frac{A_2}{A_1} = \frac{c_2}{c_1}$ or $c_2 = \frac{A_2}{A_1} \times c_1$

The unknown concentration (c_2) can be calculated.

Example 1: ***When an incident beam of wavelength 3000 Å was allowed to pass through 2 mm thick Pyrex glass, the intensity of the radiation was reduced to one-tenth of its initial value. What part of the same radiation will be absorbed by 1 mm, thick glass.***

Solution: We are given that $\frac{I_t}{I_0} = \frac{1}{10}$ and $x = 2 \text{ mm} = 0.2 \text{ cm}$

According to Lambert's Law, we have:

$$\log \frac{I_t}{I_0} = -\alpha x \quad \text{or} \quad \log \frac{1}{10} = -\alpha \times 0.2$$

or

$$\alpha = \frac{1}{0.2} = 5$$

Again, we are given that $x = 1 \text{ mm} = 0.1 \text{ cm}$ and $K = 5$

Applying Lambert's law, again

$$\log \frac{I_t}{I_0} = -\alpha x = -5 \times 0.1 = -0.5$$

$$\frac{I_t}{I_0} = \text{antilog}(-0.5) = \text{antilog}(\bar{1}.5) = 0.3162$$

We know

$$I_\alpha = I_0 - I_t$$
$$= I_0 - 0.3162 I_0,$$
$$I_\alpha = 0.6838 I_0$$
$= \mathbf{68.38\%}$ of the intensity of incident light.

Example 2: ***A 1 molar solution in a cell of 0.200 mm thickness transmits 10% of a beam of light. Neglecting the absorption by a solvent and cell windows, what percentage of the initial intensity will be transmitted by a 2 molar solution in a cell of 0.150 mm thick.***

Solution: According to Beers law, we have

$$\log \frac{I_t}{I_0} = -\epsilon' c \cdot x$$

It is given that; Transmittance $T = \frac{10}{100}$

$$\Rightarrow \quad \frac{I_t}{I_0} = \frac{1}{10},$$

$$c = 1 \text{ mol dm}^{-3} \quad \text{and} \quad x = 0.20 \text{ mm} = 0.02 \text{ cm}$$

$$\therefore \quad \log \frac{1}{10} = -\in' \times 1 \times 0.2$$

or $$\in' = \frac{1}{0.2} = 50 \text{ dm}^3 \text{ mol}^{-1} \text{ cm}^{-1}$$

Applying Beer's Law again, we have

$$\log \frac{I_t}{I_0} = -\in' c \cdot x$$

We know $c = 2$ M and $x = 0.15$ mm $= 0.015$ cm

$$\log \frac{I_t}{I_0} = -50 \times 2 \times 0.015$$

$$\therefore \quad \log \frac{I_t}{I_0} = \text{antilog } (-15) = \text{antilog } \bar{2}.5 = \mathbf{0.0316}$$

or $I_t = 0.0316 I_0 =$ **3.16%** of the intensity of incident light.

Example 3: ***What is the absorption coefficient of a solute which absorbs 90% of a certain wavelength of light beam passed through 1 cm cell containing 0.25 M solution?***

Solution: Given that 90% of the incident light is absorbed

or $I_0 = 100$ and $I_\alpha = 90$

We know $I_a = I_o - I_t$ or $90 = 100 - I_t$ or $I_t = 100 - 90 = 10$

According to Lambert–Beer's law;

$$\ln \frac{I_t}{I_0} = -\in' cx \quad \text{or} \quad \log \frac{I_t}{I_0} = \frac{\in'}{2.303} cx$$

where $\in'$ is absorption coefficient, c the molar concentration and x is the thickness of the solution.

$$\therefore \quad \log \frac{10}{100} = -\frac{\in'}{2.303} \times 0.25 \times 1$$

$$\therefore \quad \in' = \frac{2.303}{0.25} \times \log \frac{100}{10}$$

Hence $\in' = 9.21$ Absorption coefficient $=$ **9.21.**

Example 4: ***In an absorption cell, the transmittance of 0.1 M solution of a substance X is 80% and that of 0.1 M solution of another substance Y is 60% at a given wavelength. What is the transmittance of solution that is simultaneously 0.1 M in X and 0.1 M in Y?*** **(P.T.U. Dec. 06, May 2007)**

Solution: The transmittance of any cell is given by the absorbance which are additive. Hence

Total absorbance (A) = Absorbance due to X + Absorbance due to Y

i.e., $$A = Ax + Ay$$

$$= \log\left(\frac{I_0}{I}\right)_1 + \log\left(\frac{I_0}{I}\right)_2 = \log \frac{100}{80} + \log \frac{100}{60}$$

$$= 0.0969 + 0.2219 = 0.3188$$

$$A = \log \frac{I_0}{I}$$

$$0.3188 = \log \frac{I_0}{I} \qquad \therefore \frac{I_0}{I} = 2.083$$

Hence $$T = \frac{I}{I_0} = \frac{1}{2.083} = 0.48 = \mathbf{48\%}.$$

Example 5: ***If 2 millimeter thick plate material transmits 70% of the incident light, what percentage will be transmitted by 0.5 mm thick plate?***

Solution: In the first case, We know $$\frac{I_t}{I_0} = \frac{70}{100} = 0.7; x = 2 \text{ mm}$$

In the second case let $$\frac{I_t}{I_0} = A \text{ (say)} \quad \text{where} \quad x = 0.5 \text{ mm}$$

According to Lambert's Law $$\ln \frac{I_t}{I_0} = -\alpha x \quad \text{or} \quad \log \frac{I_t}{I_0} = \frac{-\alpha x}{2.303}$$

In the first case, $$\log 0.7 = -\frac{-\alpha}{2.303} \times 2 \qquad ...(i)$$

In the second case, $$\log A = -\frac{-\alpha}{2.303} \times 0.5 \qquad ...(ii)$$

Dividing (*i*) by (*ii*), $$\frac{\log 0.7}{\log A} = 4$$

$$\therefore \quad \log 0.7 = 4 \log A \quad \text{or} \quad \bar{1}.8451 = 4 \log A$$

$$\therefore \quad \log A = \frac{\bar{1}.8451}{4} = \frac{(\bar{4} + 3.8451)}{4} = \bar{1}.9613$$

$$\therefore \quad A = \text{Antilog } \bar{1}.9613 = 0.9147 \quad \text{or} \quad = \mathbf{91.47\%}.$$

5.7 INSTRUMENTATION

Principle and Instrumentation

It may be noted that emission and absorption spectroscopy gives the same information about the energy level separations. However, the practical considerations decide as to which technique should be used. *Generally absorption spectroscopy is used as absorption spectra are easier to interpret than emission spectra*. Emission spectroscopy is used when the visible or ultraviolet incident radiations are used. In this chapter, we shall, therefore, mainly concentrate on the absorption spectroscopy.

Principle: All organic compounds interact with electromagnetic radiation, that is, they absorb energy. When a molecule absorbs energy, a transformation occurs that may be either temporary or permanent. Lower energy radiation may cause a molecular rotation or a bond vibration. However, energy radiation may cause the promotion of electrons to higher energy levels or bond cleavage.

Whether the transformation involves molecular rotation, bond vibration or electronic transition, the molecule absorbs only the wavelength of radiation with exactly the energy necessary for the transition.

The absorption of wavelengths of radiation is selective for a particular transition which depends on the structure of the molecule. By measuring the absorption spectra of known compounds, we can correlate the wavelengths of energy absorbed with characteristics structural features. This information is then used to determine the structure of unknown compounds.

Spectrophotometer: The instrument used to measure the amount of electromagnetic radiation absorbed by an organic molecule is called spectrophotometer or spectrometer. It consists of a source of radiation, with a prism that can select the desired wavelengths, which are passed through a sample of the compound being investigated. The radiation that is absorbed by the sample can be detected, analyzed and recorded. The spectrum obtained in this manner is known as **absorption spectra**. It is always preferred over the **emission spectra** because in that case we have to irradiate the sample with suitable radiations which is not easy job. The recording is called the **spectrum** of the compound.

5.7.1 Components of a Spectrophotometer

The basic components of spectrophotometer are as given Fig. 5.5. The important parts of spectrophotometer are:

(i) **Source of electromagnetic radiation:** A spectrometer can be made to operate in the ultraviolet, visible and infrared regions by using suitable source of radiant energy.

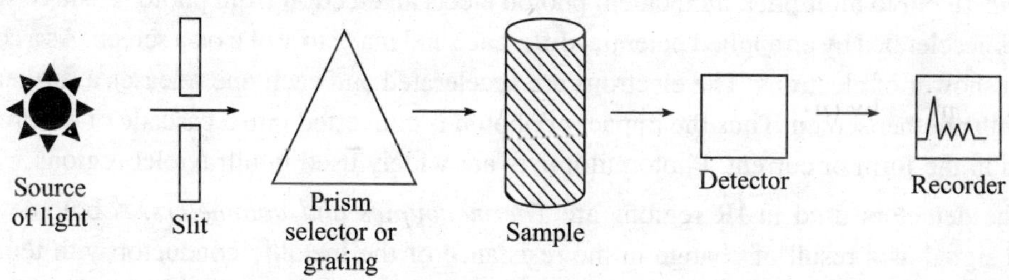

Fig. 5.5 Basic components of spectrophotometer

Source of Electromagnetic Radiation in Various Spectral Regions

Radio waves: Radio waves are generated from alternating electronic currents of high frequencies.

Micro waves: Micro waves are generally produced by an electromagnetic oscillator such as the klystron or an amplifier such as the travelling wave tube.

Infrared Radiation: Radiations in the infrared region are emitted from a filament which is maintained at red or white heat by an electric current. Two common sources are the Nernst filament, and the globar filament. The Nernst filament requires to be pre-heated before it will conduct electricity, but once the red heat is reached, the temperature is maintained by the current.

Visible Radiation: In this region, a tungsten filament lamp is most widely used.

Ultraviolet Radiation: In ultraviolet region, the most commonly used radiation sources are hydrogen discharge lamps, the Xenón discharge lamps and mercury arcs.

X-Rays: X-rays are produced when an electrically charged particle of sufficient kinetic energy is rapidly de-accelerated. Electrons are usually used for this purpose, radiation being produced in an X-ray tube which contains a source of electrons and two metal electrodes.

γ-Rays: These rays are emitted from nuclei of radioactive atoms by radioactive disintegration going on spontaneously in nature.

(*ii*) Monochromator: After passing through sample the radiation is separated into its constituent frequencies by dispersing device such as a prism or grating. The prism used for ultraviolet radiation is a quartz prism or grating; for Visible radiations glass prism, and for infrared radiations rock salt prism is used. Generally diffraction grating is preferred to a prism for use as monochromator. A grating consists of a glass or ceramic plate into which several fine grooves are cut lying about 1000 nm apart, then covered with reflective aluminium coating. The various wavelengths get separated due to constructive interference at various angles in the reflected waves.

(*iii*) Sample Holder: It is used to hold the sample in the instrument. In spectrophotometer a test tube is used or holding the sample. In UV and visible spectroscopy, a quartz cell is used. In IR spectroscopy, infrared transparent sodium chloride pallets are used for holding the samples. In NMR spectroscopy, a glass tube called NMR tube is used.

(*iv*) Photodetector and Receptor unit: This unit detects the radiant energy and converts it into measurable electrical signals. A detector converts the incident radiation into an electric current for the appropriate signal processing or plotting e.g. *photomultiplier, thermocouple detector, semiconductor devices etc.* In photo multiplier, an incident photon ejects an electron from photo sensitive surface. This electron is accelerated by an applied potential difference and made to strike on a screen. As a consequence, it ejects a shower of electrons. The electrons are accelerated and each one releases a further shower on impact with another screen. Thus the impact of photon is converted into a cascade of electrons which is measured in the form of current. Photomultipliers are widely used in ultraviolet regions.

The detectors used in IR regions are *Thermocouples and bolometers*. A bolometer gives an electrical signal as a result of change in the resistance of the metallic conductor with temperature. A *microwave detector* is typically a crystal diode consisting of a tungsten tip in contact with a semi-conductor such as germanium, silicon or gallium arsenide. The detector for visible or ultraviolet radiation is a photomultiplier tube or a photographic plate.

(*v*) Amplification and Recording unit: The signal obtained from detector is amplified and recorded on computer.

5.7.2 Single and Double Beam Spectrophotometer

Spectrophotometer can be single beam or double beam type. In single beam, the beam of radiation is passed through sample solution and reference solution one by one Fig. 5.6 (*a*). In case of double beam spectrophotometer, the beam of radiation is split into two identical halves and simultaneously passed through sample and reference solution Fig. 5.6 (*b*). Thus the instrument can compare the intensities of the two beams at each wavelength.

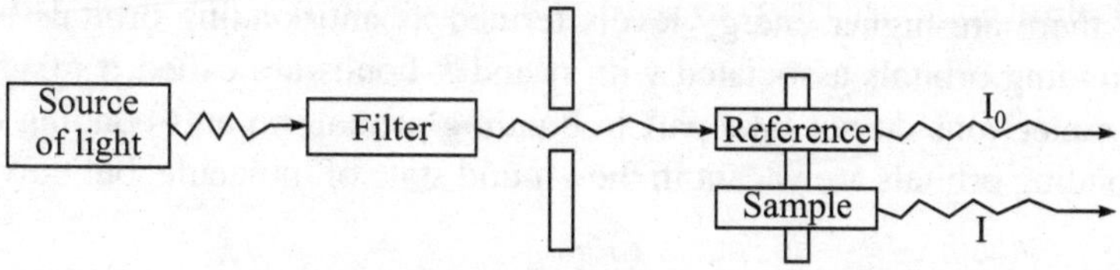

(*a*) Single beam spectrophotometer

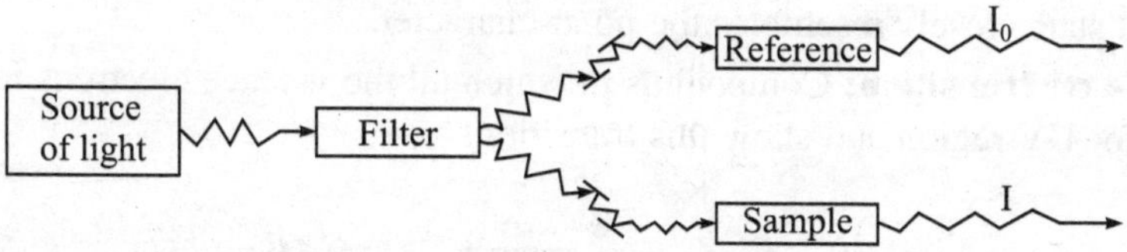

(*b*) Double beam spectrohotometer

Fig. 5.6

Suppose the intensity of beam coming out of sample solution is I and that for reference solution is I_0. It may be noted that I is always less than I_0. Therefore

$$\text{Absorbance A} = \log \frac{I_0}{I}$$

Utility: The absorbance (A) or tansmittance (T) for a compound is directly measured by spectrometers at various wavelengths. The curve showing a variation of A or T with wavelength for a given concentration of solution is known as **absorption spectrum** of the compound. It is a characteristic property of the compound. No two compounds can have the same absorption spectrum. The wavelength corresponding to maximum absorption is termed as λ_{max} and is also the characteristic property of the compound. A careful analysis of the absorption spectrum of an unknown compound and its composition with that of a compound of known structure helps in confirming the structure of the unknown compound.

5.8 THEORY OF ELECTRONIC SPECTROSCOPY

Electronic Transitions

The spectra arise in a molecule from electronic transitions due to absorption of radiations falling in the visible and ultra violet regions. In the visible region, the spectra occur in the spectral range of 12,500 to 25,000 cm^{-1}, whereas in the ultra violet region, the spectral range is 25,000 – 70,000 cm^{-1}. Since electronic transitions in a molecule are accompanied by vibrational and rotational transitions, the electronic spectra of molecules is highly complex. Even then, this spectra is studied in preference to other types because these spectra occur in visible and ultra violet region and can be examined easily. A diagram showing the various kinds of electronic excitation that may occur in organic molecules is shown below. Of the six transitions outlined, only the two lowest energy ones are achieved by the energies available in the 200 to 800 nm spectrum. The origin of bands in the electronic spectra (ultraviolet and visible spectra) is due to transitions in the electronic energy of valence electrons. These valence electrons exist in atomic orbitals in atoms. When a molecule is formed, then valence electrons taking part in bond formation exist in *bonding orbitals*. Electrons which form bonds are called σ-electrons and those which from π-bonds are called π-electrons. Some unshared electrons in a molecule do not take part in any formation and exists in *non-bonding orbitals*. Such electrons are called *n*-electrons.

Apart from these, there are higher energy levels termed as antibonding orbitals by molecular orbital theory. The antibonding orbitals associated with σ and π-bonds are called σ* (sigma star) and π* (pi star) orbitals. As *n*-electrons do not take part in bonding orbital, no anti-bonding orbital is associated with them. Antibonding orbitals are vacant in the ground state of molecule, but they are filled in excited states.

A schematic arrangement showing the energy levels of all these orbitals is shown in Fig. 5.7.

(*i*) *Transition between bonding orbitals and antibonding orbitals,* e.g. σ → σ* or π → π*. This excited state closely resembles the polar character.

(*a*) **σ → σ* transition:** Compounds in which all the valence electrons form π bonds absorb in for UV region and show this transition.

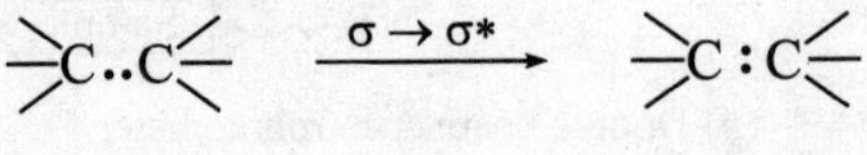

σ → σ* transition requires highest energy. In u.n, this energy is not of much importance as oxygen (O_2) starts aborping radiation.

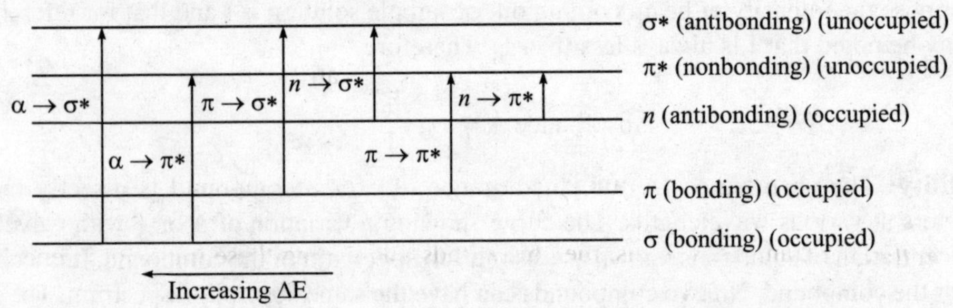

Fig. 5.7 Relative energies required for various types of electronic transitions

(*b*) **π → π* transition:** Molecules having unsaturation show this kind of transition and absorb in ultraviolet or visible region.

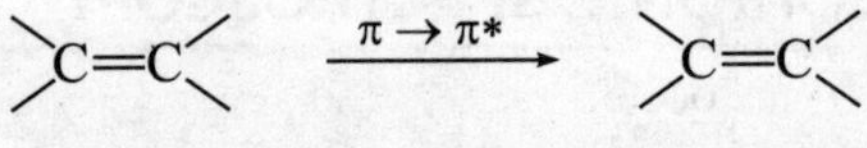

(*ii*) *Promotion of non-bonding electron* (*i.e.*, unshared pair, *n*) into antibonding sigma or antibonding π, *i.e. n* → π*. These involve transitions of non-bonding lone pair of hetro atom (*e.g.* O,N) to an empty anti-bonding molecular orbital. The absorption bands for such transitions occur in longer wavelength with low intensity.

(*iii*) *Transition between bonding (occupied)* σ *and* π *orbitals to antibonding (unoccupied)* σ* and π* *orbitals i.e.,* σ → π* and π → σ* can also take place. According to selection rules these radiations are forbidden and therefore have weak intensities.

All the above six type of electronic transitions have been summed up Fig. 5.7. The relative energies required for these transitions have also been shown. The energy required is maximum for σ → σ* transition and minimum for *n* → π* transitions. The energy requirements in the decreasing order of these transitions is given below:

$$\sigma \rightarrow \sigma^* > \sigma \rightarrow \pi^* > \pi \rightarrow \sigma^* > n \rightarrow \sigma^* > \pi \rightarrow \pi^* > n \rightarrow \pi^*$$

5.9 FORMATION OF ABSORPTION BANDS—THEIR DESIGNATION AND APPEARANCE (k-BAND, B-BAND, R-BAND AND E-BAND)

When a molecule absorbs in ultraviolet and visible regions, then the transitions that occur in electronic energy levels are

(*i*) $\sigma \rightarrow \sigma^*$ (*ii*) $n \rightarrow \sigma^*$

(*iii*) $\pi \rightarrow \pi^*$ (*iv*) $n \rightarrow \pi^*$

The electronic transition which are commonly observed and are used for identification of compound are characteristic of the structure of the molecule and are designated by special symbols.

It is important to note that transitions between electronic energy levels are also accompanied by both vibrational and rotational level transitions. Hence, when spectra is recorded in gaseous sample, a fine structure consisting of closely lines appear, whereas when sample is taken in solution, then due to solvent—solute interaction, a broad electronic spectrum is observed.

(*i*) $\sigma \rightarrow \sigma^*$ transitions: Should be mentioned.

(*ii*) $n \rightarrow \sigma^*$ Transitions: Those saturated compounds which contain singly bonded groups with atoms having non-bonding electrons undergo $n \rightarrow \sigma^*$ transitions. The atoms containing non-bonding electrons are generally O, N, S and halogens. As $n \rightarrow \sigma^*$ transitions require lesser energy than $\sigma \rightarrow \sigma^*$ transitions (Fig. 5.7), the absorption takes places at comparatively higher wave lengths in ultraviolet region. For example, such transitions occur in trimethylamine, methyl iodide and methyl alcohol and the corresponding λ_{max} values are 227 nm, 258 nm, and 183 nm respectively.

(*iii*) $\pi \rightarrow \pi^*$ Transitions (K-band): From Fig. 5.7 $\pi \rightarrow \pi^*$ transitions are of intermediate energies, *i.e.*, the energy required for these transitions is lesser than required in $n \rightarrow \sigma^*$ transitions, but more than required in $n \rightarrow \pi^*$ transitions. Thus, the compounds undergoing these transitions generally absorb in ultraviolet or visible regions known as *k*–Band (German-Konjugierte). In compounds containing conjugated double bond, the $\pi \rightarrow \pi^*$ transitions takes play with high molar absorptivity. For example, $\pi \rightarrow \pi^*$ transitions cause absorption at, λ_{max} = 170 to 190 nm and 180 nm in aldehydes, ketone, and ethylene as shown in Fig. 5.7. Further, $\pi \rightarrow \pi^*$ transitions in organic compounds may produce the following two types of bands.

B (Benzenoid) and E (Ethylenic) bands: These bands are characteristic of spectra of aromatic or hetero aromatic molecules. The Benzenoid system show **E and B bands** representing $\pi \rightarrow \pi^*$. For example E_1 and E_2 bands in benzene occur near 180 and 200 nm respectively, and their molar absorptivity varies between 2000 and 14,000.

(*iv*) $n \rightarrow \pi^*$ Transitions: R–Band. In a molecule if an atom has *n* electrons and also an unsaturated linkage with another atom, then $n \rightarrow \pi^*$ transitions take place. These transitions require least energy (Fig. 5.7), hence the bands are observed towards longer wavelengths, *i.e.*, in near ultraviolet or visible regions. These bands are called **R-bands.** Many chromophores like >C=O, —N—N—, >C=S exhibit $n \rightarrow \pi^*$ transitions, e.g., acetone and nitrosobutane absorbing at 277 nm and 665 nm respectively produce R-bands. These bands are characterized by low ε_{max} values usually less than 100. When additional bands like *k* and B bands appear in a compound. R band is shifted to longer wavelengths.

5.10 SELECTION RULES

We know that atomic spectra as well as molecular spectra are obtained due to transitions taking place between energy levels. To make the spectra study simpler and easy interpretation of the data obtained from it, certain restrictions are imposed on quantum numbers during transitions, which are known as '**Selection Rules**'. If a spectral transition is as per the selection rule, it is called as allowed transition.

If these rules are not followed, the transitions cannot take place and it is called a **forbidden** *transition. Example of forbidden transitions are from s to s orbital and from p to p orbital. For these transitions* $\Delta l = 0$*. In general, allowed transition are intense strong where as* **forbidden transitions** *are weak.*

The selection rules to be followed depend upon the type of transition (in case of molecular spectra).

The selection rules are generally expressed in terms of changes in quantum number for the allowed transitions. For example:

(*i*) In general $\Delta l = \pm 1$, where l is called azimuthal quantum number. Transitions from *s* to *p* orbital and form *p* to *d* orbital are allowed transitions, because $\Delta l = \pm 1$.

(*ii*) For the pure rotational transition, the selection rule is $\Delta J = \pm 1$, where J represents rotational quantum number, ($\Delta J = +1$ corresponds to absorption and $\Delta J = -1$ corresponds to emission).

(*iii*) For pure vibrational transition, the selection rule is $\Delta v = \pm 1$, where *v* represents the vibrational quantum number.

5.11 LINE WIDTHS AND INTENSITIES OF THE SPECTRAL LINES

For convenient analysis of various spectra, it is always desirable to get sharp and intense signals but generally broad signals are recorded in the spectrum of a molecule.

In any type of molecular spectroscopy, the two important characteristic of any spectral lines are:

(*i*) How sharp (or broad) is the line? This is decided by the widths of the line.

(*ii*) How strong (or weak) is the line? This is decided by the intensity of the line.

5.11.1 Width of the Spectral Line

If a spectral line is sharp, it will have no width as shown in Fig. 5.8 (*a*). On the other hand, if the line is not sharp, it will have a certain width as shown in Fig. 5.8 (*b*) *i.e.,* it spreads over a range of frequencies (or wave numbers).

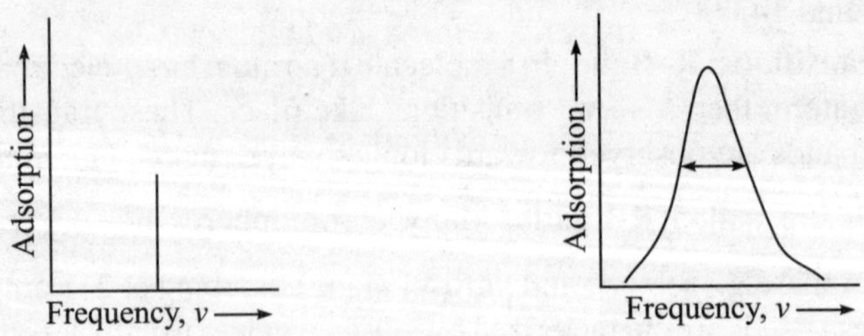

(*a*) A sharp spectral line having no width (*b*) A spectral line having a width

Fig. 5.8 Width of spectral lines

Factors Leading to Broadening of signals are: The various factors which lead to broadening of signals are discussed below.

(*i*) Natural Line Width Broadening: The natural line width of a spectral line is determined by the Heisenberg uncertainty principle.

$$\Delta E \times \Delta t \geq \frac{h}{4\pi} \qquad \ldots(5.20)$$

Where ΔE is the uncertainty in the energy and Δt is the uncertainty in the lifetime of the energy level.

We also know that for a photon, $E = h\nu$ or $\Delta E = h\Delta\nu$, where h is Plank's constant and $\Delta\nu$ is the change in frequency *i.e.,* line width.

From these two expression, the line width ($\Delta\nu$) is given as:

$$\Delta\nu \geq (4\pi\Delta t)^{-1} \quad \text{or} \quad \frac{1}{4\pi\Delta t} \qquad \ldots(5.21)$$

where $\Delta\nu$ is the width line.

We know that $\nu = \frac{c}{\lambda} = c\bar{\nu}$, where $\bar{\nu}$ *i.e.,* wave number is equal to $\frac{1}{\lambda}$.

In terms of wave number, equation (5.21) may be written as.

$$hc\Delta\bar{\nu} \times \Delta t = \frac{h}{4\pi}$$

$$\Delta\bar{\nu} = \frac{h}{4\pi\Delta t\, hc} = \frac{1}{4\pi\Delta t c}$$

Putting the value of $\pi = 3.14$ and $c = 3 \times 10^{10}$ cms^{-1}, we get

$$\Delta\bar{\nu} = \frac{1}{4 \times 3.14 \times 3 \times 10^{10} \times \Delta t}$$

$$= \frac{5.3 \times 10^{-12}}{\Delta t} \text{ sec cm}^{-1}$$

$$= \frac{5.3}{\Delta t} \text{ p.s cm}^{-1} \qquad (\because 10^{-12} = 1 \text{ pico sec } i.e., \text{ ps})$$

The broadening of signals also depends on the lifetime of excited state. A sharp spectral line is only observed if the lifetime of state involved in transitions is high.

***(ii)* Collision broadening:** The collision broadening is responsible for broadening of spectral line while recording ultraviolet and visible spectra. When UV or visible light is thrown on a molecule, the transitions take place between outer shells of the molecule. When molecules in the liquid or gaseous phase collide with one-another, they deform the electron clouds of each other. When transition takes place between such perturbed energy levels, a broad signal is recorded. *Thus broad signals in UV or visible spectroscopy are due to collisions between the perturbed energy levels of the colliding molecules.*

***(iii)* The Doppler broadening:** This type of broadening is generally observed in the spectra when recorded for gaseous molecules. As the molecules of a gas are in random motion when we observe the relative velocity of gas molecules with respect to instrument, then Doppler's broadening is observed. It can be minimized by working at low temperatures.

If a molecule is moving with velocity u towards the measuring instruments, then the frequency ν' of radiation observed by the instrument is given by

$$v' = v\left(1 + \frac{u}{c}\right) \qquad \text{...(5.22)}$$

where v is the radiation frequency and c is the velocity of light, u is velocity of molecule. On rearranging, we get.

$$\frac{v - v'}{v} = \frac{\Delta v}{v} = \frac{-u}{c}$$

where Δv is the Doppler's broadening.

Similarly, if the molecule is moving away from the measuring instruments, then the frequency of the radiation observed by the instrument is given by:

$$v' = v\left(1 - \frac{u}{c}\right) \qquad \text{...(5.23)}$$

On rearranging, we get

$$\frac{v - v'}{v} = \frac{\Delta v}{v} = \frac{u}{c}$$

Thus the observed frequency of radiation becomes higher or lower depending upon the direction of motion of molecule. The observed frequency is higher than the actual frequency, if the molecule is moving towards the instrument and lower if it moves away from the instrument.

5.11.2 Intensity of the Spectral Line

The strength or intensity of a spectral line depends upon

(i) **Boltzmann Population of states:** *The intensity of the spectral line is determined by the Boltzmann population of the energy level.*

According to Boltzmann, if N_0 is the number of molecules in the ground state, then the number of molecules in the excited state (N) at a temperature (T) is given by

$$N = N_0 e^{-\Delta E/kT} \qquad \text{...(5.24)}$$

where ΔE is the energy difference between the ground and excited states and k is the Boltzmann constant. The relative population at equilibrium is, thus given by

$$N/N_0 = e^{-\Delta E/kT} \qquad \text{...(5.25)}$$

Evidently, if ΔE is large N/N_0 is small, *i.e.,* the number of molecules in the excited is less than that the number of molecules in ground state would always be greater than any other excited state. Therefore, the transition from ground state to first excited state will be stronger than any transition from first excited state to any other state.

It may be noted that at room temperature, the molecules are in the ground state. So the spectral lines originating from the ground state *i.e.,* E_0 state to any higher state (say second state *i.e.,* E_2) would be intense than these originating from first excited state (E_1) to second excited state (E_2).

***(ii)* The probability that a transition will take place between the energy levels is called transition probability:** The term transition probability means the chances of a transition between two given energy levels. We know that absorption of a photon by a molecule will lead to transition but the molecule would not make a transition of its choice.

The probability of transition is decided by the "Selection rules". If the transition is allowed, a strong (or intense) spectral line will be obtained. If it is forbidden, no spectral line (or a weak spectral line) will be obtained. The details have already been discussed in Article 5.5.

***(iii)* The strength of the incident radiation:** Similarly, greater is the intensity of the incident radiation, greater is the rate at which the transitions are induced and hence stronger is the absorption by the sample.

5.12 CHROMOPHORES

In Greek chromophores means **colour carrier**. It has been recognized since long that the colour of a molecule is due to one or more unsaturated groups present in it. These groups are called chromophores. They exhibit characteristic absorption in ultraviolet and visible regions. In simple words, chromophores are the functional groups which absorb electromagnetic radiations and may or may not impart colour to compound. Typical examples of chromophores are $>C=C<$, $-C\equiv N$, $>C=O$, $-N=N-$, $>C=S$, and $-N=O$ in the order of increasing power of imparting colour. Simple chromophores undergo $\pi \rightarrow \pi^*$ transitions in the short wavelength region of ultraviolet radiations. Some common chromophores along with the type of excitation and λ_{max} value have been summarized in the table below:

Chromophore	*Example*	*Excitation*	*λ_{max} in nm*	*Molar Absorptivity*	*Solvent*
C = C	Ethene	$\pi \rightarrow \pi^*$	171	15,000	hexane
C ≡ C	1-Hexyne	$\pi \rightarrow \pi^*$	180	10,000	hexane
C = O	Carbonyl	$n \rightarrow \pi^*$	290	15	hexane
		$\pi \rightarrow \pi^*$	180	10,000	hexane
N = O	Nitromethane	$n \rightarrow \pi^*$	275	17	ethanol
		$\pi \rightarrow \pi^*$	200	5,000	ethanol
C – X; X = Br	Methyl bromide	$n \rightarrow \sigma^*$	205	200	hexane
X = 1	Methyl Iodide	$n \rightarrow \sigma^*$	255	360	hexane

From the chart above, it should be clear that the only molecular species likely to absorb light in the 200 to 800 nm region are pi-electron functions and hetero atoms having non-bonding valence-shell electron pairs. Such light absorbing groups are referred to as **chromophores**. The oxygen non-bonding electrons in alcohols and ethers do not give rise to absorption above 160 nm. Consequently, pure alcohol and other solvents may be used for spectroscopic studies.

The presence of chromophores in a molecule is best documented by UV-visible spectroscopy, Fortunately, conjugation generally moves the absorption maxima to longer wavelength, as in the case of isoprene, so conjugation becomes the major structural feature identified by this technique. Molar absorptivities (ε) may be very large for strongly absorbing chromophores (> 10,000) and very small if absorption is weak (10 to 100). The magnitude of ε reflects both the size of the chromophore and the probability that light of a given wavelength will be absorbed when it strikes the chromophore.

5.13 AUXOCHROMES

Those groups which do not impart colour to a compound by themselves, but increase the colouring of a chromophore when attached to them are called **auxochromes**. These groups show characteristic absorption below 2000 Å by themselves. When they are attached to a chromophore, they usually cause a shift in the absorption towards longer wavelength. Auxochromes also increase the intensity of absorption. Typical example of auxochromes are —OH, —NH_2, —NR_2, —OR, —CH_3 and —Br. All auxochromes contain unshared pair of non-bonding electrons. The action of an auxochrome is due to the electronic transitions involving these non-bonding electrons, when conjugated with chromophores *e.g.*, Nitrobenzene has only a tinge of colour but when —NH_2 group is attached at para position then conjugation leads to bathochromic shift *i.e.*, towards longer wavelength and darkening of colour occur.

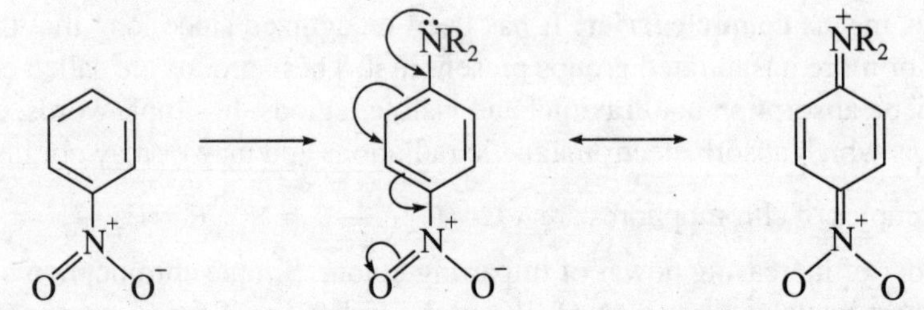

Nitrobenzene   *p*-amino nitrobenzene   Darkening of colour occur due to quinoid structure

5.14 FACTORS AFFECTING λ_{max} INTENSITY OF SPECTRAL LINES

The effect of substituents and solvent on $\pi \rightarrow \pi^*$ transitions have been extensively studied. A most suitable solvent is one which does not itself absorb in the region under investigation. Hexane and other hydrocarbons can be used as these are less polar and which have least interactions with the molecule under investigation. **For ultra-violet spectroscopy, ethanol, water and cyclohexane serve the best purpose.**

The position and the intensity of absorption maximum is shifted for a particular chromophore by changing the polarity of the solvent. In general, the absorption maximum for the non-polar compounds is the same in alcohol (polar) as well as in hexane (non-polar) solvent. The absorption maximum for the polar compounds changes with change in polarity of the solvent. The various effects are as follows:

1. Bathochromic Shift: *It is an effect by virtue of which the absorption maximum is shifted towards longer wavelength due to the presence of an auxochrome or by change of solvent.* If an electron donor group like OH, CH_3, NH_2 etc., is substituted in a compound, then the absorption band is shifted to longer wavelengths. Such shifts of absorption bands towards longer wavelengths due to substitution or solvent effect are known as bathochromic or **red shifts.** For example, the λ_{max} for C_6H_6 and $C_6H_5CH_3$ are 254 mm and 261 mm. Thus the substitution of CH_3 group in benzene causes bathochromic shift in the absorption band. Conjugation of double bonds also cause bathochromic shift.

For example: Ethylene ($H_2C = CH_2$) absorbs at λ_{max} = 175 nm in comparison to 1-butene (H_2C = CH—CH—CH_3) at λ_{max} = 185 nm or isobutene $H_2C = C\ (CH_3)$ at λ_{max} = 188 nm. The bathochromic shift is more pronounced as the number of alkyl group increases.

2. Hypsochromic shift: *It is an effect by virtue of which the absorption maximum is shifted towards shorter wavelength.* If an electron acceptor group like acetyl group is substituted in a compound, then the absorption band is shifted to shorter wavelength. Such shifts of absorption band towards shorter wavelength due to substitution of solvent effect are known as *hypsochromic or* ***blue shifts***. If CH_3 group is introduced in an unsaturated chain of an organic compound, then conjugation is destroyed and absorption band of the compound is shifted to shorter wavelengths. Hydrogen bonding also produces Hypso–chromic shifts as more energy is required for breaking Hydrogen bond hence less wavelength.

For example:

Compound	λ_{max}*(nm)*	*Compound*	λ_{max}*(nm)*	*Compound*	λ_{max}*(nm)*
H—C(=O)—H	290	H—C(=O)—Cl	235	H—C(=O)—NH_2	205

Even change in solvent can also bring the Hypsochromic shift. For example, CH_3—C(=O)—CH_3 (acetone) absorbs at λ_{max} = 279 nm in hexane and at λ_{max} = 264.5 nm when dissolved in water.

3. Hyperchromic Effect: *It is an effect due to which the intensity of absorption maximum increases i.e., ε_{max} increases.* If a substituent group causes increase in the intensity of band, then the effect is called *hyperchromic effect*. The increase in the intensity of absorption band is due to the fact that the substituent group causes the increase in the molar extinction coefficient ε of the compound.

For example: Benzene shows absorption at λ_{max} = 254 with ε_{max} – 200 whereas toluene shows absorption at λ_{max} = 261 with λ_{max} – 300, showing the hyperchromic effect.

4. Hypochromic Effect: If a substituent group causes the intensity of the band to decrease, then the effect is called hypochromic effect. As an example, ε_{max} for C_6H_6 and C_6H_5Cl are 200 and 170 respectively. Thus, substitution of Cl causes hypochromic effect.

5.15 FRANCK-CONDON PRINCIPLE

The transition between two electronic states take place according to this principle. The vibrational structure of electronic spectra can be investigated with the help of well known **Franck-Condon principle.**

It states that an electronic transition takes place so rapidly that a vibrating molecule does not change its inter-nuclear distance to any appreciable extent during the transition because the electrons move much faster than the nuclei hence the nuclei do not change their positions and velocities.

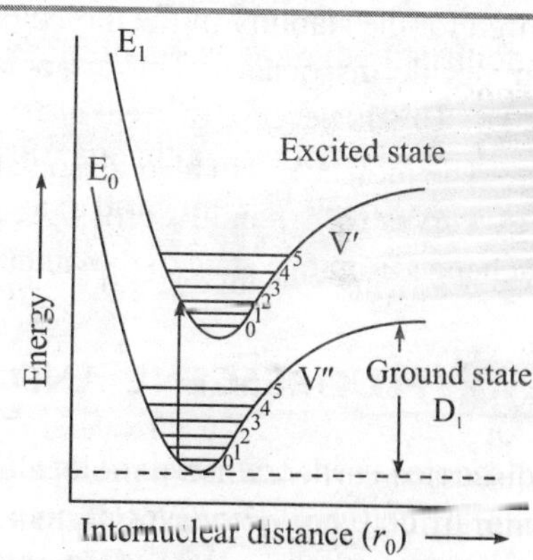

Fig. 5.9 Electronic transition in diatomic molecule

This principle for the electronic transition of a diatomic molecule has been explained in Fig. 5.9.

The potential energy curve (Fig. 5.9) indicate that at equilibrium atoms of the diatomic molecule do vibrate *i.e.*, their internuclear distance keep on changing.

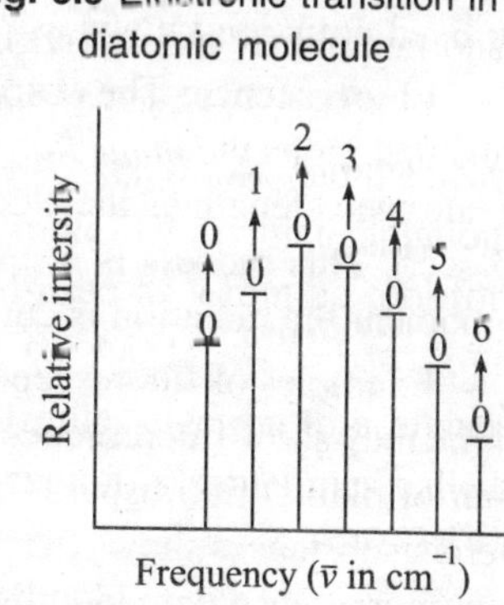

Fig. 5.10 Relative probability of electronic transitions

Two potential energy curves for the molecule in the ground state represented by E_0 and in the first excited electronic state represented by E_1 are shown. The minimum in the potential energy curve E_1 occurs at slightly greater internuclear distance than the corresponding minimum in the ground state curve E_0. *This is because the bonding in the excited state is weaker than in the ground state.* The electronic transition is represented by a vertical line drawn between the energy levels of the two curves. From quantum mechanics it is known that the molecule is in the center of the ground vibrational level of the ground electronic state. *When a photon falls on the molecule, the most probable electronic transition, according to Franck Condon principle, takes place from v″ = 0 to v′ = 2.* The probability of other vibrational level transitions is low. As a result the relative intensities of these transitions are smaller than the intensity of the 0-2 transitions as shown in Fig. 5.11.

5.15.1 Bond Dissociation Energy

Let us consider a diatomic molecule which has a potential energy curve E_0 in the ground state and two potential energy curves E_1 and E_2 in the excited states. The curve E_2 does not have a minimum (Fig. 5.11) and the state is unstable for all inter nuclear separations.

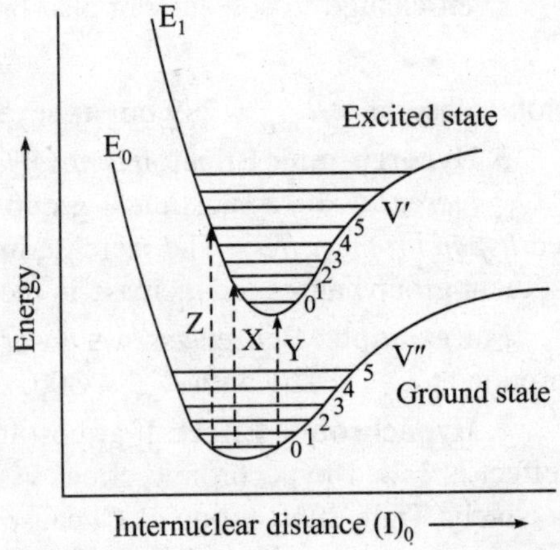

Fig. 5.11 Potential energy curves during electronic transitions in a diatomic molecule

Since the bond is weaker in the excited state, the equilibrium internuclear distance is longer in E_1 than in E_0. According to Franck-Condon principle, the transitions represented by arrow X *i.e.* (0 → 2) have maximum intensity. Two other possible transitions are represented by arrows Y and Z. The transition Y ends in the lowest vibrational level of the first excited electronic state, where the molecule is still held firmly. However, the transition Z promotes the molecule to a point where its energy is much higher than the potential energy required for the stability of the molecule. Consequently the molecule dissociates. The dissociation energy D_2 of curve II is lesser than that of curve I *i.e.*, $D_2 < D_1$. This is because of weaker bonding in excited state than in the ground state.

The molecule will also dissociate if it is excited to a level above the point C where the potential energy curves for E_1 and E_2 cross. At an internuclear distance corresponding to point C, the molecule may change from one state to the other without energy change and continue to dissociate into two parts.

5.16 FLUORESCENE AND PHOSPHORESCENCE

As discussed earlier, when a molecule is exicted by light, it can lead to a chemical reaction or energy transfer by collision. However, if both these processes do not occur then the molecule will return to the ground state with the release of energy which is either a short lived emission known as fluorescence or long lived emission known as phosphorescence.

Fluorescence: The absorbed photo-energy sends electron from lower to higher energy level and the excited molecule *immediately or instantaneously* re-radiates or emits part of the absorbed energy at a greater wavelength as the electron returns to or reverts back to its normal state within a short time (= 10^{-8} sec.). This process is instantaneous and starts immediately after the absorption of light and stops the moment the radiation is cut off and is called "fluorescence".

Examples of fluorescence are solutions of fluorescein and eosin. When their solution are placed in light, they show fluorescence from green to violet colour; uranyl sulphate, chloropyll (green colouring matter of plants) fluorospar (CaF_2) and petroleum also show this phenomenon. It may be pointed that different substances shows the phenomenon of fluorescence with different wavelengths. Thus, fluorospar show fluorescence with blue light, chlorophyll with red light, uranium glass with green light and so on.

The phenomenon of fluorescence finds wide application such as:

(*i*) For lighting purposes in fluorescent tubes. Mercury arc producing large proportion of ultra violet light is used in a tube coated with fluorescent salts which gives visible light.

(*ii*) By mixing fluorescent dyes with coloured paints, the fluorescence of the dye helps the light reflected by the paint to produce extra-ordinary brightness and luster. These materials are used in road signs. For example, Brilliant Sulpho Flavine FF and Rhodamine 6G dispersed in a special kind of plastic are used for this purpose.

(*iii*) In industry for testing and identifying materials, e.g. in rubber industry.

(*iv*) In analysis—concentration of riboflavin (Vitamin B_2) in chloroform has been examined.

(*v*) In television—the cathode stream of the photo-electric effect is made visible in the cathode ray tube by adding ZnS to which a little Ni is added to cut off phosphorescence which otherwise makes the picture blurred.

(*vi*) Use of fluorescent microscopes and, fluoroscope used in X-ray diagnosis help in testing the condition of food stuff and detecting ring worms etc.

(*vii*) A fluorescent dye is used as whitener for washing cloth when mixed with washing powder.

Phosphorescence: In this process the absorbed light energy raises the electron to a higher and the excited electron revert back to the original state after some time lag and consequently absorbed light is emitted after some time. *Thus, phosphorescence may be regarded as slow fluorescence.* In this phenomenon, the emission of light of different wavelength continues even after the source of radiation has been cut off.

The familiar examples of phosphorescence are BaS or SrS containing about 2.5% alkali chlorides (NaCl or KC1) and a trace of heavy metal sulphide. Such a mixture is generally used for painting watch dials, electric switches etc.

Many dyestuffs which are fluorescent in aqueous solution become phosphorescent when dissolved in glycerol and cooled.

5.16.1 Mechanism of Fluorescence and Phosphorescence

In **Fluorescence** light absorbed by molecule is emitted almost instantaneously with in 10 seconds in one or more steps. The emission of light ceases with the removal of source of light.

The absorption of radiation takes the molecule to singlet excited state. As the excited molecule collides with the surrounding molecules, it gives up energy and steps down the ladder of vibrational levels of singlet excited state (radiationless decay). As the transition occurs from lowest vibrational energy level of singlet excited state to singlet ground state, radiation is emitted and fluorescence is observed.

In **Phosphorescence** light absorbed by molecule is given out slowly and even long after the removal of source of light.

The absorption of radiation takes the molecule to singlet excited state. **As** excited molecules collides with the surrounding molecule it gives up energy and steps down the ladder of vibrational levels of singlet excited state. The initial steps are common in fluorescence and phosphoresce, till the vibrational energy levels of triplet excited state. At this stage, singlet and triplet excited states share a common point. At this point intersystem crossing occurs *i.e.*, switch from singlet to triplet state. *As the return from a triplet state to singlet state is spin forbidden, so triplet state act as a slowly radiating reservoir.*

5.17 JABLONSKI'S DIAGRAM

Fluorescence and **Phosphorescence** can be more conveniently shown with the help of **Jablonski's diagram.** In order to understand **Jablonski diagram**, we should know the term spin multiplicity.

Spin multiplicity: Spin multiplicity of a state is given by (2S + 1) where S is the total electron spin.

If S = 0, then spin multiplicity = 1, Such an electronic state is called **Singlet state**.

If S = 1, then spin multiplicity = 3, Such an electronic state is called **Triplet state.**

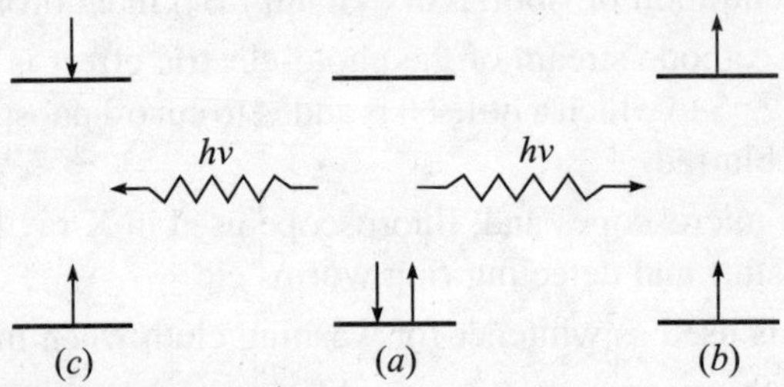

Fig. 5.11 (a) when two electrons are present in lower energy level and their spins are anti parallel (b) One electron is promoted to upper energy level and spins of both electrons are parallel (c) One electron is promoted to upper energy level and spins of both electrons are antiparallel.

Consider an energy level, containing two electrons as shown in Fig. 5.11 (*a*). If a radiation of suitable frequency falls on it, one of the paired electrons is promoted to upper energy level. If the spin of two electrons are parallel Fig. 5.10 (*b*), then total spin $S = S_1 + S_2 = ½ + ½ = 1$

⇒ **spin multiplicity** = (2S + 1) = 3

As **spin multiplicity** is 3, so the molecule is in **Triplet excited state.**

If the spins of two electrons are antiparallel, then total spin $S = S_1 + S_2 = ½ - ½ = 0$

⇒ **spin multiplicity** = (2S + 1) = 1

As **spin multiplicity** is 1, so the molecule is in **Singlet excited state.**

Since the electron can jump to any of the higher electronic states depending upon the energy of photon absorbed, we get a series of Singlet excited states represented as S_1, S_2, S_3 S_4... and are known as first singlet excited state, second singlet excited state, third singlet excited states, and so on. Similarly T_1, T_2, T_3 are known as first triplet excited state, second triplet excited state, third triplet excited state and so on.

For each singlet excited state there is corresponding triplet excited state and as the energy of singlet excited state is higher than the triplet state. Accordingly the energy sequence may be written as

$$E_{S_1} > E_{T_1} ; E_{S_2} > E_{T_2} ; E_{S_3} > E_{T_3} ;$$

Fluorescence and phosphorescence are shown in Jablonski diagram in Fig. 5.12

The transitions which involve the return of excited states (S_3, S_2 or T_3, T_2) to the first excited state (S_1 or T_1) do not involve the emission of any radiations and are thus referred to as Non-radiative or radiationless transitions. When the energy of activated molecule is dissipated in the form of heat through molecular collisions, it is called **Internal** conversion (IC) it occurs in less than 10^{-11} second.

The molecule may lose energy by intersystem crossing (ISC). ISC involves the transition from singlet state to corresponding triplet state.

For example from S_2 to T_2 or S_1 to T_1. ISC like IC is also a non-radiative or radiationless transition. ISC is a spin forbidden process and hence occurs at slow rate.

The transition from S_1 to S_0 state is an allowed transition and occurs in 10^{-8} second. The emission of radiation in this transition is called Fluorescence. The transition from the triplet state T_1 to the ground state S_0 is slow and is spin forbidden. The emission of radiation in this transition is called phosphorescence. The life time of phosphorescence is larger and varies from 10^{-3} second to 100 seconds.

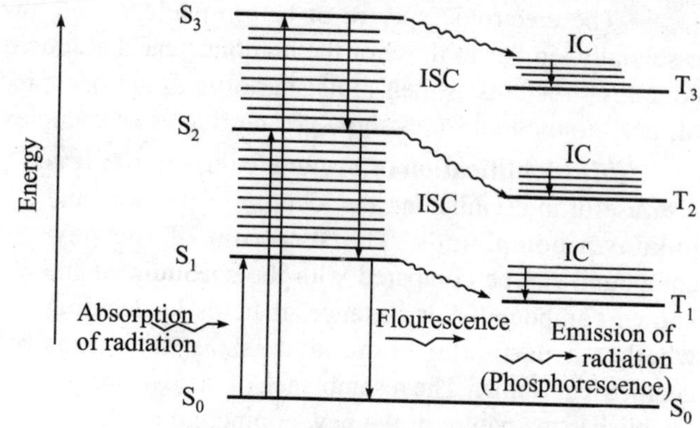

Fig. 5.12 Jablonski diagram showing Fluorescence and Phosphorescence

Both fluorescence and phosphorescence emit radiations of shorter frequencies or large wavelength than the incident radiation or exciting radiation because some part of radiation absorbed by the molecule is dissipated in the form of heat during the non-radiative transitions.

Difference between Fluorescence and Phosphorescence

1. Phosphorescence has much longer decay period (10^{-4} to 100s) than fluorescence (10^{-6} to 10^{-9} s).
2. Phosphorescence spectrum is not the mirror image of the absorption spectrum where as fluorescence spectrum is the mirror image of the absorption spectrum.
3. Phosphorescence is the radiation emitted in a transition between states of different multiplicity whereas fluorescence is the radiation emitted in a transition between states of same multiplicity.
4. Phosphorescence in not observed in solutions at room temperature whereas fluorescence can be observed in solutions at room temperature.
5. Phosphorescence is rarely observed in gases. Fluorescence is exhibited by some elements in the vapour state, *e.g.*, sodium, iodine and mercury vapours.

5.18 APPLICATIONS OF ULTRAVIOLET AND VISIBLE SPECTROSCOPY

There is extensive application of ultraviolet and visible spectroscopy in different fields of chemical interest. However, a few important applications are given here.

***(i)* Determination of dissociation energy:** The dissociation energy for a molecule can be determined with great accuracy from this spectrum. The dissociation energy is calculated from the wavelength that separates the discontinuous part of the spectrum from the continuous part of the spectrum resulting from the dissociation of the molecule into two parts in varying amount of kinetic energy.

***(ii)* Calculation of moment of inertia, vibrational frequency and interatomic distances of diatomic molecules:** This spectra also provide information regarding vibrational frequencies, moment of intertia, interatomic distances etc. In case of homo-nuclear diatomic molecules such as H_2 and N_2, the electronic transition takes place from the highest occupied molecular orbital (HOMO) to the lowest unoccupied molecule orbital (LUMO). This is in the short wavelength side of the ultraviolet (UV) region. The behaviour of O_2 molecule is, however, exceptional, which shows a complex spectra.

The electronic spectra of hetero-nuclear diatomic molecules such as CO, HCl, NO etc. are essentially similar to those of the homonuclear diatomic molecules, although they have comparatively complex structures as their molecular orbitals are occupied by various electrons. The electronic spectra of, polyatomic molecules show greater degree of complexity.

***(iii)* Identification of organic compounds:** It is also useful in establishing the identity of the new and unknown compounds. The spectrum of the new compound can be compared with the spectrum of the known compound. For instance, it helps in deciding whether a particular compound is cis or a trans geometrical isomer. The resemblance of the two specti a establishes the nature of the new compound.

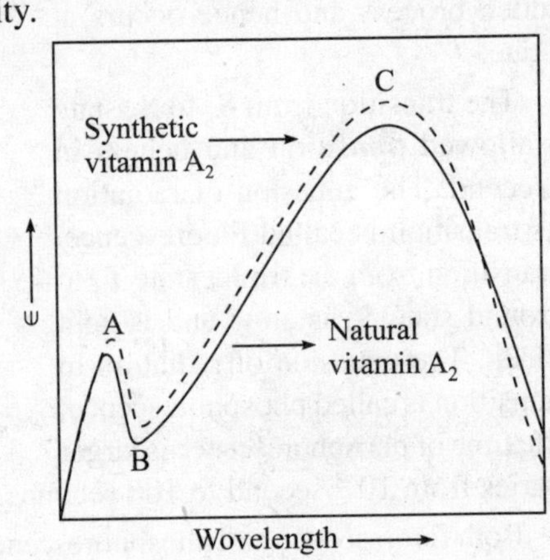

Fig. 5.13 Absorption spectrum of vitamin A_2 in Ultraviolet Region

The absorption spectrum of an unknown compound is compared with those of a number of known compounds. If the absorption spectrum of the unknown compound tallies with a particular known compound, then the structure of the two will also be similar. The absorption spectra of synthetic (unknown) and natural vitamin A_2 are shown in Fig. 5.13. The similarity between the two spectra proved helpful in the identification and assignment of structure of synthetic vitamin A_2. The technique in which the spectrum of an unknown compound is tallied with those of known compounds, is known as *finger printing*. The points A, B and C marked in the Fig. 5.12, where the value of λ_{max} is the same for two compounds are called isobestic points.

***(iv)* Presence of impurities:** It can be used to detect the presence of impurities in a substance. The spectrum of the compound is compared with the spectrum of pure compound. The impurities will certainly cause extra absorption bands, characteristic to their structures.

Ultraviolet and visible spectrophotometry is used for the control of purification of a compound. If a compound is transparent in ultraviolet and visible regions, then its purification is continued until it gives a minimum value of absorbance. On the other hand, if a compound absorbs in ultraviolet and visible regions, then its purification is continued until its molar extinction coefficient attains a constant maximum value.

***(v)* Determination of structure of organic compounds:** The fact that compounds having similar structures show similar absorption spectra can be utilized in deciding the constitutions of compounds where more than one possibility arise. For example, two possible structures for isatin have been suggested viz.

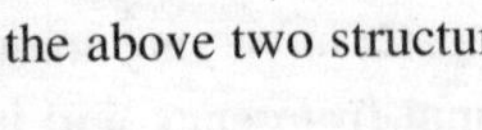

I

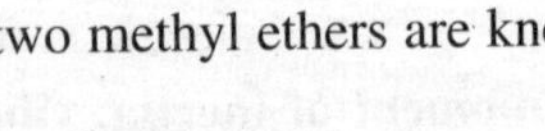

II

Corresponding to the above two structures, two methyl ethers are known

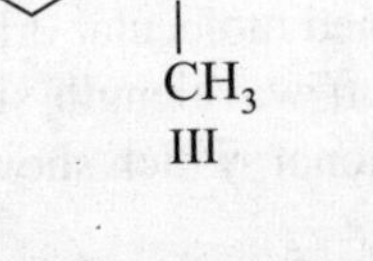

III

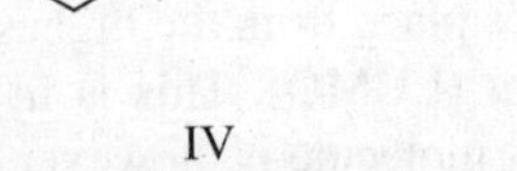

IV

Comparison of the spectrum reveals that the absorption spectrum of isatin is similar to that of N-methyl ether, structure III. Thus, structure I is assigned to isatin.

***(vi)* Confirmation of cis and trans hydrogen:** Determination of Geometrical isomer *i.e.* whether compound is cis or trans, is also established with the help of uv - visible spectroscopy, e.g.

Cis–Cinnamic acid :

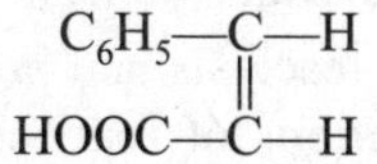

Trans–Cinnamic acid :

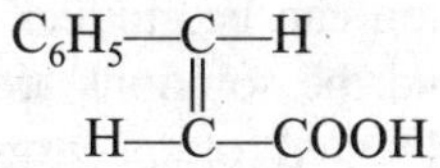

In case of cis isomer, due to steric repulsion, coplanarity is lost and thus conjugation diminishes. It absorbs at shorter λ *i.e.*, 268 nm as compared to trans which absorb at 272 nm.

***(vii)* Determination of extent of conjugation:** It is very useful technique to know the extent of multiple bonding in a molecule. It can also be used to distinguish between conjugated diene and conjugated triene, conjugated and non-conjugated molecule, α, β unsaturated ketones and β, γ unsaturated ketones (these differ in the position of carbon-carbon double bond). For example, the values of absorption measurement (λ_{max}) as well as extinction coefficient for conjugated and unconjugated alkenes can be compared from their spectra shown in Fig. 5.14. The values of λ_{max} and ε_{max} are more for extended conjugated moieties as compared to those for an unconjugated alkenes. A bathochromic as well as hyperchromic effect are observed when the spectrum of conjugated triene is compared to that of conjugated diene.

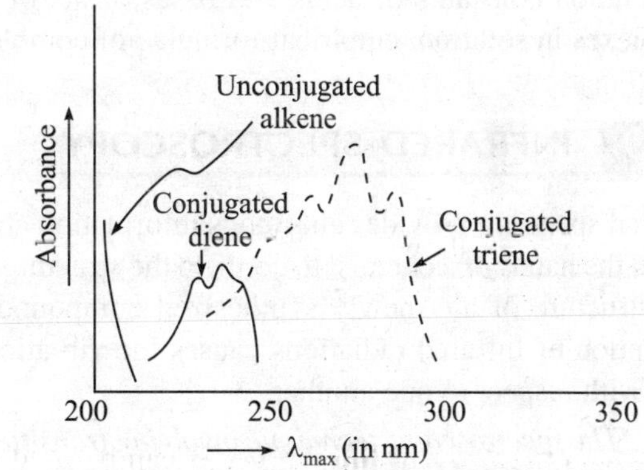

Fig. 5.14 Absorption in conjugated and uncongugated system

***(viii)* Quantitative Analysis:** Ultraviolet and visible spectrophotometric methods have been used to determine the unknown concentration of a compound in solution. The essential condition to develop such a method is that Beer's law, $A = \varepsilon c x$, must be obeyed in the range of concentration to be used. The absorbance A is measured directly from spectrophotometer at various concentrations *c* of the solution. A graph is then plotted between A and *c* as shown in Fig. 5.14. Absorbance A′ of the solution of unknown concentration is then determined by spectrophotometer and its concentration *c′* is then determined from the graph (Fig. 5.15). The extinction coefficient ε can also be evaluated from the slope of the graph.

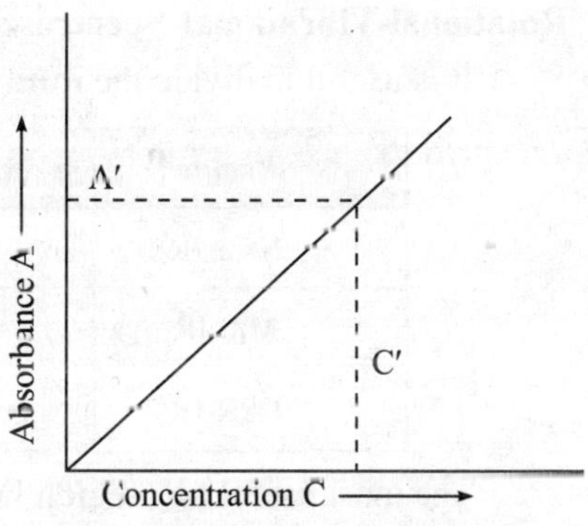

Fig. 5.15 Graph between Concentration and Absorbance

***(ix)* Determination of molecular weight:** Let M be the molecular weight of the compound. First, absorbance (A), *i.e.*, the absorbance of 1% solution of compound in a 1 cm thick cell, is determined. In this case, the molar concentration, c = 10/M mole per lit., cell thickness, x = 1 cm and absorbance = A. Putting these values in Beer's law equation Absorbance (A) = $\varepsilon c x$.

We get

$$A = \varepsilon\frac{10}{M} \quad \text{or} \quad M = \frac{10\varepsilon}{A}$$

Molecular weight M of the compound can be evaluated from the above relation.

***(x)* Study of Kinetics:** If the λ_{max} values of reactants and products of a chemical reaction are quite different, then the change in the concentrations of reactants or products can be followed spectrophotometrically and kinetics of the reaction can be studied. Spectrophotometric methods are particularly useful when the reactions are fast and the solutions are very dilute, e.g., photochemical transformation of ergosterol to vitamin D_2 is well suited for spectrophotometric method.

***(xi)* Other Physico-chemical Studies:** Apart from these, ultraviolet and visible spectrophotometry has been very useful in other physico-chemical studies, such as determination of dissociation constants of acids and bases, heats of formation of molecular addition compounds and complexes in solution, empirical formulae of complexes in solutions etc.

5.19 INFRARED SPECTROSCOPY

Infrared spectrum provides enormous information about the structure of a compound. It is an important tool in the hands of a chemist to confirm the structure of known as well as unknown organic compounds. The structure of any newly synthesized compound is established using this unique technique. The absorption of Infrared radiations causes the vibration of various bonds in the molecule to stretch and bend with respect to one another.

The spectroscopic technique involving transitions in the vibrational energy levels due to absorption of Infrared radiations by molecules is defined as IR Spectroscopy. lt is called vibrational spectroscopy because the transitions take place between the vibrational energy levels. It is called IR spectroscopy since the frequency of the radiations absorbed lie in the IR region of electromagnetic spectrum.

The Infrared radiations are capable to induce transitions between vibrational and rotational levels as each vibrational level is associated with rotational level of a molecule. Hence it is also called **"Rotational-Vibrational Spectroscopy".**

It is useful to divide the infra red region into three sections; *near, mid and far infrared*:

Region	*Wavelength range (mm)*	*(Wavenumber range (cm^{-1})*
Near IR	0.78 – 2.5	12800 – 4000
Mid IR	2.5 – 50	4000 – 200
FAR IR	50 – 1000	200 – 10

The most useful I.R. region lies between 4000 – 670 cm^{-1}. Photon energies associated with this part of the infrared (from 1 to 15 kcal/mole) are not large enough to excite electrons, but may induce vibrational excitation of covalently bonded atoms and groups. The covalent bonds in molecules are like stiff springs that can be stretched and bent. Hence, molecules experience a wide variety of vibrational motions, characteristic of their component atoms. Consequently, virtually all organic compounds will absorb infrared radiation that corresponds in energy to these vibrations.

Infrared spectra may be obtained from samples in all phases (liquid, solid and gaseous). Liquids are usually examined as a thin film sandwiched between two polished salt plates of NaCl which is transparent. If solvents are used to dissolve solids, care must be taken to avoid abscuring important spectral regions by solvent absorption. Alternatively, solids may either be incorporated in a thin KBr disk, prepared under high pressure, or mixed with a little non-volatile liquid and ground to a paste (or mull) that is smeared between salt plates.

5.20 THEORY OF INFRARED ABSORPTION

Absorption of IR is restricted to compounds with small energy differences in the possible vibrational and rotational states. For a molecule to absorb IR, the vibrations or rotations with in a molecule must cause a net change in the dipole moment of the molecule. The alternating electrical field of the radiation interacts with fluctuations in the dipole moment of the molecule. If the frequency of the radiation matches the vibrational frequency of the molecule then radiation will be absorbed, causing a change in the amplitude of molecular vibration.

Types of Molecules Exhibiting Infrared Spectra

This type of spectra is given by diatomic molecules with ***permanent dipole moment*** *and polyatomic molecules with and without dipole moment.*

For example, molecule like H_2, N_2, do not possess electric dipole, hence they do not give rise to IR absorption spectra. The polyatomic molecules having zero or no dipole moment of their own may be made to undergo *vibrational-rotational transitions* because exciting radiations can induce an oscillating dipole moment in the molecules.

When the molecule having dipole is kept in the electric field *i.e.,* exposed to Infrared radiations, this field exert force on the electric charges in the molecule and tends to increase or decrease the separation of charges. When these charged atoms vibrate, they absorb I.R. radiation from the radiation source. Depending upon the rate of vibration of charged atoms in the molecule, intensity of the bond varies.

5.21 FORCE CONSTANTS AND THEIR CALCULATIONS

The molecular vibrations are due to changes in the bond length or bond angle of a molecule. For diatomic molecules vibrations are simple oscillation of two atoms along the line of their centres around their mean position within the molecule resulting in compression or extension of the bond. To understand the vibrations in a diatomic molecule, bond is considered an elastic like spring. The **Vibrating Diatomic Molecule** is viewed as.

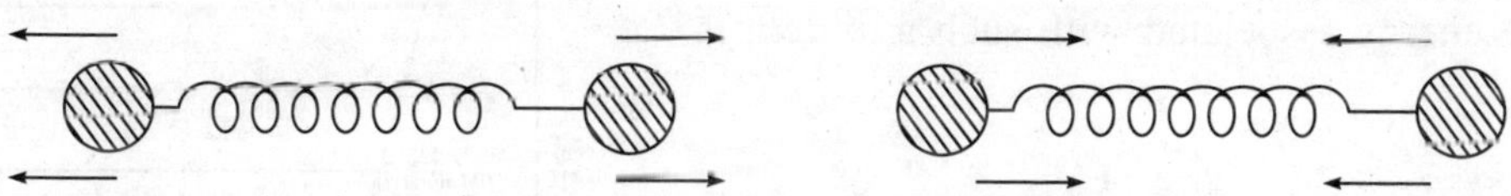

Fig. 5.16 Vibratory motion of a diatomic molecule

These vibrations are simple harmonic in nature and obey **Hooke's law.** According to this law *the restoring force (F) is directly proportional to the distance for displacement of atoms from their mean or equilibrium positions, but it acts in a direction opposite to that of the displacement of atoms.*

i.e.,

$$F = -kx \quad \text{....(5.26)}$$

where k is the force constant for the chemical bond in the molecule and F is the restoring force and x is the displacement of an atom from its equilibrium position in the molecules and is equal to the difference between internuclear distance (r) *i.e.*, the distance to which the atoms have been stretched at any time during oscillation and the internuclear distance between the atoms (r_e) *i.e.*,

$$x = r - r_e$$

If $$x = 1 \text{ cm}, k = -\text{F}$$

*The **force constant** may be defined as the restoring force per unit displacement of a harmonic oscillator. It is a measure of bond stiffness or bond strength.*

The fundamental vibrational frequency of a simple harmonic oscillator is given by

$$\nu = \frac{1}{2\pi}\left(\frac{k}{\mu}\right)^{\frac{1}{2}} s^{-1} \quad \text{or} \quad \nu = \frac{1}{2\pi c}\sqrt{\frac{k}{\mu}} \text{ cm}^{-1} \qquad \text{...(5.27)}$$

where μ, is the reduced mass of the system

We know $\mu = \dfrac{m_1 \times m_2}{m_1 + m_2}$ where m_1 and m_2 are the masses of oscillating atoms.

Force constant k for diatomic molecules can be calculated when frequency and the masses of the atoms of the diatomic molecules are known.

Force constant helps in predicting the multiplicity of the bonds. Force constant have been found to increase almost directly with the multiplicity of the bond as shown below:

Bond	**Force constant (dynes/cm)**
C – C	4.6×10^6
C = C	9.5×10^6
C ≡ C	15.8×10^6

C = C stretching is expected to absorb at higher frequency than C—C stretching. It is due to the higher bond strength (value, of k) of the double bond compared to the single bond. Similarly, O—H stretching absorbs at higher frequency compared to C—C bond. It can be explained on the basis of the smaller value of reduced mass for O—H compared to with C—C bond.

The behaviour of chemical bonds can be interpreted in terms of a fundamental frequency, ν_0 given by equation (5.28)

$$\nu_0 = \frac{1}{2\pi}\sqrt{\frac{k}{\mu}} \qquad \text{...(5.28)}$$

The energy associated with such a system is given by

$$E_{vib} = \left(v + \frac{1}{2}\right) h\nu_0 \qquad \text{...(5.29)}$$

For example for third vibrational level ($v = 3$) energy is calculated as

$$E = \left(v + \frac{1}{2}\right) h\nu_0$$

$$E = \left(3 + \frac{1}{2}\right) h\nu_0 = \frac{7}{2} h\nu_0$$

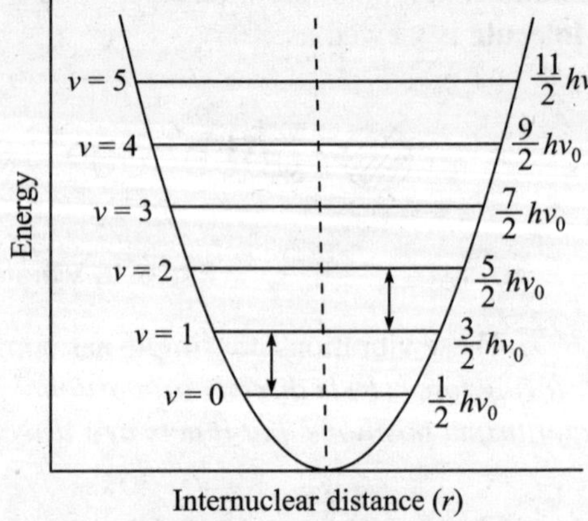

Fig. 5.17 Allowed Vibrational levels

Thus the allowed vibrational energies for a diatomic molecule, as given by equation (5.29) are shown schematically in Fig. 5.17, where energy is plotted against internuclear distance.

In terms of wave number

$$\bar{v} = \frac{v}{c} = \frac{1}{2\pi c}\sqrt{\frac{k}{\mu}}$$

To energy levels are seen to be equally spaced with a spacing hv_0.

Types of Transitions: Normally transitions occur corresponding to the selection rule $\Delta v = \pm 1$. The lines corresponding to the transitions form $v = 0 \rightarrow v = 1$ are very intense [because ground level is the most populated]. These are called fundamental transitions. Sometimes we may observe transitions corresponding to $v = 0 \rightarrow v = 2$ or $v = 0 \rightarrow v = 3$. These are called as overtones. They are less intense than fundamental transitions.

At higher temperature there is probability of transitions from $v = 1 \rightarrow v = 2$ or $v = 1 \rightarrow v = 3$. These are called hot bands. [Because they occur at high temperature.]

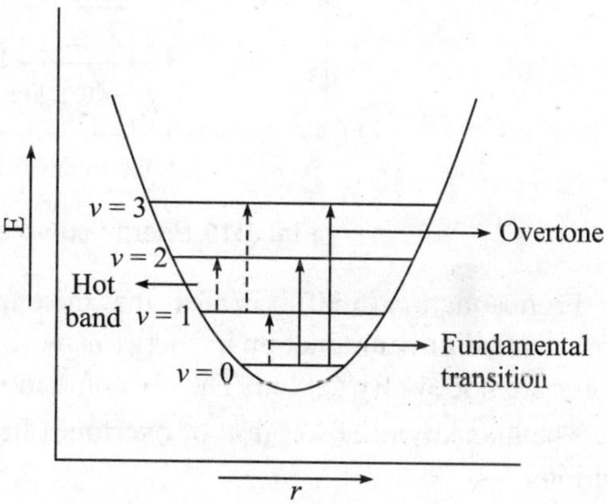

Fig. 5.18 Various types of Transiions

What is zero point energy?

The energy associated with lowest vibrational level ($v = 0$) will be as follows:

$$E(0) = \left(v + \frac{1}{2}\right) hv_0 = \left(0 + \frac{1}{2}\right) hv_0 = \frac{1}{2} hv_0$$

It means the molecule possess energy even at zero vibrational level, this is called zero point energy.

5.22 ANHAROMIC OSCILLATOR MODEL

At higher quantum levels there is a deviation from simple harmonic oscillator and the vibrations become anharmonic. The simple harmonic motion model does not explain the behaviour of real molecules. At larger distortions they deviate from the Hooke's law.

The energy curve for typical diatomic real molecule which behaves as anharmonic oscillator and undergoes extensions and compressions. In the same figure, a dotted parabola is there which is due to ideal diatomic molecule obeying simple harmonic motion.

To account for the anharmonicity in molecular oscillations, the expression for vibrational energy is modified as

$$E_v = \left(v + \frac{1}{2}\right) hv_0 - \left(v + \frac{1}{2}\right)^2 hxv_0 \qquad \ldots(5.30)$$

where x is anharmonic constant

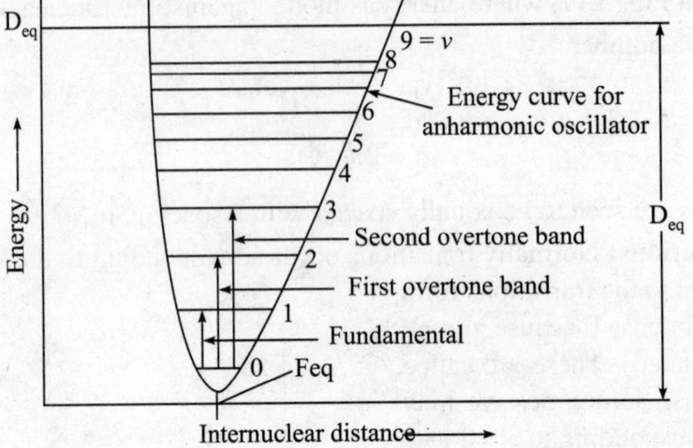

Fig. 5.19 Energy curve of an anharmonic oscillator

From equation (5.30) it is clear that the energy of every vibrational level is reduced because of anharmonicity. Further reduction in energy is more for higher levels. So the energy levels of anharmonic oscillator are not evenly spaced. They become more and more closely spaced as v increases.

The intensity of absorption of overtones lines will be much less than that of fundamental ones.

Transitions:

(*i*) $v = 1$ to $v = 0$ gives fundamental band

(*ii*) $v = 2$ to $v = 0$ gives first overtone (second anharmonic)

(*iii*) $v = 3$ to $v = 0$ gives second overtone (third anharmonic etc.)

The energy change (ΔE) when a transition results from any two, an upper level to a lower level of anharmonic oscillator is given by

$$\Delta E = E_{v+1} + 1 - E_v = h\nu_0 \, [1 - 2x_e \, (v + 1)] \quad ...(5.31)$$

The energy of transition ΔE_v for transition from $v = 0$ to $v = 1$ is

$$\Delta E = E_{v=1} - E_{v=0} = h\nu_0 \, [1 - 2x_e] = hc\overline{\nu}_1$$

or $\quad \overline{\nu}_2 \, (\text{cm}^{-1}) = h\nu_0 / hc \, (1 - 2x_e)$

$$= \nu_0 \, (1 - 2x_e)$$

The above expression gives the wave number for the line corresponding to fundamental transition. The frequency of anharmonic oscillator is less than that of harmonic oscillator.

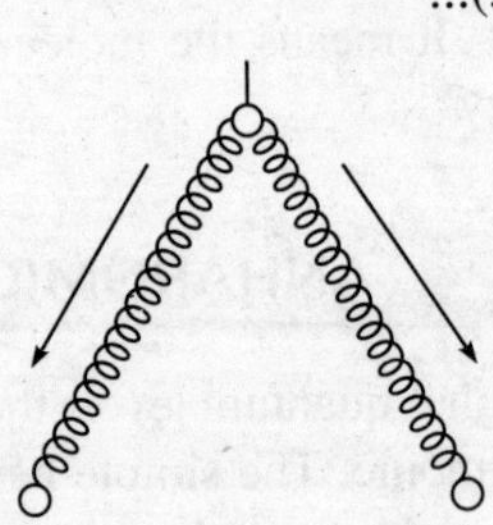

Fig. 5.20 Stretching

5.23 MOLECULAR VIBRATIONAL FREQUENCY

It may be noted that the atoms in a molecule are not held rigidly. The molecule may be visualized as consisting of balls of different sizes tied with springs of varying strengths. Here balls and springs correspond to atoms and chemical bonds respectively. When Infrared light is passed through the sample, the vibrational and the rotational energies of the molecules are increased which results in excitation of bond deformations: stretching and bending.

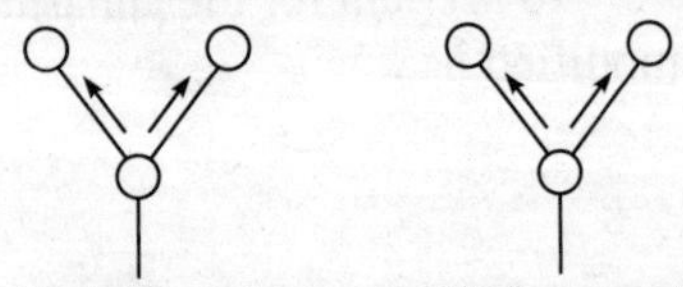

(*i*) Symmetric (*ii*) Asymmetric

Fig. 5.20 (*a*) Types of stretching vibrations

(*a*) Stretching: In this type of vibration the distance between the two atoms increases or decreases but atoms remains in the same bond axis. (Fig. 5.20)

Types of stretching vibrations: There are two types of stretching vibrations.

(*i*) Symmetric stretching: In this type, the movement of the atoms with respect to a particular atom in a molecule is in the same direction [Fig 5.20(*i*)].

(*ii*) Asymmetric stretching: In these vibrations, one atom approaches the central atom while the other departs from it [Fig. 5.20(*ii*)].

(*b*) Bending: In this type of vibrations, the positions of the atoms change with respect to the original bond axis (Fig. 5.21). We know that more energy is required to stretch a spring than that required to bend it. Thus, we can say that stretching absorptions of a bond appear at high frequencies as compared to the bending absorptions of the same bond.

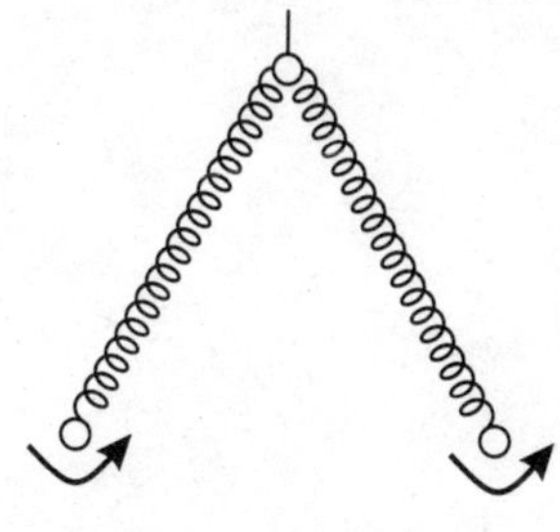

Fig. 5.21 Bending

The various stretching and bending vibrations of a bond occur at certain quantised frequencies. When Infrared radiation is passed through the substance, energy is absorbed and the amplitude of that vibration is increased. From the excited state, the molecule returns to the ground state by the release of extra energy by rotational, translational processes. As a result, temperature of the sample under investigation increases.

Types of bending vibrations *(Deformations)*

Bending vibrations are of four types (Fig 5.22):

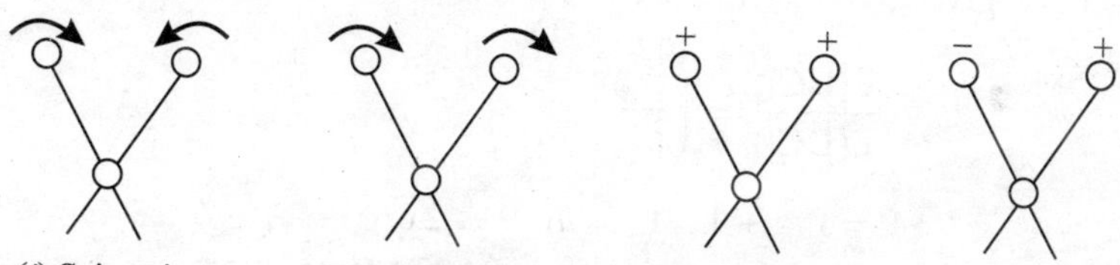

(*i*) Scissoring (*ii*) Rocking (*iii*) Wagging (*iv*) Twisting

Fig. 5.22 Types of bending vibrations

(*i*) Scissoring: In this type, two-atoms approach each other.

(*ii*) Rocking: In this type, the movement of the atoms takes place in the same direction.

(*iii*) Wagging: Two atoms move up and below the plane with respect to the central atom.

(*iv*) Twisting: In this type, one of the atoms moves up the plane while the other moves down the plane with respect to the central atoms.

Note: *Bending vibrations require lesser energy and hence occur at higher wavelengths or lower wave numbers than stretching vibrations.*

NUMERICAL PROBLEMS FOR TUTORIALS

Example 6: ***The wave number of centre of the fundamental bond for CO is 2140 cm^{-1}. Calculate the force constant of the bond between C and O.***

Solution: Fundamental frequency $\overline{v}_0 \rightarrow 1$ is given as:

$$\overline{v}_0 \rightarrow 1 = \overline{v} = \sqrt{\frac{k}{\mu}}$$

or

$$v = 2140 \text{ cm}^{-1} = 2.140 \times 10^5 \text{ m}^{-1}$$

$$\mu = \frac{m_c \times m_0}{m_c + m_0} = 1.1385 \times 10^{-26} \text{ kg}$$

$$k = (\overline{v})^2 \times p^2 c^2 \text{ m}$$

$$= (2.140 \times 10^5 \text{ m}^{-1})^2 \times 4 \times (3.14)^2 \times (3 \times 10^8 \text{ ms}^{-1})^2 \times 1.1385 \times 10^{-26}$$

$$= \mathbf{1851\ N\ m^{-1}} (1\text{N} = 1 \text{ kg ms}^{-2})$$

Example 7: ***Calculate the fundamental Vibrational frequency of HCl if that of DCl^{35} is 2080 cm^{-1}. Assume the force constant of HCl and DCl to be same.***

Solution:

$$\overline{v}\text{ (HCl)} = \frac{1}{2\pi c}\sqrt{\frac{k_h}{\mu_{HCl}}}$$

$$\overline{v}\text{ (DCl)} = \frac{1}{2\pi c}\sqrt{\frac{k_h}{\mu_{DCl}}}; \quad m_{HCl} = 1.6261 \times 10^{-27} \text{ kg}$$

$$\frac{\overline{v}\text{(HCl)}}{\overline{v}\text{(DCl)}} = \left(\frac{\mu_{HCl}}{\mu_{DCl}}\right)^{\frac{1}{2}}$$

$$\therefore \quad \overline{v}\text{ (HCl)} = 2080 \text{ cm}^{-1} \times \left(\frac{3.1618}{1.6261}\right)^{\frac{1}{2}} = \mathbf{2900.4\ cm^{-1}}$$

Example 8: ***The reduced mass of a diatomic molecule is 5×10^{-26} kg and its vibrational frequency is 1500 cm^{-1}. Calculate its force constant in (Nm^{-1}).***

Solution: Fundamental frequency $\overline{v}_c$ is given by the expression

$$\overline{v}_0 = \frac{1}{2\pi c}\sqrt{\frac{k}{\mu}}$$

when the molecule is assumed to act as simple harmonic.

The force constant

$$k = 4\pi^2 c^2 \overline{v}_0^2 \mu$$

Now

$$\mu = 5 \times 10^{-26} \text{ kg.}$$

$$\overline{v}_0 = 1500 \text{ cm}^{-1} = 1500 \times 10^2 \text{ m}^{-1}$$

$$\therefore \qquad k = 4\times\left(\frac{22}{7}\right)^2 \times (3 \times 10^8\ \text{ms}^{-1})^2$$

$$\times (1500 \times 10^2\ \text{m}^{-1})^2 \times 5 \times 10^{-26}\ \text{kg}$$

$$= \mathbf{4000.4\ Nm^{-1}} \qquad (\because 1\ \text{N} = 1\ \text{kg ms}^{-2})$$

Example 9: ***Calculate the force constant for the bond in HCl from the fact that the fundamental vibration frequency is 8.667 × 10^13^ s^–1^.***

Solution: Fundamental vibration frequency is related to force constant as under

$$v_0 = \frac{1}{2\pi}\sqrt{\frac{k}{\mu}}$$

$$k = 4\pi^2 v_0^2 \mu \text{ kg s}^{-2}, \text{ where } v_0 \text{ is in s}^{-1}, \mu \text{ in kg.}$$

$$\mu \text{ for HCl} = \frac{m_1 m_2}{m_1 + m_2} = \frac{1\times 35.5}{1+35.5} \text{ amu}$$

$$= \frac{35.5}{36.5} \times 1.66 \times 10^{-27}\ \text{kg}$$

$$k = 4\pi^2 v_0^2 \mu$$

$$= 4\times\left(\frac{22}{7}\right)^2 \times (8.667 \times 10^{13})^2 \times \frac{35.5}{36.5} \times 1.66 \times 10^{-27}$$

$$= \mathbf{48.31\ kg\ s^{-1}} \text{ or } \mathbf{Nm^{-1}}$$

Example 10: ***The reduced mass of diatomic molecule is 2.5 × 10^–26^ kg and its vibrational frequency is 2900 cm^–1^. Calculate its force constant.***

Solution: We know that fundamental vibration frequency

$$\overline{v}_0 = \frac{1}{2\pi c}\sqrt{\frac{k}{\mu}}$$

when the molecule behaves as simple harmonic oscillator

Given $\qquad \mu = 2.5 \times 10^{-26}$ kg

$$\overline{v}_0 = 2900\ \text{cm}^{-1} = 2900 \times 10^2\ \text{m}^{-1}$$

$\therefore$ Force constant $\qquad k = 4\pi c^2\ \overline{v}_0^{\,2}\ \mu$

$$= 4\times\left(\frac{22}{7}\right)^2 \times (3 \times 10^{10}\ \text{ms}^{-1})^2$$

$$\times (2900 \times 10^2\ \text{m}^{-1})^2 \times 2.5 \times 10^{-26}\ \text{kg}$$

$$= \mathbf{7476.3\ Nm^{-1}} \qquad (\because 1\ \text{N} = 1\ \text{kg ms}^{-2})$$

5.24 FUNDAMENTAL MODES OF VIBRATION AND ITS TYPES

Polyatomic molecules may exhibit more than one fundamental vibrational absorption bands. The number of these fundamental bands is related to the degrees of freedom in a molecule. The number of degrees of freedom is equal to the sum of the co-ordinates necessary to locate all the atoms of a molecule in space. Each atom has three degrees of freedom corresponding to the three space coordinates. An isolated atom has only translational degrees of freedom. When atoms combine to form a molecule, no degrees of freedom are lost, *i.e.*, the total number of degrees of freedom of a molecule will be equal to $3n$ where n is the number of atoms in a molecule. A molecule which is made up of *rotational, vibrational and translational degrees of freedom. Thus,*

$3n$ degrees of freedom = Translational + Rotational + Vibrational.

Since we are concerned only with the number of fundamental vibrational modes of a molecules, so we calculate only the number of vibrational degrees of freedom of a molecule. **A molecule has always three translation degrees of freedom.**

(*i*) In case of **linear molecule,** there are only two degrees of rotation. Thus, for a linear molecule of n atoms.

Total degrees of freedom = $3n$, Translational degrees of freedom = 3

Rotational degrees of freedom = 2,

Therefore, Vibrational degrees of freedom = $3n - 3 - 2 = 3n - 5$

Each vibrational degree of freedom corresponds to the fundamental mode of vibration and each fundamental mode corresponds to a band.

Hence, theoretically there will be $3n - 5$ possible fundamental bands for the linear molecules.

In a linear molecule of carbon dioxide (CO_2), the number of vibrational degrees of freedom can be calculated as follows:

Number of atoms (n) = 3, Total degrees of freedom = $3n = 3 \times 3 = 9$

Translational = 3, Rotational = 2

Vibrational degree of freedom = $9 - 3 - 2 = 4$

Hence, for CO_2 molecule, theoretical number of fundamental bands should be equal to four.

The four modes of vibration of CO_2 are shown in Fig. 5.23. One of these is symmetric stretching, second is asymmetric stretching vibration, third and fourth are bending and stretching. The frequencies associated with third and fourth modes are identical (*i.e.*, $\overline{v}_3 = \overline{v}_4$). These motions are identical in all respects except direction.

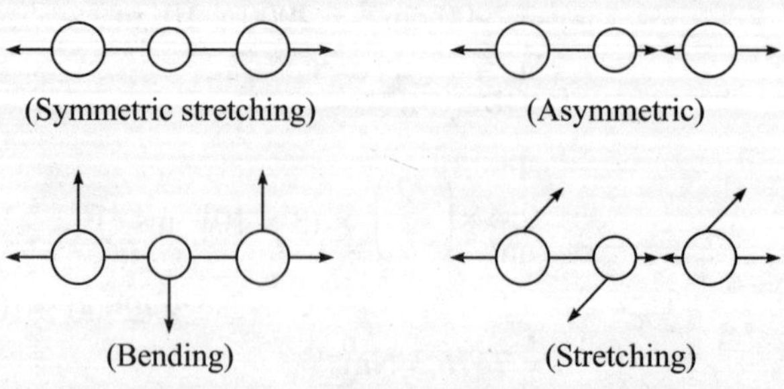

Fig. 5.23 Normal vibrational modes of CO_2 molecules

(*ii*) In the case of non-linear molecule, there are three degrees of rotation as the rotation about all the three axes will result in a change in the position of the atoms. So, for a non-linear molecule of n atoms, vibrational degrees of freedom can be calculated as follows:

Total degrees of freedom = $3n$

Translation degrees of freedom = 3

Rotational degrees of freedom = 3

Vibrational degrees of freedom = $3n - 3 - 3 = 3n - 6$

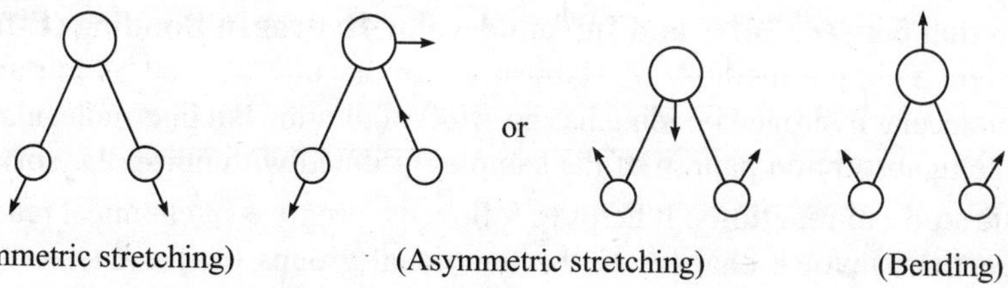

Fig. 5.24 Normal vibrational models of H_2O molecules

Two of these ($\overline{\nu}_1, \overline{\nu}_2$) are symmetric and of stretching type and third ($\overline{\nu}_3$) is asymmetric and is of bending type.

The exact frequency at which a given vibration occurs is determined by the strength of the bonds involved and the mass of the component atoms. In practice, infrared spectra do not normally display separate absorption signals for each of the $3n - 6$ fundamental vibrational modes of a molecule. The number of observed absorptions may be increased by additive and subtractive interactions leading to combination tones and overtones of the fundamental vibrations, in much the same way that sound vibrations from a musical instrument interact. Furthermore, the number of observed absorptions may be decreased by molecular symmetry, spectrometer limitations, and spectroscopic selections rules.

One selection rule that influences the intensity of infrared absorptions, is that a change in dipole moment should occur for a vibration to absorb infrared energy. Absorption bands associated with C = O bond stretching are usually very strong because a large change in the dipole takes place in that mode.

Some General Trends:

(*i*) Stretching frequencies are higher than corresponding bending frequencies.

(*ii*) Triple bonds have higher stretching-frequencies than corresponding double bonds which in turns have higher than single bonds.

5.25 APPLICATIONS OF INFRARED SPECTROSCOPY

1. Identification of an organic compound: The infrared spectrum of an organic compound is the most reliable method to establish its identity. The region of the spectrum from 1400 cm^{-1} to 700 cm^{-1} is complex because it contains, apart from fundamental stretching and bending vibrations, many bands resulting their vibrational frequencies. Thus, this part of the spectrum is characteristic of a compound and is called **fingerprint region**. The molecules containing the same functional groups may have similar absorptions above 1500 cm^{-1} but their spectra differ significantly in the fingerprint region.

2. Structure determinations: Infrared spectroscopy is used to establish the structure of a compound. For example, the infrared spectra of amino acids exhibit bands for ionised carboxylic acids and amine salts but no band for free amino and carboxylic groups is observed. Thus, we conclude that the amino acids exist as zwitter ions shown below:

$$H_3N^+—CH_2—COO^-$$

For the same molecular formula, different isomers can be proposed and their identity can be established with the help of IR spectra.

3. Detection of Impurities: The use of IR spectroscopy for the detection of impurities is also used although IR bands/peaks are not very sharp and intense even for pure sample. The presence of impurities can be detected if they absorb in the region where the sample does not show any absorption.

4. Distinction between Inter and Intramolecular Hydrogen Bonding (Effect of dilution): IR spectroscopy offers a simple method to establish intermolecular as well as intramolecular hydrogen bonding. Intra-molecular hydrogen bonding has no effect of dilution but intermolecular hydrogen bonding shows the change in absorption pattern of the sample if diluted with non-polar solvents.

5. Kinetic study of reactions: It helps to follow the progress of chemical reaction. Since most of the chemical reactions involve changes in the functional groups of the reactants, therefore periodic withdrawal of sample may be sufficient to reveal the change in the reaction in terms of functional group.

6. Others: It also helps to study the structure of complex molecules, keto-enol tautomerism, stability of conformers and optical isomerism etc.

Table 5.2: Characteristic IR Absorption Frequencies of Organic Functional Groups

Functional Group	*Type of Vibration*	*Characteristic Absorptions (cm^{-1})*	*Intensity*
Alcohol			
O – H	Stretch, H-bonded	3200 – 3600	strong, broad
O – H	Stretch, free	3500 – 3700	strong, sharp
C – O	Stretch	1050 – 1150	strong
Alkane			
C – H	stretch	2850 – 3000	strong
–C – H	bending	1350 – 1480	variable
Alkene			
=C – H	stretch	3010 – 3100	medium
=C – H	bending	675 – 1000	strong
C = C	stretch	1620 – 1680	variable
Alkyl Halide			
C – F	stretch	1000 – 1400	strong
C – Cl	stretch	600 – 800	strong
C – Br	stretch	500 – 600	strong
C – I	stretch	500	strong
Alkyne			
C – H	stretch	3300	strong, sharp
–C ≡ C–	stretch	2100 – 2260	variable, not present in symmetrical alkynes

Amine			
N – H	stretch	3300 – 3500	medium (primary amines have two bands; secondary have one band, often very weak
C – N	stretch	1080 – 1360	medium–weak
N – H	bending	1600	medium
Aromatic			
C – H	stretch	300 – 3100	medium medium-weak,
C = C	stretch	1400 – 1600	multiple bands
Analysis of C–H out-of-plane bending can often distinguish substitution patterns			
Carbonyl 1			
C = O	stretch	1670 – 1820	strong
(conjugation moves absorptions to lower wave numbers)			
Ether			
C – O	stretch	1000 – 1300 (1070 – 1150)	strong
Nitrile			
CN	stretch	2210 – 2260	medium
Nitro			
N – O	stretch	1515 – 1560 and 1345 – 1385	strong, two bands

5.26 NUCLEAR MAGNETIC RESONANCE

Nuclear magnetic resonance involves the interaction between oscillating electro-magnetic forces and the magnetic properties of certain magnetic nuclei notably the nucleus of hydrogen atom—**the proton.**

Examples of magnetic isotopes are ^{13}C, ^{1}H, ^{19}F, ^{14}N, ^{17}O, ^{31}P, and ^{33}S. If a nucleus is not magnetic, it can't be studied by nuclear magnetic resonance spectroscopy.

NMR was independently discovered in 1946 by the American physicists F. Bloch and E.M. Purcell; they shared the 1952 Physics Nobel Prize. The first application of NMR to chemistry was made in 1951 when NMR spectrum of ethanol was recorded (details would be discussed later on in the chapter). Since then NMR technique has almost been completely monopolized by chemists, especially organic chemists and biochemists for structural elucidation.

NMR spectroscopy is based on the measurement of absorption of electromagnetic radiations, in the radio frequency region from roughly 4 to 900 MHz. Absorption of radio waves in the presence of magnetic field is accompanied by a special type of nuclear transition, and for this reason, such type of spectroscopy is known as Nuclear magnetic resonance (NMR) spectroscopy.

Since the early sixties, the growth of NMR has been nothing short of an explosion and has led to highly sophisticated instrumental techniques which in turn have led to complex organic structural investigations and the elucidation of mechanism and stereochemical details. Some important applications of NMR are:

(*i*) *Determining structure of molecules.*

(*ii*) *Detection of number of protons present in a molecule.*

(*iii*) *Detection of hydrogen bonding*

(iv) *Identification of structural isomers*

(v) *Distinction between cis-trans isomers*

(vi) *Detection of aromaticity*

(vii) *Detection of electronegative groups*

5.26.1 Nuclear Spin

Just as electrons spin about their axes, it is believed that the protons and neutrons present in the nuclei of atoms also spin about their axes. Like electron, each proton or neutron is assigned a spin quantum number equal to ½.

The spin quantum number for the nucleus can then have values 0, $\frac{1}{2}$, 1, $1\frac{1}{2}$, 2 etc. depending upon the number of neutrons and protons having parallel and antiparallel spins.

It is usually represented by I, For example, deuterium nucleus which contains one proton and one neutron is expected to have *nuclear spin quantum number* equal to 1 or 0 depending upon whether the neutron and proton have parallel spins (↑↑) or antiparallel spins (↑↓). Deuterium nucleus is actually found to have I = 1 in the ground state showing that the proton and neutron have parallel spins. In general, it is observed that:

(i) *For nuclei in which the sum of protons and neutrons is even as well as the number of each of them is also even* (*i.e.*, mass number as well as atomic number is even) I = 0 (because there is pairing). For example, nuclear spin I *is zero* for $^{4}_{2}He$, $^{12}_{6}C$, $^{16}_{8}O$, $^{32}_{16}S$.

(ii) *For nuclei in which the sum of protons and neutrons is even but the number of protons is odd* (*i.e.*, mass number is even but atomic number is odd), I = 1,2, 3 etc. For example, for $^{2}_{1}D$ and $^{14}_{7}N$, I = 1, while that of $^{10}_{5}B$ = 3. *The spin nuclear is integral.*

(iii) *For nuclei in which the sum of protons and neutrons is odd (i.e., mass number is odd),* I is always odd integral multiple of $\frac{1}{2}$ *i.e.*, I = $\frac{1}{2}, \frac{3}{2}, \frac{5}{2}, \frac{7}{2}$ ... For example for $^{1}_{1}H$ and $^{19}_{9}F$, I = $\frac{1}{2}$, for $^{23}_{11}Na$, I = $\frac{3}{2}$, for $^{35}_{17}Cl$, I = $\frac{5}{2}$. *Thus nuclear spin is half integral multiples.* These rules be summarized in the form of table as shown below:

S.No.	*Mass Number*	*Atomic Number*	*Nuclear Spin (I)*
(i)	Odd	Odd or Even	$\frac{1}{2}, \frac{3}{2}, \frac{5}{2}, ...$
(ii)	Even	Even	0
(iii)	Even	Odd	1,2,3 ...

Since a nucleus possesses an electric charge, *the spinning nucleus gives rise to a magnetic field whose axis coincide with the axis of spin.* Thus, each nucleus can be considered as a tiny magnet having a magnetic moment. Each nucleus with I ≠ 0 has magnetic moment. However, nuclei with I = 0 do not possess magnetic moment, *i.e.*, they are non-magnetic, hence they do not absorb in NMR.

5.27 PRINCIPLE OF NUCLEAR MAGNETIC RESONANCE SPECTROSCOPY

The nuclei of certain atoms behave as if they were spinning charges. Any spinning charge creates a magnetic field and behaves as if it were a tiny bar magnet. Of the three nuclei most common in organic compounds (C, H and O), only the hydrogen nucleus (proton) behaves in this manner. When a proton in an organic molecule is placed in a strong magnetic field, it can align with the field or against it. In the more stable low energy state, it is aligned with the magnetic field. If energy is supplied in the form of radio waves of exactly the right frequency, radiation will be absorbed and the nucleus will 'flip' and align against the applied magnetic field in the higher energy state. At some value of applied magnetic field strength, the energy required to flip the proton matches with the energy of the radiation. The sample is then said to be in resonance with the incident frequency. At this stage absorption occurs and a signal is observed. Such a spectrum is called a nuclear magnetic resonance (NMR) spectrum.

NMR spectroscopy is based upon the fact that an electromagnetic radiation of suitable energy can induce transitions between the spin level of the nucleus placed in a fixed magnetic field.

Nuclei are characterized by **nuclear spin number,** I, which has the values, $\frac{1}{2}, 1, \frac{3}{2}, 2, \ldots$ expressed in the units of $\frac{h}{2\pi}$. For a given nucleus with spin number I, there are (2I + 1) possible orientations of the nucleus in a magnetic field.

Nuclear spin is the result of the spin angular momentum. The resultant angular momentum of all the protons and neutrons contained in the nucleus is called *nuclear spin,* which is given as

$$\text{Nuclear spin} = \sqrt{I(I+1)}\ h/2\pi r \qquad \ldots(5.32)$$

where I is the spin quantum number of nucleus.

(Just as the spin angular momentum of the electrons is given $\sqrt{s(s+1)}\,\frac{h}{2\pi}$, where s is the electron spin quantum number)

A nucleus with spin ½ will have two possible orientations. In the absence of an external magnetic field, these orientations are of equal energy. If a magnetic field is applied, then the energy levels split. Each level is given a *magnetic quantum number, m.*

Nuclear Magnetic Spectroscopy is based on the fact that when population of magnetic nuclei is placed in an external magnetic field, the nuclei become aligned in a predictable and finite number of orientations. For ^1H there are two orientations. In one orientation the protons are aligned with the external magnetic field (north pole of the nucleus aligned with the south pole of the magnet and south pole of the nucleus with the north pole of the magnet) and in the other where the nuclei are aligned against the field (north with north, south with south). The alignment with the field is also called the **"alpha"** orientation and the alignment against the field is called the **"beta"** orientation.

When the nucleus is in a magnetic field, the initial populations of the energy levels are determined by the Boltzmann distribution that **the lower energy level will contain slightly more nuclei than the higher level.** It is possible to excite these nuclei into the higher level with electromagnetic radiation. The frequency of radiation needed is determined by the difference in energy between the energy levels.

It is useful to consider the NMR of a bare nucleus. From quantum mechanics, it is known that the energy of interaction of nuclear magnetic moment *m* with an external magnetic field B_z, applied in the z-direction, is given by $E = -\vec{\mu}\vec{B}_z$

substituting the value of $\vec{\mu} = g_n \mu_N \vec{I}$

$$E = -g_N \mu_N \vec{B}_z \vec{I}$$

As nuclear spin I is quantized in the presence of the magnetic field and the energy of nuclear spin is defined only by us component m_I

$$E_{m_I} = -g_N \mu_N B_z m_I$$

Consider a bare proton ^1H with I = 1/2 so that $m_1 = \pm 1/2$.

Thus, for $m_1 = 1/2$, the energy.

$$E_{1/2} = -\frac{1}{2} g_N \mu_N B_z \quad \text{and}$$

for $m_1 = -\frac{1}{2}$, the energy $$E_{-1/2} = \frac{1}{2} g_N \mu_N B_z$$

It should be noted that in the absence of the external magnetic field, the two components of the proton spin have the same energy (*i.e.*, they are degenerate). However, in the presence of the magnetic field, the components have different energies. The magnetic field lifts or removes the degeneracy of the spin components. This is called Zeeman splitting of energy levels. The energy difference is given by

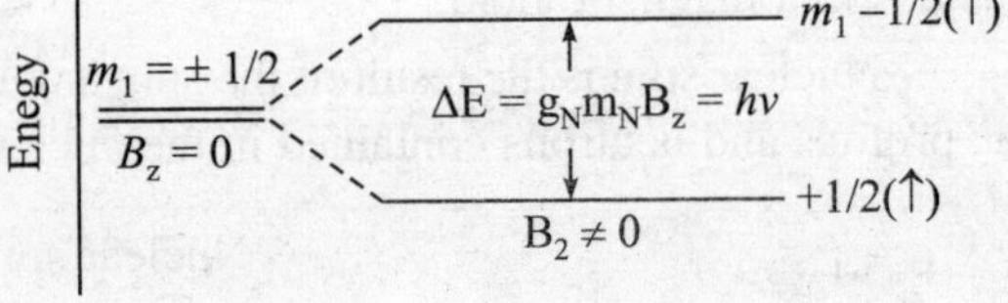

Fig. 5.25 Splitting of nuclear energy levels of a proton in a magnetic field (Magnetic field is kept constant and frequency is varied)

$$\Delta E = E_{-1/2} - E_{1/2} = g_N \mu_N B_z \quad \text{...(5.33)}$$

In order to induce transitions between the two energy levels $E_{-1/2}$ and $E_{1/2}$, *i.e.*, in order to let the nuclear spin 'flip' from the upward direction (↑) to the downward direction (↓), apply an oscillating radiofrequency field perpendicular to the direction of B_z. *The nuclear spin flips, provided the energy (hν) of the radiofrequency photon of frequency ν is exactly equal to ΔE, i.e.,*

$$\Delta E = g_N \mu_N B_z = h\nu \quad \text{...(5.34)}$$

This is called the **Bohr frequency condition.** Thus, the NMR frequency is given by

$$\nu = \frac{\Delta E}{h} = \frac{g_N \mu_N B_z}{h} \quad \text{...(5.35)}$$

It may be noted that g_N can be positive or negative but ν must be positive.

Example 11: ***Calculate the NMR frequency (in MHz) of the proton (^1H) in a magnetic field of intensity 1.4092 tesla, given that g_N = 5.585 and μ_N = 5.05 × 10^{-27} JT^{-1}.***

Solution: From Eq. (5.34)

$$\nu = \frac{g_N \mu_N B_z}{h}$$

$$= \frac{(5.585)\times(5.05\times10^{-27}\ JT^{-1})\times(1.4092T)}{6.626\times10^{-34}\ Js}$$

$$= 60 \times 10^6\ s^{-1} = 60 \times 10^6 = \mathbf{60\ MHz}$$

This frequency lies in the radiofrequency region

Example 12: ***Calculate the magnetic field strength required to observe the NMR frequency for fluorine (^{19}F) at 60 MHz. Given g_N for $^{19}F = 5.257$ $\mu_N = 5.05 \times 10^{-27} JT^{-1}$.***

Solution: Since from Eq. (5.34), $v = g_N \mu_N B_z/h$, we have

$$B = \frac{hv}{g_N \mu_N} = \frac{(6.626\times10^{-34}\ Js)\,(60\times10^6\ s^{-1})}{(5.257)\,(5.05\times10^{-27}\ JT^{-1})}$$

$$= \mathbf{1.49739T}$$

5.27.1 The Absorption of Radiation by a Nucleus in a Magnetic Field

Imagine a nucleus (of spin 1/2) in a magnetic field. This nucleus is in the lower energy level (*i.e.*, its magnetic moment does not oppose the applied field). The nucleus is spinning on its axis. In the presence of a magnetic field, this axis of rotation will process around the magnetic field.

The frequency of precession is termed as the *Larmor frequency as precessional frequency,* which is identical to the transition frequency. The energy of the processing nucleus is given by:

$$E = \mu B \cos\theta$$

where θ is the angle between the direction of the applied field and the axis of nuclear rotation. If energy is absorbed by the nucleus, then the angle of precession, θ, will change. For a nucleus of spin 1/2, absorption of radiation "flips" the magnetic moment so that it opposes the applied field (the higher energy state.) Fig. 5.26 and 5.27.

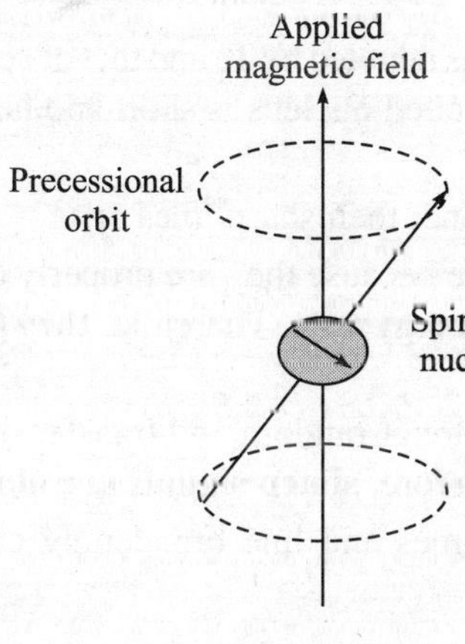

Fig. 5.26

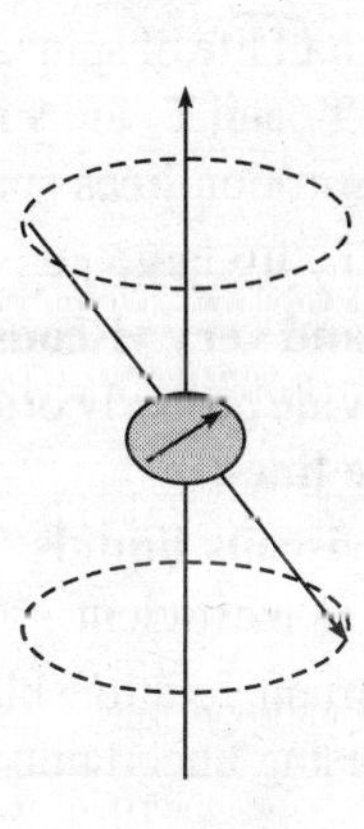

Fig. 5.27

5.27.2 Relaxation Processes

It is important to realize that only a small proportion of "target" nuclei are in the lower energy state (and can absorb radiation). There is the possibility that by exciting these nuclei, the *populations* of the higher and lower energy levels will become equal. If this occurs, then there will be no further absorption of radiation. The spin system is saturated. The possibility of saturation means that we must be aware of the *relaxation processes* which return nuclei to the lower energy state.

How do nuclei in the higher energy state return to the lower state? Ideally, the NMR spectroscopist would, like relaxation rates to be fast—but not too fast. If the relaxation rate is fast, then saturation is reduced. If the relaxation rate is too fast, *line-broadening* in the resultant NMR spectrum is observed.

There are two major relaxation processes:

- Spin-lattice relaxation
- Spin-spin relaxation

(*i*) Spin-lattice relaxation: Nuclei in an NMR experiment are in a sample. The sample in which the nuclei are held is called the *lattice*. Nuclei in the lattice are in vibrational and rotational motion, which creates a complex magnetic field. The magnetic field caused by motion of nuclei within the lattice is called the lattice field. This lattice field has many components. Some of these components will be equal in frequency and phase to the *Larmor frequency* of the nuclei of interest. These components of the lattice field can interact with nuclei in the higher energy state, and cause them to lose energy (returning to the lower state). The energy that a nucleus loses increases the amount of vibration and rotation within the lattice (resulting in a tiny rise in the temperature of the sample).

The relaxation time, T, (the average lifetime of nuclei in the higher energy state) is dependant on the magnetogyric ratio of the nucleus and the mobility of the lattice. As mobility increases, the vibrational and rotational frequencies increase, making it more likely for a component of the lattice field to be able to interact with excited nuclei. However, at extremely high mobilities, the probability of a component of the lattice field being able to interact with excited nuclei decreases.

(*ii*) Spin-spin relaxation: Spin-spin relaxation describes the interaction between neighbouring nuclei with identical processional frequencies by differing magnetic quantum states. In this situation, the nuclei can exchange quantum states; a nucleus in the lower energy level will be excited, while the excited nucleus relaxes to the lower energy state. There is no net change in the populations of the energy states, but the average lifetime of a nucleus in the excited state will decrease. This can result in line-broadening.

The mean half life of spin lattice relaxation process is denoted by T_1 and that of spin-spin relaxation process by T_2. If T_1 and T_2 are small, then life time of excited nucleus is short and hence this give rise to very broad absorption lines spectrum.

If T_1 and T_2 are large *i.e.*, approximately one second, then sharp lines arise.

In solids and very viscous liquids, T_1 is very large because they are properly oriented, but T_2 is short, as they provide properly oriented nucleus in lower energy state. Therefore, **they normally display broad resonance lines.**

For non-viscous liquids, molecular orientations are at random and transfer of energy by spin-lattice relaxation is inefficient , T_1 and T_2 are large. Therefore, **sharp signals are obtained in NMR.**

The important relationship between relaxation times and line broadening can be understood qualitatively by using uncertainty principle in the form.

$$\Delta E \cdot \Delta t = \frac{h}{2\pi} \qquad (\Delta E = h\Delta v)$$

$$\therefore \qquad h\Delta v \cdot \Delta t \approx \frac{h}{2\pi} \qquad \Delta v \cdot \Delta t \approx \frac{1}{2\pi} = \text{constant}$$

It Δt is large, then Δv will be small. There is little spread in frequency and line width is narrow *i.e.,* sharp peak is obtained and vice-versa.

5.28 INSTRUMENTATIONS OF NMR SPECTROPHOTOMETER

The schematic arrangement of apparatus used is shown in Fig. 5.28. It consists of the following parts:

(*a*) **An electromagnet** with poles M and M′ whose strength field can be varied by varying the feeding voltage.

(*b*) **A sample tube S** placed between the poles of the electromagnet and containing the sample under study.

(*c*) **A radio frequency transmitter T** which sends out constant frequency around the sample tube through the coil B.

(*d*) **A receiver coil C** to pick up the frequency emitted by the sample.

(*e*) **A receiver and amplifier** (A) to receive and amplify the frequency picked up by the receiver coil C.

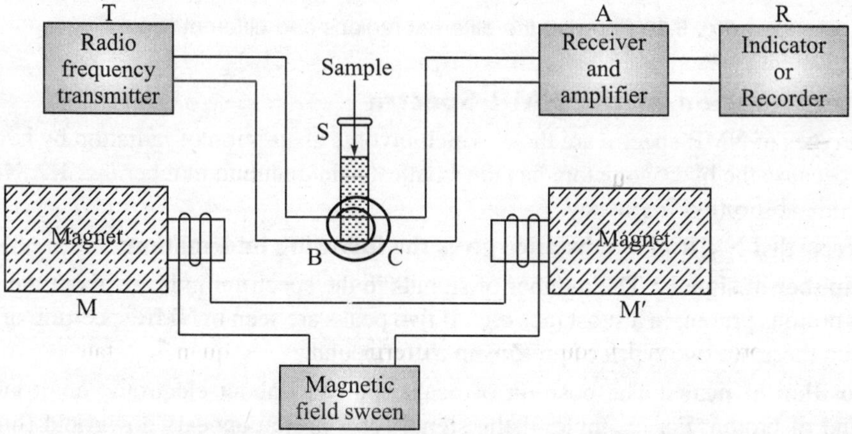

Fig. 5.28 Schematic arrangement of NMR spectrometer

(*f*) **An indicator *i.e.*, a cathode ray oscilloscope or a recorder R** to display or record the amplified signals.

There can be two different ways to bring about the nuclear transitions. These are:

(*i*) Keeping (B_0) applied Magnetic Field constant, and ν may be varied gradually. No energy is absorbed till ν reaches a value at which it is satisfied *i.e.,* the energy supplied is equal to the nuclear energy level spacing. The sample is then said to be in resonance with the incident frequency.

(*ii*) Keeping ν constant, B_0 may be varied, till it is satisfied. This procedure is found to be experimen tally more convenient and satisfactory and is briefly explained below:

The procedure consists in supplying some constant frequency (usually 60 megacycles/sec or 60 MHz) to the sample from the radio frequency transmitter. A sweeping magnetic field is then applied on the sample. When this energy is absorbed, the nuclei are said to be in the excited state. They then start re-emitting the energy. The energy thus emitted is picked up by the receiver coil C, amplified by the amplifier A and recorded on the recorder R. At this stage, the sample is said to be in resonance with the applied frequency.

When magnetic field is increased further, no absorption of energy takes place till it again reaches a value at which the sample is again in resonance with the applied frequency. In this way, a plot of the type shown in Fig. 5.29 is obtained. Such a plot is called an **NMR spectrum.**

When magnetic resonance condition is reached, the energy absorbed by the sample from the radio frequency oscillator is measured electrically and displayed on the chart of a recorder or on an oscilloscope.

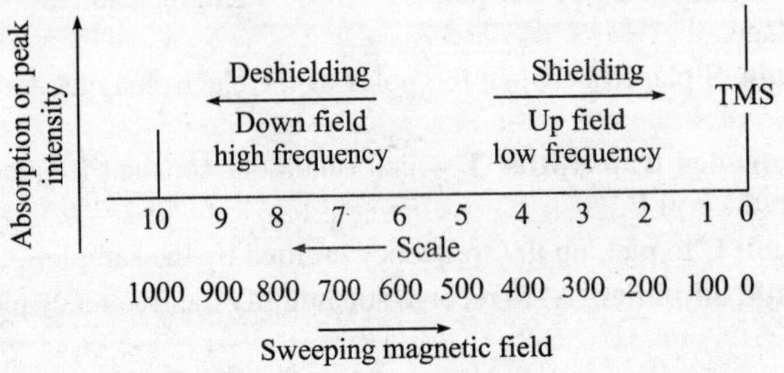

Fig. 5.29 Showing the different regions and different scales

5.28.1 Interpretation of the NMR Spectra

The simplest types of NMR spectra are those which involve absorption of radiation by hydrogen nuclei only. This is because the hydrogen atom has the smallest spin quantum number *i.e.,* 1/2. Hence it would require only one absorption frequency.

The recorded NMR of a substance gives the following information:

(i) **Number of signals:** The number of signals in the spectrum indicate about the various types of protons present in a substance *e.g.,* If two peaks are seen in NMR spectrum of a compound, then these are two types of protons in a compound.

(ii) **Position of peaks:** The position of peaks indicates about electronic environment of each kind of proton. For example, if the signal for a proton appears downfield (higher δ value) then the proton is attached to carbon which has an electronegative atom.

(iii) **Intensities of peaks:** The intensities (relative height) of signal indicate number of protons of each kind. For example, if two signals are recorded on NMR spectrum and ratio of heights of signal is 3 : 2, then there are 3 protons of one type and 2 protons of the other type.

(iv) **Splitting of signals in the form of multiplets:** The splitting of signals indicates number of protons on the neighbouring carbon.

If the proton have same chemical shift, then they are called **chemical shift equivalents** and are designated by the same letter. In other words, protons with the same environment absorb at the same applied field strength and such a set of protons are called equivalent protons. The equivalent protons can be judged by looking at the molecular structure of a molecule *e.g.,* Ethyl chloride has two types of protons labeled ⓐ and ⓑ.

$$\underset{\text{(a)}}{CH_3}—\underset{\text{(b)}}{CH_2}—Cl \quad \text{2 NMR signals}$$

5.28.2 Number of Signals

The number of signals in the NMR spectrum tell the number of different sets of equivalent protons in a molecule. Each signal corresponds to a set of equivalent protons. **It may be noted that magnetically equivalent protons are chemically equivalent protons.** Let us find the various sets of equivalent protons (signals) in the following types of compounds:

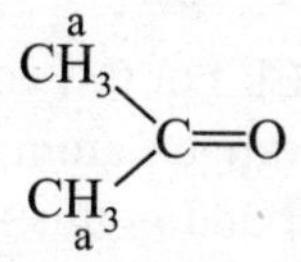

Acetone (one signal)

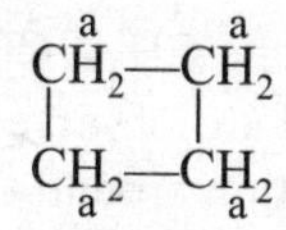

Cyclobutane (one signal)

In acetone, all the six protons are in exactly similar environment. Therefore, only one signal is observed. Similarly, we see only one signal for cyclobutane, because all the eight H atoms in it have similar environment.

(*b*) Some compounds showing more than one signal are as follows:

(*i*) $\overset{b}{CH_3}—\overset{a}{OH}$
Methyl alcohol
2 NMR signals

(*ii*) $\overset{c}{CH_3}—\overset{b}{CH_2}—\overset{a}{NH_2}$
Ethyl amine
3 NMR signals

(*iii*) $\overset{c}{CH_3}—\overset{b}{CH_2}\overset{a}{CH_2}Cl$
Propyl chloride
3 NMR signals

(*iv*) $\overset{b}{CH_3}—\overset{c}{CH}(\overset{a}{OH})—\overset{b}{CH_3}$
Isopropyl alcohol
3 NMR signals

(*c*) Strictly speaking, chemically equivalent protons must also be stereo-chemically equivalent *i.e.*, a particular set of protons are said to be chemically equivalent only if they remain in exactly similar environment when the stereo-chemical formula of the molecule under consideration is written.

Consider the case of 2-chloropropene, $CH_3—C(C1)=CH_2$. From this structure, one can expect two sets of equivalent protons (two peaks). But stereo-chemical formula reveals three sets of protons in it and correspondingly three signals are obtained in its NMR spectrum.

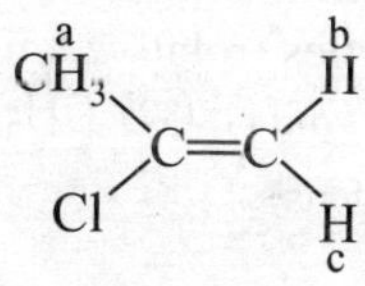

H^b and H^c are not exactly similar environment.

Now consider methyl cyclo-propane. For this compound, one can expect three sets of protons. But actually four signals are observed. There are four types of different H atoms marked *a, b, c* and *d* and correspondingly there are *four signals.*

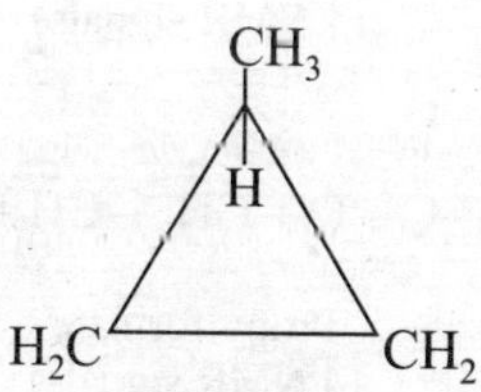

4 NMR signals
(Methyl cyclopropane)

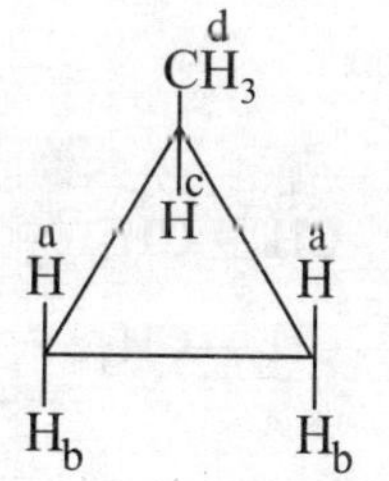

Methyl cyclopropane

Consider 1, 2-dichloropropane $\overset{3}{C}H_3—\overset{2}{C}H(Cl)—\overset{1}{C}H_2Cl$. For this compound one can expect three signals. But actually, four signals are observed as two hydrogen atoms attached with C_1 are not in exactly similar environment. Rotation around C—C single bond in this molecule cannot bring similar environment for the said hydrogen atoms.

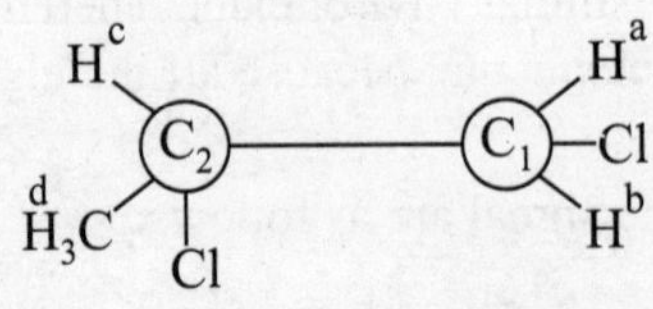

1, 2 dichloropropane

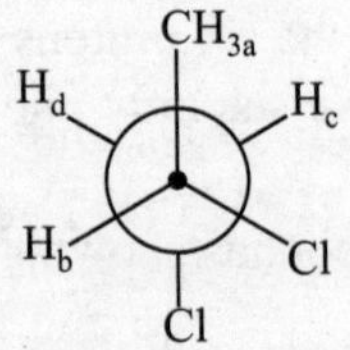

Let us note the number of NMR signals for the following compounds:

(i) 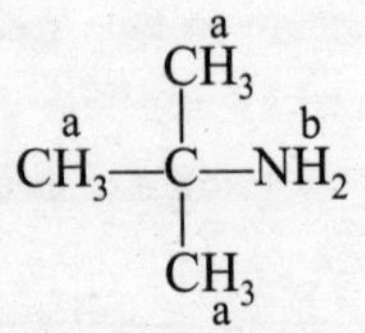

Tert-butylamine
(2 NMR signals)

(ii) 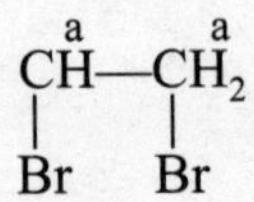

1, 2-diromoethane
(1 NMR singnal)

(iii) $\overset{b}{CH_3}\overset{a}{C}HBr_2$

1, 1-dibromoethane
(2 NMR signals)

(iv) 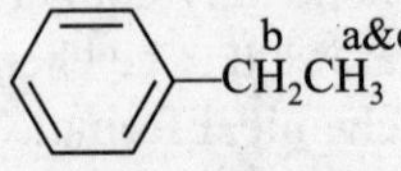 (ring hydrogens)

Ethyl benzene
(3 NMR signals)

(v) 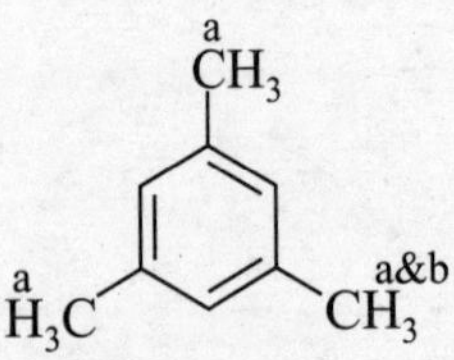 (ring hydrogens)

Mesitylene
(2 NMR signals)

(vi) $\overset{b}{CH_3}—\overset{a}{CH_2}—O—\overset{a}{CH_2}—\overset{b}{CH_3}$

Diethyl ether
(2 NMR signals)

(vii) $\overset{c}{CH_3}\overset{b}{CH_2}\overset{a}{CHO}$

Propanal
(3 NMR siganls)

(viii) 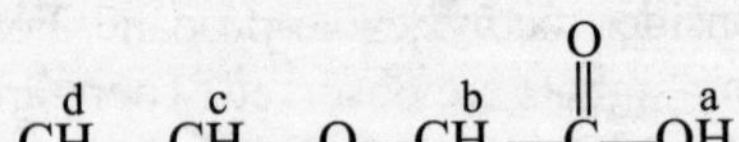

Ethoxy acetic acid
(4 NMR signals)

(ix) 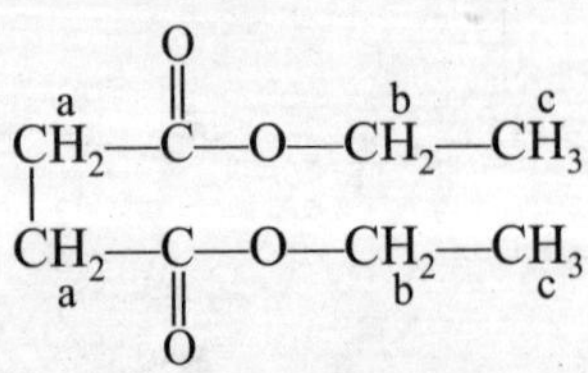

Diethyl succinate
(3 NMR signals)

(x) 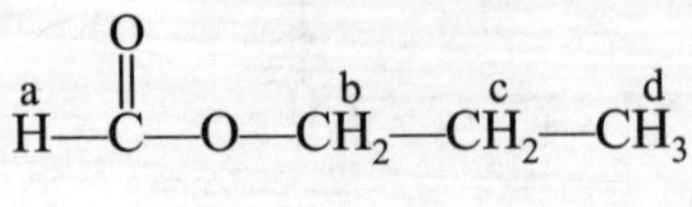

Propylformate
(4 NMR signals)

5.29 SHIELDING AND DESHIELDING EFFECTS

Hydrogen nuclei in a molecule are surrounded by the electronic charge which shields the nucleus from the influence of the applied field. Thus, to overcome the shielding effect and to bring the protons to resonance, greater external field is required. Clearly, greater the electron density around the proton, greater will be induced secondary magnetic field (Local diamagnetic effect) which opposes the applied field and thus, greater external field will cause proton absorption. When absorption occurs, the field H felt by the proton is represented as:

$$H_{eff} = \underset{\text{(applied field)}}{H_0} - \underset{\text{(induced field)}}{\sigma H_0}$$

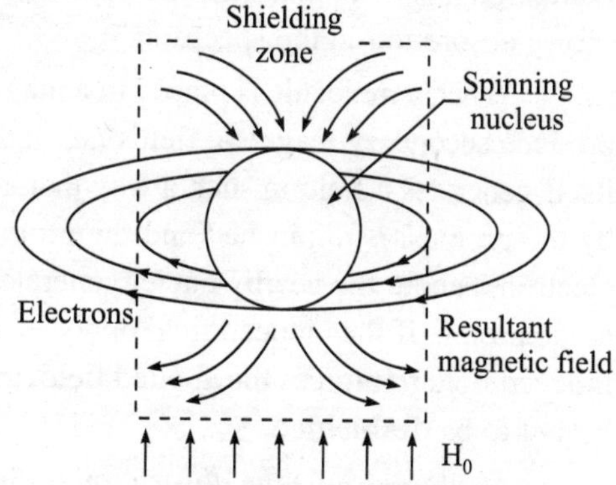

Fig. 15.30 Diamagnetic shielding about the nucleus

where H_0 is the applied field strength in Gauss. Clearly the field felt by the proton does not correspond to the applied field.

Greater the value of shielding, greater will be the value of applied field strength which has to be applied to get the effective field required for absorption and vice versa.

$H_0 \propto \nu$, $\therefore$ ν will be different. Hence in different structural environment extent of shielding will be different, therefore, we get different signals.

$$H = H_0 (1 - \sigma) \quad \text{...(5.36)}$$

where σ = shielding parameter

Moreover $$\nu = \frac{\gamma H}{2\pi} \quad \text{...(5.37)}$$

(Here, H is the magnetic field experienced by proton in Gauss and r is called gyromagnetic ratio *i.e.*, $r - \frac{2\pi\mu}{hI}$, where μ is magnetic moment of spinning nuclei, I is spinquantum number of nuclei and h is Planck's constant).

From Eqns. (5.36) and (5.37), $\nu = \frac{\gamma H_0 (1 - \sigma)}{2\pi}$

From this relation, it is clear that the protons with different electronic environments or with different shielding parameter can be brought into resonance in two ways:

(*i*) The strength of the external field is kept steady and the radio-frequency is constantly varied.

(*ii*) The radio-frequency is kept steady and the strength of the applied field is constantly varied.

Clearly, at constant radio-frequency, shielding shifts the absorption upfield in the molecules when, there is spherical distribution of electrons around the proton. It is called **positive shielding.**

5.30 POSITIONS OF SIGNALS (CHEMICAL SHIFT)

The position of the signals in the spectrum help us to know the nature of protons viz., *aromatic, aliphatic, acetylenic, vinylic,* adjacent to some electron attracting or electron releasing groups etc. Each of these types of protons will have different electronic environments and thus, they absorb at different applied field strengths. It is important to note that it is the electronic environment which tells where a proton shows absorption in the spectrum.

When a molecule is placed in a magnetic field, its electrons are caused to circulate and thus, they produce secondary magnetic fields *i.e.,* induced magnetic field. Rotation of electrons about the proton itself generates a field in such a way that at the proton, it opposes the applied field. Thus, the field felt by the protons, is diminshed and the protons is said to be **shielded**. Rotation of electrons (specially π electrons) about the nearby nuclei generates a field that can either oppose or reinforce the applied field at the proton. If the induced field opposes the applied field, then proton is said to be shielded. But if the induced field reinforces the applied field, the proton feels a higher field strength and thus, such a proton, is said to be **deshielded.**

Shielding shifts the absorption ***upfield*** *and deshielding shifts the absorption* ***downfield*** *to get an effective field strength necessary for absorption. Such shifts in the positions of NMR absorptions which arise due to the shielding or deshielding of protons by the electrons are called* ***chemical shifts***.

For measuring chemical shifts or various protons in a molecule, the signal for **tetramethyl silane (TMS)** is taken as a reference. The structure of TMS *i.e.,* $(CH)_4$ Si is given as:

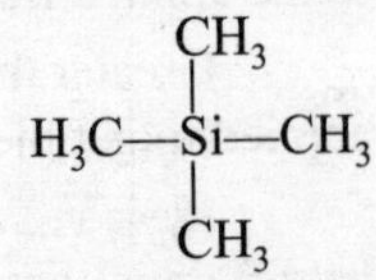

Structrue of TMS

Due to the low electronegativity of silicon, the shielding of equivalent protons in tetramethyl silane is greater than most of the organic compounds. Therefore, NMR signal or tetramethyl silane is taken as a reference and chemical shift for different kinds of protons are measured relative to it. Clearly, the NMR signal for a particular proton in a molecule will appear at different field strengths as compared to a signal from TMS. **This difference in the absorption position of the proton with respect to TMS signal is called chemical shift (δ-value).**

Tetramethyl silane is the most convenient reference and has the following characteristics:

(*i*) All the 12 protons present in it are equivalent *i.e.,* they have the same environment. Hence it gives a single well defined absorption peak in the NMR spectrum.

(*ii*) The protons are highly shielded by their electrons from the external magnetic field so that it shows NMR at a very high external magnetic field strength as compared to other protons.

(*iii*) It is chemically, inert and miscible with almost all organic substances.

(*iv*) Its boiling point is low (34°C) so that it evaporates off quickly and the sample compound is easily recovered.

The environmental effect on any type of hydrogen atom is expressed quantitatively in terms of a quantity called chemical shift. It is generally represented by the letter δ and is defined by the equation,

$$\boxed{\delta = \frac{H_s - H_r}{H_0} \times 10^6 \text{ ppm}} \qquad \ldots(5.38)$$

where H_s is the magnetic field strength at which the maximum (peak) appears for a particular type of protons of the sample taken, while H_r is the field strength at which absorption occurs for some reference substance.

It may be pointed out that different samples show absorption or resonance at field strengths that differs only by the order of miligausses when the applied magnetic field is 14000 gauss. Hence the factor 10^6 in expressing δ values is introduced to make the values convenient size. Thus chemical shift is expressed in terms of parts per million *i.e.*, ppm.

As H ∝ ν, equation (5.38) can also be written as:

$$\boxed{\delta = \frac{v_s - v_r}{v_0} \times 10^6} \qquad \text{...(5.39)}$$

where v_0 = frequency of the instrument.

It is not measured in gauss but is measured in equivalent frequency units which is then divided by the frequency of the spectrometer used. This gives the value of δ (delta). Protons with the same chemical shift are called equivalent protons. Non-equivalent protons have different chemical shifts.

The values of δ for a substance with respect to TMS can be obtained by measuring $v_s - v_{TMS}$

where v_s = Resonance frequency of the sample

v_{TMS} = Resonance frequency of TMS

Keeping the radio-frequency constant, an NMR signal for a particular set of proton in the sample will appear at different field strength than the signal obtained from TMS. Thus, the value for $v_s - v_{TMS}$ can be expressed in the corresponding field strength in gauss. Since, the operating frequency of the instrument is directly proportional to the strength of the magnetic field, we can define δ as

$$\delta = \frac{v_{\text{sample}} - v_{\text{reference}}}{\text{operating frequency in megahertz (MHz)}}$$

$$= \frac{\Delta v}{\text{operating frequency in megahertz (MHz)}}$$

where Δv is the frequency shift. The value of δ is expressed in parts per million (ppm). Most chemical shifts have δ value between 0 and 10. In the τ scale, signal for the standard reference, TMS, is taken as 10 ppm.

δ(Delta) or τ (Tau) scales are commonly used. In majority of organic compounds, Protons resonate at a lower field than the proton of tetramethyl silane (TMS). Thus assigning delta (δ) value for TMS equal to zero, a scale can be defined in which most protons resonances are of the same sign. Any proton or set which absorb at a field lower than TMS is given a positive value for δ.

The two scales are related by the expression τ = 10 – δ

NMR signal is usually plotted with magnetic field strength increasing to the right. Thus, the signal for TMS (highly shielded) appears at the extreme right of spectrum with δ = 0 ppm. **Greater the deshielding of protons, larger will be the value of** δ. For most of the organic compounds, signals appear downfield to the left of TMS signal.

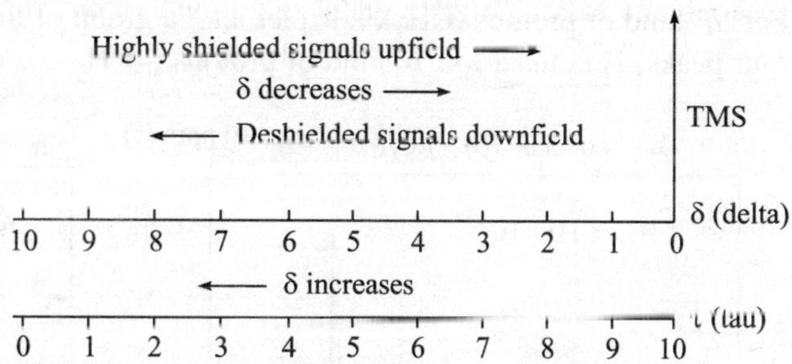

Fig. 5.31 Two scales for chemical shifts

Table 5.3: Chemical shift for various types of protons

Types of protons		*Chemical shifts in ppm* δ *(Delta)*	τ *(Tau)*
TMS		0(zero)	10
Cyclo-propane		0.2	9.8
Primary	R—CH_3	0.9	9.1
Vinylic	C=C—H	4.9 – 5.8	5.1 – 4.2
Acetylenic	C≡C—H	2 – 3.5	8 – 6.5
Aromatic	Ar—H	6 – 9.0	4 – 1.0
Fluorides	H—C—F	4 – 4.5	6 – 5.5
Chlorides	HC—Cl	3 – 4	7 – 6
Alcohols	H—C—OH	3.4 – 4	6.6 – 6
Ethers	H—C—OR	3.3 – 4	6.7 – 6
Esters	H—C—COOR	2 – 2.2	7 – 7.8
Acids	H—C—COOH	2 – 2.5	8 – 7.5
Aldehydic	RCHO	9 – 10	1 – 0
Hydroxy	R—OH	1 – 5.5	9 – 4.5
Phenolic	Ar—OH	4 – 12	6 to –2
Enolic	C≡C—OH	15 – 17	–5 to –7
Carboxylic	RCOOH	10.5 – 12	–0.5 to –2

These days most commonly used scale for measuring chemical shift is δ scale but in some old books τ scale is also given $\tau = 10 - \delta$.

5.30.1 High Resolution Spectra: Spin-Spin Splitting

It is already pointed out that each signal in an NMR spectrum represents one kind or one set of protons in a molecule. It is found that in certain molecule, a single peak (singlet) is not observed, but instead, a multiplet (group of peaks) is observed. Consider a molecule of $\overset{a}{CH_3}—\overset{b}{CH_2}—Br$, ethyl bromide. This molecule has two kinds of protons in it and thus, two signals are expected in its NMR spectrum. It has been observed that for each kind of protons, we do not get singlets but a group of peaks are observed. For '*a*' kind of protons (CH_3), a triplet, *i.e.*, a group of three peaks is observed and a quartet (group of four peaks) is noticed for '*b*' kind of protons ($–CH_2–$) (Fig. 5.32).

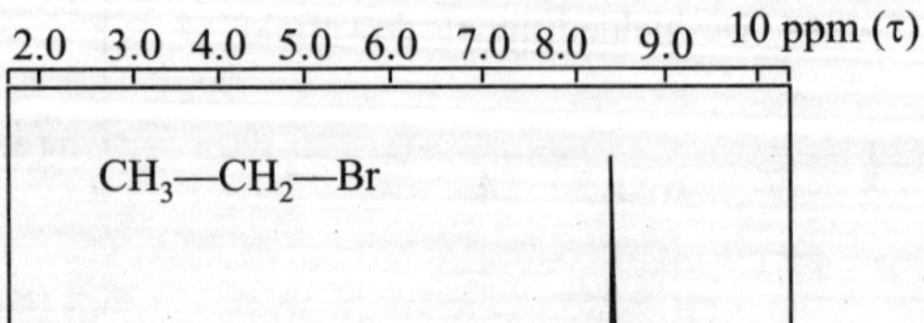

Fig. 5.32 NMR spectrum of ethyl bromide

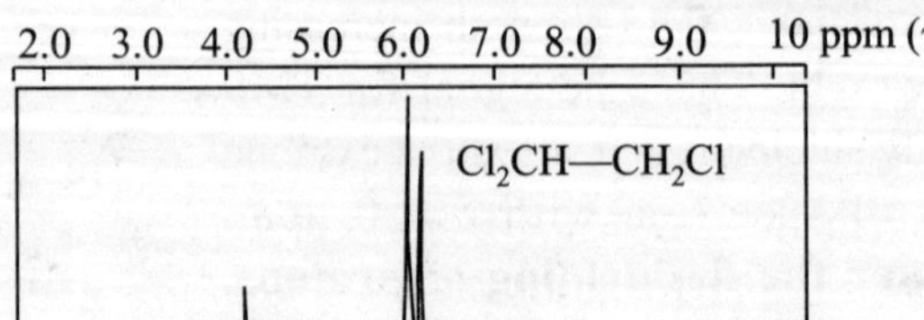

Fig. 5.33 NMR spectrum of 1, 1, 2-trichloroethane

Signals and their absorption positions

(*i*) a three proton triplet $\delta = 1.65$ ppm

(*ii*) a two proton quartet $\delta = 3.4$ ppm

Similarly, in the NMR spectrum of 1, 1, 2-trichloroethane $(Cl_2\overset{b}{CH}.\overset{a}{CH_2}Cl)$, two signals are observed. For '*a*' kind of protons, a doublet (a group of two lines) is observed while a triplet (a group of three lines) is noticed for '*b*' kind of protons.

Signals and their absorption positions

(*i*) a two proton triplet $\delta = 3.95$ ppm

(*ii*) a one proton quartet $\delta = 5.8$ ppm

Cause of splitting: The splitting of signals is related to number of protons in the neighbouring atoms (*n*)

e.g.,

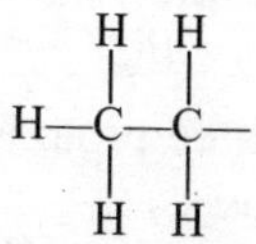

(A) (X)

A = Tertiary proton
X = Secondary porton

A-protons (Tert Proton) lies in magnetic environment of X–proton (Sec. proton) and this Proton-A can have two orientation as shown in Fig. 5.34 (*a*)

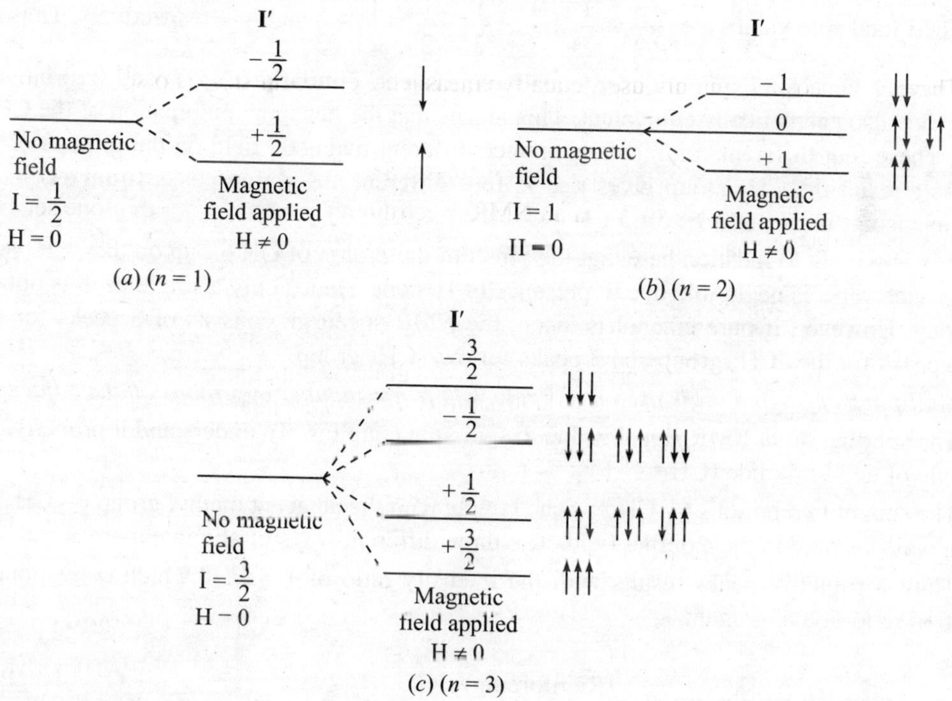

Fig. 5.34 Splitting of the energy levels in a magnetic field by neighbouring proton (*n*)

Effective magnetic field experienced by sec. proton is now modified.

(*a*) If A-proton lies **parallel** to the H_0 then absorption of X-proton will be **downfield**

(*b*) If A-proton lies **antiparallel** then absorption of X-proton will be **up field.**

The chances to align parallel or antiparallel are equal. Therefore, the absorption signal will split into peaks, called **doublet** with equal intensities *i.e.,* 1 : 1.

Likewise, with respect to a proton of the CH_3 group, the two protons of the CH_2 group can align themselves in the following four different ways [Fig. 5.34 (*b*)].

↑↑	↑↓	↓↑	↓↓
(*i*)	(*ii*)	(*iii*)	(*iv*)

Thus their total spin values are 1, 0, 0 and –1 respectively. As (*ii*) and (*iii*) have same total spin, they are energetically equivalent. Therefore, the net spin combinations of CH_2 group are three. These three different combinations induce different magnetic fields on the protons of the CH_3 group. As a result, CH_3 group gives rise to three different absorption peaks **(triplet)** having the relative intensities in the ratio **1 : 2 : 1.**

Similarly with respect to a proton of the CH_2 group, the three protons of CH_3 group can align themselves in eight different possible ways [Fig. 5.34 (*c*)] as follows:

↑↑↑	↑↑↓	↑↓↑	↓↑↑	↓↓↑	↓↑↓	↑↓↓	↓↓↓
(*i*)	(*ii*)	(*iii*)	(*iv*)	(*v*)	(*vi*)	(*vii*)	(*viii*)

Their total spin values are $+\frac{3}{2}, +\frac{1}{2}, +\frac{1}{2}, +\frac{1}{2}, -\frac{1}{2}, -\frac{1}{2}, -\frac{1}{2}$ and $-\frac{3}{2}$ respectively. Thus (*ii*), (*iii*) and (*iv*) having same total spin are energetically equivalent. Similarly (*v*), (*vi*) and (*vii*) having same total spin are also energetically equivalent. This means that the net spin combination of the CH_3 group are four. These four different combinations induce different magnetic field on the protons of the CH_2 group. As a result the CH_2 group gives rise to four different absorption peaks **(quartet)** having the relative intensities in the ratio **1 : 3 : 3 : 1.**

It is important to mention here that the effect of the proton of OH group on the $-CH_2$ and $-CH_3$ groups or vice versa is negligible due to presence of H^+ ions. Hence only a single peak is obtained for –OH group. However, if pure ethanol is taken, the NMR spectrum consists of 3 peaks for the –OH group, 8 peaks for the $-CH_2$ group and 3 peaks for the $-CH_3$ group.

Multiplicity of signal is given by n + 1 rule, where n = number of protons on the adjacent atoms.

The splitting of an NMR signal is due to spin-spin coupling. To understand it properly, consider a molecule of ethyl bromide (CH_3CH_2Br).

The spin of two protons ($—CH_2—$) can couple with the adjacent methyl group ($—CH_3$) in three different ways relative to the external field. The three different ways of alignment are:

Thus, a triplet of peaks results with the intensity ratio of 1 : 2 : 1 which corresponds to the distribution ratio of a alignment.

(*i*)	↑↑	(Reinforcing)	↑ External field (Applied magnetic field)
(*ii*)	↓↑ ↑↓	(Not effecting)	
(*iii*)	↓↓	(Opposing)	

Similarly, the spin of three protons (CH_3—) can couple with the adjacent methylene group (—CH_2—) in four different ways relative to the external field.

(*i*)	↑↑↑	(Strongly reinforcing)	
(*ii*)	↑↑↓ ↑↓↑ ↓↑↑	(Weakly reinforcing)	External field (Applied magnetic field) ↑
(*iii*)	↓↓↑ ↓↑↓ ↑↓↓	(Weakly opposing)	
(*iv*)	↓↓↓	(Strongly opposing)	

Thus, quartet of peaks results with an intensity ratio of 1 : 3 : 3 : 1 which corresponds to the distribution ratio of all the alignment. The relative intensities of the individual lines of a multiplet correspond to the numerical coefficient of the lines in the binomial expression:

$$(1 + x)^n = 1 + x \text{ if } n = 1$$

If $n = 2$, then $(1 + x)^2 = 1 + 2x + x^2$. Thus, the lines of the triplet have relative intensities 1 : 2 : 1.

If $n = 3$, then $(1 + x)^3 = 1 + 3x^2 + 3x + x^3$. Thus, the lines of the quartet formed due to the influence of three equivalent protons will have relative intensities 1 : 3 : 3 : 1.

Similarly, the lines of the pentet (quintet) formed will have relative intensities 1 : 4 : 6 : 4 : 1.

Hence, the splitting of a signal is due to the different environment of the absorbing proton not with respect to electrons but with respect to the nearby protons (protons attached to the adjacent atom).

It is important to remember that no splitting is caused in the following cases:

(*i*) ***When all the protons in a molecule are equivalent.*** For example, in case of cyclobutane, benzene etc., we observe a singlet as all the protons in each of the above compounds are exactly equivalent.

(*ii*) When a set of equivalent protons do not have another set of protons attached with the adjacent carbon atom. For example, in the NMR spectrum of 2-butanone ($\overset{c}{C}H_3\overset{b}{C}H_2—CO—\overset{a}{C}H_3$), a singlet appears for protons '*a*', since there is no hydrogen atom attached with the adjacent carbon atom.

The peak intensities of the various multiplet can be described by Pascal's triangle as under.

Peak intensities of various multiplies (Pascal's triangle)

Peak area ratio	*Signal*
1	Singlet
1 : 1	Doublet
1 : 2 : 1	Triplet
1 : 3 : 3 : 1	Quartet
1 : 4 : 6 : 4 : 1	Quintet
1 : 5 : 10 : 10 : 5 : 1	Sextet

It is important to note that the peak area depends upon the number of the absorbing protons and the multiplicity of a signal depends upon the number of neighbouring protons.

Let us consider a few examples:

$$\begin{array}{ll} CH_2 — & O \\ | & | \end{array}$$

(*a*) In the spectrum of $CH_2—CH_2$, two signals are observed;

(*i*) A four proton triplet downfield (Area = 2)

(*ii*) A two proton quintet upfield (Area = 1)

(*b*) In the spectrum of $\begin{matrix} CH_3 \\ CH_3 \end{matrix} \!\!> CH—C{\equiv}N$, two signals are observed;

(*i*) A one proton septet downfield (Area =1)

(*ii*) A six proton doublet Upfield (Area = 6)

(*c*) In the NMR spectrum of 2-Butanone, $\overset{a}{CH_3}—CO—\overset{b}{CH_2}—\overset{c}{CH_3}$ three signals are observed '*a*' proton appears as a singlet and the multiplicities of protons '*b*' and '*c*' are:

(*i*) A two proton quartet ($—CH_2—$), downfield (Area =2)

(*ii*) A three proton triplet ($—CH_3$), upfield (Area = 3)

From the above examples, it is clear that the chemical shifts for various protons are exactly in accordance with the factors on which the values of chemical shifts depend. The presence of electronegative atoms or groups near the proton, bring the absorption downfield and the effect diminishes with the increasing distance of the said atom or group from the concerned proton.

Important note: *Spin-spin coupling takes place between non-equivalent neighbouring (or on adjacent carbon atom) protons. Non-equivalent protons are those which have different chemical shifts.*

- The multiplicity of a multiple is given by number of equivalent **protons** in **neighbouring** atoms plus one, *i.e.* then $n + 1$ rule.
- Equivalent nuclei do not interact with each other. The three methyl protons in ethanol cause splitting of the neighbouring methylene protons; they do into cause splitting among themselves.
- The coupling constant is not dependant on the applied field. Multiplets can be easily distinguished from closely spaced chemical shift peaks.

Splitting of signals in other cases

If the protons responsible for spin-spin coupling are not in exactly similar environment, the number of lines for a particular multiplet will be equal to $(n + 1)(n' + 1)(n'' + 1)$ where n, n' and n'' are the number of protons in different environments. A few examples are listed below:

(***a***) Consider a compound, 1, 1-dibromo-3, 3-dichloropropane $Br_2\overset{a}{CH}—\overset{b}{CH_2}—\overset{c}{CHCl_2}$. In this compound, we see three sets of protons. Protons '*a*' and '*c*' are non-equivalent (with different chemical shifts) and thus, will influence the methylene ($—CH_2—$ *i.e.*, protons '*b*') protons differently. Thus the signal for $—CH_2—$ protons appear as a multiplet and consists of $(n + 1)(n' + 1) = (1 + 1)(1 + 1) = 4$ lines where n and n' are the number of protons of the kinds '*a*' and '*c*'.

(***b***) Consider a compound, 1, 1-dibromo-3, 3-dichloropropane $Br_2\overset{a}{CH}—\overset{b}{CH_2}—\overset{c}{CHCl_2}$. The signal for the central methylene ($—CH_2—$) protons will clearly be a multiplet consisting of $(n + 1)(n' + 1) = (2 + 1)(2 + 1) = 9$ peaks where n and n' are the number of protons '*a*' and '*c*' respectively. The central $—CH_2—$ proton being under the influence of two kinds of protons (two in each set) with different chemical shifts. Thus, for central methylene, asymmetrical pattern results.

5.30.2 ^{1}H-NMR Spectra of Ethanol

The way in which the proton NMR spectra help in the study of molecular structure is explained below taking the example of *acidified* ethyl alcohol (CH_3CH_2OH).

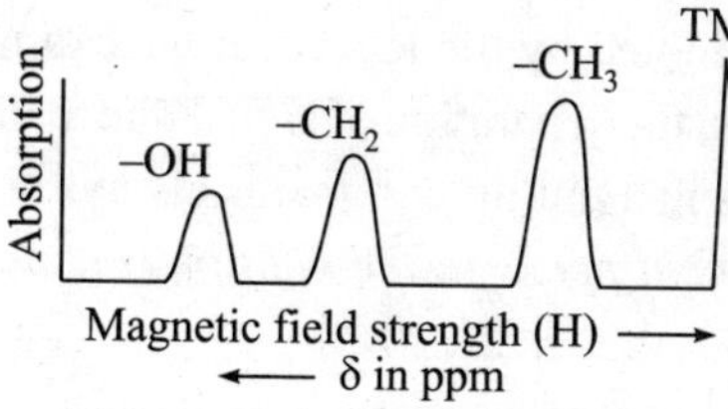

(*a*) Low resolution NMR spectrum or ethyl alcohol

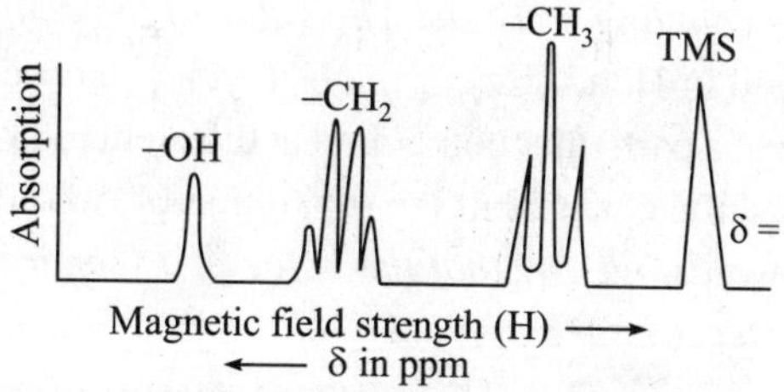

(*a*) High resolution NMR spectrum or ethyl alcohol

Fig. 5.34 NMR spectrum of ethyl alcohol

5.31 SPIN - SPIN COUPLING

Methylene ↓

Consider the structure of ethanol $CH_3—CH_2—OH$

↑ Methyl

The ^{1}H-NMR spectrum of ethanol (below) shows the methyl peak has been *split into* three peaks (a *triplet)* and the methylene peak has been split into four peaks (a *quartet).* This occurs because there is a *small interaction (coupling)* between the two groups of protons. The spacings between the peaks of the methyl triplet are equal to the spacings between the peaks of the methylene quartet. This spacing is measured in Hertz and is called the coupling constant (J).

To see why the methyl peak is split into triplet, let's look at the methylene protons. There are two of them, and each can have one of two possible orientations (aligned with or opposed against the applied field). This gives a total of four possible states;

In the first possible combination, spins are paired and opposed to the field. This has the effect of reducing the field experienced by the methyl protons; therefore a slightly higher field is needed to bring them to resonance, resulting in an up field shift. Neither combination of spins opposed to each other has an effect on the methyl peak, The spins paired in the direction of the field produce a downfield shift. Hence, the methyl peak is split into three, with the ratio of areas 1 : 2 : 1. Similarly, the effect of the methyl protons on the methylene protons is such that there are eight possible spin combinations for the three mehtyl protons.

Out of these groups, there are two groups of three magnetically equivalent combinations. The mehtylene peak is split into a quartet. The areas of the peaks in the quartet have the ratio 1 : 3 : 3 : 1. In a *first-order* spectrum (where the chemical shift between interacting groups is much larger than their coupling constant), interpretation of splitting patterns is quite straight forward.

The magnetic field experienced by a proton is influenced by various structural factors. Since the magnetic field strength dictates the energy separation of the spin states and hence the radio frequency of the resonance, the structural factors mean that **different types of proton** will occur at **different chemical shifts.** This is what makes NMR so useful for structure determination, otherwise all protons would have the same chemical shift.

5.32 COUPLING CONSTANT

The distance between the centres of the two adjacent peaks in a multiplet is usually constant and is called the coupling constant. The value of the coupling constant is independent of the external field. It is measured in Hertz (Hz) or in cps (cycles per second). It is denoted by the letter J. If we examine the spectrum of a particular compound at different radio-frequencies, the separation of signal due to different chemical shifts change but the separation of two adjacent peaks in a multiplet remains always constant. *In other words, we say that the value of J (separation of adjacent peaks in the multiplet) remains the same whatever be the applied field.*

From the value of coupling constant, one can distinguish between the two singlets and one doublet and also a quartet from two doublets. It can be done by simple recording the spectrum at two different radio-frequencies. If the separation (in Hz) between the lines (value of J) does not change, then the signal is a doublet. On the other hand, if the separation between the lines increases with increasing frequency, then the signal in fact, will be two singlets. The value of 'J' generally lies between 0 and 20 Hz. Similar explanation can be given to distinguish a quartet from the doublets.

Now let us consider a compound,

$>\overset{b}{CH}-\overset{a}{CH_2}-$. In this compound two signals are expected in the NMR spectrum. Under the influence of two equivalent protons '*a*', the signal for proton '*b*' will appear as a triplet. The distance between any two adjacent peaks in a multiplet will be exactly the same. The triplets formed due to spin coupling as shown in Fig. 5.36.

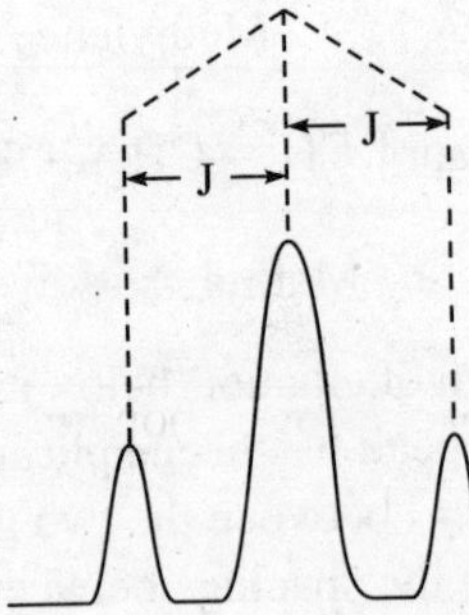

Splitting of proton 'b' signal w.r.t. Equivalent proton 'a'

Fig. 5.36 A triplet

Similarly, consider a compound, $>\overset{b}{CH}-\overset{a}{CH_3}$. In the NMR spectrum of this compound, proton '*b*' is under the influence of three equivalent protons '*a*'. Thus, due to spin-sin coupling, the signal for proton '*b*' will appear as quartet with intensity ratio 1 : 3 : 3 : 1. The quartet for '*b*' can be represented as in Fig. 5.37.

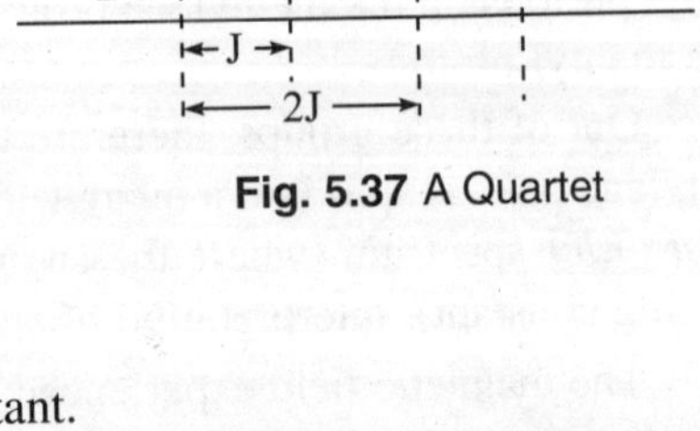

Fig. 5.37 A Quartet

From the constant value of 'J', we say that the multiplets formed are of symmetrical nature. It has been noted that the departure from the symmetrical nature of the multiplet results when the absorption position (chemical shifts) of the interacting groups are close enough.

In the NMR spectrum of 1,1,2- trichloroethylene (Fig. 5.38) two-multiples are observed.

The value J in each of its multiplets is found to be constant.

But if a signal is split due to the influence of two different sets of protons (protons with different chemical shifts), then the peak separations in the multiplet observed will not be equidistant.

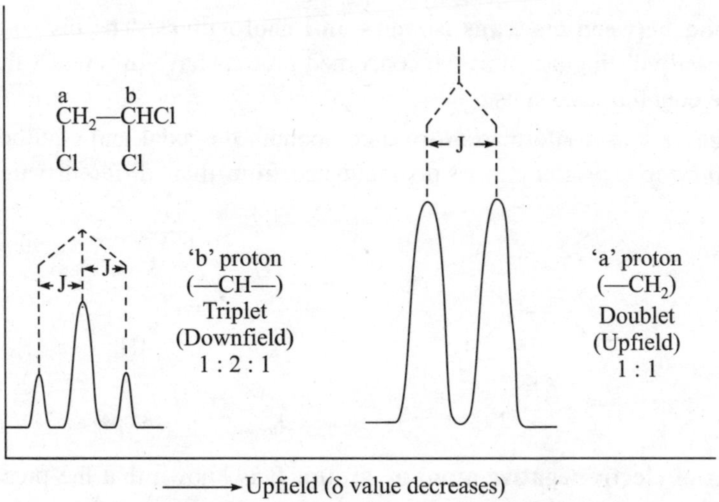

Fig. 5.38 NMR Spectrum of 1, 1, 2-trichloroethane

5.33 APPLICATIONS OF NMR SPECTROSCOPY

1. Detection of hydrogen bonding: Intermolecular hydrogen bonding shifts the absorption for a concerned proton downfield. The extent of hydrogen bonding varies with the solvent, concentration of the solution and the temperature. Intra-molecular hydrogen bonding also shifts the absorption downfield. The two types of hydrogen bonding can be distinguished as the intra molecular hydrogen bonding is not concentration dependent.

2. Identification of structural isomers: The distinction between the following isomers can be easily made from their NMR spectra

(*i*) $CH_3CH_2CH_2Cl$ (*a*)

$CH_3—CH(Cl)—CH_3$ (*b*)

In the Isomer '*a*' three signals are observed whereas we see only two signals in the spectrum for '*b*' which is a clear distinction between the above isomers. The three signals for isomer '*a*' in order of decreasing tau values are:

(*i*) A three proton triplet (CH_3—)

(*ii*) A two proton sextet (—CH_2—) and

(*iii*) A two proton triplet (—CH_2Cl).

For Isomer (*b*), two signals have their multiplicities as:

(*i*) doublet (6H)—upfield and

(*ii*) septet (IH)—downfield.

3. Detection of aromaticity. Protons attached to the benzoyl, polynuclear and heterocyclic compounds whose π-electrons follow Huckel's rule [*i.e.*, $(4n + 2)$ π-electrons where $n = 1, 2, 3$ ... (whole number)] are extremely deshielded due to the circulation sextet (ring current) of π-electrons. As a result of this, the signal for the aromatic protons appear at a very low field than that observed even for benzene. From this, the aromatic character of the compound under investigation can be predicted.

4. Distinction between cis-trans isomers and conformers: The cis and transisomers of a compound can be easily distinguished as the concerned protons have different values of the chemical shifts as well as the coupling constants.

Similarly, the various conformations of a compound, the axial and equitorial positions of the proton or group carrying a proton can be distinguished from their different values of the coupling constants.

(*cis*) $J_{cis} = 7 - 12$ cps (*cis*) $J_{trans} = 13 - 18$ cps

5. Detection of electronegative atom or group: It is known that the presence of an electro-negative atom or group in the neighbourhood of the proton cause deshielding and the signal is shifted downfield. Greater the electro-negativity of the adjacent atom, smaller is the tau value of absorption for the concerned proton. Fluorine causes more downward shift as compared to oxygen and oxygen in turn causes more downward shift as compared to nitrogen and so on.

6. Detection of some double bond character due to resonance: In some compounds, the molecule acquires a little double bond character due to resonance. Due to this, two signals can be expected for apparently equivalent protons. It is due to the hindered rotation which changes the geometry of the molecules. Consider N, N dimethyl formamide, It can be written in the following resonating structures:

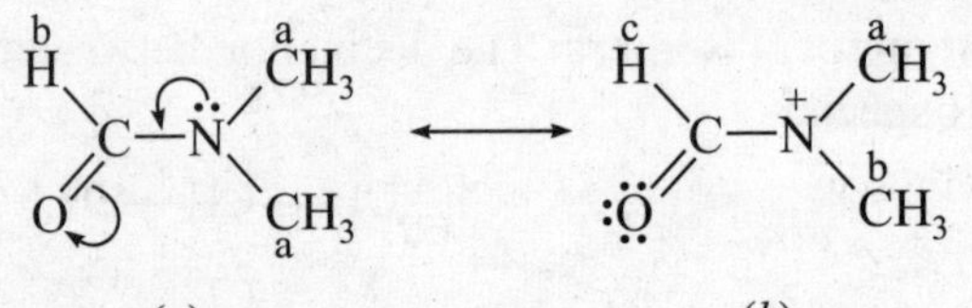

(*a*) (*b*)

For structure (*a*), two signals (singlets) should be expected with peak areas 6 : 1 as the two methyls are exactly equivalent.

In structure (*b*), the presence of double bond restricts rotation and now the two methyl groups remain no longer equivalent (Geometrical isomers). For this structure, two signals appear for two methyl groups.

7. Importance in quantitative analysis: NMR spectroscopy is gaining importance for the quantitative analysis of the compounds. Equilibrium mixtures can be analysed when the proton signals of the components are well separated. In the *NMR* spectrum of pure ethanol (CH_3CH_2OH), a triplet is formed for the OH proton but when water is added in alcohol, then due to proton exchange, the triplet collapses to a singlet. The position of this singlet depends upon the water content in alcohol. From the values of the chemical shift, the ratio of water and alcohol can be estimated by comparing with the known results.

8. Keto-enol Tautomerism: Keto-enol tautomerism has also been studied in the same manner. Consider acetyl acetone in equilibrium with the enolic form.

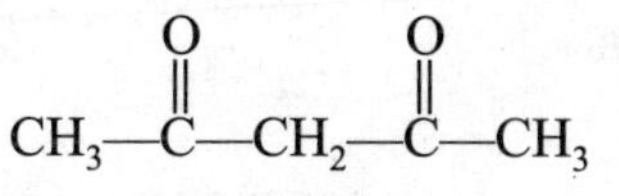

(Keto)

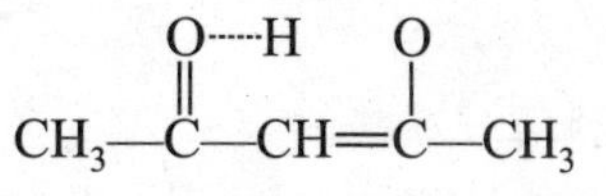

(Enol)

The *NMR* spectrum of pure acetyl acetone shows signals for the following equivalent protons:

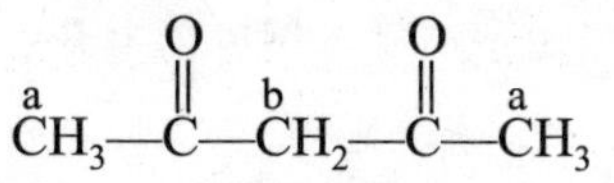

(Ketonic)

a = 7.86 τ
b = 6.35 τ [ketonic]

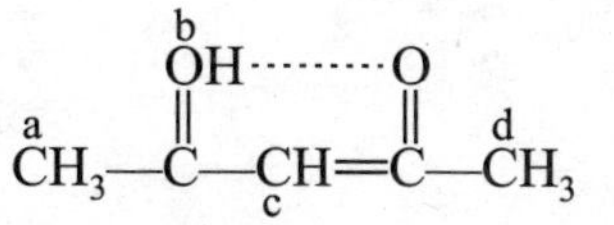

a = 8.03 τ [enolic]
b = –4.3 τ
c = 4.4 τ
d = 7.87 τ

A distinct two proton singlet at 6.35 τ appears due to the ketonic form whereas all the eight protons in the enolic form are observed as four signals. Quantitatively, the number of enol protons relative to the keto protons yields the ratio as under :

Enol protons : keto protons = 8 : 2 or 4 : 1

Thus, we say that 80% of enolic form and 20% of the ketonic form are present in the equilibrium mixture of acetyl-acetone.

PRACTICE PROBLEMS

1. What values can be assigned to the nuclear spin quantum number (I)?
2. What inference can be drawn when the value of I is (*i*) integral (*ii*) fractional?
3. Which of the followings would exhibit nuclear magnetic resonance?
   (*i*) Hydrogen, (*ii*) Deuterium,
   (*iii*) Carbon, (iv) Sodium,
   (*v*) Fluorine, (*vi*) Oxygen.
4. Write expression for (*i*) nuclear spin and (*ii*) Magnetic moment, in terms of the spin quantum number.
5. In which of the following cases the molecule will show splitting of energy levels in a magnetic field?
   (*i*) I = 0 (*ii*) I = 0.5
   (*iii*) I =1 (*iv*) I =1.5
   (*v*) I = 2
6. Write expressions for the energy of different levels of a molecule in magnetic field when the value of spin quantum number (I) is 1.5.
7. What is the expression for the resonance frequency (*v*) in NMR spectroscopy?
8. Write an expression for the chemical shift in NMR, giving the significance of the terms involved.
9. Why three peaks appear in the NMR spectrum of ethyl alcohol?
10. Name the commonly employed reference substance in NMR spectroscopy.

Answers

1. 0, 0.5, 1, 1.5, 2, 2.5 etc:
2. (*i*) The sum of Neutrons and Protons in the nucleus is an even number.
   (*ii*) Neutrons + Protons = Odd number.
3. Hydrogen, sodium and fluorine.
4. Nuclear spin = $\sqrt{I(I+1)}\ h/2\pi$ and $\mu = g_n\mu_n \sqrt{I(I+1)}$ where g_n is the Lande's splitting factor and m_n is the nuclear magnetic factor.
5. (*ii*), (*iii*), (*iv*) and (*ii*) all will show splitting as $I \neq 0$ for them.
6. $-1.5\ g_n\mu_n$ H, $-0.5\ g_n\mu_n$ H, $0.5\ g_n\mu_n$ H and $1.5\ g_n\mu_n$ H
7. $\nu = \pm\dfrac{g_n\mu_n\ H}{h}$
8. $\delta = \dfrac{H_s - H_{ref}}{H_{ref}} \times 10^6$ where H_s and H_{ref} are the field strengths at which the sample and the reference substance are in resonance for a given frequency.
9. It is due to the presence of H atom in three different environment.
10. Tetramethyl Silane (TMS).

FOLLOW-UP SOLVED PROBLEMS

Problem 1: ***Determine which of the following molecules will show spin-spin coupling: in their NMR spectra. If the coupling or splitting is observed, give the multiplicity of each kind of proton, i.e., give the number of lines arising from each proton:***

(*a*) $ClCH_2CH_2Cl$ **(*b*) $ClCH_2CH_2I$** **(*c*)** 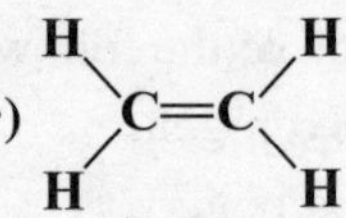

(*d*) 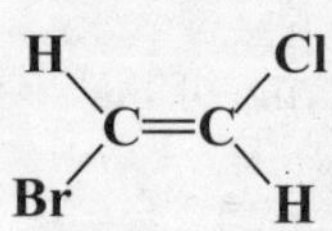 **(*e*)** 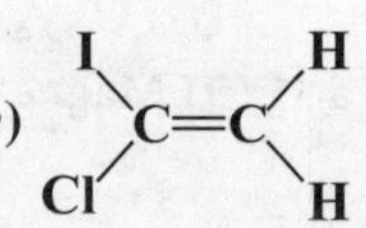 **(*f*)** 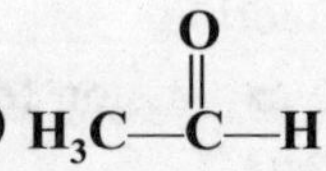

(g) 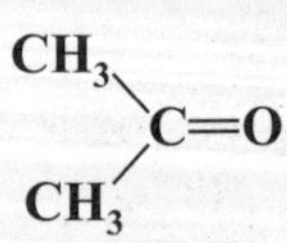 **(*h*) $CH_3—C(Cl_2)—CH_3$**

Solution: Spin-spin splitting is not observed in (*a*), (*c*), (*g*) and (*h*) which have only equivalent protons. The protons of the CH_2 groups in (*b*) are non-equivalent each being split into a 1 : 2 : 1 triplet. In (*d*), the two protons are not equivalent and each is split into a doublet. The vinyl protons in (*e*) are not equivalent since one is cis to Cl and the other is cis to I so that each is split into a doublet. Notice that in this case the protons undergoing spin-spin interaction are on the same carbon. In (*f*) the protons of CH_3 and the proton of the —CH=0 group are not equivalent;^ the CH_3 resonance is split into a doublet and the —CH=0 resonance is split into a 1 : 3 : 3 : 1 quartet.

Problem 2: ***How do you account for the fact that nuclear spin-spin splitting is observed in 2 methyl propene but not in l-chloro-2,2-dimethyl-propane?***

Solution: In 2-methyl propane, 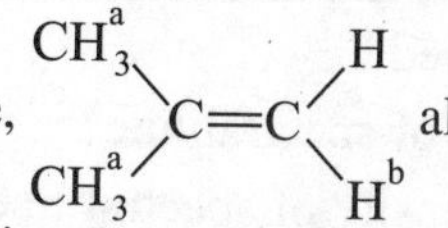 although H^a and H^b are not on adjacent carbon atoms, they undergo spin-spin splitting because the coupling is transmitted through the π-electrons. In,

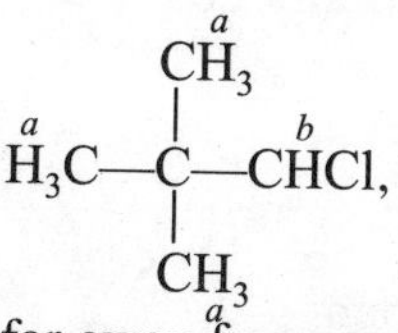 , H^a and H^b do not couple because they are not on adjacent carbons atoms; they are too far away from one another.

Problem 3: ***How can the NMR spectrum distinguish between the isomers p-xylene and ethyl benzene?***

Solution: The isomers are:

p-xylene 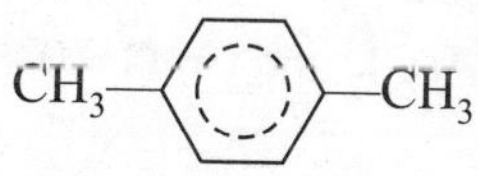 *i.e.*, *p* $CH_3^a\, C_6H_4^b\, CH_3^a$

Ethyl benrene, 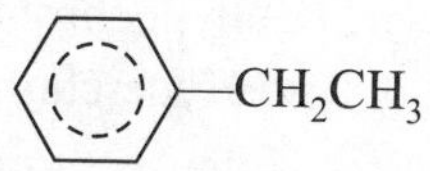 *i.e.*, $C_6H_5^a\, CH_2^b\, CH_3^c$

$\overset{a}{CH_3}\,\overset{b}{C_6H_4}\,\overset{a}{CH_3}$ has two kinds of protons and its NMR spectrum consists of two singlets.

$C_6H_5^a\, CH_2^b\, CH_3^c$ has three kinds of protons and its NMR spectrum consists of a singlet for H^a, a quartet for H^b and a triplet for H^c.

Problem 4: ***Predict the high-resolution proton-NMR spectra of CH_3CH_2Cl and H_3CHDCl.***

Solution. Deuterium (D) does not give a signal in the proton NMR spectrum, nor does it split the nearby protons. Thus, we can ignore its presence in the molecule. The NMR spectrum of $CH_3^a\, CH_2^b\, Cl$ consists of 1 : 2 : 1 triplet for H^a and, more down-field, a 1 : 3 : 3 : 1 quartet for H^b.

The NMR spectrum of $CH_3^a\, CH^b DCl$ has a doublet for H^a and, more downfield, 1 : 3 : 3 : 1 quartet for H^b.

Problem 5: ***Predict the proton NMR spectra of H_2, CH_4, C_2H_6 and C_6H_6***

Solution: All the four molecules contain equivalent protons which do not show spin-spin fraction. Hence, their proton NMR spectra will consist of one signal only, *i.e.* a singlet.

REVISION QUESTIONS

1. What is molecular spectroscopy? How does it differ from atomic spectroscopy?
2. Draw the molecular energy levels of the first two electronic levels. What are vibrational and rotational quantum numbers?

3. Briefly explain the terms: Absorption and Emission spectroscopy. Why absorption spectroscopy is preferred? Give its experimental set-up.
4. Name the different types of molecular spectra. Explain each of them briefly. In which region of the electromagnetic spectrum are they obtained?
5. (*a*) What are the factors that affect the width of spectral line?
   (*b*) What are the factors which affect the intensity of spectral line?
6. Show vibrational, rotational and electronic energy levels on an energy level diagram. In which region of electromagnetic spectrum do rotational, vibrational and electronic spectra lie?
7. Explain the following terms.
   (*i*) Spin-spin Coupling (*ii*) Coupling constant
   (*iii*) Shielding-Deshielding Effect
8. Explain the principle and working of spectrophotometer, giving its neat diagram.
9. What is an absorption spectra? Write what do you know about $n \rightarrow \pi^*$ and $n \rightarrow \sigma^*$ transitions? Explain by giving examples.
10. Why molecules absorb in UV-visible region? What are the type of electronic transitions that can occur in a molecule? Discuss giving examples.
11. Predict the kind of electronic transitions in (*a*) Cl_2 and (*b*) C = 0 group. Also give their intensity.
12. (*a*) Define the following terms
    (*i*) Bathochromic shift (*ii*) hypochromic shift
    (*iii*) A chromophore (*iv*) Hyperchromic effect
    (*b*) Which solvent in generally used in ultra-violet spectroscopy and why?
13. Explain any two of the following in respect of UV-visible spectroscopy:
    (*i*) Bathochromic shift (*ii*) Blue shift
    (*iii*) Hyperchromic effect
14. Calculate the force constant for HCl^{35} when it's fundamental vibrational frequency is 8.667×10^{13}/sec. How does the force constant of HCl^{37} compare with it?
15. (*a*) What do you understand by nuclear spin? When will a nuclei show nuclear magnetic resonance?
    (*b*) Which of the following nuclei do not show NMR?
    1H, 2H, 12C, 13C, 14N, 15N, 16O, 19F, 31P, 32S.
16. (*a*) Briefly explain the principle of NMR spectroscopy.
    (*b*) How is the frequency involved in the nuclear transitions related to the strength of magnetic field?
17. (*a*) What is meant by nuclear magnetic resonance? Describe a method for determination of NMR spectrum.
    (*b*) What information regarding the structure of molecules can be obtained from NMR spectroscopy?
18. Write a short note on chemical shift. What information do you get from intensity of the absorption peaks in an NMR spectrum?
19. Briefly explain the term chemical shift. Why is the factor 10^6 introduced in expressing its values? Name the most commonly used compound as a reference and give its advantages.
20. (*a*) Draw structural formulae of the following compounds which give only one NMR signal.
    (*i*) C_3H_6 (*ii*) C_4H_6
    (*iii*) C_2H_6O (*iv*) $C_2H_4Br_2$

(*b*) Draw structural formulae of the following compounds having two singlets in their NMR spectra.

(*i*) $C_3H_6O_2$ (*ii*) $C_3H_8O_2$

21. (*a*) Draw the low and high resolution spectra of C_2H_5OH.

(*b*) What type of molecules gives pure rotational spectra? Derive expressions for frequency of rotational lines.

(*c*) What type of nuclei will give NMR spectra? Give two examples of each type.

(*P.T.U. July 2001*)

22. (*a*) Draw a block diagram of a typical instrument for making UV/Visible absorption measurement. Briefly describe the various components of the instrument. (*P.T.U. Jan. 2002*)

(*b*) Explain the following in respect of UV-Visible spectroscopy:

(*i*) Red shift or bathochromic effect, (*i*) Blue shift or hypsochromic effect.

(*iii*) Hyperchromic effect (*iv*) Isobestic point

(*P.T.U. June 2003*)

23. (*a*) State and explain the principle of NMR spectroscopy. (*P.T.U. May 2004*)

(*b*) Give the high resolution H-NMR spectrum of Ethanol. (*P.T.U. May 2004*)

(*c*) How many kinds of signals are obtained in the NMR spectrum of CH_3CU_2OH?

(*P.T.U. Dec. 2004*)

24. Describe a UV-Visible double beam spectrophotometer. (*P.T.U. May. 2004*)

25. (*a*) What do you understand by I.R. Spectra? (*P.T.U. Dec. 2004*)

(*b*) On the basis of infrared spectroscopy, how can you distinguish among *l*-hexyne, *l*-hexene and hexane? (*P.T.U. Dec. 2004*)

(*c*) How many NMR signals are observed in the spectrum of

(*i*) CH_3COOCH_3 (*ii*) $CH_3—CH_2—$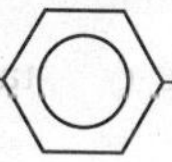$—CH_2—CH_3$

(*iii*) CH_3OCH_3 (*iv*) $CH_3CH_2CH_3$

(*P.T.U. Dec. 2005*)

26. (*a*) How can one establish whether a carbonyl group is part of an aldehyde, ester or ketone from UV spectroscopy? (*P.T.U. May 2006*)

(*b*) What is the principle behind nuclear magnetic resonance spectrum? (*P.T.U. May 2006*)

(*c*) Define the terms bathochromic shift and hypsochromic shift. What structural feature may produce bathochromic or a hypsochromic shift in an organic compound.

(*P.T.U. Dec. 2006, May 2007*)

27. (*a*) What is the importance of IR spectroscopy in finger print region?

(*P.T.U. Dec. 2006, May 2007*)

(*b*) Explain the Chemical shift of NMR spectroscopy: (*P.T.U. Dec. 2002,2007*)

28. (*a*) Why is methanol a good solvent for UV but not for IR determination? (*P.T.U. Dec. 2007*)

(*b*) Differentiate between bathochromic and hypsochromic shifts. (*P.T.U. Dec. 2008*)

(*c*) Explain UV-Transitions. (*P.T.U. Dec. 2008*)

(*d*) How do you distinguish between different types of transitions involved in UV-visible spectrophotometry? (*P.T.U. Dec. 2008*)

29. (*a*) Indicate the number and type of ^{1}H-NMR signals expected in $CH_3—CH=CH_2$. (*P.T.U. Dec. 2008*)

(*b*) Write notes on: Chemical shift in NMR. (*P.T.U. May 2008*)

30. (*a*) What type of molecules exhibit vibrational-rotational spectra and why? (*P.T.U. Dec. 2009*)

(*b*) What do you understand by term spin-spin coupling? Is the coupling constant independent of the applied field or depend on it. (*P.T.U. Dec. 2009*)

(*c*) Cause of chemical shift in NMR. (*P.T.U. Dec. 2009*)

31. (*a*) Give the high resolution ^{1}H-NMR of methyl ethyl ether. (*P.T.U. May. 2010*)

(*b*) What are different kinds of electronic transitions? Explain chromophores and auxophores. (*P.T.U. May. 2010*)

❑❑❑

6 Fuels and Combustion

INSIDE THIS CHAPTER

6.1 INTRODUCTION

Fuel may be defined as: *"Any combustible substance which on burning in air gives large amount of heat that can be used economically for domestic and industrial purposes."* The important common fuels are: Wood, coal, petrol, kerosene, diesel oil, coal gas, oil gas, water gas, producer gas etc.

Fuels are primarily used for heating purposes and heat energy is the main source of power, but they also perform several other important functions. For example, coal in locomotive engines and petrol in internal combustion (I.C.) engines is used to do some mechanical work. Similarly, coke is used in blast furnace in the manufacture of iron, it provides not only heat, but also acts as a reducing agent.

The main sources of fuel are coal and petroleum oils, but due to rapidly growing industries these are reducing day by day. These **are fossil/stored fuels** available in earth's crust, in the near future, the primary source of fuels may get exhausted with the progressive consumption and great demand all over the world. Thus, attempts are being made to use uranium (nuclear fuel), sea water and solar energy etc. as fuels.

It is row well established that combustion is not necessary for a fuel to give out energy. For example, electrical energy which is used as a source of heat does not involve combustion but it is also called fuel. Nuclear energy may also be used as a source of heat. Chemical compounds, such as power alcohol, hydrazine, ammonia etc. are used as fuels.

According to a modified definition, *a fuel is any fissionable material, chemical or reactant which produce energy in any form that can be used for producing power.*

For example, in electrochemical cells or fuel calls, the chemical energy of the reactants is converted into electrical energy by a suitable device. Similarly, in a nuclear reactor fissionable materials like U-235 can be broken into smaller fragments and the loss in mass appears in the form of energy. The huge amount of energy thus produced can be controlled and used for steam generation which in turn can produce electricity. The materials used in these devices which act as a source of energy in any form are nothing else but fuels. A fuel in the broader sense of the term is not essentially a combustible substance, it should be a source of energy through combustion or even otherwise.

6.2 CLASSIFICATION OF FUELS

Fuels may be classified on the basis of (*i*) their occurrence or preparation and (*ii*) physical state. Thus on the basis of the occurrence or preparation fuel may be of two types:

(*a*) **Natural or primary fuels:** Fuels which occur in nature are called *natural fuels.* Common examples are wood, coal, peat, petroleum and natural gas.

(*b*) **Artificial or secondary fuels:** Fuels which are manufactured artificially generally from primary fuels are called *artificial fuels.* Common examples are coke, kerosene oil, petrol, coal gas, producer gas etc.

On the basis of physical state, fuel may be above types, *viz.* ***solid, liquid and gaseous.*** All the three types of these fuels are further divided into primary (natural) and secondary (artificial) fuels. The important examples of primary and secondary solid, liquid and gaseous fuels are tabulated below:

Table 6.1: Classification of Fuels

Type of fuel	*Natural or Primary*	*Artificial or Secondary*
Solid	Wood, peat, lignite, bituminous coal and anthracite coal	Coarcoal, charred peat, coke an petroleum coke
Liquid	Crude petroleum	Various fractions of petroleum, *viz.* kerosene oil, petrol, diesel oil, fuel oil etc.
Gaseous	Natural gas	Liquefied Petroleum Gas (LPG), coal gas, oil gas, water gas, blast furnace gas etc.

6.2.1 Calorific Value of a Fuel

Calorific value is the important property of fuel. **It is defined as "the total quantity of heat liberated in calories by the complete combustion of unit weight or volume of the fuel in air/oxygen".** There are two types of calorific values *i.e.,* higher and lower calorific values, also known as gross calorific values and net calorific values respectively.

(*a*) **Higher calorific value (HCV) or gross calorific value (GCV):** Almost all fuels contain hydrogen and when the calorific value of such fuel is determined hydrogen is converted in to steam. If the products of combustion are condensed to the room temperature (25°C or 77°C) the latent heat of condensation of steam gets also included is then called higher or gross calorific value'. It may be defined as *the total amount of heat produced when one unit (weight or vol.) of the fuel is burnt completely and the products of combustion have been cooled to room temperature.*

(*b*) **Lower calorific value (LCV) or net calorific value (NCV):** In actual practice the water vapour and moisture etc. present in fuel are not condensed and escape as such along with flue gases.

Hence a lesser amount of heat is available. The amount of heat, so, available, is called lower or net calorific value. It may be defined *as the net heat produced when unit mass (or volume) of the fuel is burnt completely and the combustion products are allowed to escape.*

Thus, NCV = GCV – Latent heat of water vapours formed.

Since 1 part by weight of hydrogen gives 9 parts by weight of water, the above equation becomes NCV = GCV – Wt. of hydrogen × 9 × Latent heat of steam.

The latent heat of steam is 587 kcal/kg at room temperature (*i.e.,* at 25°C)

6.2.2 Units of Calorific Value

The units of calorific value of fuel can be expressed in the following ways:

(*i*) **C.G.S. System**

In this system, the unit of heat is calorie and it is defined as the amount of heat required to raise the temperature of 1 g of water from 14.5° – 15.5°C *i.e.,* by 1°C. Since, solids and liquids are measured in terms of their masses, unit of calorific value is cal/g.

(*ii*) M.K.S. System

In this system, the unit of heat is kcal which is defined as the quantity of heat required to raise the temperature of 1 kg of water from 14.5° – 15.5°C by 1°C." Thus,

1 kcal = 1000 cal.

∴ Units of calorific value: kcal/kg (M.K.S. system for solid and liquid fuels)

1 kcal/m^3 (M.K.S. system for gaseous fuels)

J/m^3 (S.I. system for gaseous fuels)

(*iii*) F.P.S. (or British System):

In this system, the unit of heat is B.Th.U., which is defined as the amount of heat required to raise the temperature of 1 product/I of water by 1°F.

Unit of calorific valuye: B.Th.U./1b ...... (For solid and liquid fuels)

B.Th.U/ft^3 ...... (For gaseous fuels)

(*iv*) S.I. System

Unit of calorific value = Joule/kg

The inter conversion of units are as follows:

1 kcal/kg = 1.8 B.Th.U/1b

1 kcal/m^3 = 0.1077 B.Th.U./ft^3

1 kcal = 4.18687 kJ

6.3 CHARACTERISTICS OF A GOOD FUEL

A good fuel should have the following characteristics:

(*i*) Calorific value: *"It is amount of heat liberated when a unit mass or volume of fuel is brunt completely"*. A good fuel must have a high calorific value, since the amount of heat obtained from a definite weight of the substance depends upon this factor.

(*ii*) Moderate velocity of combustion: The highest temperature achieved by combustion of a fuel depends upon the calorific value and the velocity of combustion. For a continuous supply of heat, fuel must burn with a moderate velocity.

***(iii)* Proper ignition point:** A good fuel must have a moderate ignition point, since low ignition temperature of fuels is mostly accompanied by fire hazards, while high ignition point makes the fuel safe for transportation and storage but not favourable for starting a fire.

***(iv)* Low contents of non-combustible material:** A good fuel must have low contents of non-combustible material as the non-combustible matter is left in the form of ash which will naturally decease the calorific value of the fuel. Moreover, disposal of ash is itself a big problem and it increases the cost of operation.

***(v)* Highest pyrometric effect:** The highest temperature obtained with the fuel on is burning known as *pyrometric burning effect.* A good fuel should have a high pyrometric effect. Gaseous fuels have the highest pyrometric effect.

***(vi)* Low moisture content:** The moisture content of the fuel should be low, because moisture reduces its calorific value. Since moisture in fuel (coal) evaporates and is removed as steam during its burning and some of the heat liberated is utilized during the conversion of moisture into steam. Therefore, the presence of moisture in fuel (coal) reduces its calorific value. Moreover, moisture also increases the transport charges.

***(vii)* Products of combustion should not be harmful:** Fuels on burning should not produce any objectionable and harmful gases which may act as physiological poisons for human lives in nearby area. It should produce minimum amount of smoke. In other words, the gaseous products of combustion of fuel should not pollute the atmosphere. Gases like carbon monoxide (CO), sulphur dioxide (SO_2), hydrogen sulphide (H_2S) and phosphine (PH_3) etc. are some of the harmful gases.

***(viii)* Low cost:** A good fuel should be readily available in bulk at a low cost. Its supply must be reliable *i.e.,* a fuel should not be adulterated.

***(ix)* Easy to transport:** A good fuel is that which can be easily handled, stored and transported at a low cost. Solid and liquid fuels can easily be transported from one place to other. On the other hand, transportation of gaseous fuels is costly and can even cause fire hazards.

(x) The storage cost of fuel in bulk should be low.

(xi) A fuel should burn in air with efficiency, without producing much smoke.

(xii) The combustion of fuel should be easily controllable.

(xiii) A fuel should not undergo spontaneous combustion. The spontaneous ignition can cause fire hazards.

(xiv) In case of solid fuels, size should be uniform, so that combustion is regular.

6.4 COAL

Coal is a highly carbonaceous complex substance, which originated from various plants and wood material that existed on the surface of the earth several million years ago. It mainly consists of carbon, hydrogen, nitrogen, oxygen and to a lesser extent sulphur and phosphorus, besides some non-combustible inorganic matter like oxides of alkali metals, alkaline earth metals, silicone etc. The plants were caught up in the swamps, buried underground and slowly changed due to heat, pressure and complex chemical and biochemical processes. Now it is found as beds or layers in between clay or sediments, which were deposited in course of time. Coal is an important fuel. The fact that on proper processing it gives gases, oils, fertilizers and chemicals increases its usefulness. As it is very easy to mine and handle, its use will continue even against severe competition from mineral oils, hydro-electro power and nuclear fuels.

6.4.1 Origin of Coal

Two different theories have been put forward for coal formation. Millions of years ago, when the temperature was moderate and rainfall was heavy, vegetation was thick and grew at far greater rate then it does at present. This was so especially in low lying wet areas. According to the first theory, known as *in situ* theory, coal formation took place where the vegetation grew and accumulated.

The second theory, known as *drift* or *transportation* theory holds that trees and plants etc. were uprooted and drifted or transported to big vallies and other such depression and deposited there. Thus huge files of wood collected and caught buried underground. There are arguments for and against each of these theories but it is now accepted that in some cases, the first theory and in other cases, the second theory is more applicable. For example, the great purity of many coal streams can be better explained by the *in situ* theory and great thickness of coal steams appears more reasonable on the basis of the *drift* theory.

Anyway after the plants were water logged, some silt was also deposited with them. It was joined by some mineral water at some stage. The initial transformation of wood is considered to be mainly due to bacterial action. In the early stages, aerobic bacteria caused some oxidation but latter on the degradation of organic matter and removal of oxygen was mainly done by anaerobic bacteria. Acid was produced by bacterial action and it prevented further action and formation of coal. In some situation, the soil was alkaline which neutralized the acidity and hence further degradation of organic matter was possible. There is still some difference of opinion regarding the mechanism of bacterial action but it has now been accepted that bacterial action took place only in early stages for breaking of organic matter into simpler molecules. The degraded matter gave rise to heat with passage of time. After this, heat, pressure and time were responsible for further transformation resulting in evolution of carbon dioxide and methane and loss of water and volatile matter. This process of coal formation can be stopped at any stage and therefore, we get coal of different ranks.

6.4.2 Classification of Coal

Most common method of classification of coal is on the basis of rank. Rank represent the extent to which the process of coal formation has taken place. Various stages of conversion of wood material to coal can be summarized as follows:

Wood material → peat → lignite → brown coal → sub-bituminous coal → bituminous coal → semi-bituminous coal → anthracite

Each species is more mature or is of higher rank than the preceding one in the series. Hence lignite ranks better than peat. As we move from peat to anthracite, moisture and oxygen become progressively less and the percentage of carbon and hence the calorific value are increased.

Table 6.2: Approximate composition and calorific value of some typical of coals

Coal	*% moisture*	*% carbon*	*Calorific value (kcal/kg)*
Peat	25–30	55–60	4500–5900
Lignite	20–25	60–70	6500–7100
Sub-bituminous	10–15	75–85	7000–7500
Bituminous	5–10	85–90	8000–8500
Semi-bituminous	< 5	90–95	8500–8600
Anthracotic	1–3	92–98	8600–8700

6.5 ANALYSIS OF COAL

The composition of coal may vary form mine to mine. Hence it is important to analyse and be able to interpret correctly the result from the point of view of fixation of price for each grade and also for proper utilisation of coal. Coal analysis is of two types.

(A) Proximate analysis (B) Ultimate analysis

6.5.1 Proximate Analysis

It is the simplest type of analysis of coal and it tells us about the practical utilization. It involves the determination of (*a*) Moisture content, (*b*) Volatile matter, (*c*) Ash content and (*d*) Fixed carbon.

This analysis is quite empirical in nature since the results may vary with the procedure adopted. However, it gives fairly good idea about the quality of the coal for the requirement of the industries.

Significance of proximate analysis: The results of proximate analysis have no absolute significance. However, if the observations are made in accordance with standard specifications, reproducible results can be obtained. This analysis informs about the relative qualities of the coal samples with regard to its cost and the probability of its use for a particular industry.

1. Moisture: Coal contains two types of moisture:

(*i*) Free content, Adsorbed or Surface moisture: The moisture of this type is lost on air drying. It depends upon the treatment given to coal after mining the rank, size and surface, nature of the coal. It reduces the calorific value of the sample. A lot of heat is wasted in evaporating the free moisture during combustion. It also increases the cost on transport of the coal.

(*ii*) Inherent or Internal Moisture: It is that type of moisture which is obtained from air dried coal. It is an indicator of the rank of the coal *i.e.,* the degree of decomposition of vegetable matter to carbon content of coal. It decreases with increasing rank of the coal. In other words *the lower the moisture content, the better is the quality of coal.* The sum of the weight loss on air drying and Inherent moisture content gives the total moisture content of the coal sample.

Determination of moisture content: About one gram of finely powdered, air dried coal sample is taken in a silica crucible. It is heated in an electric hot air oven maintained at about 105–110°C for one hour. Then the crucible is taken out from the oven and cooled in a desicator and finally weighed. The loss in weight is due to the loss of moisture from the coal.

Suppose, the weight of empty crucible = W_1 g

Weight of crucible + coal sample taken = W_2 g

Weight of crucible + coal sample after heating in oven = W_3 g

$$\text{Percentage of moisture} = \frac{\text{Loss in weight of coal}}{\text{Weight of coal taken}} \times 100 = \frac{W_2 - W_3}{W_2 - W_1} \times 100$$

Significance of moisture in coal: Moisture present in coal evaporates during burning of the coal and takes away some portion of heat liberated, in the form of latent heat of evaporation. Therefore, higher the % of moisture in a coal sample, lower is the effective calorific value of the coal. Moreover, it quenches the fire and increases the cost since water is bought and transported at the fuel price. However, coal containing 5–10% free moisture improve the yield and quality of metallurgical coke formed. It helps in forming uniform fuel bed and reduces the amount of "fly ash".

2. Volatile Matter: It consists of a complex mixture of gaseous and liquid products resulting from the thermal decomposition of coal. It is not a constituent of coal as such. Its magnitude depends upon the condition of heating particularly the temperature. So, the experimental conditions must be maintained during its determination. High volatile matter content of a coal gives high flames, greater smoke and relatively low heat content, the larger is the combustion space required. Moreover, it determines the proportion of coal, which would be converted into tarry matter on heating. It includes the combustible gases like CO, H_2, CH_4 etc. and incombustible gases like N_2, CO_2, etc. The presence of incombustible gases is undesirable as they do not have any heating value. *The lower the volatile matter content, the better would be the quality of coal.* Coking quality of coal also depends upon its volatile matter content. Coking power decreases with increase of volatile matter.

Determination of volatile matter content: After determining moisture content, the dried sample of coal taken in the crucible with lid is heated in a muffle furnace maintained at 925°C ± 25°C for exactly seven minutes. The crucible and the lid are removed from the furnace with the help of long legged tongs carefully. It is first placed on a cooled iron plate and the warm crucible is then transferred in a desicator. The contents of the crucible are then cooled at room temperature and finally the crucible is weighed. Loss in weight is reported as volatile matter content on percentage basis.

Suppose the weight of crucible + coal after heating in muffle furnace and cooling = W_4 g

$$\text{Percentage of volatile matter} = \frac{\text{Loss in weight due to volatile manner}}{\text{Weight of moisture free coal taken}} \times 100$$

Significance of volatile matter: The volatile matter present in coal distills over as gas or vapour, as the fuel is burnt in air, without adding to the heat value of the coal. *Therefore, higher the % of volatile matter in coal, lower is the relative calorific value.*

A high volatile matter in the coal sample means high % of combustible or non-combustible gases in it. The presence of non-combustible gases is always undesirable, as they increase the volume of the furnace without adding to the heat liberated. Therefore, furnace with small volume is not suitable for burning fuels with high volatile matter content since large portion of fuel goes unburnt. On the other hand, burning of low volatile coals requires forced drought. As the % of volatile matter decreases, the intensity of drought needs to be increased.

Coals containing high volatile matter burn with long flame and high smoke. Therefore, they are not suitable for domestic purposes. Since the heat produced by such coals is dissipated over large space (calorific intensity is low). Such coals are not suitable for purposes like steam generation.

However, coals with high volatile matter content are suitable for production of coal gas. Moreover, coals with medium amount of volatile matter (around 20%) are suitable for producing hard and strong coke. Coals with very high or very low volatile matter content are not suitable for the purpose.

3. Ash Content: The non-combustible inorganic matter which is left behind after burning off the organic matter in coal is called ash. It is of two types:

Free or extraneous ash: The ash which occurs in different layers of the coal is known as free ash or extraneous ash. It can be removed by washing.

Fixed or inherent ash: The ash which is intimately interspersed with in the mass of coal is called fixed or inherent ash.

Determination of ash content: The residual, coal which is free from moisture and volatile matter is placed in a muffle furnace in the crucible without lid and is heated at 700 to 750°C for half an hour. The matter in the crucible is occasionally stirred with platinum wire in between. After half an hour the crucible is taken out, cooled firstly in air and then in a desicator and finally weighed. The process of heating, cooling and weighing is continued in a same way till a constant weight is obtained *i.e.,* whole of the carbonaceous matter is completely burnt. The final weight of the residue left is reported as the weight of the ash is the coal sample.

Suppose weight of residue after ignition = W_5 g = $\dfrac{W_5 - W_1}{W_2 \ W_1} \times 100$

Percentage of ash (Air dried basis) = $\dfrac{\text{Weight of ash formed}}{\text{Weight of air dry coal taken}} \times 100 = \dfrac{W_5 - W_1}{W_2 - W_1} \times 100$

Significance: Ash is a totally useless, non-combustible matter present in the coal, which does not add to the heating value of the coal. *Therefore, more the % of ash in coal sample, lesser is the relative calorific value*. Moreover, it also hinders the flow of air and heat. Therefore, it decreases the flame temperature and causes trouble during firing by forming clinkers which block the inter spaces of the grating on which the coal is burnt. As a result, the combustion becomes irregular. The presence of ash causes early wear of furnace and also increases the cost of transportation, handling and storage of the coal. It also increases the cost in ash disposal. The presence of ash also demands increase in furnace size, which is also undesirable.

4. Fixed carbon: The percentage of fixed carbon can be obtained by subtracting the percentages of moisture, ash and volatile matter from 100.

$$\% \text{ of fixed carbon} = 100 - \% \text{ (moisture + volatile matter + ash)}$$

$$\text{Alternatively, } \% \text{ of ash} = \frac{W_4 - W_5}{W_2 - W_1} \times 100$$

Significance: Carbon present in the coal burns in air to give carbon dioxide.

$$C + O_2 \longrightarrow CO_2 + \text{Heat}$$

It is highly exothermic process and hence greater the % of fixed carbon in a coal sample, greater is the heat liberated and hence greater is its calorific value. Moreover, greater the % of fixed carbon, smaller is the % of volatile matter. It also shows the importance of higher % of carbon in a coal sample. Moreover the % of fixed carbon helps in deciding the shape and size of the furnace to be used.

NUMERICAL PROBLEMS ON PROXIMATE ANALYSIS

Example 1: ***A sample of coal was analysed as follows: 1.000 g of an air dried coal sample was weighed in a silica crucible. After heating for 1 hour at 105–110°C, the dry coal residue weighed 0.980 g. The crucible was then covered with a vented lid and then heated strongly for exactly 7 minutes at 950 ± 20°C. The residue weighed 0.800 g. The crucible was then heated strongly in air, until a constant weight was obtained. The last residue was found to weigh 0.100 g. Calculate the proximate analysis.***

Solution: Given wt. of coal sample, $(W_1) = 1.000$ g

Weight of residue after heating at 105–110°C, $(W_2) = 0.980$ g

Percentage of moisture $= \dfrac{\text{loss of wt.}}{\text{wt. coal}} \times 100 = \dfrac{1.000 - 0.980}{1.000} \times 100 = \mathbf{2\%}$

Again, weight of residue after heating at 950 ± 20°C, $(W_3) = 0.800$ g

∴ Percentage of volatile matter $= \dfrac{W_2 - W_3}{W_1} \times 100 = \dfrac{0.980 - 0.800}{1.000} \times 100 = \mathbf{18\%}$

Final weight of residue, $(W_4) = 0.100$ g

Percentage of ash $= \dfrac{W_4}{W_1} \times 100 = \dfrac{0.100}{1.000} \times 100 = \mathbf{10\%}$

Percentage of dixed carbon $= 100 - \%$ (moisture + volatile matter + ash)

$= 100 - (2 + 18 + 10) = \mathbf{70\%}$

Alternatively, percentage of fixed carbon

$$= \frac{W_3 - W_4}{W_1} \times 100 = \frac{0.800 - 0.100}{1.000} \times 100 = \mathbf{70\%}$$

Example 2: ***A sample of coal was analysed as follows: Exactly 2.500 g was weighed into a silica crucible. After heating for one hour at 110°C, the residue weighed 2.415 g. The crucible next was covered with a vented lid and strongly heated for exactly seven minutes at 950 ± 20°C. The residue weighed 1.528 g. The crucible was then heated without the cover, until a constant weight was obtained. The last residue was found to weigh 0.245 g. Calculate the percentage results of the above analysis.***

Solution: Mass of moisture in coal sample = 2.500 – 2.415 = 0.085 g

Mass of volatile matter = 2.415 – 1.528 = 0.887 g

Mass of ash = 0.245 g

$$\text{Percentage of moisture} = \frac{0.085 \times 100}{2.500} = \mathbf{3.40}$$

$$\text{Percentage of volatile matter} = \frac{0.0887 \times 100}{2.500} = \mathbf{35.48}$$

$$\text{Percentage of ash} = \frac{0.245 \times 100}{2.500} = \mathbf{9.80}$$

Percentage of fixed carbon = 100 – (3.40 + 35 48 – 9 80) = **51.32.**

6.5.2 Ultimate analysis

It involves the determination of

(*a*) Carbon and hydrogen
(*b*) Nitrogen
(*c*) Sulphur
(*d*) Oxygen and
(*e*) Ash

This analysis is exact and can be employed in calculating the calorific value of coal, theoretically using the ***Dulong's formula*** *i.e.*,

$$\text{Gross or higher calorific value} = \text{HCV} = \frac{1}{100} \times \left[8080\,\text{C} + 34500\left(\text{H} - \frac{\text{O}}{8}\right) + 2240 \times \text{S} \right] \text{kcal/kg}$$

where C, H, O and S represent the % ages of carbon, hydrogen, oxygen and sulphur in coal. The calorific values of the fuel constituents namely C, H, S are 8080, 34500, and 2240 kcal/kg respectively. In this formula, oxygen is assumed to be present in combination with hydrogen as water. It means hydrogen is present in fixed form in water as H_2O. So, the amount of hydrogen available for combustion is:

Available or combustible hydrogen = Total mass of H in fuel – fixed hydrogen

$= \text{Total mass of H in fuel} - \frac{1}{8}(\text{mass of O in the fuel})$

It is because 8 parts of oxygen combine with one part of hydrogen to form 9 parts by weight of H_2O. The lower calorific value is given by the formula.

$$LCV = \left[HCV - \frac{9}{100} H \times 587 \right] \text{kcal/kg}$$

$$= [HCV - 0.09\, H \times 587] \text{ kcal/kg}$$

Thus, ultimate analysis is useful in calculating the higher as well as lower calorific value of the coal sample. Further this analysis is used for calculating the amount of air required for complete combustion of given coal sample and the composition of the products of combustion.

(*a*) **Determination of carbon and hydrogen:** A known weight (1-2 gm) of coal sample is burnt in a combustion chamber in free supply of dry air. As a result carbon and hydrogen present in the coal are converted into carbon dioxide and water respectively.

$$\underset{12}{C} + \underset{32}{O_2} \longrightarrow \underset{44}{CO_2} \qquad \ldots(6.1)$$

$$\underset{2}{H_2} + \underset{16}{\frac{1}{2} O_2} \longrightarrow \underset{18}{H_2O} \qquad \ldots(6.2)$$

The gasses coming out from the combustion chamber after purification are passed through firstly anhydrous $CaCl_2$ and then through KOH test tubes of known weight. These tubes absorb respectively H_2O and CO_2 produced in the combustion. The increase in weights of these tubes is noted.

$$2KOH + CO_2 \longrightarrow K_2CO_3 + H_2O$$

$$CaCl_2 + 7H_2O \longrightarrow CaCl_2.7H_2O$$

Suppose the increase in weights of KOH and $CaCl_2$ tubes are x and y g respectively. This implies that x and y g of CO_2 and H_2O have been produced by combustion respectively.

From equation (6.1)

44 parts by weight of CO_2 produced = 12 parts by weight of C in coal sample

$\therefore$ x gms of CO_2 produced $= \frac{12}{44} \times x$ g of C in the coal sample

$$\text{Percentage of carbon} = \frac{\text{Weight of C in coal}}{\text{Weight of coal sample}} \times 100 = \frac{12}{44} \times \frac{x}{W} \times 100$$

Where W = weight of coal taken for combustion and x is the weight of CO_2 formed.

Similarly, $\quad$ percentage of H $= \frac{2}{18} \times \frac{y}{W} \times 100$

Where y represents the weight of H_2O formed and W is the weight of coal sample taken.

Significance: Carbon present in the given coal sample changes into CO_2 when coal in burnt in air. This is highly exothermic reaction. Therefore, *more the % of carbon in coal, higher is the heat liberated and thus higher is the calorific value of coal sample*. Moreover, the % of carbon in coal sample also determines the size and shape of the furnace to be used for combustion.

Hydrogen is present in two forms in a coal sample (1) free form (2) associated with oxygen as H_2O *i.e.,* moisture. Hydrogen present in free form changes to H_2O on combustion. It is again highly exothermic reaction and higher the % of free hydrogen in coal sample more is the heat produced and hence more is the calorific value of coal sample. On the other hand, the hydrogen present as moisture, takes away some amount of heat produced during combustion to get evaporated. Therefore, *it is desirable to have more % of free hydrogen and lesser % age of associated hydrogen or in other words lesser percentage of oxygen in the coal sample.*

(*b*) Determination of nitrogen: A known weight of coal sample is finely powdered and heated with concentrated H_2SO_4 in presence of K_2SO_4 and a little $CuSO_4$ in a Kjeldahl's flask till solution becomes clear. K_2SO_4 raises the boiling point of the content and heating becomes more effective, $CuSO_4$ acts as a catalyst.

C, H and S present in the sample get converted to their respective oxides (CO_2, H_2O and SO_2) and N is quantitatively converted to ammonium sulphate

$$2N + 6H + H_2SO_4 \longrightarrow (NH_4)_2SO_4 \quad ...(6.3)$$

The solution so obtained is treated with excess of KOH solution. It leads to liberation of ammonia.

$$(NH_4)_2SO_4 + 2KOH \longrightarrow 2NH_3 + K_2SO_4 + 2H_2O \quad ...(6.4)$$

The ammonia so liberated is absorbed in a known excess of standard acid solution.

$$2NH_3 + H_2SO_4 \longrightarrow (NH_4)_2SO_4 \quad ...(6.5)$$

The unreacted acid solution is then determined by back titration with standard solution of NaOH.

$$H^+ + OH^- \longrightarrow H_2O \quad ...(6.6)$$

Calculations

Suppose N_1 is the normality and V_1 is the volume of acid, neutralised by ammonia.

$$V_1 \text{ ml of } N_1 \text{ acid used} \equiv V_1 \text{ ml of } N_1\text{-}NH_3 \text{ produced}$$
$$\equiv 17\, N_1 V_1 \text{ mg of } NH_3 \text{ produced}$$
$$\equiv 14\, N_1 V_1 \text{ mg of N in coal sample}$$

From equation (6.3) and (6.4), $2N \equiv 2NH_3$

Moreover, one normal solution contains one equivalent of the substance in a litre of the solution *i.e.*, one milligram equivalent in one milliliter (ml) of the solution.

$$\text{Percentage of nitrogen} = \frac{\text{Weight of N in coal}}{\text{Weight of coal taken}} \times 100$$

$$= \frac{14\, N_1\, V_1 \text{ mg}}{\text{Wg}} \times 100$$

$$= \frac{14\, N_1\, V_1 \text{ in grams}}{1000 \times \text{W in grams}} \times 100$$

$$= \frac{1.4 \times \text{Normality of acid used} \times \text{vol. of acid}}{\text{Weight of coal}}$$

The above formula can also be derived by an alternate method as under:

Let the weight of sample taken = Wg

Volume of standard acid taken in excess = V cm^3

Normality of the acid = N_1

Volume of the standard alkali required for the unused acid = V_2 cm^3

Normality of the alkali = N_2

Applying the normally equation *i.e.,* $N_1V_1 = N_2 V_2$

V_1 ml of N_1 unused acid $= V_2$ ml of (Acid) of N_2 alkali

$$V_1 \text{ ml of unused acid} = \frac{V_2 \times N_2}{N_1} = V \text{ cm}^3$$

Hence volume of N_1 acid used for neutralizing $NH_3 = (V - v)$ cm³.

But $(V - v)$ cm³ of N_1 acid $= (V - v)$ cm³ of N_1 ammonia solution

i.e. Ammonia evolved $= (V - v)$ cm³ of N_1 solution.

We know 1000 cm³ of one normal ammonia solution contains 17 g of ammonia (NH_3) or 14 g of nitrogen.

$$\text{Weight of Nitrogen in } (V - v) \text{ cm}^3 \text{ of } N_1 \text{ } NH_3 \text{ Solution} = \frac{14 \times (V - v) \times N_1}{1000}$$

Hence percentage of Nitrogen in the compound

$$= \frac{14 \times (V - v) \times N_1}{1000} \times \frac{100}{W}$$

$$= \frac{14 \times N_1 \times (V - v)}{W}$$

Significance: Nitrogen is present in very small amounts in coal (say 1% only) and therefore does not contribute much to the properties to coal. Moreover, when coal is burnt in air, temperature is too low for nitrogen to get changed to its oxides and hence it escapes as such in air without adding to the heating value of coal. Hence its presence in coal is undesirable.

(*c*) Determination of sulphur: Sulphur present in coal sample is determined from bomb calorimeter washings. In this method, a known weight of coal is taken in a platinum crucible and is placed in a steel bomb. Around 5 to 10 ml of distilled water is also put in the bomb. The bomb is closed tightly and oxygen is passed in it slowly to a pressure of nearly 25 atm. Now the bomb is placed in water and the sample is ignited by making use of electric current. After about 10 minutes, the bomb is removed and pressure is released at a uniform rate. Now it is gently opened and the cover is washed with water. If some unburnt sample is left in the crucible, the whole process is repeated again. Now the contents of the bomb are transferred to beaker and washed several times with water. About 5 ml of Br_2 is also added to oxidize any sulphide to sulphate. The washings are finally treated with $BaCl_2$ solution, when $BaSO_4$ gets precipitated. The precipitates so obtained are filtered, washed and dried to a constant weight.

$$S \longrightarrow SO_2 \longrightarrow SO_3^{2-} \longrightarrow S_4^{2-} \xrightarrow{BaCl_2} BaSO_4$$

$$S(32) \equiv BaSO_4(137 + 32 + 4 \times 16 = 233)$$

Therefore the percentage of Sulphur

$$= \frac{32}{233} \times \frac{\text{Weight of } BaSO_4}{\text{Weight of coal}} \times 100$$

Significance: Sulphur present in the coal changes to SO_2 and SO_3, when it is burnt in air. These reactions are exothermic and therefore, the presence of sulphur in coal increases its calorific value. However, its presence is undesirable because SO_2 and SO_3 produced on combustion of sulphur causes environmental pollution and in moist conditions, gives H_2SO_3 and H_2SO_4 which cause corrosion of the metallic components in their vicinity. Hence, the presence of sulphur is highly undesirable in coal used for making of coke, since it badly affects the quality and properties of steel used.

Determination of oxygen: Percentage of oxygen is determined as follows:

$$\% \text{ of O} = 100 - \% (\text{C} + \text{H} + \text{N} + \text{S} + \text{Ash})$$

Significance: Oxygen is never present alone in coal. It is always present with hydrogen as H_2O in the form of moisture. This moisture takes away some amount of heat produced when coal is burnt in air. Hence, higher the % of oxygen in coal, lower is the relative calorific value of coal. On average, 1% increase in percentage of oxygen decreases the calorific value by 7–8%. Hence, its presence is undesirable.

NUMERICAL PROBLEMS ON ULTIMATE ANALYSIS

Example 3: ***0.25 g of coal sample on burning in a combustion chamber in the current of pure and dry O_2 was found to increase weight of anhydrous $CaCl_2$ U-tube by 0.075g and of KOH U-tube by 0.52 g. Find the % of C and H in coal.***

Solution: Wt. of coal taken = 0.25 g

Wt. of H_2O formed = 0.075 g

Wt. of CO_2 formed = 0.52 g

$$\% \text{ of C} = \frac{12}{44} \times \frac{\text{Wt. of } CO_2 \text{ formed}}{\text{Wt. of coal taken}} \times 100 = \frac{12}{44} \times \frac{0.52}{0.25} \times 100 = \mathbf{56.73}$$

$$(2) \% \text{ of H} = \frac{2}{18} \times \frac{\text{Wt. of } H_2O \text{ formed}}{\text{Wt. of coal taken}} \times 100 = \frac{2}{18} \times \frac{0.075}{0.25} \times 100 = \mathbf{3.33}$$

Example 4: ***0.45 of coal sample was kjeldahlised. NH_3 gas evolved upon distilling kjeldahlised solution with 50% Naoh solution, neutralised 7.15 ml of 0.12 N HCl. Calculate the % of N in the above sample.***

Solution: Wt. of coal taken (W) = 0.45 g

Volume of 0.12 N HCI neutralized by

NH_3 = 7.15 ml

$$\therefore \quad \% \text{ of N} = \frac{1.4 \times \text{N} \times \text{V}}{\text{W}} = \frac{1.4 \times 0.12 \times 7.15}{0.45} = \mathbf{2.67\%}$$

Example 5: ***In Kjeldahl's nitrogen estimation, the NH_3 evolved from 0.14 g of coal sample was absorbed in 50 ml of 0.1 N H_2SO_4. The unused acid required 25 ml of 0.1 N NaOH for neutralisation. Calculate the percentage of N in the above sample.***

Solution: Weight of coal taken, (W) = 0.14 g

Vol. of H_2SO_4 consumed by NH_3 = 50 – 25 = 25 ml

Normality of H_2SO_4, (N) = 0.1 N

$$\therefore \quad \% \text{ of N} = \frac{1.4 \times \text{Normality of } H_2SO_4 \times \text{Vol. of } H_2SO_4 \text{ consumed}}{\text{Wt. of coal sample}}$$

$$= \frac{1.4 \times 0.1 \times 25}{0.14} = \mathbf{25}$$

Example 6: ***In an estimation of sulphur by ultimate analysis 0.2175 g of the coal sample gave 0.5825 gm of $BaSO_4$. Calculate the % of S in coal sample.***

Solution:

$$\text{Wt. of coal sample} = 0.2175 \text{ g}$$

$$\text{Wt. of } BaSO_4 \text{ formed} = 0.5825 \text{ g}$$

$$\therefore \quad 233 \text{ g } BaSO_4 \text{ contains S} = 32 \text{ g}$$

$$\therefore \quad 0.5825 \text{ g } BaSO_4 \text{ will contains} = \frac{32 \times 0.5825}{233} \text{ g}$$

$$\therefore \quad 0.2175 \text{ g of coal} \equiv 0.5825 \text{ g } BaSO_4 \text{ or } \frac{32 \times 0.5825}{233} \text{ g}$$

$$\therefore \quad 100 \text{ g coal} = \frac{100 \times 32 \times 0.5825}{233 \times 0.2175} = 36.78$$

$$\therefore \quad \% \text{ of S in Coal sample} = \mathbf{36.78}$$

Example 7: ***0.35 g of a coal sample was kjeldahlised and ammonia evolved was absorbed in 100 cm³ of N/5 sulphuric acid. The residual acid required 154 cc of N/10 NaOH for neutralization. Calculate the percentage of nitrogen in the sample.***

Solution:

$$\text{Weight of the sample taken} = 0.35 \text{ g}$$

$$\text{Volume of N/5 } H_2SO_4 \text{ taken} = 100 \text{ cc}$$

$$\text{Since, 100 cc of N/5 } H_2SO_4 = 200 \text{ cc of N/10 } H_2SO_4$$

Or

$$\text{Volume of N/10 } H_2SO_4 \text{ taken} = 200 \text{ cc}$$

$$\text{Volume of N/10 NaOH required for residual acid} = 154 \text{ cc}$$

$$\therefore \quad \text{Volume of N/10 } H_2SO_4 \text{ actually neutralised by } NH_3 = 200 - 154 = 46 \text{ cc}$$

$$\text{Hence, } \% \text{ of N} = \frac{1.4 \times \text{normality of acid} \times \text{volume of acid used}}{\text{weight of the sample}}$$

$$= \frac{1.4 \times 1 \times 46}{10 \times 0.35} = \mathbf{18.4}$$

Example 8: ***Calculate the gross and net calorific values of a coal having following composition: C = 80%, H = 7%, O = 3%, S = 3.5%, N = 2.1% and ash = 4.4%.***

Solution: According to Dulong's formula,

$$\text{Gross calorific value} \quad (GCV) = \frac{1}{100}\left[8080\,C + 34500\left(H - \frac{O}{8}\right) + 2240\,S\right] \text{ kcal/kg}$$

$$GCV = \frac{1}{100}\left[8080 \times 80 + 34,500\left(7 - \frac{3}{8}\right) + 2240 \times 3.5\right] \text{ kcal/kg}$$

$$= \frac{1}{100}[646400 + 228562.5 + 7840] \text{ kcal/kg}$$

$$= \frac{1}{100}[882802.5] \text{ kcal/kg} = 8828.025 \text{ kcal/kg}$$

$$\text{Net calorific value} \quad (NCV) = (GCV - 0.09\,H \times 587) \text{ kcal/kg}$$

$$= (8828.025 - 0.09 \times 7 \times 587) \text{ kcal/kg}$$

$$= (8828.025 - 369.81) \text{ kcal/kg} = 8458.215 \text{ kcal/kg}$$

$$= \mathbf{8458.215 \text{ kcal/kg}}$$

PROBLEMS FOR PRACTICE

1. 2.500 g weighed in coal sample was a silica crucible. After heating for one hour at 110°C the residue weighed 2.415. The crucible next was covered with a vented lid and strongly heated for exactly seven minutes at 950 ± 20°C. The residue weighed 1.528 g. The crucible was then heated without the lid, until a constant weight was obtained. The last residue was formed to weigh 0.245 g. Calculate the percentage result of the above analysis.
2. 1.0 g of a sample of coal was used for nitrogen estimation by kjeldahl's method. The evolved ammonia was collected in 25 ml N/10 sulphuric acid. To neutralise exess acid, 15 ml of 0.1 N sodium hydroxide was required. Determine the percentage of nitrogen in the given sample of coal.
3. 1.56 g of the coal was kjeldahlised and NH_3 gas thus evolved was absorbed in 50.0 ml of 0.1 N H_2SO_4. After absorption, the excess (residual) acid required 6.25 ml of 0.1 N NaOH for exact neutralisation. 2.6 g of the coal sample in a quantitative analysis gave 0.1755 g of $BaSO_4$. Calculate the percentage of N and S is the coal sample.
4. 1.525 g of a coal sample was kjeldhlised and then distilled with excess KOH. The evolved NH_3 was absorbed in 50 cm^3 of NH_2SO_4. This was diluted to 750 cm^3. 25 cm^3 of diluted solution required 11.43 cm^3 of decinormal KOH for neutralization. Calculate the percentage of nitrogen.
5. 0.5 g of a sample of coal was used in a bomb calorimeter. The ash formed in the bomb calorimeter was extracted with acid and the acid extract was heated with barium nitrate solution and a precipitate of barium sulphate was obtained. The precipitate was filtered, dried and weighed. The weight of the precipitate was found to be 0.05 g. Calculate the percentage of sulphur in the coal sample.
6. Calculate the gross and net calorific values of coal having the following composition: carbon = 85%, hydrogen = 8%, sulphur = 1%, nitrogen = 2% and ash = 4%. Latent heat of stem = 587 kcal/kg.
7. A coal has the following composition by weight. C = 90%, O = 3.0%, S = 0.5%, N = 0.5% and ash = 2.5%. Net calorific value of coal was found to be 8490.5 kcal/kg. Calculate the percentage of hydrogen and higher calorific value of coal.

Answers

1. Moisture = 3.4%, Volatile matter = 35.48%, ash = 9.8% and fixed carbon = 51.32%
2. 1.4%
3. N = 3.93% and S = 0.93%
4. 14.3%
5. 1.37%
6. GCV = 9650.4 kcal/kg, NCV = 9227.8 kcal/kg.
7. H = 4.58%; HCV = 8731.8 kcal/kg

6.6 PETROLEUM FUELS—CRUDE PETROLEUM AND ITS REFINING

Petroleum or crude oil as obtained from oil wells is a dark coloured viscous liquid with unpleasant smell. It is a complex mixture of hydrocarbons (chiefly aliphatic with lesser amounts of aromatic hydrocarbons) along with varying amounts of organic compounds of sulphur and nitrogen.

The crude oil is purified by subjecting it to fractional distillation. During this process, the impurities are removed and the crude oil is separated into different useful fractions.

The *process of separating petroleum into different useful fractions having different boiling ranges with the simultaneous removal of undesirable impurities is called* refining.

Refining of petroleum: The crude oil is first of all neutralised by washing it with an acidic or basic solution depending upon whether the impurities are basic or acidic in nature. This is then subjected to fractional distillation. For this purpose, the crude oil is heated to 650 K in coiled pipes in a furnace and the pressure reduced. The vapours, thus obtained, are admitted into the bottom of a fractionating tower called the *bubble tower* (Fig. 6.1). It is a tall cylindrical steel structure fitted with horizontal trays. Each tray is provided with a number of holes and an overflow pipe having a *bubble cap*. As the vapours of the oil rise up, the high boiling fractions condense in the lower portions of the tower. The lower boiling fractions rise up the tower and condense at different levels depending upon their boiling ranges. In this way, a number of fractions are collected. Each fraction, thus collected, is a mixture of hydrocarbons having a definite boiling range. The uncondensed gases, however, pass out of the tower at the top.

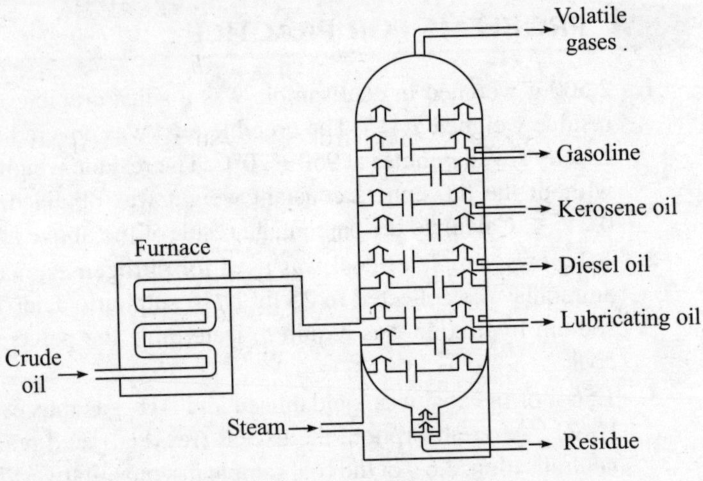

Fig. 6.1 Fractional distiallation of crude oil

The gasoline obtained by this process is called *straight run gasoline*. Some of the important fractions along with their approximate composition, boiling ranges and important uses are given in Table 6.3.

Table 6.3: Some important fractions of petroleum

	Fraction	*Approximate Composition*	*Approximate Percentage*	*Boiling Range (K)*	*Uses*
1.	Gaseous	C_1–C_4	2	113–303	As a fuel, production of carbon black, H_2 and CO, used to make NH_3, methanol and gasoline.
2.	Petroleum ether (ligroin)	C_5–C_7	2	303–363	As a solvent for rubber, oils, fats, varnishes and, for dry cleaning clothes.
3.	Gasoline or petrol	C_7–C_{12}	32	343–473	As a motor fuel, for dry cleaning and preparation of petrol gas.
4.	Kerosene oil	C_{12}–C_{15}	18	448–548	As an illuminant fuel, jet engine fuel and for making oil gas.
5.	Gas oil, fuel oil, and diesel oil	C_{15}–C_{18}	20	573–673	Furnace fuel, for diesel engine, used in cracking.
6.	Lubricating oils, greases and petroleum jelly	C_{16} and up		623 and up	As a lubricant.
7.	Paraffin wax	C_{20} and up		melts between 325–330	In the manufacture of candles, waxed papers and for water-proofing.
8.	Petroleum coke	C_{30} onwards		Left as residue	Used as artificial asphalt, fuel and for electrodes.

6.6.1 Petroleum Cracking

The process of decomposition of less volatile higher hydrocarbons into more volatile lower hydrocarbons with the application of heat and catalysts is known as cracking.

As the yield of petrol from the distillation of petroleum is small (about 16%), the petrol or gasoline is prepared by cracking of less useful and cheap straight run oils like heavy oils and residual oils. In fact approximately 50% of gasoline is now prepared by this method. The cracking, besides increasing the yield of petrol, also improves the quality of gasoline in terms of antiknock properties.

The fractions like heavy oil or residual oil contain alkanes of higher hydrocarbons having C_{12} or more carbon atoms. When they are cracked these large hydrocarbon molecule break up to give hydrocarbons with lesser number of carbon atoms. This decrease in size or number of carbon atoms is accompanied with a change in the boiling range or volatility of the resulting product.

$$\underset{\substack{\text{Dodecane} \\ \text{b.p. } 216\,^\circ\text{C}}}{C_{12}H_{26}} \xrightarrow{\Delta} \underset{\text{etc.}}{\begin{Bmatrix} C_7H_{16} + C_5H_{10} \\ C_6H_{14} + C_6H_{12} \\ C_5H_{12} + C_7H_{14} \end{Bmatrix}} \underset{30\text{–}150^\circ C}{\text{Boiling range}} \quad \underset{\text{Hexdecane}}{C_{16}H_{34}} \xrightarrow{\Delta} \underset{\text{etc.}}{\begin{Bmatrix} C_8H_{18} + C_8H_{16} \\ C_9H_{20} + C_7H_{14} \\ C_{10}H_{22} + C_6H_{12} \end{Bmatrix}} \underset{30\text{–}150^\circ C}{\text{Boiling range}}$$

The nature of products formed depend on the following :

(*i*) *Structure of the hydrocarbon.*

(*ii*) *Temperature and pressure applied in the cracking.*

(*iii*) *Presence or absence of catalyst.*

Changes occurring during cracking: Besides the decrease in the size of the molecule which results in increased volatility or lower boiling fractions, the other important changes taking place are as follows:

(*i*) Cracking results in formation of a mixture of large number of hydrocarbons—all having lower b.pt. than initial alkane.

(*ii*) Straight chain alkanes are converted to branched chain hydrocarbons (isomerization).

(*iii*) Saturated hydrocarbons are converted to unsaturated hydrocarbons.

(*iv*) Ring closure or cyclization of aliphatic alkanes may occur.

Types of cracking: There are two main processes used for cracking: (*a*) Thermal cracking, and (*b*) Catalytic cracking.

(*a*) Thermal cracking: When the cracking is carried out simply by the application of heat and pressure, it is known as thermal cracking. The furnace used for thermal cracking is made of special chromium nickel alloy and the design is such that it allows high oil velocity in the tubes and very low residence time. The reaction products are then passed through fractionating towers to separate gasoline from other fractions formed in cracking. Thermal cracking is done in either liquid phase or vapour phase.

(*i*) *Liquid phase cracking:* In this process the cracking of the stock (consisting of either heavy oil or residual oil) is done at a temperature of 475–530°C under high pressure, usually 7 to 10 atmospheric pressure, to keep the reaction product in liquid state. The yield of petrol on fractionation is approximately 60–65% and the octane number is in the range 65 to 70.

(*ii*) *Vapour phase cracking:* In this process the stock is first converted to vapour state by heating to about 600°C and then subjected to 3.5 to 10.5 atmospheric pressure for a very short time for cracking. The yields are good (approx. 70%) but the octane number is again below 70. However, it is disadvantageous to subject heavy oil or residual oil to this process because they cannot be vaporized completely.

(*b*) **Catalytic cracking:** When the cracking is done in the presence of catalysts, high pressure and temperature are usually not necessary. The feedstock used is generally in the boiling range of 350–550°C. The catalytic cracker may be fixed bed, moving bed or fluid bed type. Nowadays generally the latter types are used.

The product mix obtained from a catalytic cracker depends on a number of variable *viz.* nature of feedstock, catalyst, cracking conditions, etc.

(*i*) **Fixed bed catalytic cracking:** In this process the catalyst is used in the form of a fixed bed and cracking is carried out at around 500°C. A number of cracking furnaces are used, so that if the catalyst is spent in one furnace, because of carbon, deposition, the other one is used while the spent catalyst is being regenerated by burning the carbon in a current of air.

Process: The oil vapours are heated in a preheater to cracking temperature (420–450°C) and then forced through a catalytic chamber (containing artificial clay mixed with zirconium oxide) maintained at 425–450°C and 1.5 kg/cm^2 pressure. During their passage through the tower, about 40% of the charge is converted into gasoline and about 2–4% carbon is formed.

Carbon gets absorbed on the catalyst bed. The catalyst, stops functioning, due to the deposition of black *layer of carbon,* formed during cracking. This is *re-activated* by burning off the deposited carbon. *During the re-activation interval, the vapours are diverted through other catalyst chamber.* The vapours produced are then passed through a *fractionating column,* where heavy oil fractions condense. The vapours are then led through a *cooler,* where some of the gases are condensed along-with gasoline and uncondensed gases move on. The gasoline containing some dissolved are is then sent to a *'stabilizer'*, where the dissolved gases are removed and pure gasoline is obtained (*see* Fig. 6.2).

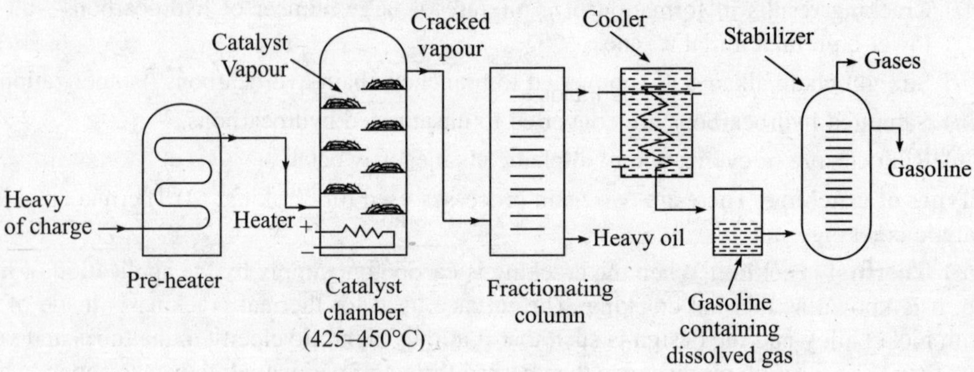

Fig. 6.2 Fixed-bed catalytic cracking.

(*ii*) **Fluid bed (or moving bed) catalytic cracking:** To carry out catalytic cracking continuously, fluidized catalyst is used. The catalyst is suspended in a stream of oil vapours. The spent catalyst continuously flows out of the bottom of the furnace and can be regenerated while fresh amounts of catalysts are continuously added from the top. The temperature is maintained between 450 to 550°C and the pressure is between 1 to 2 atmosphere.

Process

In the fluid catalytic cracking, oil preheated to about 300°C is sprayed from the bottom into the reactor containing the catalyst (Fig. 6.3). Formerly natural zeolites and now synthetic molecular sieves in the powdered form are used catalysts. In the reactor the oil mixes with the powdered catalyst to produce a reaction temperature of about 530°C. At this stage steam is introduced, which converts the catalyst and the oil mixture in a fluidized state into the reactor. The oil undergoes cracking in less than a contact time of 3 second. The products are then transferred into a fractionating column not shown in the figure. The deactivated (spent) catalyst is conveyed into a regenerator where the excess oil is stripped by steam and the carbon deposited on the deactivated catalyst is burnt off using hot air. The regenerated catalyst is again introduced into the reactor by a current of air. The advantage of fluidized cracking process is that a high degree of mixing is achieved and consequently a good contact is established between the catalyst and the vapour. This results not only into a higher yield, but a gasoline of better quality with higher octane number.

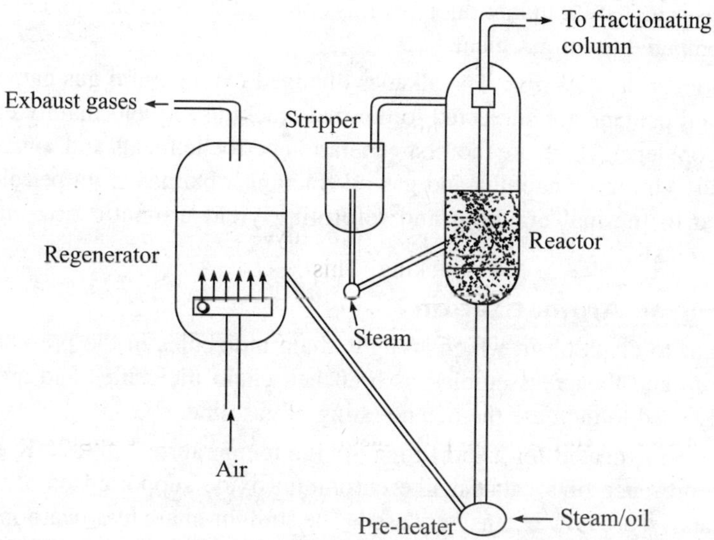

Fig. 6.3 Fluidized catalytic cracking

Most of the gasoline petrol that is being used is obtained by the cracking process rather than by direct refining of Petroleum. It also gives several gaseous hydrocarbons employed in the manufacture of petrochemicals. A comparison of catalytic and thermal cracking processes is given below:

Process	*Temperature*	*Pressure*	*Quality of Gasoline*	*Compositions of flue gases*
Thermal cracking	550°C	5 atm	Octane No. 65–68 It contains mainly normal alkanes and alkenes	mainly C_1–C_2
Catalytic cracking	500°C	0.1 atm	Octane No. 80–85 It contains mainly branched chain alkanes and aromatics	C_3–C_5

The yields of gasoline by catalytic cracking are generally higher and the octane number, also above 80. The resulting product is fractionated in a tower to give gasoline and other heavy fractions which are again subjected to cracking.

Applications of Cracking

1. *Obtaining petrol:* The most important single application is the preparation of gasoline or petrol by converting high boiling fractions into petrol. Roughly more than 50% of petrol

produced is by cracking processes, as it also given as better quality petrol in terms antiknocking properties.

2. *Preparation of oil gas and petrol gas:* Oil gas is obtained by cracking of kerosene oil by dropping it over red hot iron retorts. The kerosene is converted to a mixture of gaseous hydrocarbons like methane, ethane, propane, butane etc., which are the main constituents oil gas, together with small amounts of other hydrocarbons and hydrogen. The gas used in laboratory burners is generally oil gas which is stored in iron gas holders over water and supplied to laboratory through pipe lines.

   The petrol gas is prepared by cracking of petrol by passing it through electrically heated coils or retort. The petrol is converted to a mixture of simple gaseous hydrocarbons. The petrol gas is also used in laboratory as it is convenient to prepare and the cost of the plant is low a compared to oil gas plant.

3. *Obtaining petrochemicals :* The alkanes obtained from natural gas namely ethane propane, butane and pentane are subjected to thermal cracking to yield mainly ethylene an to some extent propylene, which are the first generation petrochemicals and source of man important chemicals. Similarly, naphtha and gas oil fractions obtained from petroleum refining when subjected to thermal cracking and reforming yield aromatic raw materials (BTX) for petrochemicals.

6.6.2 Reforming or Aromatization

It is a process similar to cracking in which straight chain molecules in the presence of a catalyst are effectively broken up and then reassembled as branched chain molecules and aromatic compounds. Reforming is widely used to increase the octane rating of gasoline.

In the process, gasoline is heated for a short time at high temperature 725–825 K under high pressure (10–20 atm) in the presence of a catalyst like chromium oxide supported on aluminium oxide. But aluminium-based catalyst are now extensively used. The straight-chain hydrocarbons may be converted to branched chain hydrocarbons by isomerization or alkylation, alkanes to cycloalkanes by cyclization and cycloalkanes to aromatic hydrocarbons by dehydrogenation or aromatization. Thus, *n*-hexane changes into benzene, *n*-heptane changes into toluene.

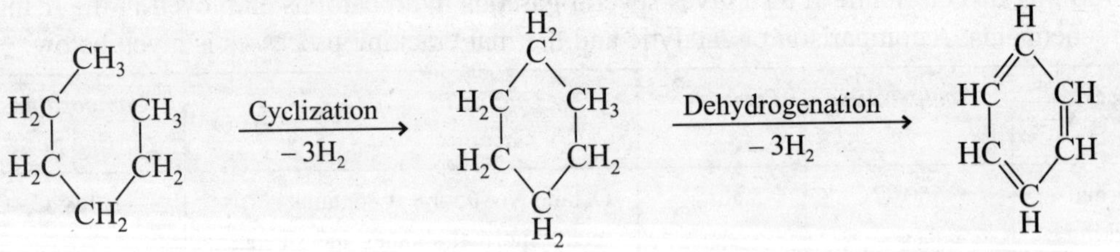

n-Hexane   Cyclohexane   Benzene

Aromatization of a gasoline increases its octane number because unsaturated hydrocarbons are better anti-knock compounds.

Reforming carried out in the presence of hydrogen is known as *hydroforming*.

6.6.3 Gasoline or Petrol

It is a mixture of hydrocarbons such as C_5H_{12} (Pentane) to C_8H_{18} (Octane) and this fraction is obtained between 40–120°C. It calorific value is about 11250 kcal/kg.

It is obtained from crude petroleum by fractional distillation, by cracking of heavier fraction of crude petroleum or by synthetic process such as polymerisation of gaseous alkenes and alkylation of iso-butane. It is the most important vehical fuel having the following characteristics.

Characteristics of Ideal Gasoline

(*i*) It must be cheap and readily available.

(*ii*) It should have high calorific value.

(*iii*) It should not pre-ignite easily.

(*iv*) It must burn clean and produce no corrosion in the engine.

(*v*) It should not decrease the volumetric efficiency of engine.

(*vi*) It should readily mix with air and should be easy to vapourise.

(*vii*) It must be knock resistant.

6.6.4 Fuel Rating—Octane Number (Knocking and Anti-knocking)

Gasoline (or petrol) is used as a fuel in the internal combustion engines of scooters, cars and aeroplanes. For this purpose, the vapours of gasoline are mixed with air. The mixture is compressed to a small volume and then ignited by the spark from the spark plug. The mixture explodes and the petrol burns to form a large number of gaseous products which pushes back the piston. This thrust provides the power for moving the automobile.

For maximum efficiency, the fuel mixture must be highly compressed before firing in the cylinder. But in some cases, due to over compression of the fuel mixture, pre-ignition of fuel takes place in the cylinder before sparking which causes a violent jerk to the piston. The violent jerk **gives** rise to a metallic sound or rattle known as **knocking.** This causes damage to the piston and decreases the efficiency of the engine.

Knocking is an undesirable property. However, it depends on nature of fuel. Straight-chain compounds have a greater tendency of knocking than branched-chain compounds which are said to have an *anti-knocking quality*

Quality of Petrol Octane Number

The anti-knock quality of the gasoline used as a fuel is determined by its octane rating or octane number. It has been found that the hydrocarbon *n-heptane* produces maximum knocking when used as the fuel. Due to its very poor antiknock quality, it has been arbitrarily assigned an octane number of zero. On the other hand, 2, 2, 4-trimethylpentane (incorrectly called iso-octane) produces minimum knocking and has been assigned an octane number of 100.

$$CH_3—CH_2—CH_2—CH_2—CH_2—CH_2—CH_3$$

n-Heptane

(Octane number-zero)

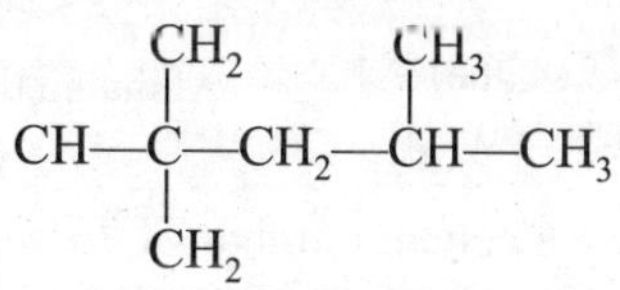

2, 2, 4-Trimethylpentane

(Octane number –100)

This means that n-heptane is the worst and iso-octane the best fuel. All other fuels are rated between zero and 100. Thus, the octane number or octane rating of a given fuel may be defined *as the percent of iso-octane in the mixture of iso-octane and n-heptane which has the same anti-knocking qualities as the fuel being tested under the similar experimental conditions.* For instance if the octane number of a fuel is 80, it means that this behaves like a mixture of 80 percent iso-octane and 20 percent *n*-heptane.

Generally, gasolines having an octane number of 74 are used in motor cars, scooters and motorcycles. Such gasolines are known as regular gasolines. Fuels having octane number lower than this value are called *third grade gasolines.* Gasolines having an octane number of 81 and above are known as *premium gasolines.* Aviation gasoline (gasoline used in aeroplane) has an octane number of 100. Gasolines with rating even higher than 100 have been produced. For example, the octane number of 2, 2, 3-trimethylbutane (Triptane) has an octane number of 125.

The octane number changes with structure of the hydrocarbons as follows:

Straight chain alkanes < Branched chain alkanes < Olefins < Cycloalkanes < Aromatics

(*i*) Straight-chain alkanes have low octane number. As the chain length increases, the octane number decreases, *e.g.,* octane number of *n*-non-one is less than zero.

(*ii*) Branched-chain alkanes have higher octane numbers than the isomeric straight chain alkanes, *e.g.,* Octane numbers of *n*-heptane and 2, 2, 3-Trimethylbutane are zero and 125 respectively.

(*iii*) Octane number of cycloalkanes is higher than that of corresponding open-chain alkanes.

(*iv*) Alkenes and aromatic hydrocarbons have higher octane numbers than alkanes.

For the same reason, the gasoline obtained through the process of catalytic cracking or reforming which contains higher percentage of alkenes, branched-chain alkanes and aromatics has higher octane number than the gasoline obtained by direct distillation of petroleum.

6.6.5 Diesel Oil

It is the fraction of petroleum distillation which is collected in the temperature range of 300–350°C. It is heavier than gas oil and lighter than the residual oil. Its specific gravity lies between 0.86–0.95. Its calorific value is about 11,000 kcal/kg. It is used as a fuel in diesel engines. Diesel fuel is used in engines where the fuel is directly ignited into the cylinders and is atomised before it is ignited and burnt. In a diesel engine the fuel is exploded by the application of heat and pressure and not by a spark. The main characteristic of a diesel engine fuel is that, it should easily ignite below compression temperature. Chemically, diesel oil consists of straight-long chain hydrocarbons with minimum amount of branched chain and aromatic hydrocarbons.

$$H_3C{-}(CH_2)_{14}{-}CH_3$$

Cetane (*n*-Hexadecane)
(Centane No. = 100)

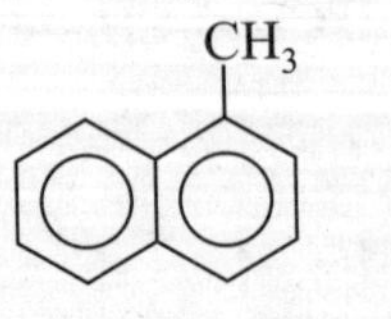

Alpha methylnaphthalene

For example, a diesel oil having a cetane number of 75 would have the same ignition properties as a mixture of 75% cetane and 25% alpha methylnaphthalene.

6.7 GASEOUS FUELS

Gaseous fuels are considered to be best fuels because they possess several advantages over solid and liquid fuels.

The composition of gaseous fuels depends upon the nature of the gases. An examination of the constituents present in various fuel gases shows that there are large variations in the percentage composition. CO, CO_2, O_2, H_2 and hydrocarbons are generally the main constituents of important gaseous fuels.

The most important commercial gaseous fuels are

(A) Primary fuels: Natural gas and Liquefied Petroleum Gas (LPG)

(B) Secondary Fuels: Coal gas, water gas, producer gas, oil gas, gobar gas etc.

6.7.1 Water Gas or Blue Gas

It is blue gas because it burns with a blue flame. Blue gas consists mainly of carbon monoxide and hydrogen. It is produced by passing steam over red hot coke.

Construction: The gas generator consists of steel cylinder lined inside with refractory bricks. It is about 3 m in dia and 4 m high and fuel is introduced in it through a cup and cone feeder provided at the top. There are two tucyers near the bottom to pass air and steam separately through the coal bed and an outlet for removing ash.

Method of preparation: It consists of two steps repeated alternatively.

1. Air is passed though coke bed, when carbon reacts with oxygen of air to give carbon monoxide and carbon dioxide. Both the reactions are highly exothermic. Therefore the temperature of the bed rises to around 1100°C.

$$C + CO_2 \longrightarrow CO_2 + 97 \text{ kcal}$$
$$2CO + O_2 \longrightarrow 2CO_2 + 52.8 \text{ kcal}$$

This step is called **hot blow**.

2. In this step, steam is passed through the coke bed. It reacts with red hot coke to give water gas.

$$\underset{\text{Redhot coke}}{C} + \underset{\text{Steam}}{H_2O} \longrightarrow \underset{\text{Water gas}}{CO + H_2} - 41.9 \text{ kcal.}$$

The reaction is endothermic and hence the temperature of the bed falls. This step is called **cold blow**. As the temperature falls below 1000°C, hot blow is again repeated. Therefore the water gas is produced and collected during alternative blows.

Composition: CO = 40%, H_2 = 50%, N_2 = 5%, CO_2 = 5%

Calorific Value: Its calorific value is 2700 kcal/m^3.

Uses:

(*i*) It is used as a reducing agent in metallurgical operations.

(*ii*) as a source of hydrogen gas for various purposes.

(*iii*) as a fuel in industries.

(*iv*) as an illuminating gas.

The calorific value of the gas can be increased by mixing water gas with oil vapours. This is done by firstly heating a checker work of fine bricks by burning a portion of the gas.

Then kerosene oil is sprayed on this heated brick work when it gets vapourised and partly cracked. Water gas is passed through it. The mixture so obtained is called **carbureted water gas.** A simple sketch to explain the working of **carbureted water gas generator** is shown in Fig. 6.5.

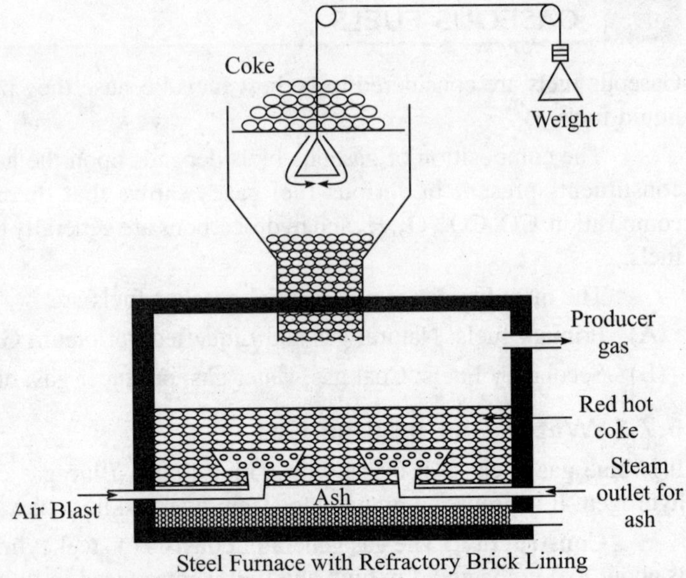

Fig. 6.4 Manufacture of water gas

The furnace proper consists of a cylindrical vessel made of steel. The coke is introduced through an opening at the top. There are separate inlets for blowing steam and air (alternately) from below and an exit near the top for the gases to escape.

The water gas from the furnace passes through the ***carburetter*** which is a big tower packed with heated bricks into which a spray of oil is injected. The oil gets volatilized and the vapours mixed with the gas, enter the ***superheater,*** which is another similar big tower packed with bricks heated to redness by the burning of a mixture of producer gas and air.

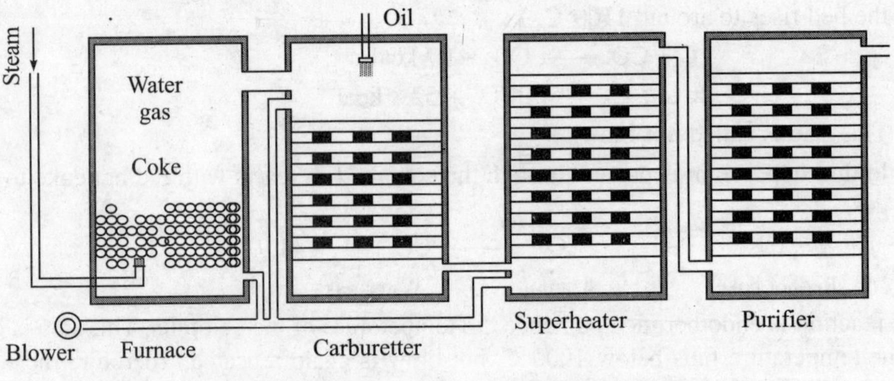

Fig. 6.5 Preparation of carburetted water gas

The oil vapours get cracked at the high temperature into simple gaseous hydrocarbons (saturated and unsaturated) such as ethane, propane propylene, ethylene, etc., which burn with a luminous flame.

The ***carbureted water** gas* is then passed through the ***purifiers*** which contain hydrated ferric oxide where hydrogen sulphide (which originates from the sulphur present in the coke) is removed by adsorption on ferric hydroxide gel.

It consists of approx. 25% CO, 35% H_2, 17–20% saturated hydrocarbons, 10–15% unsaturated hydrocarbons, 2% CO_2 and about 5% N_2. It has a calorific value of 4500 kcal/m^3.

6.7.2 Bio Gas

The most common example of biogas is gobar gas. Gobar gas is obtained by the fermentation of gobar (dung) in absence of air by anaerobic bacteria.

Manufacture: The essential parts of gobar gas plant (Fig. 6.6) are:

Digester: A Well constructed measonary work, built usually below the ground level (3.7 m deep and 1.7 m wide) called as digester or fermentation well.

(*a*) ***Gas-holder:*** A drum made of mild steel sheets of about 1.2 m height and 1.5 m diameter is placed in inverted position on the fermentation well and making air tight. It can be made up and down with the help of pulleys.

(*b*) ***Pipe line:*** There are three pipe lines connected to the plant:

(*i*) Feeding pipe used for feeding in the dung in the form of slurry in the fermentation well.

(*ii*) Outlet pipe for spent slurry used for collecting the spent slurry in outlet tank for manure.

(*iii*) Exit pipe for gas: used for supply of gas to kitchen from the top of the drum.

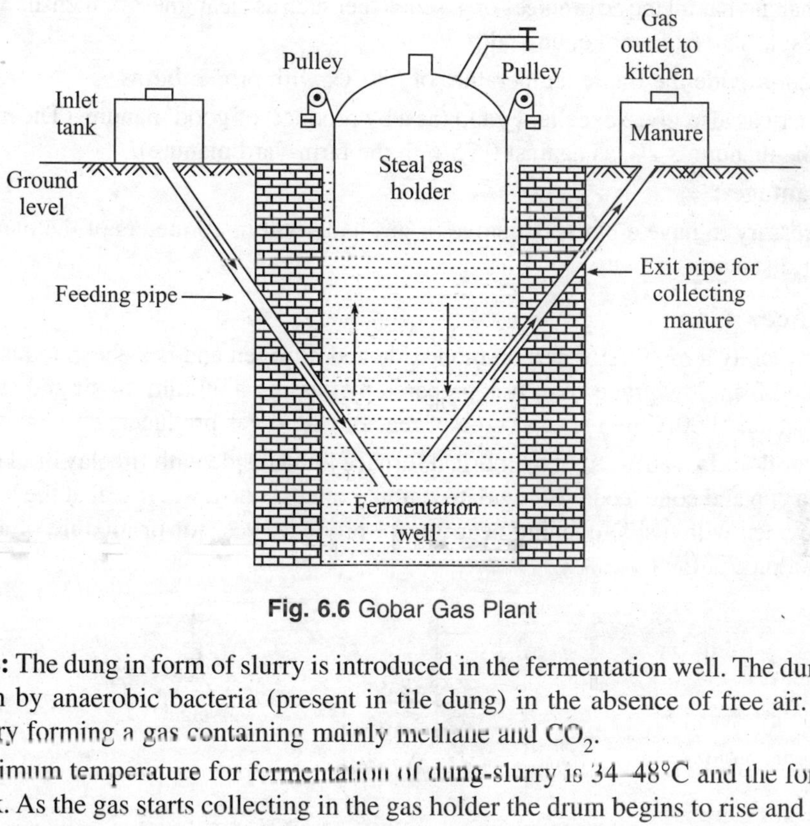

Fig. 6.6 Gobar Gas Plant

Process: The dung in form of slurry is introduced in the fermentation well. The dung is subjected to fermentation by anaerobic bacteria (present in tile dung) in the absence of free air. Such bacteria digest this slurry forming a gas containing mainly methane and CO_2.

The optimum temperature for fermentation of dung-slurry is 34–48°C and the formation of gas starts in a week. As the gas starts collecting in the gas holder the drum begins to rise and floats over the slurry. The gas is taken out from the exit provided at tile top of the drum, from where it is supplied through pipe line to the kitchen.

Composition: The average composition of the gobar gas is:

Methane (CH_4) = 55%, Hydrogen (H_2) = 8%, Carbon dioxide (CO_2) = 35%, Nitrogen = 2% and traces of H_2S

Its average gross calorific value is 5300 kcal/m^3.

Uses:

(*i*) It is used as a domestic fuel particularly in villages where plenty of dung is available (about 45 kg of dung per day is required to produce 2.8 m^3 of the gobar gas which is sufficient for an average family).

(*ii*) It can also be used for lighting and power purposes like running water pumps, chaff cutting machines etc.

(*iii*) The manure derived from the spent slurry is a richer manure than the original gobar.

Advantages and Disadvantages of Gobar Gas

Advantages:

1. The production of gobar gas is highly economical than other fuel gases. It has been found that 4.25 kg of fresh cattle dung (or 1 kg dry cattle dung) gives nearby 160 liters of gobar gas, which can supply 188 kcal of heat. On the other hand, 1 kg of dry dung on direct burning gives only 23.4 kcal of effective heat.
2. It does not contain the poisonous 'CO-gas' as an ingredient.
3. Gobar gas has all the advantages of gaseous fuel such as cleanliness of utensils and surrounding areas, absence of smoke, dust, dirt etc.
4. It can provide the flame temperature of 630°C, with proper burns.
5. Gobar gas also gives excellent yield (as a by product) of good manure. (The nitrogen content of the manure is 2% as against 0.75% in the farm-yard manure).

Disadvantages:

It is necessary to have a burner or stove or gas lamp within 10 meters of the plant. It cannot be used for longer distances.

6.7.3 Producer Gas

Producer gas is mainly a mixture of carbon monoxide and nitrogen and is a cheap industrial fuel, since it can be obtained from low grade coal. It is prepared by passing a mixture of air and steam over a red hot coal bed at about 1000°C in a special type of reactor called gas producer.

Construction: It is an air tight cylindrical furnace lined inside with fireclay bricks. It is provided at the top with a cup and cone feeder of regulated supply of coal and a water seal at the base for removal of ash. It is provided with a side opening near top for producer gas. Air or mixture of air and steam is blown through a pipe at the base.

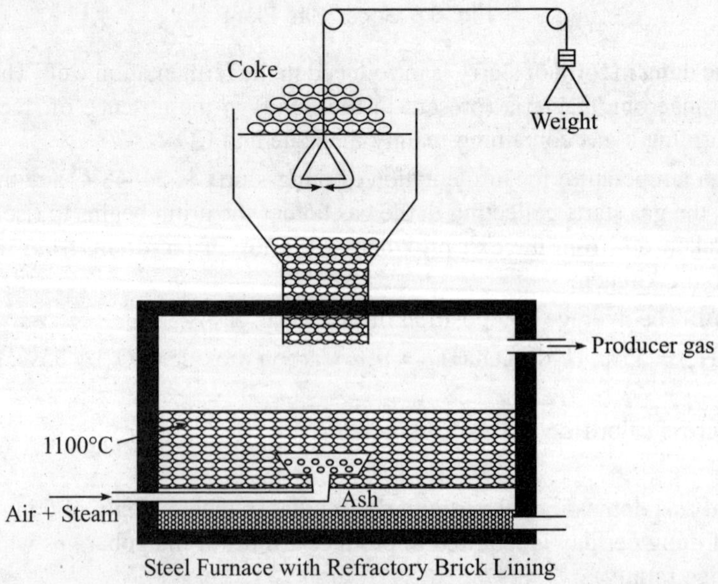

Fig. 6.7 Manufacture of Producer gas

Method of preparation: The producer is filled with coal and is ignited. Air is forced in through the hot fuel in the lower zone known as **oxidation zone.** Carbon burns to give carbon mond oxide and carbon dioxide, the reactions are strongly exothermic. The temperature of this zone is about 1100°C.

$$2C + O_2 \longrightarrow 2CO + 52.8 \text{ kcal}$$

$$2CO + O_2 \longrightarrow 2CO_2 + 135.3 \text{ kcal}$$

Carbon dioxide thus produced moves upwards through the bed of coal and is reduced to carbon monoxide. The reaction is endothermic. This zone is called **reduction zone** where temperature falls to 1000°C.

$$CO_2 + C \longrightarrow 2CO - 41.2 \text{ kcal}$$

In the uppermost part of the fuel bed, there is fresh coal which is at a much lower temperature and no reaction takes place. But the coal gets preheated and volatile matter distills. This part is called **distillation zone.** The temperature of this zone is 400–800°C. Gas goes out through the outlet and ash is removed through water seal. The lowest parts called **ash zone** consisting mainly of ash, which protects the grate of the plant from the intense heat of combustion.

When air is passed through carbon in lower zone, it gets heated and clinkers are formed which are difficult to remove. The fire burns on which the coal rests and the fractory lining of the plate also gets heated with the result that their lives are reduced. Moreover, the gas coming out is at very high temperature and if it is transported to distant place through pipes, much of its sensible heat is lost.

To overcome these difficulties about 0.35 kg of steam per kg of coal is blown along with the air. In the reduction zone, steam reacts with carbon producing carbon monoxide, which is an endothermic process.

$$C + H_2O \longrightarrow CO + H_2 - 41.9 \text{ kcal}$$

Hence, the temperature remains low.

Composition: Since the nitrogen of the air remains unaffected and passes as such with the mixture of gasses, producer gas mainly consists of carbon monoxide and nitrogen

$$CO = 22\text{–}30\%,\ N_2 = 52\text{–}55\%,\ H_2 = 8\text{–}12\%,\ CO_2 = 3\%$$

Calorific value: Since it consists of mainly non-combustible gasses like N_2 and CO_2, hence it has low calorific value i.*e.* about 1300 kcal/m^3.

Uses: Since it has low calorific value and comes out at high temperature, it is used on the spot and is not transported to distant places.

1. It is used in open hearth furnaces for manufacture of steel and glass.
2. It is used as a reducing agent in metallurgical operations.
3. It makes possible the utilization of low grade coal.

In the manufacture of coal gas, it is used for heating iron retorts.

6.7.4 Coal Gas as a Fuel

Coal gas *is obtained, when coal is carbonized or heated in the absence of air at about 1,300°C in either coke ovens or gas-making retorts.* Coal is fed in closed *silica retorts,* which are then heated to about 1,300°C (by burning *producer* gas and *air* mixture).

$$\text{COAL} \xrightarrow[\text{(In the absence of air)}]{1{,}300^\circ\text{C}} \underset{\text{(Residue in retort)}}{\text{COKE}} + \text{COAL GAS}\uparrow$$

The gas (coming out from the retort) is first scrubbed by passing through a *hydraulic main.* Much of the tar is then removed by cooling the gas in a huge water-cooled heat exchanger, called *condenser.* Any remaining tar and ammonia, present in the gas, are removed by scrubbing with water in a *scrubber.* The cooled gas is then scrubbed with *creosote oil,* which dissolves benzol, naphthalene, etc. The gas is then freed from hydrogen sulphide by passing it over *moist ferric oxide,* contained in a *purifier.* Any hydrogen sulphide gas, if present, is removed here (*see* Fig. 6.8).

$$2Fe(OH)_3(s) + 3H_2S(g) \longrightarrow Fe_2S_3(s) + 6H_2O(l)$$

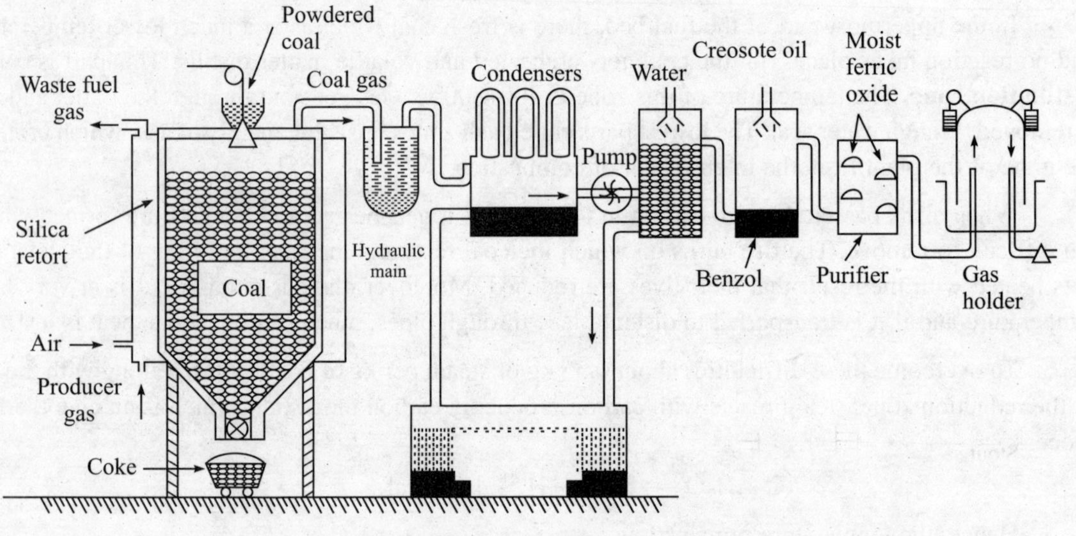

Fig. 6.8 Manufacture of coal gas.

When the iron oxide of the purifier is *exhausted,* it is taken out from the purifier and *exposed to air* when it gets oxidized to ferric oxide.

$$2Fe_2S_3(s) + 3O_2(g) \longrightarrow 2Fe_2O_3(s) + 6H_2S(g)$$

The purified coal gas is finally *stored over water in gas holders.*

Coal gas is a *colourless* gas having a *characteristic odour*. It is *lighter* than air and burns with a *long smoky flame.* Its average composition is: H_2 = 40%; CH_4 = 32%; CO = 7%; C_2H_2= 2%; C_2H_4 = 3%; N_2 = 4%; CO_2 = 1% and rest = 4%. Its calorific value is about *4,900 kcal/m³*.

Uses. It is used as: (*i*) *illuminant* in cities and towns; (*ii*) *a fuel,* and (*iii*) in metallurgical operations for providing *reducing atmosphere.*

6.7.5 Oil Gas

Composition:

The approximate composition of the oil gas is:

Methane (CH_4) 25–30% Hydrogen (H_2) 50–55%
Carbon monoxide (CO) 10–15% Carbon dioxide (CO_2) 3%

The calorific value of oil gas is about 4500–5400 kcal/m.

Production: Oil gas is obtained by cracking of kerosene oil. It consists of mixture of hydrocarbons.

$$\underset{\text{(Kerosene oil)}}{C_{12}H_{26}} \xrightarrow{\text{Cracking}} \underset{\text{(Mixture of Lower hydrocarbons)}}{CH_4 + C_2H_6 + C_2H_4 + C_2H_2 + C_3H_8 + C_3H_6} + \text{Tar}$$

The equipment used for preparing oil gas is shown in Fig. 6.9.

Cracking of kerosene oil is done in strong cast iron retort, enclosed in a coal-fired furnace. At the mouth of the retort, a bonnet is fitted through a molten, lead seal. The other end of the bonnet is connected to a hydraulic main, through a water seal A pipe from the hydraulic main leads to the gas holder. This pipe has a testing tap. From where the samples of gas can be taken for testing.

The proper cracking of the oil is estimated from the colour of the gas. A good oil gas should have a golden colour. Finally, the gas is stored over water in gas holders. Yield of the gas is about 40 to 50%. From the gas holders, gas is conveyed to laboratory through pipes, whenever required.

Applications: Oil gas is used as laboratory gas. It is also used to improve the calorific value of water gas and the mixture of the two gases is called carburetted water gas (discussed under water gas)

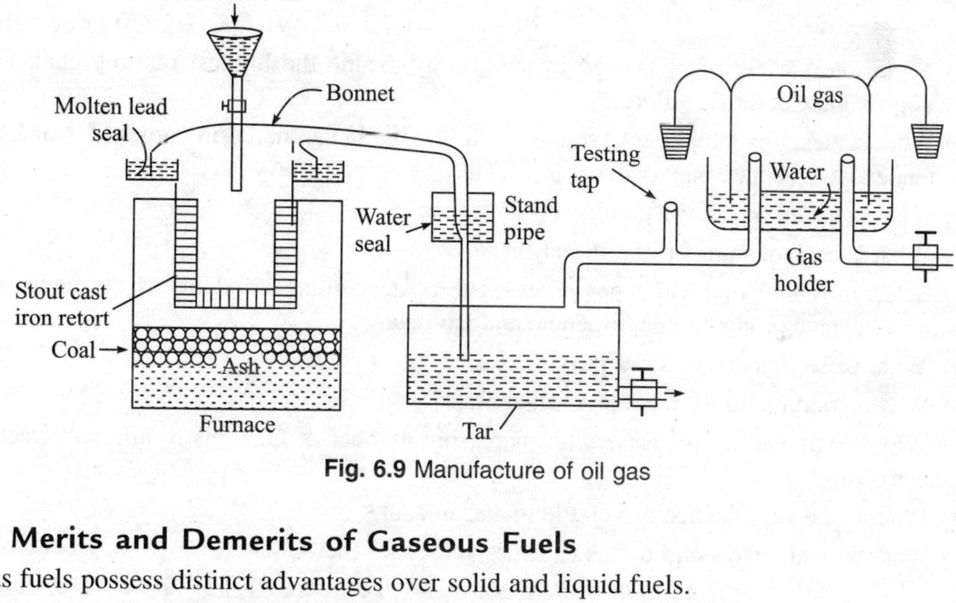

Fig. 6.9 Manufacture of oil gas

6.7.6 Merits and Demerits of Gaseous Fuels

Gaseous fuels possess distinct advantages over solid and liquid fuels.

(*i*) They are ashless and smokeless.

(*ii*) They are clean in use and economic in labour.

(*iii*) They have high calorific values and hence they produce high temperature.

(*iv*) They can be transported easily through pipes to the place without any manual labour.

(*v*) They can be lighted at a movement notice.

(*vi*) Due to their perfect combustion, the loss of heat is practically negligible.

(*vii*) Their flame can be made oxidizing, reducing or neutral as needed by adjusting the air supply.

(*viii*) The supply of gas to the burners can be easily controlled.

(*ix*) They burn in slight excess of air.

(*x*) They can be preheated by the heat of waste gases and hence their use is economical as far as heat is concerned.

(*xi*) They are efficient in converting heat into work and hence can be used in internal combustion (I.C.) engine for the generation of power.

Disadvantages:

(*i*) Since gases occupy a large volume, they require large storage tanks for carrying.

(*ii*) They are more costly as compared to solid and liquid fuels.

(*iii*) They are highly inflammable, hence chances of explosion (fire hazards) are more during their used.

REVISION QUESTIONS

1. (*a*) Define a chemical fuel. Give three examples of artificial fuels.

(*b*) Why should an ideal fuel have moderate ignition temperature? (*JNTU, Dec., 04*)

2. (*a*) What are the main varieties of coal formed inside the earth?

(*b*) What are the different types of fuels. What are the characteristics of a good quality fuel? (*GG SIP, Dec. 2K, Tripura 2K, SV, Aug. 01*)

(*c*) Name the variety of coal used in metallurgy. Name the highest ranking coal. Give its approximate calorific value.

3. (*a*) What is meant by knocking? Arrange the following in the increasing order of their knocking tendency: *n*-octane, naphthalene, iso-octane.

4. (*a*) What is meant by octane-number of a gasoline?

(*b*) What is centane number of a diesel fuel?

(*c*) Select the compound which possesses highest octane number and highest cetane number out of *n*-heptane, *n*-hexdecane, *n*-octane and iso-octane.

5. (*a*) What is the main purpose of proximate analysis of coal?

(*b*) Why a good fuel must have low ash content?

(*c*) Why does volatility decrease with increasing number of C atoms in different fractions of petroleum?

(*d*) What is the significance of volatile matter in coal?

6. (*a*) What do you understand by the calorific value of a fuel? (*VT, Aug 2K; March. 2001, Aug. 2001; RGPV, May 2001; UPT, Jan. 2001*)

(*b*) Distinguish between gross and net colorific value of a fuel (*CGSIP, March 2K; Anna July 2K; PT, Dec. 2K; UPT, Jan. 2001; MK, April 2001*)

7. (*a*) What is the importance of determining fixed carbon in coal?

(*b*) What do you understand by proximate and ultimate analysis of coal? Give the importance of ultimate analysis of coal.

8. (*a*) What are chemical fuels? Give complete classification of fuels with examples. (*Tripura, 2K; PT, Dec. 2K; UPT, Jan. 2001*)

(*b*) Distinguish between solid, liquid and gaseous fuels. (*Dibrugrh, Dec. 2K*)

9. (*a*) Differentiate between proximate and ultimate analysis of coal.

(*b*) How is the ranking of coal based on ultimate analysis? (*Dibrugrh, Dec. 2K*)

10. (*a*) Explain what do you understand by the term octane number. What is its significance?

(*b*) What do you understand by cetane number? What is its significance?

11. What is water gas? Give its composition. How is it prepared on large scaled? What are its uses? (*Anna, June, 2005, Dec. 2006*)

12. (*a*) What is LPG? What are the advantages of LPG over gaseous fuels?
(*b*) What is gobar gas? Describe the manufacture of gobar gas with a neat diagram. (*U.P.T. May 2007*)

13. How will you determined nitrogen and sulphur contents in the coal sample.

14. (*a*) Explain merits and demerits of liquid fuels.
(*b*) Give the advantages and disadvantages of gaseous fuels.

15. Explain the principle and calculations of kjeldahls method for estimation of nitrogen.

16. What do you mean by ultimate and proximate analysis of coal? Discuss their significance. (*R.G.P.V., Feb. 2005*)

17. Write short note on the following:
(*a*) water gas (*b*) gobar gas (*U.P.T., May 2003; Dec. 2007*)

18. Write notes on:
(*a*) Estimation of sulphur in coal (*b*) Knocking in IC engine
(*c*) Advantages of gaseous fuels

19. Write informative notes on:
(*a*) water gas (*b*) octane number (*c*) cetane number

20. (*a*) Discuss how the following constituents in coal are estimated:
(*i*) Sulphur (*ii*) Volatile matter (*iii*) Moisture

21. Describe the methods used for estimation of carbon, oxygen, nitrogen and sulphur in coal.

22. (*a*) What is ultimate analysis of coal? Explain the significance of each constituent.
(*b*) Explain the role of anti-knocking agents. (*Mumbai, May, 2006*)

23. (*a*) Why is producer gas used immediately after its production?
(*b*) Why is $C_2H_4Br_2$ added, when TEL is used as an anti-knock. (*RCPV,* May 2001)
(*c*) Give the different types of fuels. What are the requirements of fuel to be used in industry?

24. (*a*) Describe the refining of petroleum with various fractions obtained. (*Mumbai May 2007*)
(*b*) How is producer gas manufactured? State its composition and uses. (*Anna, June 2006*)
(*c*) Explain the proximate analysis of coal. (*Anna June, 2005, Dec. 2005, 2006*)
(*d*) Why is coke preferred to coal in metallurgical processes. (*Anna, June 2006, JNTU, Dec. 2004*)

25. (*a*) Define the terms CV, HCV and LCV.
(*b*) Discuss the parameters used in determinations of proximate analysis of coal.
(*c*) Why ethylene dibromide is added with TEL while using as antiknocking agent? Explain with chemical reactions involved in A. (*RGPV, June 2007*)

26. (*a*) Explain the fluidised bed catalytic cracking of petroleum fractions. (*VT, Feb. 2006*)
(*b*) Write a note on knocking in IC engines. (*VT, Aug. 2005*)

❑❑❑

SECTION D

Chapter 7: Polymers

Chapter 8: Composite Materials

7 Polymers

INSIDE THIS CHAPTER

7.1 INTRODUCTION

The word polymer is the combination of two Greek words—*poly* meaning several and *meros* meaning parts. Thus, polymers are high molecular weight substances consisting of large molecules of colloidal dimensions (*i.e.*, $10^{-4} - 10^{-7}$ cm), predominantly organic (but to a lesser extent inorganic) consisting of long chain macromolecules. The chains in them are made up of similar or different repeating units usually on a backbone of carbon atoms or other polyvalent atoms. The smallest unit, with which a polymer is made is known as *monomer or repeating unit or building block*. The monomers in polymers are joined with one another through covalent bonding.

Polymers form the backbone of the modern civilization and are the chief products of the modern chemical industries. An important attribute of the polymers is the ability to modify their inherent physical properties by the addition of certain additives while retaining their characteristics. Most of the polymers are of recent origin but they have made an impact on our daily life, starting from our kitchen/drawing room to space.

Difference between a polymer and macromolecule: Usually one does not make much distinction between a polymer and a macromolecule because both are high molecular weight substances. A polymer consists of a large number of structurally similar or dissimilar repeating units but a macromolecule may or may not have all the repeating units structurally similar. Let us consider, two well known bio-molecules (*i*) starch (*ii*) nucleic acid. The starch should be considered both as a polymer as well as macromolecule because it is made by the condensation polymerization of a large number of only one type of monomer, *i.e.*, of a glucose molecules. However, nucleic acid should be considered as a macromolecule and not as a polymer because it is made by the condensation polymerisation of three structurally as well as chemically different molecule, *i.e.,* phosphoric acid, pentose sugar and nitrogen base. Similarly, polyvinyl choride (PVC) may be regarded both as macromolecule as well as polymer because it is made up of only one type of monometer, *i.e.,* vinyl chloride. Thus, all polymers are macromolecules but all macromolecules are not polymers. Haemoglobin present in blood is a macromolecule but not a polymer. It has no repeating unit in its structure.

7.2 CLASSIFICATION OF POLYMERS

Homopolymers: These are the polymers obtained by the polymerisation of only one type of monomer. For example, polyethylene (PE), polypropylene (PP), polyvinyl chloride (PVC), polystyrene (PS), polyisoprene (PIP), poly acrylonitrile (PAN), polyvinyl acetate (PVA), polymethyl methacrylate (PMMA), nylon 6, polybutadiene (PBD), Teflon, cellulose, starch, glycogen etc.

Copolymers: These are the polymers obtained by the polymerisation of two or more monomers. For example, nylon 66, nylon 611, bakelite, urea formaldehyde, melamine formaldehyde, styrene butadiene rubber, polyesters, epoxy resins, alkyl resins etc.

Heterochain polymers: These polymers contain two different atoms in the main chain, one of which is usually carbon atom. For example, nylon 6, nylon 610, nylon 611, urea formaldehyde, melamine formaldehyde etc. All have nitrogen as the other atom along with the carbon atoms in the main chain. Similarly, polyesters, epoxy resins, alkyl resins, cellulose, starch etc. contain oxygen as the other atom along with the carbon atoms in the main chain. Similarly, polyesters, epoxy resins, alkyl resins, cellulose, starch etc. contain oxygen as the other atom along with the carbon atoms in the main chain.

Graft polymers: These are the special type of copolymers in which the main chain is made up of one type of monomer (A) and side chain by another type of monomer (B). Thus branched chain copolymers are also known as graft polymers *i.e.,*

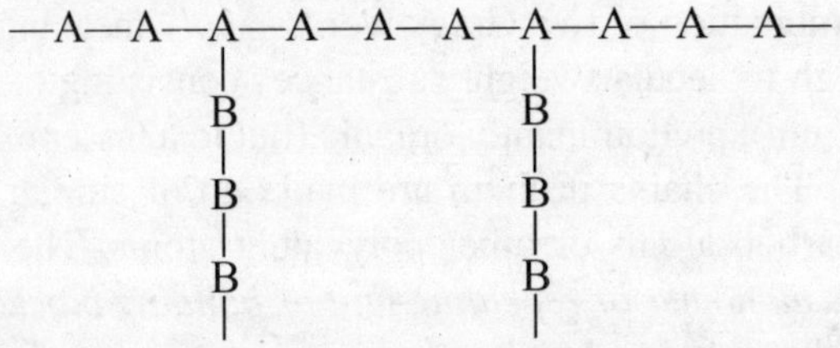

Block polymers: These are another special type of copolymers in which the main chain is made up of both types of monomers, A and B. These polymers have a long chain of homopolymer of A joined by a second block of the homopolymer of the second monomer B.

—A—A—A—A—B—B—B—B—A—A—A—A—A—B—B—B—B—

The block polymers may further have random arrangement of monomers A and B in the main chain to form *random block copolymers. i.e.,*

—A—A—B—A—A—A—A—B—B—B—B—B—A—B—A—A—B—B—A—

or in a regular alternating fashion (*alternating copolymers*) *i.e.,*

—A—B—A—B—A—B—A—B—A—B—A—B—

7.3 FUNCTIONALITY

The number of bonding sites in the monomers is known as the *functionality of monomer.* For a substance to act as a monomer, it must have atleast two bonding sites or reactive sites (functional group). In an alkene, the double bond may be considered as a site with two free valencies, thus the alkene monomer is bifunctional. Similarly, in alkadiene, the two double bonds correspond to the functionality equal to four. The compounds having reactive functional group like —OH—, NH_2, —COOH etc. on the extreme ends of the chain behave as bifunctional monomers. For example,

Ethylene glycol (HO—$(CH_2)_2$—OH), Adipic acid (HOOC—$(CH_2)_2$—COOH), mexamethylene diamine (H_2N—$(CH_2)_6$—NH_2), 6-Aminohexanoic acid (H_2N—$(CH_2)_5$—COOH) are all bifunctional monomers. However, melamine is a well known trifunctional polymer.

Depending upon the functionality of monomers, it is possible to synthesis the different types of polymers. A bifunctional monomer consisting of two binding sites at the extreme ends of chain, produces a linear chain polymer. In such polymers, all the monomers are connected with one another by covalent bonds. However, the different chains are held together by van der Waal's forces of attraction. If the monomer has polyfunctional groups, a three dimensional network is obtained by combining these monomers.

 If the bifunctional monomer is represented by a rectangular box containing two holes, a covalent bond by hard line and the secondary forces of molecular attractions by the wavy line, the formation of linear chain polymer may be shown as below in Fig. 7.1.

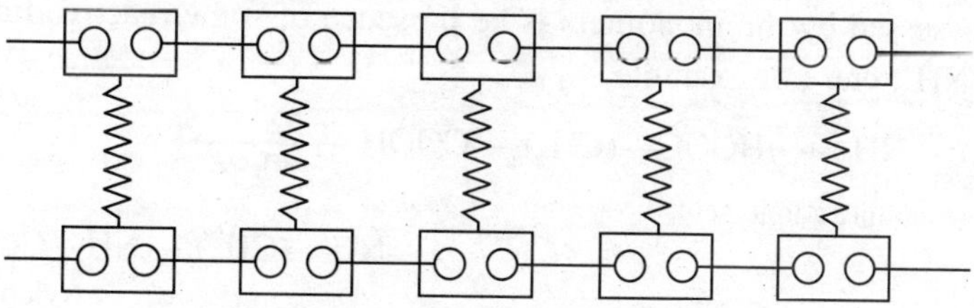

Fig. 7.1 Formation of Linear Chain Polymer

Similarly, if the trifunctional monomer is represented by a rectangular box containing three holes a covalent bond by hard line, the formation of three-dimensional cross-linked network lined by combining these monomers, may be shown as below in Fig. 7.2.

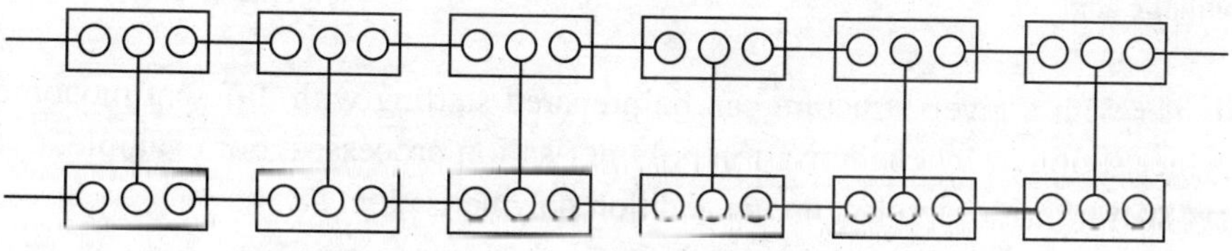

Fig. 7.2 Formation of three dimensional cross-linked network

7.4 TYPES OF POLYMERISATION

The polymerisation process is of two types (*i*) addition polymerisation and (*ii*) condensation polymerisation.

(*i*) Addition Polymerization: It is process of addition of the similar or different monomer(s) resulting in the formation of a giant product in which the molecular formula of the repeating unit is the same as that of the starting material.

Very often the molecular weight of the polymer is an integral multiple of the monomer. The polymers thus obtained are termed as addition polymers. For a monomer, to undergo the addition polymerisation, the basic requirement possessed by the monomer is the multiplicity in it *i.e.*, the presence of double or triple bond. For example,

$$n\text{CH}_2{=}\text{CH}_2 \longrightarrow (\text{—CH}_2\text{—CH}_2\text{—})_n$$

ethene (monomer) → polyethene (polymer)

$$n\text{CH}_2{=}\text{CH—Cl} \longrightarrow (\text{—CH}_2\text{—CH(Cl)—})_n$$

vinyl chloride (monomer) → polyvinyl chloride (polymer)

$$n\text{CH}_2{=}\text{CHOCOCH}_3 \longrightarrow (\text{—CH}_2\text{—CHOCOCH}_3\text{—})_n$$

vinyl acetate (monomer) → polyvinyl acetate (polymer)

***(ii)* Condensation Polymerization:** It is the process of the combination of two structurally/ chemically similar or different molecules usually with the loss of simple molecules like H_2O, NH_3, ROH etc. resulting in the formation of a high molecular weight substance. The polymers thus obtained are termed as condensation polymers. For a monomer, to undergo the condensation polymerisation, the basic requirement possessed by the monomers is be presence of some reactive functional groups like —OH, —COOH, —NH_2, etc. For example,

$$n\text{H}_2\text{N—(CH}_2)_6\text{—NH}_2 + n\text{HOOC—(CH}_2)_4\text{—COOH} \xrightarrow[-n\text{H}_2\text{O}]{\Delta} [\text{—NH—(CH}_2)_6\text{—NH—CO—(CH}_2)_4\text{—CO—}]_n$$

hexamethyl enediamine adipic acid → Nylon 66

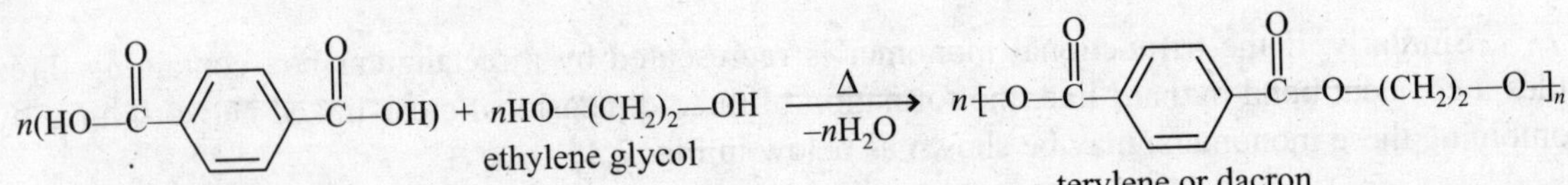

terephthalic acid + ethylene glycol → terylene or dacron

A polymer with a given structure can be prepared starting with different monomers either by addition polymerisation or by condensation polymerisation processes. For example, nylon 6 can be prepared either starting with caprolactum *via* addition polymerisation, *i.e.*,

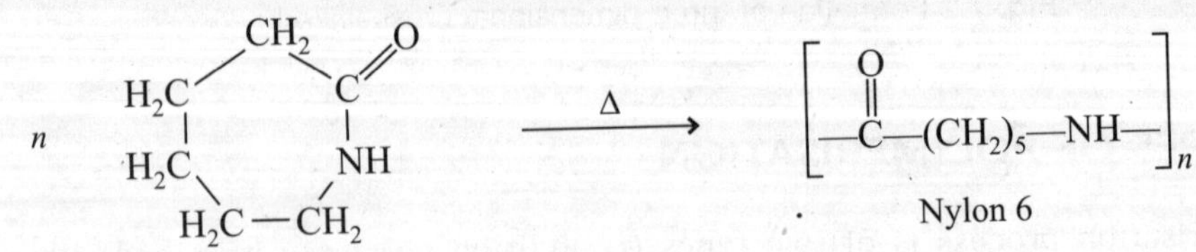

Caprolactum → Nylon 6

or starting with 6-aminohexanoic acid *via* condensation polymerization, *i.e.*,

$$n\text{H}_2\text{N—(CH}_2)_5\text{—C(=O)—OH} \xrightarrow[-\text{H}_2\text{O}]{\Delta} [\text{—NH—(CH}_2)_5\text{—C(=O)—O}]_n$$

6-aminohexanoic acid → nylon 6

7.4.1 Difference Between Addition and Condensation Polymerisation

Addition Polymerisation	*Condensation Polymerisation*
1. It requires the presence of a multiple bond in the monomer.	1. It requires two reactive functional groups to be present at both ends of the monomer.
2. There can be one or two monomers having having multiple bond or bonds.	2. There should be at least two different monomers having functional group which can re act with each other.
3. No byproduct is formed *i.e.*, no simple molecule is eliminated.	3. Generally a byproduct is formed *i.e.*, a simple molecule such as H_2O, NH_3 etc. is eliminated.
4. A homo-chain polymer, generally a thermoplastic is obtained. These polymers have only C atoms in the polymer chain.	4. Hetero-chain polymer, which may be either thermoplastic or thermosetting is obtained. These polymers have some atom than C atoms in the polymer chain.
5. The growth of chain is at one active centre only.	5. The growth of chain is at minimum two active centres.
6. It proceeds through chain growth mechanism. It. is also referred to as chain or step growth polymerisation.	6. It proceeds through step growth mechanism and is therefore, referred to as step growth polymerisation.
7. High molecular weight polymer is formed at once.	7. The molecular weight of the polymer rises steadily throughout the reaction.
8. Long reaction time give high yields but molecular weight of the polymer is affected slightly.	8. Long reaction time is essential to obtain high molecular weights.

7.5 PLASTICS

The term plastics refer to an organic macromolecule possessing the property to undergo moulding into desired shape upon subjecting to the action of heat, pressure in the presence of a catalyst.

Depending upon the source of recovery, plastics are of two types:

(*i*) Natural and (*ii*) Synthetic

(*i*) Natural Plastics: Plastics materials obtained from natural sources are called natural plastics For instance, amber and resins come from trees and casein is obtained from milk.

(*ii*) Synthetic Plastics: These are man-made plastics. For instance, Polyethene, PVC, Nylon-66, Buna-S etc.

The term 'Resins' is slightly different from plastics, but the two terms are considered synonymous to each other these days. Resins are basically binding materials which have been included in plastics as an important class.

7.5.1 Classification of Plastics

The plastic materials are classified into two categories:

(A) Thermoplastic Materials: They remain soft as long as they are heated and become hard on cooling reversibly. Since these two states namely 'soft' and 'hard' forms involve only physical changes, therefore, no chemical change occurs during heating or cooling these materials. Generally, these materials, having a long chain molecular structure are obtained via chain growth polymerisation. They are soluble in organic solvent, soft, weak and less brittle and can be reclaimed from waste materials.

Examples: Polyethylene, polypropylene, polyvinyl chloride, polyvinyl acetate, polystyrene, polymethyl methacrylate, polytetrafluoroethylene, cellulose acetate, polycarbonates, polyamides, acrylonitrile.

(B) Thermosetting Materials: These are obtained by the irreversible change of fusible, soluble materials having linear structure or non-crosslinked branched structure to produce an infusible, insoluble material which contains crosslinking between the polymer chains. The basic requirement for synthesizing thermosetting materials is that at least one of the monomers should have functionality more than two otherwise the material obtained will be linear and thermoplastic in nature. They become hard during moulding and don't undergo softening upon subjecting to heat. These materials have 3-dimensional crosslinked networks which are formed during moulding. **Prolonged heating chars these materials**. They are obtained via step growth polymerisation. Due to the presence of strong covalent bonds and crosslinking, these materials are insoluble in all the organic solvents. They are usually hard, strong and more brittle than the thermoplastic materials. They can not be reclaimed from waste materials. Examples: Bakelite, urea formaldehyde resin, polyesters, alkyl resins, epoxy resins, polyurethanes etc.

7.5.2 Difference Between Thermoplastic and Thermosetting Polymer

Basic difference between thermosetting polymers and thermoplastic polymers is that the thermoplastic polymers can be softened repeatedly by heating and hardened on cooling without change in properties. On the other hand, thermosetting plastics can be heated only once because it undergo permanent change on melting and sets into a new solid which cannot be remelted.

Thermoplastics	*Thermosetting plastics*
1. They have either linear or branched structure.	1. They have three dimensional network structure.
2. They are formed by addition polymerisation only.	2. They are formed by condensation polymerisation.
3. Neighbouring polymeric chains are held together by weak Vander Waals forces or dipole-dipole forces or hydrogen bonding. There are no crosslinks.	3. Neighbouring polymeric chains are held to gether by strong covalent bonds. The chains are crosslinked.
4. They can be softened on heating, so they can be reshaped or reused.	4. They retain their shape and structure even on heating. So they can not be reshaped or reused.
5. They are usually soft, weak and less brittle.	5. They are usually hard, strong and more brittle.
6. They can be reclaimed from wastes.	6. They can not be reclaimed from wastes.
7. These polymers are usually soluble in suitable organic solvents.	7. They are insoluble in almost all organic solvent due to strong bonds and crosslinks.
8. On strong heating they can be softened very quickly.	8. On strong or prolonged heating polymers get charred.

7.5.3 Plasticizers

To increase the plasticity and flexibility of the resin, certain chemical substances are added which are called plasticizer.

The plasticizer weakens the intermolecular forces of attraction between the macromolecules. Thus the flexibility and plasticity of the resin is increased, but at the same time the tensile strength and chemical resistance is adversely affected. The presence of small molecules of the plasticizer in between the long molecules of the resin reduce the attractive forces between the larger molecules of the resin (Fig. 7.3).

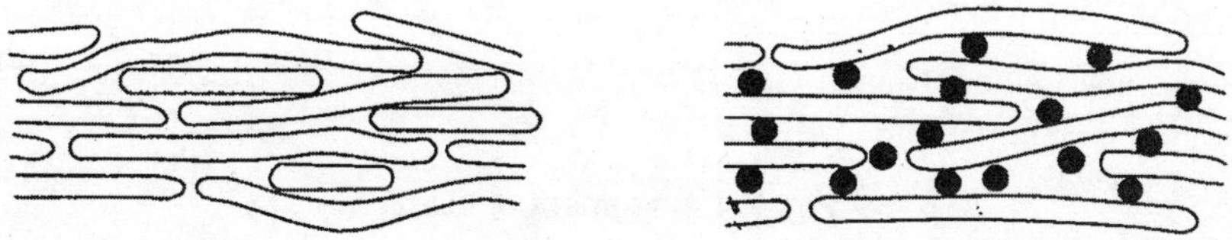

(*a*) (*b*)

Fig. 7.3 (*a*) Resin without plasticizer (*b*) with plasticizer.

Certain plastics do not soften very much on heating. These can be easily softened by the addition of some organic compounds which are called plasticizers. For example, poly vinyl chloride (PVC) is very stiff and hard but it is made soft by adding di-n-butylphthalate (a plasticizer). Some other common plasticizers are dialkyl phthalates, cresylphthalate.

Esters of higher acids like stearic acid, oleic acid or phthalic acids are commonly used plasticizer. Dioctyl phthalate (DOP), dibutyl phthalate (DBP) are some examples of these esters. Besides these, some phosphates like tributyl phosphate, Isophenyl phosphate, and tricresyl phosphate, tetra butyl phosphate are commonly used as plasticizers.

7.6 SOME IMPORTANT COMMERCIAL THERMOPLASTICS

There are very large number of commercially important thermoplastic polymers namely polyethylene (LDPE and HDPE), Teflon, PMMA, PVC, PVA etc. But keeping in view the limitations of the syllabus only a few important thermoplastics are discussed below:

7.6.1 Polythene $-(CH_2-CH-_2)_n$

It is one of the most common addition polymers and was synthesized in 1939 by imperial chemical industries from ethylene, a byproduct of petrochemical industries, via free radical polymerisation ethylene at high temperature and pressure. The monomer is a gas, which boils at –103.2°C and solid at –169.2°C. Depending upon manufacturing conditions, two types of polyethylenes are obtained.

(*a*) Low Density Polyethylene (LDPE): It is manufactured by the free radical polymerization of ethylene. The gas is first liquefied at 1500 atm and then pumped into autoclaves maintained between 180 –200°C, in the presence of oxygen and free radical initiator. The chain transfer steps result in the formation of branched polymers.

$$\underset{\text{Ethene}}{nCH_2{=}CH_2} \longrightarrow \underset{\text{Polyethene}}{-(CH_2{-}CH_2)_n-}$$

The polyethylene obtained so possesses a molecular mass of about 20,000 and low density and therefore, it is also termed as high pressure polyethylene or low density polyethylene. It has a branched structure containing both short and long chain branches. The side chain consists only upto 4–5 carbon atoms. The presence of branches in the polymer does not permit it to pack close together and hence its density is low (0.910–0.925 gcc^{-1}). The melting point of LDPE is 110–125°C and is 40–55% crystalline.

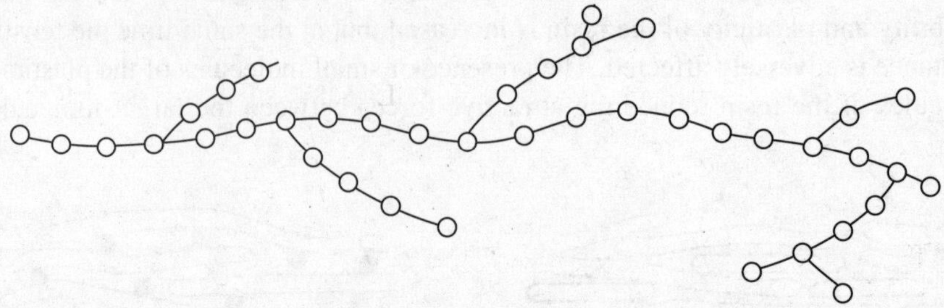

Fig. 7.4 A segment of LDPE

Properties

LDPE is a white, rigid, waxy translucent, non-polar material insoluble in common organic solvents. Strong acids, strong bases and the salt solutions don't have appreciable action on it at ordinary temperatures. However, it undergoes swelling when kept in kerosene.

Applications

It is widely used for a variety of products such as:

(*i*) Polyethylene-bottles/bags/pipes/sheets/dishes.

(*ii*) Coating on milk cartoons.

(*iii*) As a packing material in the form of thin plastic films, bags etc.

(*iv*) In high frequency insulation for electrical wires and cables.

(*v*) It is also used for making kitchen and domestic appliances, toys, etc.

The wide spread use of polyethylene is due to the low cost of both the starting monomer and the final polymer.

Limitations

(*i*) Due to its low density and low crystallinity, it has low rigidity and is not suitable for load bearing application. For instance, domestic water pipes are not made form LDPE because they may creep at anytime.

(*ii*) It is permeable to gas molecules because of its partial amorphous nature. So, LDPE is not suitable for the manufacture of pipes for gas distribution.

(*b*) High Density Polyethylene (HDPE): HDPE can be manufactured in two ways.

(*i*) Using Ziegler Natta Catalyst: In the method, ethylene is polymerized under 6-7 atmospheric pressure at 60–70°C in presence of Ziegler-Natta catalyst [a mixture of $(C_2H_5)_3Al$ and $TiCl_4$] dispersed in an inert solvent.

(*ii*) Using Mixed Oxide Catalyst: In the method, ethylene is polymerized under 35 atmospheric pressure at a temperature of 50–200°C in presence of Cr_2O_3 supported on $SiO_2 + Al_2O_3$.

HDPE obtained by above methods melts at 144–150°C and is highly crystalline (–90%). It is a linear structure containing a lower amount of short chain branching. The ratio of the side chain to the main chain is less than 1/200. These polymer molecules pack well and hence polyethylene of this type has higher density (0.941–0.965 gcc^{-1}).

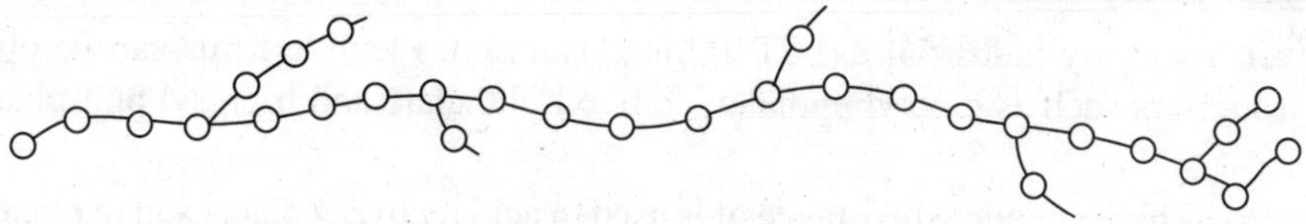

Fig. 7.5 A segment of HDPE

HDPE is comparatively more chemical inert than LDPE. It also exhibits better tensile strength (200–300 atm) than LDPE (–100 atm).

Uses of HDPE

(*i*) HDPE film is used as wrapping material for food products instead of paper because of its crisp feel and greases proof nature.

(*ii*) It is used for the manufacture of crates, food tubs, industrial containers and overhead tanks, because of its better oxidation and u.v. resistance.

(*iii*) If can also be used for domestic water and gas piping.

7.6.2 Polyvinyl Chloride (PVC)

Preparation: It is prepared from its monomer vinyl chloride, which is a gas and is commercially prepared by catalytic addition of dry HC1 to acetylene. The two gases are passed together over heated charcoal catalysts containing mercuric or other heavy metal salts at 100 to 250°C.

$$\underset{\text{Acetylene}}{HC{\equiv}CH} + HCl \xrightarrow{Hg^{2+}} \underset{\text{Vinyl chloride}}{H_2C{=}CH{-}Cl}$$

It can be also prepared starting from ethylene as follows:

$$\underset{\text{Ethylene}}{\begin{matrix} H_2C \\ \| \\ H_2C \end{matrix}} + Cl_2 \longrightarrow \underset{\text{Ethylene dichloride}}{\begin{matrix} H_2C{-}Cl \\ \| \\ H_2C{-}Cl \end{matrix}} \xrightarrow[-HCl]{+\text{ alc. KOH}} \underset{\text{Vinyl chloride}}{H_2C{=}CH{-}Cl}$$

PVC is of two Types:

(*i*) Rigid PVC (*ii*) Plasticized PVC

Rigid PVC is prepared by the free radical addition of vinyl chloride, by heating its water emulsion in the presence of small quantity of benzoyl peroxide or H_2O_2 in an autoclave under pressure.

$$n\,\underset{\text{Vinyl chloride}}{H_2C{=}\overset{\overset{\large Cl}{|}}{C}H} \xrightarrow{(H_2O_2)} \underset{\text{PVC}}{\left(H_2C{-}\overset{\overset{\large Cl}{|}}{C}H\right)_n}$$

Properties

(*i*) It is a colourless, odourless, amorphous and non-inflammable plastic.

(*ii*) Owing to the high intermolecular attraction forces present, it is a hard and stiff polymer. Its softening point is quite high (148°C).

(*iii*) It has glass transition temperature of 81°C and is soluble in cyclohexanone and tetrahydrofuran.

(*iv*) It is a polar polymer due to C—CI dipole, but its power factor is comparatively low due to immobility of the dipole at room temperature (Transition Temp. being 81°C).

(*v*) It has outstanding strength, lightness and chemical resistance.

(*vi*) It has excellent oil resistance and weathering resistance.

(*vii*) Unplasticised PVC can be extruded, calendered or press laminated.

However, for many industrial uses, PVC is plasticised. Flexible films can be obtained by the addition of plasticisers such as dioctyl phthalate, dibutyl phthalate and tricresyl phosphate.

Uses

(*i*) Due to its high chemical resistance, it is used in acid recovery plants and in plants for handling hydrocarbons. It is superior to polyolefins in this regard.

(*ii*) It is used for making pipes for drainage and guttering. Due to its excellent resistance to weather, it is replacing wood for making window frames which do not corrode.

(*iii*) PVC is less brittle, lighter and more flexible as compared to glass. It has, therefore, replaced glass for making bottles for storing consumable liquids such as edible oils, fruits squashes, vinegar, cosmetics, detergents etc. The bottles made of PVC are being used for storing mineral waters.

(*iv*) Rigid PVC sheets are used for tanks lining, light fittings, safety helmets, refrigerator components, trays, cycle and motorcycle mudguards etc.

(*v*) Plasticised PVC is very good insulator for current. It has completely replaced rubber for insulating electrical wires.

(*vi*) Plasticised PVC is used for making leather clothes, winch has greater abrasion resistance, flex resistance and washability as compared to leather. Therefore, it is used for ladies handbags, bathroom curtaining, car and kitchen upholstery.

(*vii*) It is also used for making plastic rain wears, baby pants, PVC shoes and garden hoses.

7.6.3 Polypropene

It is obtained from propene in the presence of Ziegler-Natta catalyst (AIR_3—$TiCl_4$).

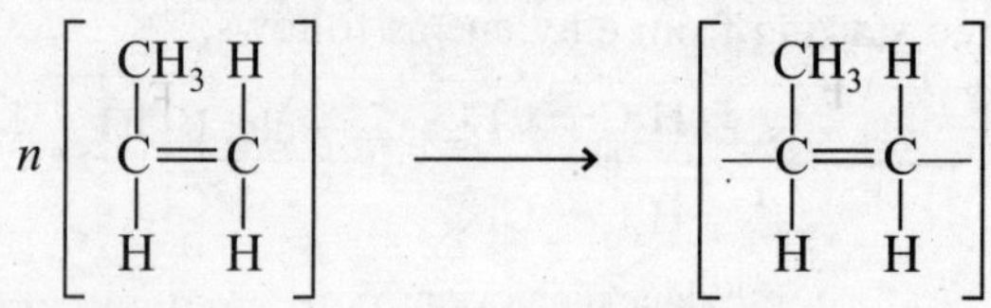

Propene Polypropene

Properties of Polypropylene

1. Polypropylene is stereoregular (isotactic) and highly crystalline polymer (m.p. 160–170 °C)
2. It possesses better hardness, strength, stiffness than polyethylene.
3. It is more resistant than polyethylene.

Uses: Polypropene is harder and stronger than polyethene. It is used for making ropes, fibres, heat-shrinkable wraps for records and seat covers.

7.6.4 Polystyrene

It is prepared by polymerization of styrene (dissolved in ethyl benzene) in presence of benzoyl peroxide catalyst.

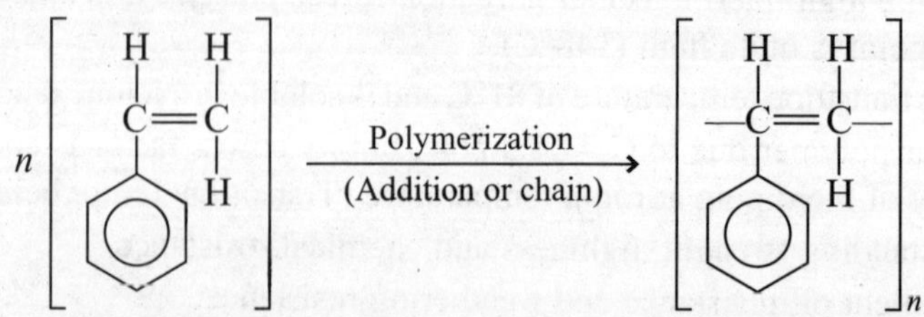

Styrene (Monomer) Polystyrene (PS)

Styrene is generally produced from benzene and ethylene.

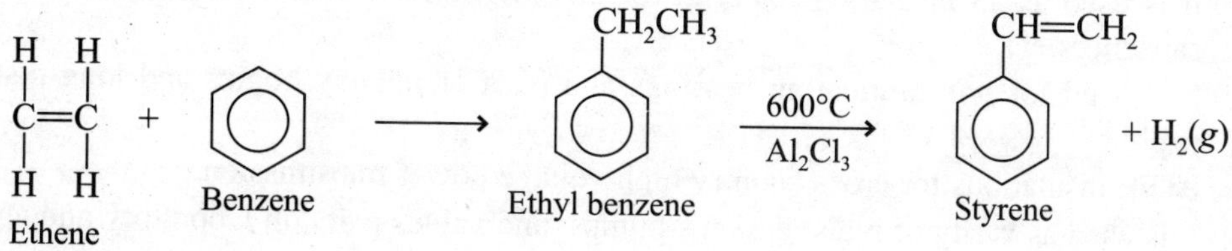

Properties of Polystyrene

Polystyrene is a transparent, light [sp. gr. 1.05 to 1.07 g/cm^3), good light stable, excellent moisture-resistant. It can be nitrated by fuming nitric acid and sulphonated by conc. H_2SO_4. At about 100 °C, it yields water-soluble emulsion. It is highly electric insulting, highly resistant to acids and good chemical resistant. But it has relatively low softening range (90–100 °C) and is brittle, However, it has the unique property of transmitting light through curved sections.

Uses: In moulding of articles like toys, combs, buttons, buckles, radio and television parts, refrigerator parts, battery cases, high frequency electric insulator, lenses, indoor lighting-panels, etc.

7.6.5 Polytetrafluoroethylene (PTFE) or Teflon

Preparation: Teflon is prepared by the polymerisation of tetrafluoro ethylene under pressure in the presence of benzoyl peroxide as catalyst.

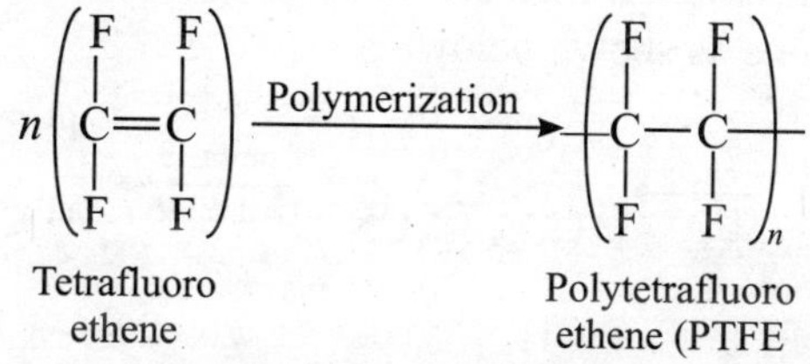

Polytetrafluoroethylene is produced from a very pure monomer which is completely free from traces of hydrogen by emulsion polymerization process using ammonium persulphate, $(NH_4)_2S_2O_8$ as initiator. The monomer is prepared by treating chloroform with hydrogen fluoride and dechlorinating the resulting chlorodiflouoromethane by heating.

$$CHCl_3 + 2HF \longrightarrow CHF_2Cl + 2HCl$$

$$2CHF_2Cl \longrightarrow CF_2{=}CF_2 + 2HCl$$

Properties

(*i*) The small atomic radius of fluorine, high electronegativity low polarizability and high linkage energy between carbon and fluorine given polytetrafluoroethylene a very high softening point (350°C) and excellent resistance to chemicals.

(*ii*) It is highly crystalline solid havng a density of 2.1–2.3 g cm^{-3}.

(*iii*) Although it is a linear polymer with no branching, yet it behaves like a thermoplastic material.

(*iv*) Its high melting point and melt viscosity (*i.e.*, low melt flow rate) and poor solubility make it impossible to fabricate by conventional means, so special methods are required.

(*v*) It is extremely resistance to the action of corrosive solvents and reagent. Because of its good weather resistance it is a very good regulator.

(*vi*) The strong C—F bond forces are responsible for its excellent toughness, heat resistance and non-adhesive characteristics.

Applications

(*i*) It is used as an insulating material for electric motors, transformers, generators, cables, capacitors etc.

(*ii*) It is used for non lubricating bearings and pads or railway bogies and long distance gas peplines.

(*iii*) PTFE insulations for cross country high voltage power transmission.

(*iv*) It is used as verity of reals, gaskets, pumps, and values parts, in laboratory and agricultural equipments.

(*v*) It is not wetted by oils or water, so its coatings have non-stick properties. It is, therefore, used for coating frying pans and other kitchen utensils.

(*vi*) It is used as laminates for printed circuitry.

(*vii*) It is used as stop cocks for burettes used in laboratories.

(*viii*) It is used for coating and impregnating glass fibres, asbestos fibres and cloth.

(*ix*) It is used for tank lining and pipe lining in which different chemicals are passed or kept for long.

7.6.6 Polymethyl Methacrylate (PMMA) or Plexiglass

It constitutes an example of polyacryloid plastics and its other trade names are **plexiglass, lucite, acrylite, perspex and crystallite.** It is prepared by free radical polymerisation of metal in presence of acetyl peroxide or hydrogen peroxide. Methyl methacrylate which is chemically an ester of methyl acrylic acid, is synthesized from acetone as shown below:

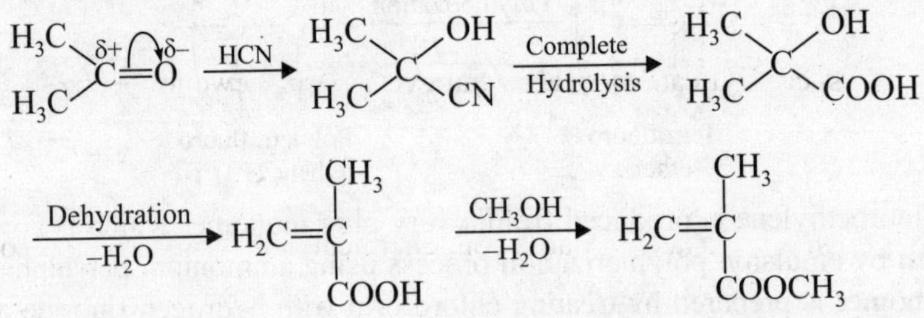

methyl methacrylate

Methyl methacrylate then undergoes polymerisation to produce PMMA

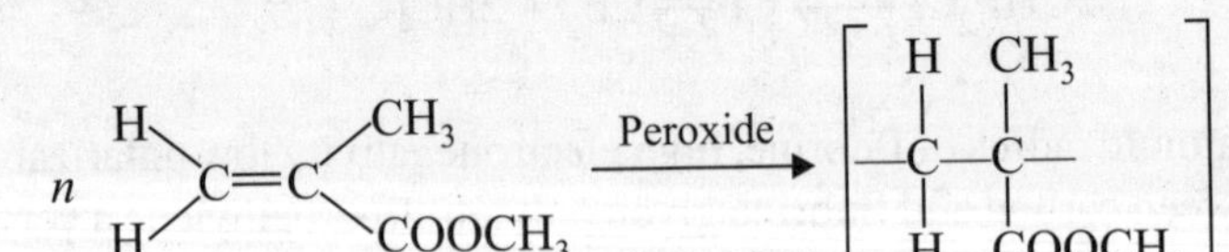

PMMA

The polymer has the structure

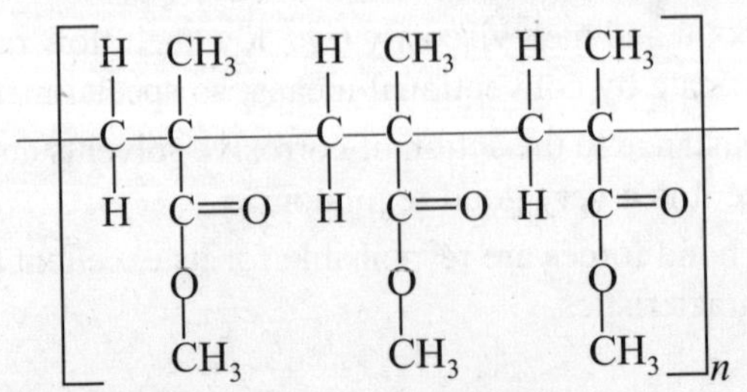

Properties

(*i*) PMMA is colourless, amorphous, transparent, hard, fairly rigid material.

(*ii*) Since it is prepared via free radical polymerization, therefore, it is atactic. The atacticity in the polymer is responsible to confer the transparent nature of polymer.

(*iii*) It becomes a rubbey substance above 65°C and softens between 130–140°C.

(*iv*) It has excellent optical properties (optical clarity) and is nearest alternative to glass owing to its non-shattering behaviour.

(*v*) It is very light substance and weighs about one-third as much as glass.

(*vi*) It can be moulded readily almost to any desired shape because of its wide range of temperature from its rigid state to viscous state.

(*vii*) It has high resistance to suhnlight and transmits light accurately even in curved sections.

(*viii*) Due to the presence of ester groups, the polymer has low resistance to hot acids and bases.

(*ix*) The polymer exhibits remarkable resistance to scratches. The scratching on PMMA can be removed by rubbing gently a tissue paper moistened with acetone over the polymer.

Uses

1. It is used for protective coating and for the manufacture of safety glass.
2. It is used for making long-lasting lenses for automobile tail lights, windscreens, TV, computer screen guards, fibre optics.
3. It can take up beautiful colours and tints and hence is used in the manufacture of emulsions, paints, adhesives, light shades, signboards, transparent domes and skylights aircraft windows, bomber noses, cockpit canopies, artificial eyes, plastic jewellery etc.

7.6.7 Polynurethanes

Polyrethanes were developed by Otto Bayer and co-workers in 1937. Because of exceptionally good polymeric properties such as resistance to abrasion, chemical attack by oxygen, ozone, hydrocarbons etc. polyurethanes can be used in a number of ways such as coatings, elastomers, fibres and foams.

(*i*) **Coatings:** Due to their resistance to impact, abrasion, greases, oils etc., they are used as coatings.

(*ii*) **Elastomers:** They have excellent abrasion resistance, good elasticity and hardness resistance to oils, greases and solvents. They are therefore, used as elastomers.

(*iii*) **Fibers:** Resilient PU fibers are very popular as *spandex*. They are used for making swim suits and foundation garments.

(*iv*) **Foams:** Both flexible and rigid foams can be prepared which are light, and tough. The foams also have good resistance to weathering, abrasion, heat and chemicals.

Polyurethanes contain the characteristics urethane linkage 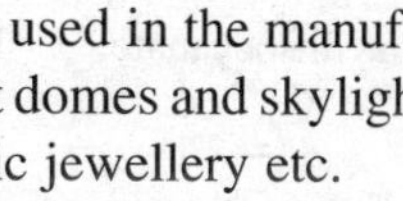 as their repeat unit.

Some important polyurethanes are discussed below:

(1) Polyurethane rubber (or **isocyanate rubber**) *is produced by reacting polyalcohols with di-isocyanates, e.g.,*

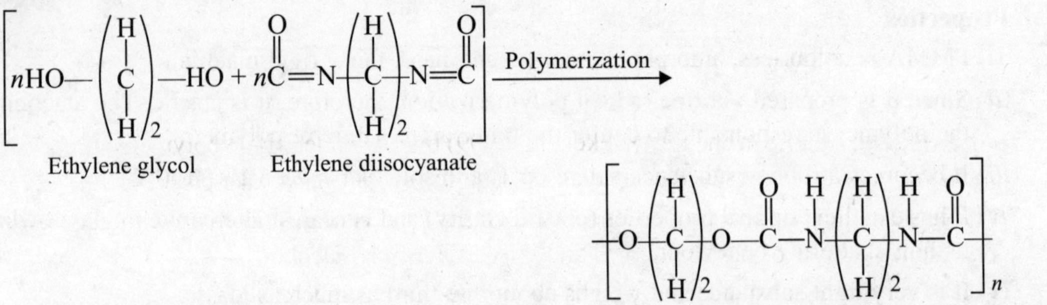

Ethylene glycol   Ethylene diisocyanate

Polyurethane rubber (or **isocyanate rubber**)

Properties: Polyurethanes are *highly resistant to oxidation,* because of their *saturated* character. They also show good resistance to many organic solvents, but are attacked by acids and alkalis, especially concentrated and hot. The polyurethane foams are light, tought and resistant to heat, abrasion, chemicals and weathering.

Uses: For surface coatings and manufacture of foams and spandex fibres.

(2) Perlon-U: A crystalline polyurethane, known as Perlon–U, is obtained by the polymerisation of 1, 4-butanediol with 1, 6-hexa methylene diisocyanate.

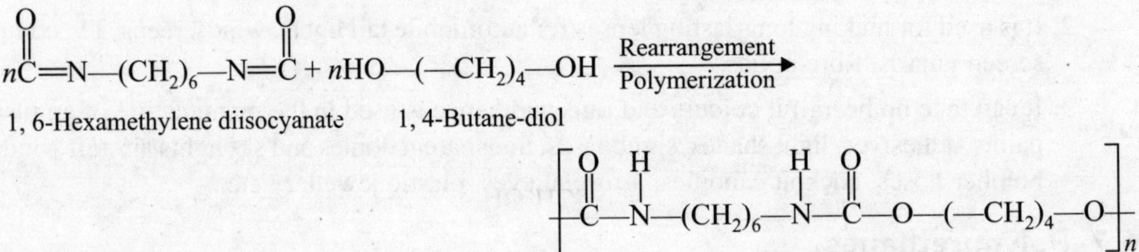

1, 6-Hexamethylene diisocyanate   1, 4-Butane-diol

Polyurethane (Perlon-U)

Applications

(*i*) Polyurethanes are used as coatings at the floor of gymnasiums and dancing halls.

(*ii*) Elastomeric polyurethanes are used as synthetic rubbers for enhancing the life of tyres.

(*iii*) They are also used for makings small industrial wheels and shoe-soles.

(*iv*) Perlon—U fibers (spandex) are use for making swim-suits and foundation garments.

(*v*) Flexible foamed polyurethanes are used for making car and furniture cushions.

(*vi*) They are also very good leather substitute and well known in market as *cor-foams,* used for making pillows, mattresses.

Properties

(*a*) They are highly resistant to oxidation by O_2, O_3 and other oxidising gases, because of their saturated character.

(*b*) In comparison with Nylons, Polyurethanes are less stable at elevated temperatures;

(*c*) They have good resistance to many organic solvents, but are attacked by concentrated and hot acids and alkalis.

(*d*) They have excellent resistance to abrasion, weather and possess good mechanical strength.

(*e*) Suitable choice of diisocyanate and diol and other additives will result in the production of polyurethanes which can be used as coatings, elastomers fibers and foams.

7.7 THERMOSETTING RESINS

7.7.1 Bakelite (Phenol Formaldehyde, PF)

It was prepared by the American chemist Backeland (1909) by the condensation polymerisation between phenol and formaldehyde in presence of acid/alkali. The following stages take place.

(*i*) In the first stage, phenol and formaldehyde in 1 : 1 molar ratio react with each other in presence of acid or base to form *o* and *p*-hydroxybenzyl alcohol.

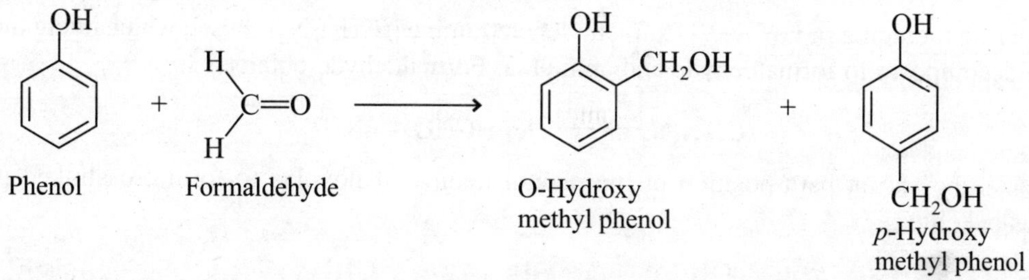

If the above ratio is kept 1 : 2, the following possible products are obtained.

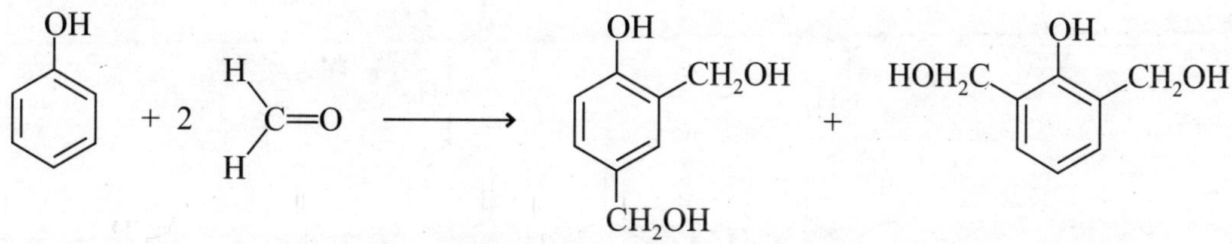

However, by adjusting P/F ratio as 1 : 3, the only product as shown below, is obtained.

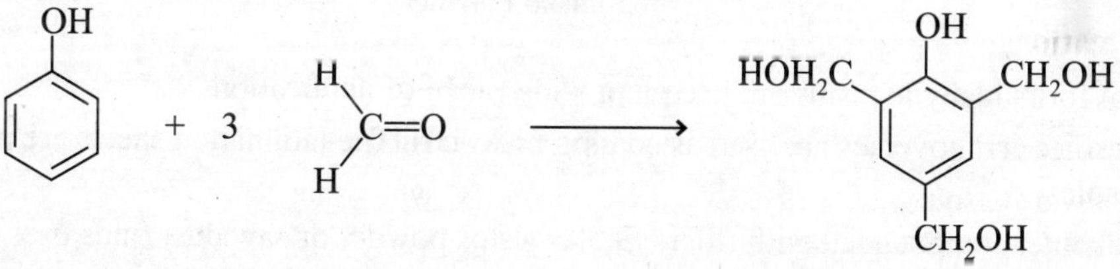

The formation of above monosubstituted, disubstituted or trisubstituted phenols is known as stage I and these products after condensing with phenol or by themselves given the product commonly known as resoles.

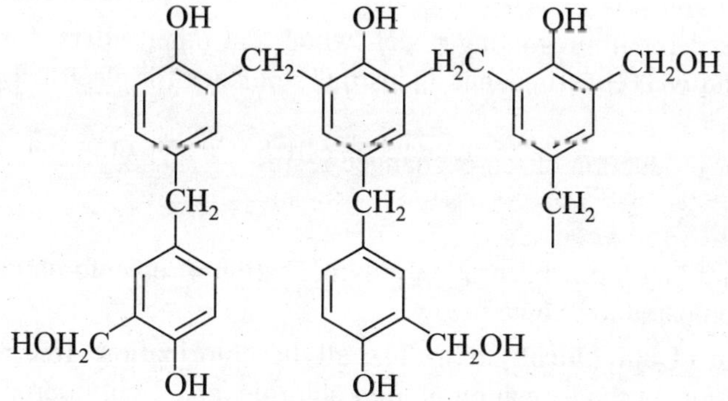

Resoles

(*ii*) In the two stage process, phenol and formaldehyde are heated with an acid catalyst like HCl, H_2SO_4, or organic acids like oxalic acid, tartaric acid etc. to form a linear polymer known as **novolac.**

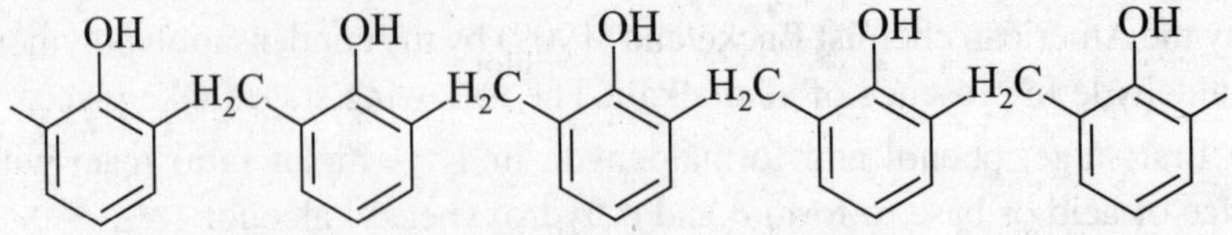

novolac

During moulding of novolac, hexamethylenetetramine $[(CH_2)_6N_4]$ is used which being thermally unstable decomposes to formaldehyde and ammonia. Formaldehyde obtained so

$$(CH_2)_4N_4 \xrightarrow{6H_2O} 6HCHO + 4NH_3$$

then attacks the vacant para position of the phenol mainly of novolac to form crosslinked polymer known as bakelite.

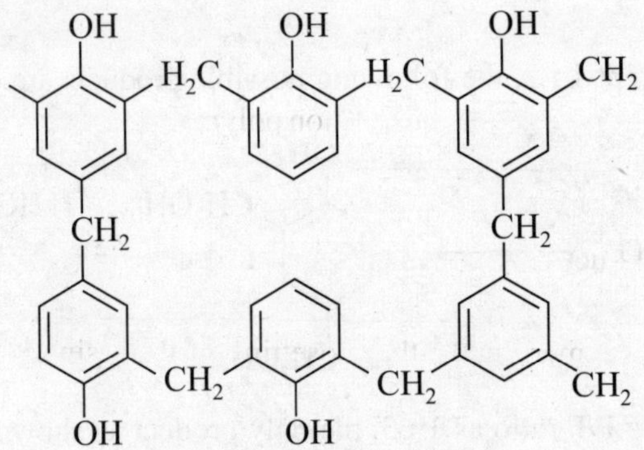

Crosslinked Polymer

Uses

Phenol formaldehyde resins are used in a wide range of applications.

(*i*) Resoles and novolacs are used as coating materials. The lamination sheets are prepared from resoles.

(*ii*) Bakelite, compounded with fillers like asbestos powder or saw dust finds uses for moulding electrical items like telephone instruments, electric insulators parts such as switches, plugs, switch boards, heater, pressure cooker handle.

(*iii*) It is also used as distributor leads of cars, adhesive for grinding wheels and brake linings.

(*iv*) It is used for varnishes, electrical insulation and protective coatings.

(*v*) It is also used for impregnating paper, wood and other fillers, for producing decorative laminates and wall coverings and industrial laminates for electrical parts including printed circuits.

(*vi*) It is used for production of ion-exchanged resins.

7.7.2 Epoxy Resins

These resins are prepared by the condensation polymerisation of epichlorohydrin and bisphenol. The starting materials are prepared as shown below:

(*i*) Preparation of epichlorohydrin: The allylic chlorination (free radical substitution) of propylene at 400°C results in the formation of allyl chloride, *i.e.*, 3-chloroprop-1-ene. The latter upon reacting with chlorine-water at 300°C produces 1, 3-dichloropropan-2-ol- which when heated with CaO at 200°C produces epichlorophydrin. All these reactions may be summarized as:

$$\underset{\text{Propylene}}{H_2C{=}CH{-}CH_3} + Cl_2 \xrightarrow[-HCl]{400°C} \underset{\text{Allyl Chloride}}{H_2C{=}CH{-}CH_2Cl} \xrightarrow[300°C]{H_2O/Cl_2}$$

$$\underset{\text{Glycerol dichlorohydrin}}{ClCH_2{-}CH(OH){-}CH_2Cl} \xrightarrow[200°C]{CaO} \underset{\text{epichlorohydrin}}{H_2\overset{\;O\;}{C{-}CH}{-}CH_2Cl} + HCl$$

(*ii*) **Preparation of Bisphenol:** Acetone and phenol (dissolved in aqueous NaOH solution) in 1 : 2 molar ratio at ~50°C form bisphenol as shown below:

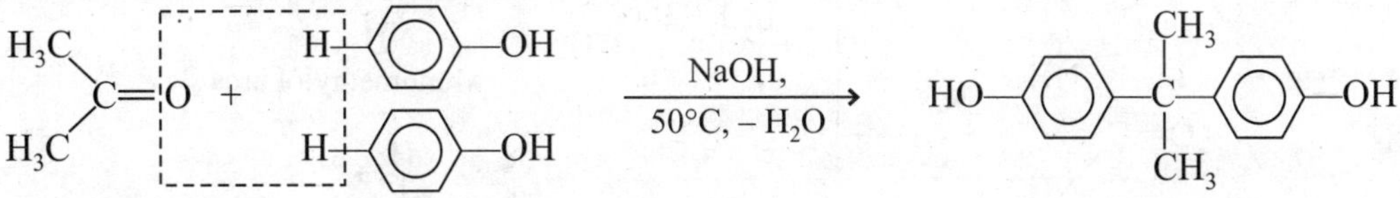

Acetone Phenol Bisphenol

(*iii*) **Condensation of ephichlorohydrin and bisphenol:** In presence of alkaline catalyst, epichlorohydrin and bisphenol undergo condensation polymerisation to form epoxy resin. In this reaction, opening of epoxide ring takes place generating OH groups on epoxide molecules. The OH groups than undergo cross-linked during moulding resulting in the formation of cross-linked polymer. At first stage, a low molecular weight product is obtained. In order to ensure that the polymer chain must terminate with the epoxide ring, an excess of epichlorohydrin is used. The excess amount of epichlorohydrin assists in cross-linked the polymer during thermosetting of the resins.

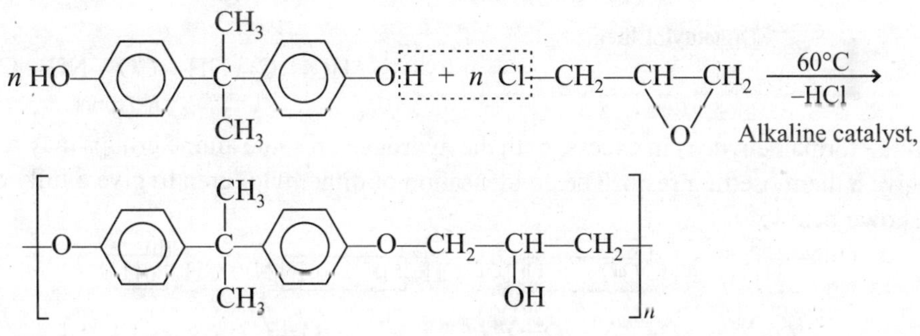

Properties

The physical nature of the polymer depends upon its molecular weight. The resin is a viscous liquid with the molecular weight 340 *i.e.*, contains one bisphenol and two epoxy moieties. However, the polymer becomes a brittle solid when its molecular weight is as high 800. The resin melts at 145–155°C. Due to the presence of stable ether linkage, the polymer shows remarkable resistance towards water and other solvents.

Applications

(*i*) Being polar, they have excellent adhesive characteristics and are used as adhesives, for glass, metals etc. and are popularly known as *araldite*.

(*ii*) They have excellent resistance to wear, they are also tough and heat resistance. So, they are used for surface coatings particularly for making skid resistant surface for high ways, road junctions and round abouts.

(*iii*) Good heat and electrical resistance, low water absorption tendency, dimensional stability make them very good materials for electronic applications particularly in mouldings containing inerts and encapsulations.

7.7.3 Urea Formaldehyde Resins (UF)

Urea reacts with formaldehyde by addition to form methylol compound depending upon the quantity of the reactants. A 1 : 1 ratio of the two reactants give monomethylol urea, whereas 1 : 2 ratio give dimethylol urea as shown below:

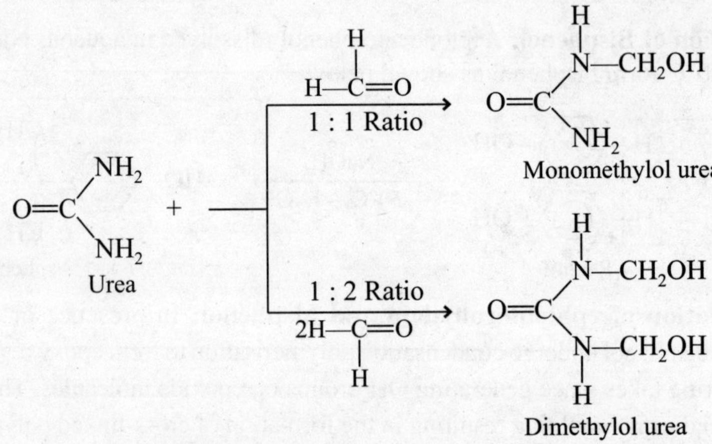

The mono and dimethylol urea may then react with more urea with the elimination of water *i.e.*, they may undergo condensation reactions with one another to give linear polymer.

$$\underset{\text{Dimethylol urea}}{HOH_2C\text{—}NH\text{—}CO\text{—}NH\text{—}CH_2OH} + H_2N\text{—}CO\text{—}NH_2 \xrightarrow[-H_2O]{} \underset{\text{A Trimer}}{HOH_2C\text{—}NH\text{—}CO\text{—}NH\text{—}CH_2\text{—}NH\text{—}}$$

But if the formaldehyde is in excess, both the hydrogen on some amine groups may react forming linkages to give a thermosetting resin. The condensation of dimethylol urea to give a fully cross-linked polymer is shown below:

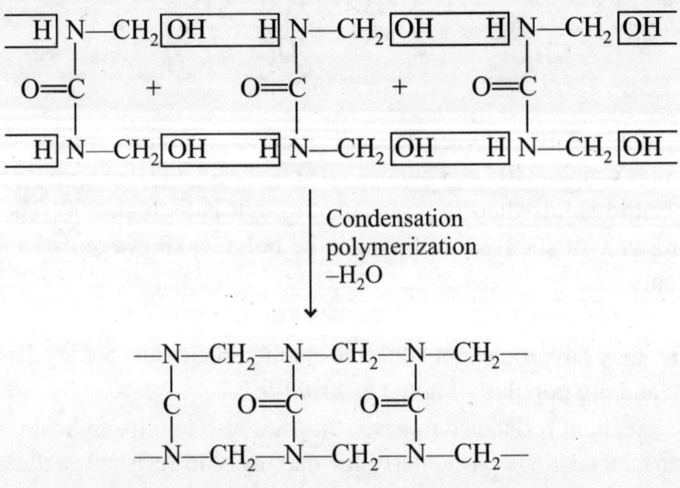

Crosslinked urea formaldehyde resin

Similarly monomethylol urea and dimethylol urea may condense to give a partially cross-link polymer.

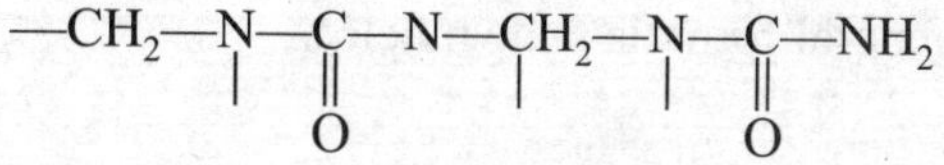

Partially crosslinked polymer

Properties

(*i*) Urea formaldehyde resins have a distinct advantage over the phenolic resins in that they are clear and colourless; and thus can be synthesised in all desired colours by adding the proper pigment.

(*ii*) The hardness and tensile strength of these resins is better than those of phenolics but their impact strength and heat moisture resistance are lower.

(*iii*) They have excellent abrasion resistance and good adhesive characteristics.

Uses

(*i*) These resins are widely used in the manufacture of buttons, bottle caps, surgical items, cosmetic closures, house appliances, coloured toilet seats and several other purposes.

(*ii*) They are also used as adhesives in the plywood industry and furniture.

(*iii*) They impart stiffness, crease resistance, fire retardation and shrinkage control to cotton fabrics and so are used for the finishing of cotton textiles.

7.8 NATURAL ELASTOMERS OR RUBBERS

The polymers that have elastic character like rubber are called elastomers. Any material like rubber that can return to its original shape after stretching is said to be elastic. In elastomers the polymer chains are held together by weak intermolecular forces. Because of the presence of weak forces, the polymers can be easily stretched by applying small stress and regain their original shape when the stress is removed. The most important example of elastomers is natural rubber. However, a few cross linked are also introduced in the chains. For example, natural rubber is a gummy material which has poor elasticity. It becomes soft and sticky when heated and becomes permanently deformed when stretched. However, when natural rubber is heated with 3 5% sulphur, it becomes non-sticky and more elastic. This process is called **vulcanization** and the product formed is called **vulcanized rubber** *which has better tensile strength, elasticity and resistance to abrasion than natural rubber.* This is due to sulphur cross links between polymer chains.

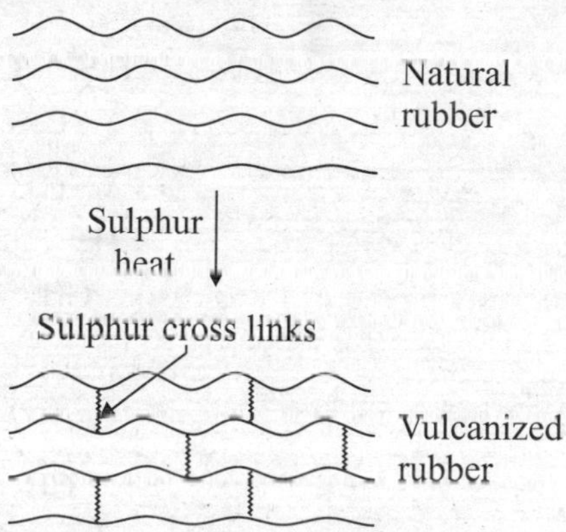

Fig. 7.6 Vulcanozation of RUbber

An elastomer is linear polymer which exhibits elasticity and other rubber like properties. Elastomers elongate on stretching and revert to the original position as soon as the stretching force is released. They have long flexible chains with weak intermolecular forces and occasional cross-links between the chains. The long chains do not fit together well enough for van der Walls forces to act properly and, therefore, the intermolecular forces are weak. Moreover, there are no polar groups or sites for hydrogen bonding. For elongation and recovery the long chains of an elastomer must be connected to each other by occasional crosslinks. These crosslinks prevent the slipping of molecules across each other, but at the same time the flexibility of the chain is preserved. For example, natural rubber is a highly soft and elastic material. It is a polymerised form of isoprene (2-mehtyl-l,3-butadiene *i.e.*, $H_2C{=}C(CH_3){-}CH{=}CH_2$ having general formula shown below:

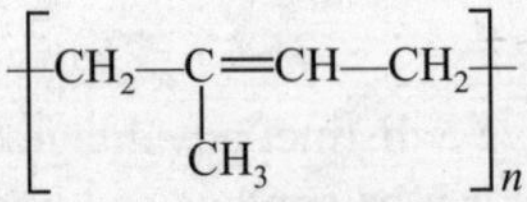

Natural rubber (*cis*-1, 4-polyisoprene) has the structure.

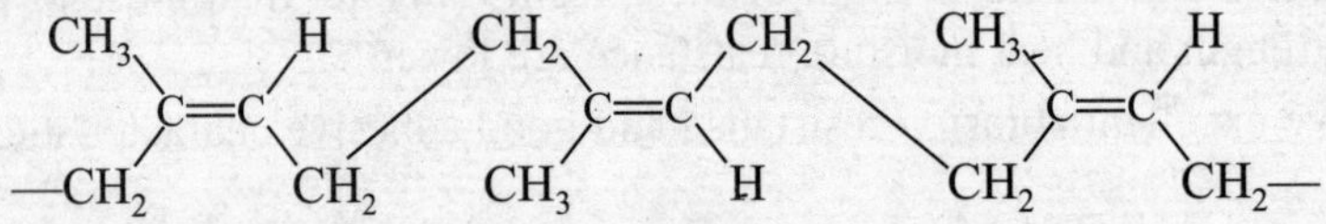

Natural rubber

Due to the all cis-configurations about the double bonds, the chains do not fit together well. Hence there are only weak vander Waal's forces. Cross-linking between chain's is done by heating with sulphur (vulcanization). This results in the formation of sulfur bridges between molecules.

Structure: It has been revealed through X-ray studies that *natural rubber is a cis isomer, whereas the natural rubber (Gutta Percha) is a trans isomer*. In the natural rubber obtained from the trea *Hevea Brazilensis,* the isoprene units are arranged in the cis form whereas in the natural rubber Gutta Percha obtained from the tree Dichopsis Gutta and Palagurm gutta the isoprene units are arranged in the trans form. This has been represented in the following (Fig. 7.7).

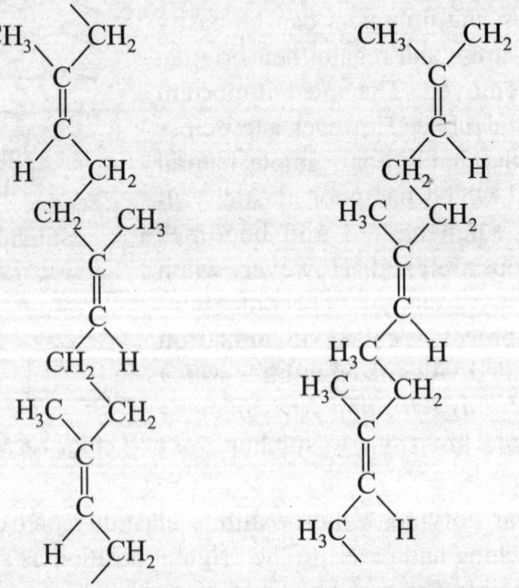

Natural rubber | Gutta percha

Fig. 7.7 The structure of natural rubber (*cis*-polyisoprene) and gutta percha (*trans*-polyisoprene).

7.9 VULCANISATION OF RUBBER

Natural rubber is a thermoplastic. There are no cross-links between the polymer chains. It becomes soft and sticky when heated. It is not hard and tough. The properties of natural rubber can be modified and improved by the process of vulcanization. To improve its properties, it is heated with Sulphur at a temperature of 373 to 415 K.

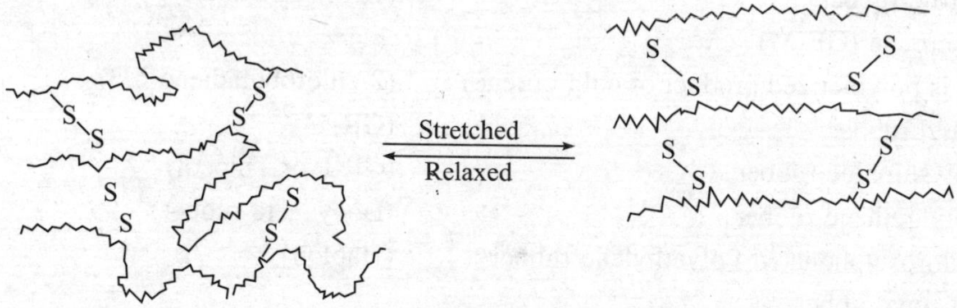

Fig. 7.8

The process of heating natural rubber with sulphur to improve its properties is called vulcanization.

Since the process is slow, some additives such as zinc oxide etc. are used to accelerate the process of vulcanization. During vulcanization, sulphur cross-links are formed. The formation of cross-links makes rubber hard, tough with greater tensile strength. The vulcanized rubber has excellent elasticity, low water absorption tendency, resistance to oxidation and organic solvents. The double bonds in the rubber molecule act as reactive sites. During vulcanization, sulphur forms cross-links at these reactive sites. In vulcanized rubber, the polymer chains are held together by polysulphide bridges or cross-links (Fig. 7.8). When the stretching force is applied, the chains straighten out but they cannot slip pass each other, because of polysulphide bridges. When the stretching force is with drawn, the chains tend to coil up again and the rubber returns to its original shape (Fig. 7.5).

The extent of hardness or toughness, however, depends upon the amount of sulphur added. Thus, about 5% sulphur is used for making tyre rubber, 20–25% S for making ebonite and 30% S for making battery case rubber. The process of vulcanization was discovered by Charles Goodyear in 1893.

The comparison of the main properties of natural rubber and vulcanized rubber are given below:

Natural rubber	*Vulcanized rubber*
1. Natural rubber is soft and sticky.	1. Vulcanized rubber is hard and non-sticky.
2. It has low tensile strength.	2. It has high tensile strength.
3. It has low elasticity.	3. It has high elasticity.
4. It can be used over a narrow range of temperature (from 40°C to 100°C).	4. It can be used over a wide range of temperature (100°C to 600°C).
5. It has low wear and tear resistance.	5. It has high wear and tear resistances.
6. It is soluble in solvents likes ether, carbon tetrachloride, petrol etc.	6. It is insoluble in all the common solvents.

7.9.1 Synthetic Rubbers

The synthetic rubber is obtained by polymerizing butadiene derivatives. These are also vulcanized by the process developed for natural rubber. The synthetic rubbers are either homopolymers of 1, 3-butadiene or copolymers in which one of the monomer is 1, 3-butadiene or its derivative so that the polymer has the availability of double bonds for its vulcanization. Synthetic rubbers are of following types:

(*i*) Styrene rubber (GR-S or Buna-S)
(*ii*) Nitrile rubber (GR-A or Buna-N)
(*iii*) Neoprene (GR-M)
(It is polymerized product of chlproprene) (2-chlorobutadiene-l, 3)
(*iv*) Butyl rubber (GR-1)
(*v*) Polysulphide rubber (GR-P or Thiocol)
(*vi*) Polyurethane rubber (Isocyanate rubber)
(*vii*) Chlorosulphonated Polyethylene rubber (Haplon)
(*viii*) Silicone rubber

Out of these, the first two are discussed below:

Styrene Rubber (Buna-S): Styrene butadiene rubber (SBR) is prepared by the copolymerisation of butadiene (75%) and Styrene (25%) is an emulsion system at 50°C in the presence of catalyst such as cumene hydroperoxide.

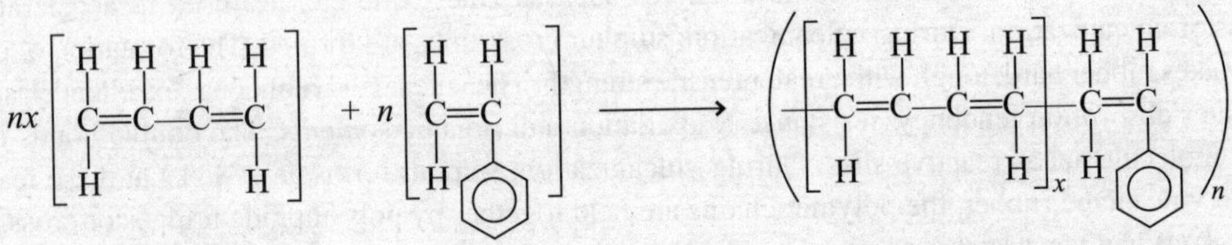

Butadiene Styrene Polybutadiene-costyrene (styrene butadiene rubber, SBR)

In addition to the above 1, 4 addition product, the polymer also contain small amount of of 1, 2-addition products. SBR is also known as Buna-S or GR-S.

Properties: It resembles natural rubber in processing characteristics as well as quality of finished products. Styrene rubber has high abrasion resistance, high load bearing capacity low oxidation resistance, swells in oil and solvents, like natural rubber. SBR is also vulcanized and produce cold rubber, which has greater tensile strength and greater abrasion resistance. However, it gets readily oxidises, especially the presence of traces of O_3 present in the atmosphere.

Uses

(*i*) It is used in the manufacture of those materials which are required to possess greater resistance to abrasion. For instance, it is used in motor tyres shoes soles, footwear components etc.

(*ii*) It is also used tanks lining, floor tiles, insulation of wire and cables, carpet backing, gasket, adhesives etc.

(*iii*) It is also used as *cold rubber*, which has greater tensile strength and resistance to abrasion. To prepare cold rubber the emulsion polymerisation of butadiene and styrene is carried out at –18°C to 5°C in the presence of an oxidation – resistance system.

(*iv*) SBR is used in motor tyres, shoes soles, footwear components, tanks lining, floor tiles, insulation of wire and cables, carpet backing, gaskets, adhesives etc.

Nitrile Rubber (Buna-N/NBR): It is a copolymer of acrylonitrile and butadiene. It is prepared by the copolymerization of acrylonitrile and butadiene in emulsion systems.

$$m\,CH_2{=}CH{-}CH{=}CH_2 \;(\text{1,3-butadiene}) \quad + \quad n(CH_2{=}CHCN) \;(\text{acrylonitrile})$$

$$\downarrow \text{copolymerization}$$

$$-\!\!(-CH_2-CH{=}CH-CH_2)_m-(CH_2-\underset{\displaystyle CN}{\underset{|}{CH}}-)_n-$$

Nitrile rubber

Properties: Nitrile rubber (GR-A) has low swelling, low solubility, good tensile strength and abrasion resistance even after immersion in gasoline or oils. It has good resistance to heat, sunlight, oils, acids, but it is less resistant to alkalies than natural rubber because of the presence of cyano groups. Vulcanized nitrile rubber is more resistant to heat and ageing than natural rubber and may be exposed to high temperatures.

Uses

(*i*) It is also used for making conveyor belts, high altitude air craft components, oil resistant foams and automobile parts etc.

(*ii*) It is also used in adhesives, for making fuel tanks, gasoline.

(*iii*) It is also used for impregnating paper, leather and textiles in the form of lates.

7.10 SYNTHETIC FIBRES

The first synthetic fibre named nylon was made by Carothers in 1938. So far only a few polymer fibres have been produced. The fibre must have sufficiently high melting point to withstand laundering and ironing. The melting point depends upon shape of the molecules. During melting the polymer chain should be able to move apart in the solid by over coming van der Waals forces between the molecules. The closer the packing of the chain, higher is the melting point. Hence fibre molecule should be capable of close packing. Thus polymers with bulky side chains are not very suitable for making fibres. Synthetic fibres belong to the following classes:

1. Polyester Fibres
2. Polyamide Fibres
3. Polyalkene Fibres
4. Polyacrylic Fibres
5. Polyminyl Fibres
6. Polyurethane Fibres

We will discuss below the important synthetic fibres belonging to the first two categories. Synthetic fibres such as Polypropylene, Polyurethanes belonging to the other categories have already been discussed in the proceeding articles.

1. **Polyesters:** These polymers contain ester linkages.

(*i*) **Polyester resin Terylene or Dacron** *are the condensation products of dicarboxylic acids with dihydroxy alcohols.* For example, *terene* (or terylene or dacron) is a *saturated polyester, formed by condensation of ethylene glycol (a diol) and terepthalic acid (a saturated diacid).*

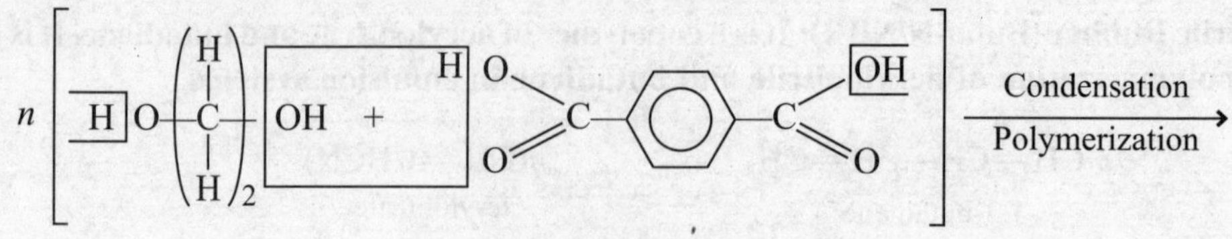

Ethylene glycol Terephthalic acid

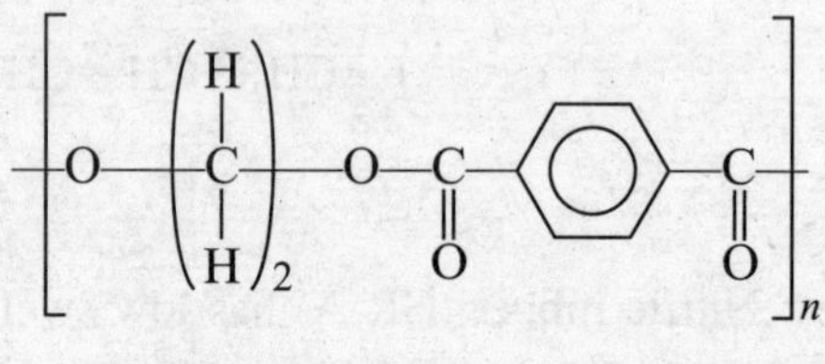

Polyester (or Terene or Dacron)

Uses

(*i*) It is mostly used for making *synthetic fibres* like terylene, dacron, etc.

(*ii*) For *blending with wool* to provide crease and wrinkle resistance.

(*iii*) As *glass reinforcing* material in safely helmets, aircrafts, battery boxes, etc.

(*ii*) Glyptal: Glyptal is a polymer prepared from ethylene glycol and phthalic acid.

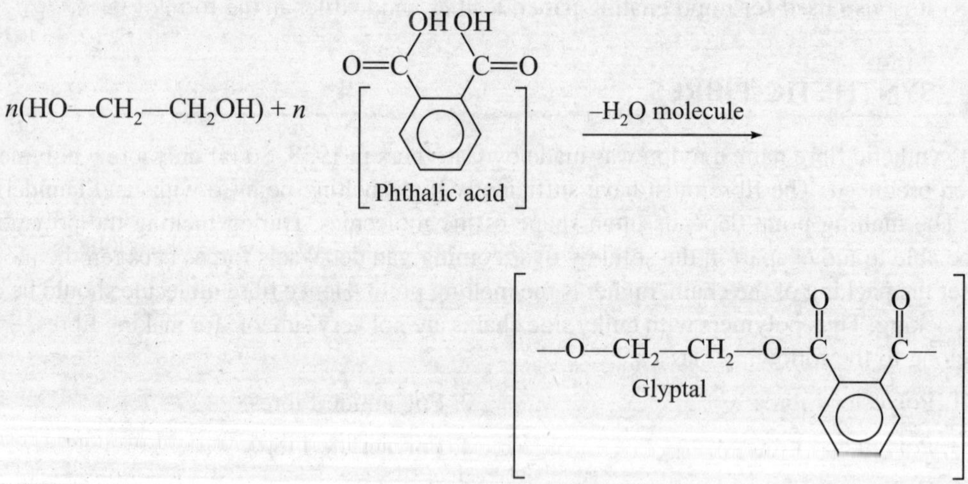

Uses: It is used in the manufacture of paints and lacquers.

2. **Polyamides** are synthetic polymers, which have recurring **amide** groups. Nylons, used mostly for making fibres, belong to this class.

   (*i*) Nylon-6 : 6 is obtained by the polymerization of adipic acid with hexamethylene diamine.

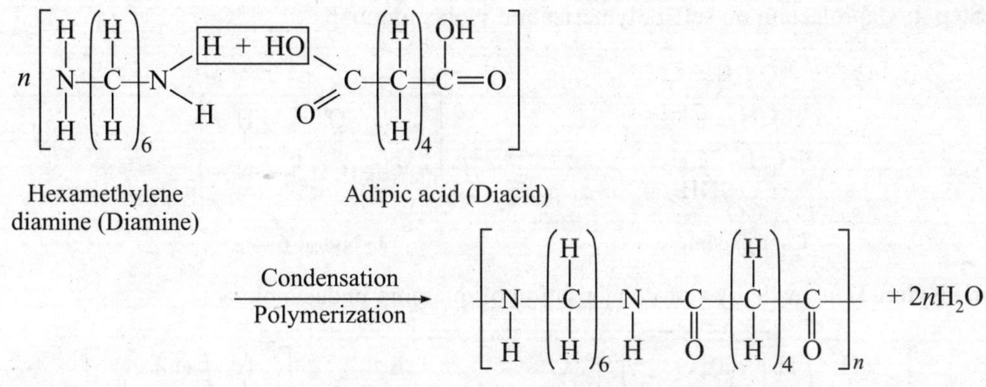

Hexamethylene diamine (Diamine) Adipic acid (Diacid)

Polyhexamethylene adipate (Nylon 6 : 6 Polyamide)

(*ii*) **Nylon-6** is produced by the self-condensation of ∈-amino caproic acid.

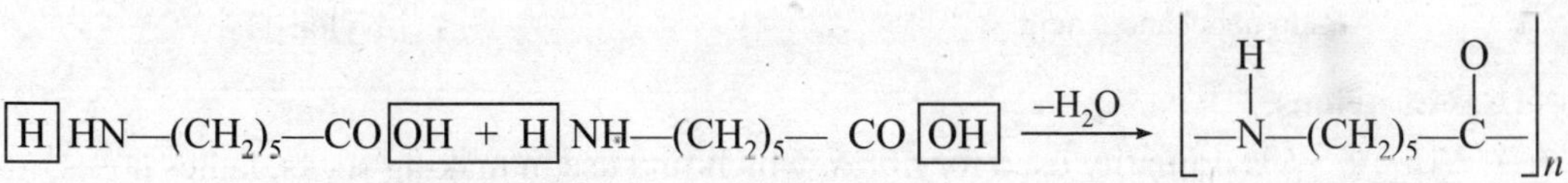

∈-Aminocarproic acid Nylon-6

Synthesis of nylon-6: Step I: Cyclohexane is oxidised to cyclohexanone.

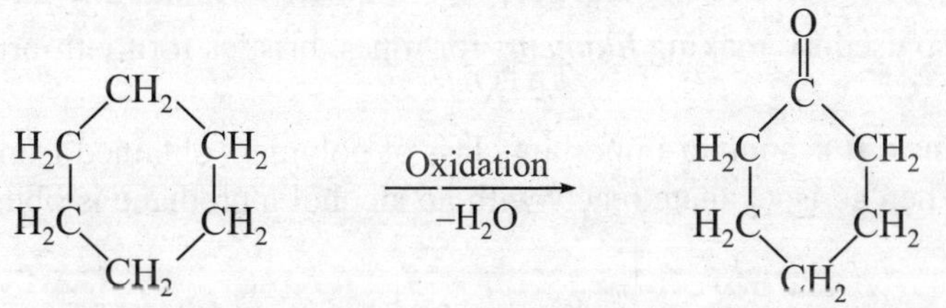

Cyclohexane Cyclohexanone

Step 2: Cyclohexanone is treated with hydroxylamine (NH_2OH) to get cyclohexanone-oxime.

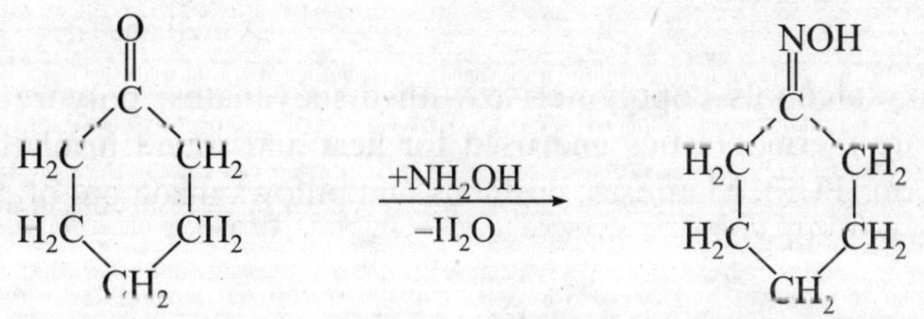

Step 3: Cyclohexanone-oxime is presence of (H_2SO_4) undergoes *Backmann rearrangement* to yield caprolactam.

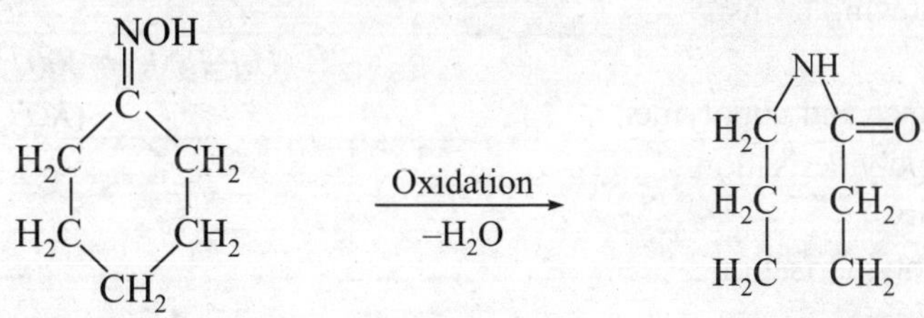

Cyclohexanone-oxime Caprolactam

Step 4: Caprolactam on self-polymerization yields nylon–6.

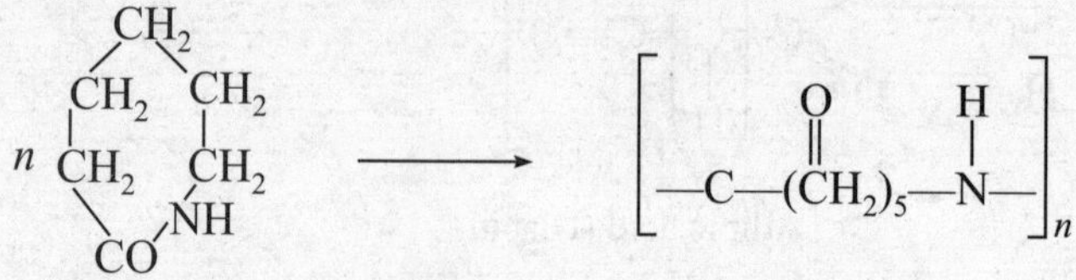

Caprolactam   Nylon-6

(iii) **Nylon-11** is made by self-condensation of ω-amino undecanoic acid.

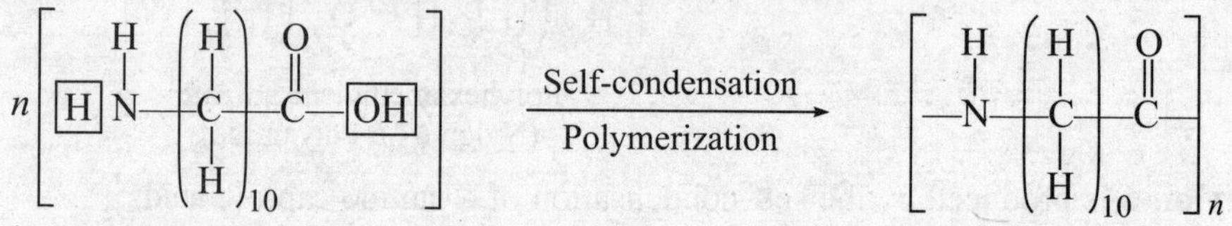

w-Amino undecanoic acid   Nylon-11

Uses of nylons:

1. *Nylon-6 : 6* is primarily used for *fibres*, which find use in making socks, ladies hoses, under-garments, dresses, carpets, etc.
2. *Nylon-6* and *Nylon-11* are mainly used for *moulding purposes* for gears, bearings, electrical mountings, etc. Nylon bearing and gears work quietly without any lubrication.
3. They are also used for making *filaments* for ropes, bristles for tooth-brushes and films, tyre-cords, etc.

3. Polyurethanes: It is another important class of polymers obtained from organic isocyanates (RNCO). When an isocyanate reacts with an alcohol a urethane is obtained.

$$\underset{\text{An isoyanate}}{R{-}N{=}C{=}O} + \underset{\text{Alcohol}}{R'OH} \longrightarrow \underset{\text{A urethane}}{R{-}\overset{H}{\overset{|}{N}}{-}\underset{O}{\underset{\|}{C}}{-}OR'}$$

When dihydroxy alcohols copolymerise with disocyanates, polyurethanes are produced. Polyurethanes are thermoplastics and used for heat and sound insulation or in the form of polyurethane form (PUF), Mattreses, cushions and pillows made out of polyurethane form are washable and long lasting.

REVISION QUESTIONS

1. Define the following terms:
   (*a*) Polymer *(U.P.T., Jan. 2001; Dibrugarh, June 2K)*
   (*b*) Homopolymer and copolymer *(K.U., May 1999; June 2K)*
   (*c*) Degree of polymerisation *(U.P.T., Jan. 2001)*
   (*d*) Functionality
   (*e*) Addition polymerisation

(*f*) Condensation polymerisation

(*g*) Monomer (*K.U., June 2K*)

2. Give reasons:

(*a*) All simple organic molecules do not produce polymers

(*b*) Bakelite can not be remoulded

(*c*) Platiciser is used during moulding of plastics.

3. What are elastomers? Explain their elasticity on the basis of structure taking natural rubber as an example. (*U.P.T., Jan 2001; Dibrugarh, June 2K*)

4. Give the structure of monomers of natural rubber. In what way the two forms of natural rubber differ. Explain giving the structure of two form's.

5. (*a*) Give the preparation, properties and uses of the following:

(*i*) Buna-S (*ii*) Buna-N

Suggest another name for each of them.

(*b*) Describe the methods of preparation, properties and applications of the following:

(*i*) PVC (*ii*) Bakelite and

(*iii*) Buna-S

6. (*a*) Discuss the preparation, properties and uses of any one thermoplastic material.

(*b*) Write a short note on GR-N elastomer. (*K.U. June 2005*)

7. (*a*) Mention preparation, properties and industrial applications of two important thermoplasts. (*M.D.U., B.E. 1 May 2005*)

(*b*) Explain the manufacture of the following polymers and mentions their use. (*V.T.U., B.E. March 2005*)

(*c*) Give the synthesis and applications of butyl rubber and epoxy resin. (*V.T.U., B.E. Jan 2003*)

8. (*a*) How are PF resins Novolacs different from resoles? Give the preparation, properties and applications of PF resins.

(*b*) Why SPR is known as cold rubber? Explain preparation, properties and technical applications of SPR. (*M.D.U. B.E. 1 Dec. 2005*)

9. (*a*) What are elastomers? Give the preparation, properties and applications of Nitrile rubber.

(*b*) What do you mean by homopolymer and copolymer? Explain the term graft copolymer. (*K.U. June 2006*)

(*c*) Give the manufacture and applications of (*i*) Polyurethane (*ii*) Epoxy Resin and (*iii*) Buna-S (*V.T.U. B.E. July/August 2004*)

10. (*a*) What are the difference between thermoplastic and thermosets? (*K.U. Jan. 2007*)

(*b*) Out of linear, branched chain and crosslinked polymers, which is the strongest and why? How is PVA prepared? Mention its important properties and uses. (*K.U. Jan. 2006, June 2006*)

(*c*) Describe the manufacture of the following

(*i*) Polystyrene (*ii*) Neoprene (*V.T.U., B.E., August 2001*)

11. (*a*) Define polymerization? Differentiate between addition and condensation polymerization. (*K.U. June 2006*)

(*b*) Why can not thermosetting plastics be reshaped and reused?

(*c*) Write preparation, properties and technical applications of PVC

12. (*a*) Write short note on: Thermoplastic and thermosetting polymers. (*M.D.U. Dec. 2007*)
(*b*) What are elastomers? Give the preparation, properties and uses of SBR and NBR. (*M.D.U. May, 2009*)

13. (*a*) What are elastomers? Give the proepartion, properties and applications of Nitrile rubber? (*M.D.U. May 2007*)
(*b*) Differentiate between addition polymers and condensation polymers. (*M.D.U. May 2008*)
(*c*) Define polymer. Why all macromolecules are not polymer? (*M.D.U. Dec. 2008*)

14. What are plastics? How are they classified? Write down the preparation, properties and industrial applications of polyester resin or phenolic resin. (*M.L.B. Tech. 2001, 2002*)

15. (*a*) Why bakelite can not be remoulded (*MG, Oct. 1998*)
(*b*) What is the role of plasticizer during moulding plastics
(*c*) Write short note on constituents of plastics (*Raj., May, 2001*)
(*d*) Write short note on vulcanization of rubber. (*Tripura 2K, RGT, May 2001, Anna, Nov. 2K*)
(*e*) Explain why natural rubber needs vulcanization. How is it carried out. (*SK June 2001, Audhra, June 2001, Raj., May 2001*)
(*f*) Mention the advantages of synthetic rubber over natural rubber (*V.T., March 1999, March 2001*)

16. (*a*) Why do all simple organic molecules not produce polymers (*U.P.T., Jan. 2001, Nagpur, May 2001*)
(*b*) What is the minimum functionality for a monomer to form a cross linked polymer? (*Dibrugarh, July 2001*)
(*c*) Differentiate between homopolymer and copolymer (*K.U., May 1999, June 2K*)
(*d*) Why is Teflon highly chamical resistant? (*U.P. Tech., B. Tech. 2000-2001*)

17. (*a*) Why is PVC stronger than polyethylene?
(*b*) Why do branched polymers have low densities and low melting points as compared to linear polymers?
(*c*) Why is natural rubber susceptible to attach by ozone but polyethylene not?
(*d*) Why are the properties of natural rubber are very different from Gutta percha although both are polyisoprenes?
(*e*) What structural changes occur when natural rubber is subjected to vulcanization?

□□□

8 Composite Materials

INSIDE THIS CHAPTER

8.1 INTRODUCTION

A composite may be defined as a multiphase material made up of two or more components., which are physically and chemically different phases separated by an interface. We see a lot of composites around us. For example, concrete is a composite. It is made up of cement, gravel and sand and has steel rods inside it to reinforce it. Some other common examples are, wood, which is a composite of cellulose fibers and lignin, packing paper impregnated with bitumen or wax and rain proof cloth in which cloth is impregnated with water proof material. As per the above definition, metallic alloys, ceramics etc. are not considered as composite materials. The composites are made from materials like metals, ceramics and polymers by adding fibers, particles etc. into them. A polymer composite, therefore, will be made from polymer along with other kind of materials. These filler materials are solid additives which modify the physical properties, particularly the mechanical properties of basic polymeric material. One of the most common types is fiber-reinforced composites. These are materials in which a fiber made of one materials is embedded in another material.

A composite shows properties entirely different from those of the ingredients. In fact, it shows extra-ordinary combination of properties like toughness and strength with low weights and high temperature resistance. For instance, compared to steel and aluminum, composites are lighter, have low coefficient of thermal expansion and have superior strength, stiffness and fatigue resistance.

Important advantages of composite materials over the conventional materials like metals, polymers and ceramics are :

(*i*) higher strength; (*ii*) higher stiffness;

(*iii*) maintainance of strength, even up to high temperatures;

(*iv*) greater toughness, impact and thermal shock-resistance;

(*v*) better creep and fatigue strength;

(*vi*) better corrosion and oxidation-resistance.

(*vii*) lower electrical conductance and thermal expansion.

Composite materials are extensively used in automobile industries, including air and sea, because of their lesser weight and higher strength. Besides, they are also used in furniture, sport goods, high speed machinery, safety equipments like ballistic protection, communication antennae, electronic circuit boards.

8.2 CLASSIFICATION OF COMPOSITES

There are two main parts i.e. the constituents of any composite material (1) body constituents (also called as matrix) and (2) structural constituents (also called as reinforcing phase)

Composite materials are classified on the basis of these constituents in to the following categories

1. ***On the basis of type of matrix***
   (*i*) Polymer matrix composites (PMC);
   (*ii*) Metal matrix composites (MMC); and
   (*iii*) Ceramic matrix composites (CMC) as shown in Fig 8.1(*a*)

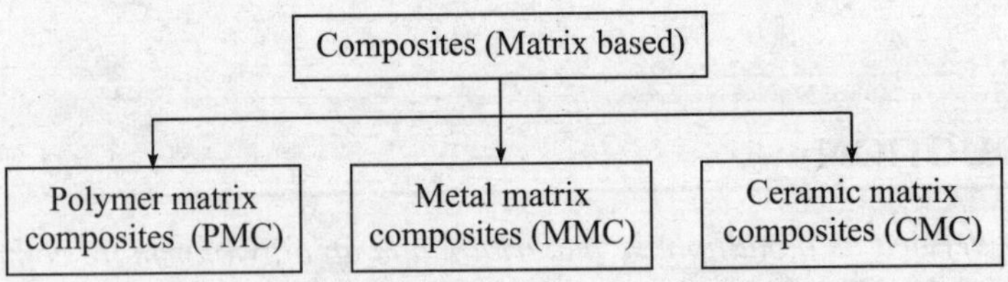

Fig. 8.1 (*a*) Matrix based classification

2. ***On the basis of type of reinforcement used in the matrix:*** The reinforced phase may be
   (*i*) Particles which may be large or small in size. The small size particles may be dispersed in the form of a dispersion.
   (*ii*) Fibers which may be of large size giving rise to continuous phase and the small fibers may give rise to discontinuous phase as shown in 8.1(*b*)

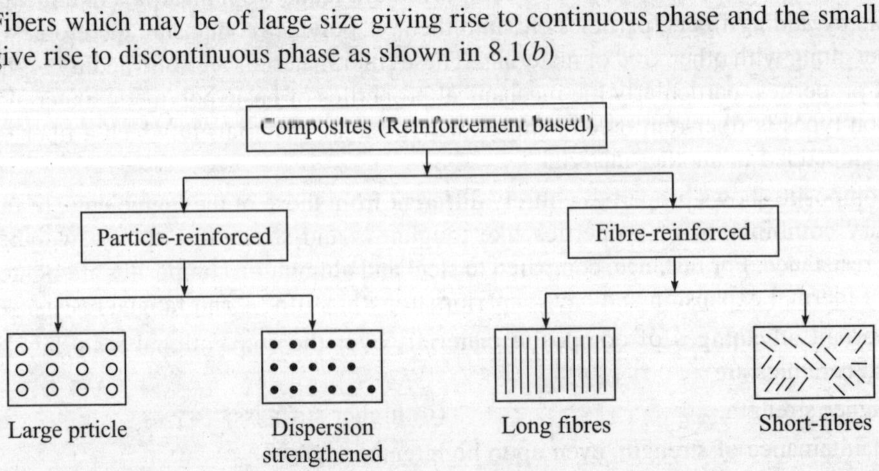

Fig. 8.1 (*b*) Reinforcement based classification

3. ***On the basis of number of layers:*** This gives rise to composite of two type (*a*) unilayer or single ply composites and (*b*) Multiply composites or laminates. Fig 8.1(*c*) The details of all the above types of composites materials will be taken up in the proceeding articles of the chapter.

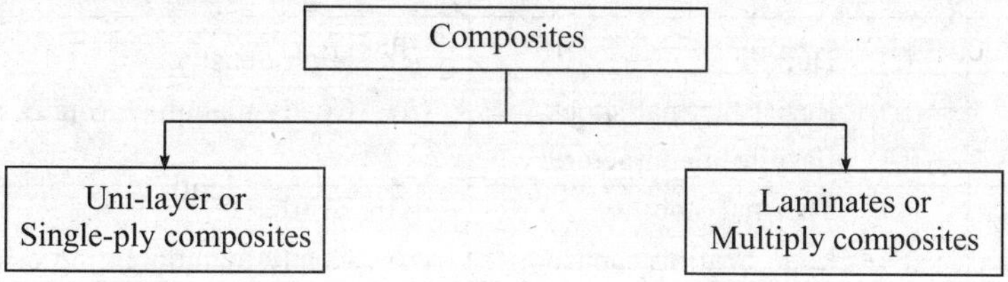

Fig. 8.1 (*c*) Layer based classification

8.3 CONSTITUENTS OF COMPOSITES

In general, composites consist of two main phases (1) Body constituents and (2) Structural constituents. The body constituent is also called matrix phase.

1. **Matrix phase** is the continuous body constituent, which encloses the composite and give it its bulk from. Matrix phase may be metal, ceramic or polymer. Composites using these matrices are known as metal matrix composites (MMC), ceramic matrix composites (CMC) and polymer matrix composites (PMC) respectively. Polymer matrix materials used in composites are epoxy, polyamides, phenolics, silicones and polysulphones.

   In polymer matrix composites (PMC), both thermoplastic and thermosetting resins can be used. Advantages of thermoplastic resins over thermosets are listed below :

   (*i*) They are tougher and have high strains to failure;

   (*ii*) They offer the potential of improved hot/wet resistance; and

   (*iii*) No chemical reactions are involved causing release of gaseous products or exothermal heat.

   Functions of matrix phase

   (*i*) It binds the fibers together by virtue of its cohesive and adhesive forces.

   (*ii*) It acts as a medium to transmit and distribute an externally applied load.

   (*iii*) It protects the individual fibers from chemical reaction with the environment and from surface damage due to mechanical abrasion.

   (*iv*) It keeps the reinforcing fibers in the proper position and orientation

   (*v*) It prevents the propagation of brittle cracks due to its relative softness and plasticity.

   (*vi*) It provides interlaminar shear strength to the composite.

   Requirements of a good matrix materials are:

   (*a*) It should be ductile;

   (*b*) The bonding strength between the matrix and fibre must be high.

   (*c*) It should have good corrosion resistance.

The properties controlled by matrix phase: Generally matrix determines the overall service temperature limitations of the composite. It also controls the environmental resistance of the composite. For example, composites based on polymers are limited to service temperatures below 300°C.

Comparason of matrix phases

The merits and demerits of different matrix phases are summarized below:

Matrix Phase	*Merits*	*Demerits*
1. Metals	(*i*) Ductile (*ii*) Resist thermal shock, (*iii*) Easy to manufacture.	(*i*) High density, (*ii*) Costly manufacturing & materials costs.
2. Ceramics	(*i*) Thermal stability, (*ii*) Load bearing capacity, (*iii*) Stiffness.	(*i*) Brittle, (*ii*) Costly manufacturing & material costs, (*iii*) Not stable against mechanical shock.
3. Polymers	(*i*) Low cost, (*ii*) Light weight, (*iii*) Corrosion resistant, (*iv*) Tough, (*v*) Easily processible.	(*i*) Lower strength and Stiffness, (*ii*) Low thermal stability, (*iii*) Low load bearing capacity

2. **Dispersed phase** *is the structural constituent, which determines the internal structure of the composite.* Important dispersed phases of composites are: (*a*) Fibers (*b*) Particulates (*c*) Flakes (*d*) Whiskers

(*a*) **Fibre** *is a long and thin filament of any polymer, metal or ceramic having high length to diameter ratio.* Its diameter nears crystal size diameter. Their main characteristics are (*i*) High length to diameter ratio (*ii*) high tensile strength, (*iii*) high stiffness, (*iv*) causes lowering of overall density of composite.

The main reinforcing fibers are discussed below:

(*i*) **Glass fibres** are obtained by forcing glass melt though an instrument having a number of small openings followed by cooling the emerging continuous filaments. Individual fibre (of diameter about 10 Nm) is called '*monofilament.*' Usually, a "sizing agent" in liquid form is applied to glass monofilaments so as to help in handling strands and protect them from abrasion.

Glass fibres are considered as high performance materials due to the following points

(*i*) ready availability, (*ii*) low cost,
(*iii*) easy to draw from molten melt, (*iv*) high specific strength,
(*v*) can be economically used to produce glass-reinforced polymer composite,
(*vi*) can be coupled with polymer matrix to get chemical inertness.

(*ii*) **Carbon fibres** *are obtained as continuous filament by the pyrolysis in an inert atmosphere of organic fibres such as cellulose, but usually of polyacrylonitrile.* The monofilaments have diameter of about 8 Nm. They are used as reinforcing materials with epoxy or polyester resins to form composites, which have higher strength/weight ratio than metals. These are high performance fibres. However, these fibres are quite costly.

Reasons for popularity of carbon fibres as reinforcing material are:

(*i*) High modulus, specific strength and stiffness even at elevated temperatures,
(*ii*) Resistant to moisture, acids, bases and a number of solvents.

(*iii*) **Aramid fibers:** They are basically aromatic polyamides (aramides) like Nomex and Kevlar. For instance, Kevlar is *p*-phenylene terephthalamide. It is spun from a nematic liquid crystalline solution using sulphuric acid as the solvent and cold water as coagulant.

The *characteristic* properties of Kevlar fibers are:

(*a*) Higher tensile strengths and tensile moduli compared to other polymeric fiber materials.

(*b*) Resistant to combustion and stability even at relatively high temperatures.

(*c*) Retain mechanical properties over a long temperature range.

(*d*) Excellent toughness, impact resistance and resistance to creep and fatigue failure.

(*e*) Weak in compression and

(*f*) Susceptible to degradation by strong bases and acids

However, the *limitations* of these fiber are:

Aramide fibers are commonly used in polymer matrix composites. The common matrix materials are polyesters and epoxies.

(*b*) **Particulate:** Between the matrix and the dispersed phase is known as interface. It plays the following two roles in determining the characteristics of a composite. Their inherent hardness, the particulates in a composite reduce the plastic deformation on the matrix phase. In other words, the matrix transfers some of the applied stress, thereby the particles bear a fraction of applied load. It has been observed that particle-matrix interactions that lead to strengthening may or may not occur on the atomic or molecular level depending upon the size of the particles. Particulate fillers are accordingly classified as *large particle and dispersion strengthened respectively*.

The properties of matrix materials which are modified by adding particulate fillers are:

(*i*) Improved performance at elevated temperatures,

(*ii*) Improved wear and abrasion resistance,

(*iii*) Modification in thermal and electrical conductivities.

(*iv*) Increase in surface hardness, and

(*v*) Reduction in (cost, shrinkage and friction)

(*c*) **Flakes** *are thin solids having a two-dimensional geometry, e.g., mica flakes*. Thus they impart equal strength in a plane compared to fibres, which reinforce in one direction. Packing of flakes can be carried out more efficiently than that of fibres or spherical particulates, Mica flakes are used in electrical and thermal insulating appliances.

(*d*) **Whiskers** *are thin strong filaments of fibres made by growing a crystal, for example*, of graphite, silicon carbide, silicone nitride and aluminum oxide. Usually, they are several mm in length and several microns in thickness. Irregularities decrease their strength, but perfect whiskers possess much higher strength than in bulk. They possess high elastic modulus, high degree of crystallinity and exceptionally high strength. However, they find limited use in composites, due to their high cost and their incorporation in matrix is a difficult process.

8.3.1 Role of Interface in Composites

The surface forming the common boundary for composites that have a good interfacial bond, the fracture mode is generally a sharp well-defined break. For composites that have a poor interfacial bond, the fracture will be a progressive one-bond failure, followed by matrix failure and finally fibre failure.

1. The transmission of stress from the matrix phase to reinforcing dispersion phase is dependent on the interfacial bond strength. If interfacial bond is strong, the failure is resisted.
2. The interface can serve as: (*i*) locus of chemical reaction, (*ii*) site for nucleation, and (*iii*) site for preferential adsorption. This is due to the difference between thermal coefficients of two

constituents of composites and cure shrinkage (in thermoset) or crystallization (in thermoplastic) when polymer is used as matrix phase.

8.3.2 Factors Controlling the Quality of Composites

It may be pointed out that the final quality of a composite depends upon:

(*i*) The properties of the constituents and their relative proportions, distribution, orientation and geometry (*e.g.*, shape, size, size distribution) of the dispersed phase.

(*ii*) **Void content:** Smaller void contents means lower susceptibility to weathering and penetration of water, higher fatigue resistance and lesser variation in strength properties. However, higher void content enhances the thermal and electrical insulating properties of the composite.

8.4 FIBER REINFORCED COMPOSITES

When fibers are used as dispersed phase for the reinforcement of matrices, the resultant composites are known as fiber reinforced composites. They involve three components, namely filament, a polymer matrix and a bonding agent, which ties fiber filaments to polymer. Glass fibres and metallic fibres are usually employed for this purpose.

The *characteristics of* fiber-reinforced composites are:

(*i*) High specific strength (*i.e.* tensile strength/ specific gravity)

(*ii*) High specific modulus (*i.e.* modulus of elasticity/specific gravity)

The characteristics of fiber-reinforced composites depend on:

(*a*) Nature and properties of fiber and matrix materials,

(*b*) Their relative volume fractions,

(*c*) Orientation and distribution of fibers in matrix phase,

(*d*) Strength of the interfacial bond between the fiber and the matrix phases,

(*e*) Length of the fibers or more appropriately aspect ratio (length diameter ratio) of fibers.

For number of fiber-reinforced composites, aspect ratio has value in between 20 and 150. In fact, for effective strengthening and stiffening of the composite, some critical fibre length is required.

Critical fibre length (l_c) is defined as: $l_c = \sigma_f d / T_c$

Where σ_f is ultimate fibre strength of the fibers.

d is diameter of fibers, and

T_c is shear yield strength of fibre-matrix bond strength.

Long (or continuous) fibers have their lengths which are about 15 times greater than the critical length but in short (or discontinuous) fibers, the lengths of the fibers are considerably lower than their critical length. That's why the reinforcement efficiency of continuous fibers is higher than that for discontinuous fibers.

8.4.1 Types of Fiber-Reinforced Composites

Some Important Types of Fiber-Reinforced Composites:

(A) ***Glass-fiber reinforced composites:*** Glass fibers (both long and short) are generally used for improving the properties of polymeric matrics like polyesters, nylon etc. These composites have the following **characteristics**:

(*i*) High impact resistance,

(*ii*) Excellent chemical and corrosion resistance.

(*iii*) Lower densities,

(*iv*) Higher tensile strengths,

But, these composites also have the following **limitations**:

(*i*) Limited service temperatures, because most polymeric matrices begin to deteriorate or flow at high temperatures.

(*ii*) These materials do not possess the desired stiffness and rigidity, particularly in their applications as structural components.

Applications of these composite materials are in: Automobiles parts, storage tanks, industrial floorings, plastic pipes and in transportation industries to reduce vehicle weight and boost fuel efficiency.

(B) ***Carbon Fiber-Reinforced Polymer Composites:*** High performance fibers are now-a-days used as reinforcement in advanced polymer matrix composites, which requires retention of desired properties even at higher temperatures and excellent resistance to corrosion along with lighter weights.

Application of these composite materials are:

(*i*) Military and commercial aircraft structural components.

(*ii*) Wing, body and stabilizer components of helicopters,

(*iii*) Sports and recreational equipments like fishing rods, golf clubs.

The uses of these materials are limited because of their higher costs.

(C) ***Aramid Fiber-Reinforced Polymer Composites****:* Aramid Fiber relate to Aromatic amides. These materials can be broadly classified into following categories:

(*i*) ***Short fibre reinforced composites.*** The short fibers of *p*-aramid have high aspect ratio and high surface area and hence they give effective reinforcement. Moreover, they have inherent toughness, strength and heat stability. They also have high wear resistance and are used in automotive brakes and clutches. In fact these composites perform better than asbestos at less than one-tenth the amount in brakes and clutches. Because asbestos causes serious health problems so its use is now-a-days restricted.

(*ii*) ***Long fibre reinforced composites.*** These composite are metal-like in being ductile, energy absorption and respond to compressive stresses without any danger. This is due to the fact that aramid fibres absorb energy. They are excellent advanced engineering and structural materials. The latest commercial aircrafts Boeing 767 and Airbus 320 contain about 3-5% of composites, much of which is *p*-aramid or hybrids of p-aramid and carbon.

The helicopter industry also utilises composites based on p-aramid motor housing and rotor blades. Even primary structures of small business air crafts are being made by using these composites

These materials are also used for manufacturing protective apparel (thermal, cuts, ballistics).

(D) ***Alumina and/or carbon fibers reinforced metal-matrix composites:*** Fibre reinforcement in ductile metal matrix may improve specific strength & stiffness, resistance to abrasion and dimensional stability. The characteristic properties and applications of two important composites of this type are tabulated below:

Matrix	*Reinforcement*	*Characteristic properties*	*Applications*
Aluminium alloys as matrix	Alumina or Carbon fibers as reinforcement	Light in weight, Resists wear and thermal distortion.	Engine component of automobiles.
Composites having nickel and cobalt based alloys as matrix	Alumina or refractory metals such as tungsten as reinforcement	Excellent high temperature, creep and rapture resistance and excellent impact strength	They are used as turbine engines.

8.5 STRUCTURAL COMPOSITES

The composite materials which are used as structural components in the form of plywood and also as stainless steel bonded on both sides with copper case etc. are known as structural composites. The properties of structural composite depends on

(*i*) The properties of the constituent materials, and

(*ii*) Their geometrical design

Structural composites are mainly of two types:

(*a*) Sandwich composites and (*b*) Laminar.

Let us discuss these two briefly

(*a*) ***Laminar Composites:*** They consist of number of two-dimensional sheets or layers that are stacked and subsequently cemented together in such a way that the orientation of reinforcement varies with each successive layer as shown in Fig. 8.2 The orientation of fibers in two adjacent layer is almost at right angle to each other as shown in Fig. 8.2(*a*).

These layers possess relatively high strength in the direction of reinforcement and hence the resulting laminar composite has high strength in number of directions. The final fabricated laminar composite is the compact form as shown in Fig. 8.2(*b*).

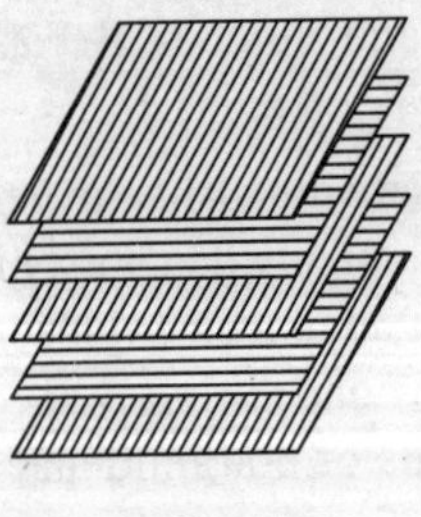

(a) Structure of laminar composites

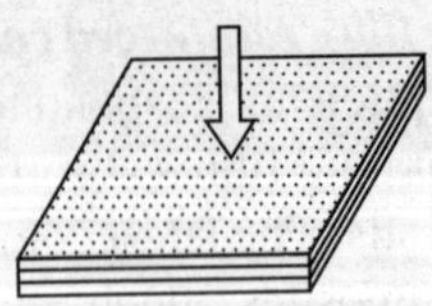

(b) Compact form of laminar composite

Fig. 8.2 Laminar Composites

(*a*) **Sandwich panels:** This types of structural composites consist of two strong outer sheets called faces separated by a layer of less-dense material called core material. Core material should be light weight, able to transfer load from one facing to the other and should have high corrosion resistance.

The structural function of the *core* is to resist any deformation at right angles to the face plane by providing certain degree of shear rigidity. The *faces* bear most of the in-plane loading, as well as any transverse bending stresses.

The stiffness of the sandwich panels increases as the core thickness increases. For instance, if we increase the thickness of the core four times, relative stiffness increases by 39 times. As the core is generally of lighter materials, so corresponding increase in weight is just 6%. A very popular core material is of following: "*Honeycomb*" structure which is made up of thin foils interlocked hexagons' cells with their axes oriented at right angle to the "face" planes. It is shown below in Fig. 8.3(*a*). The fabricated sandwich panel is shown in Fig. 8.3(*b*).

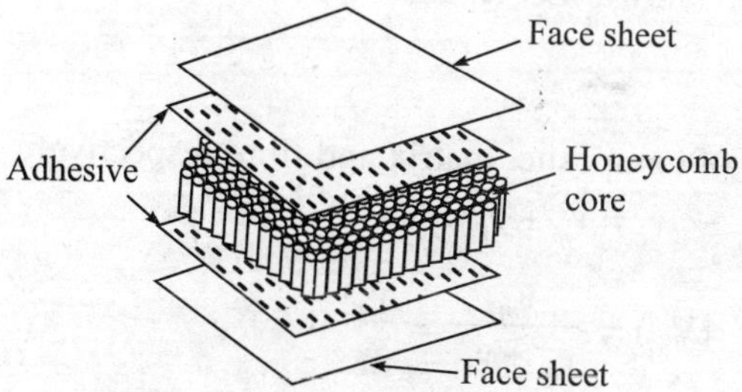

(a) Arrangement of outer sheets and core material

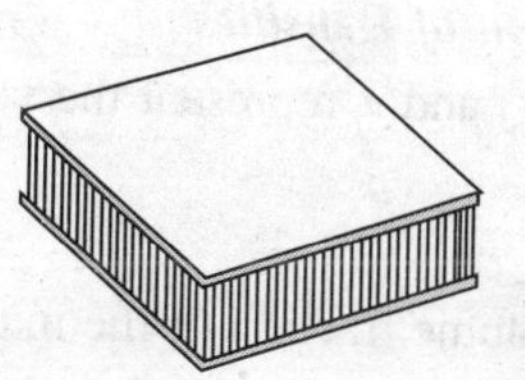

(b) Fabricated sandwich panel

Fig. 8.3 Sandwich panels

Applications of sandwich structures are listed below:

(*i*) For fabrication of wings of aircrafts,

(*ii*) For fabrication of roofs, floors and walls of buildings and

(*iii*) Design of ships, boat hulls etc.

8.6 UNDIRECTIONAL FIBRE REINFORCED COMPOSITES

The most widely used composite materials usually consist of several distinct layers of unidirectional composites. Unidirectional composites are made up of parallel fibres embedded in a matrix. Each layer of a unidirectional composite is also known as *ply or lamina*. Each lamina is generally made of same constituent materials *i.e.* matrix and fibre. But an individual lamina may differ from another lamina in the following respects.

(*a*) relative volumes of the fibre and matrix,

(*b*) orientation of fibres with respect to common reference axes and

(*c*) form of the reinforcement used such as continuous and discontinuous fibres, woven or unwoven reinforcement .

It is clear that directional properties of the individual layers are different from each other. The properties of individual layers must be known for the analysis and design of any structural element made from the fibre-reinforced

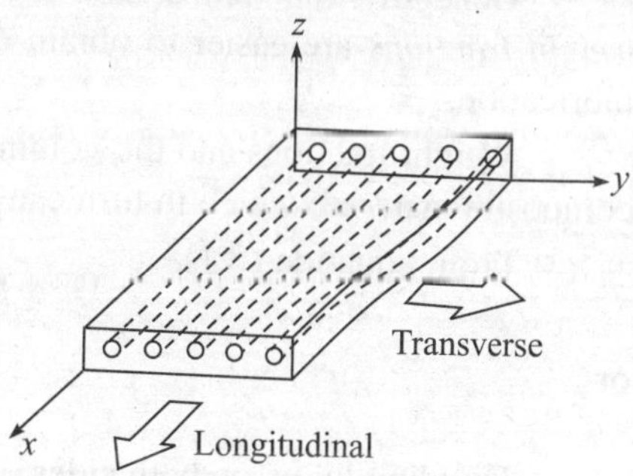

Fig. 8.4 Schematic diagram of unidirectional composite.

composites. Let us study the properties and behavior of single layer composites which are basic building blocks for laminate construction.

A typical unidirectional composite consists of parallel fibers embedded in a matrix as shown in Fig. 8.4. The direction parallel to fibers is known as *longitudinal direction* (x-axis) and the direction at right angle to the fibers is called the *transverse direction* which lies in the y-z plane.

A lamina has the strongest properties in the longitudinal direction.

8.6.1 Properties of Fibre Reinforced Unidirectional Composites

We can predict the properties of a composite, if the relative proportions of the matrix and reinforcing material and their geometrical arrangement are known. Let us discuss some of the properties of fibre-reinforced composite.

1. *Evaluation of Densities*

Let v_c, v_m, and v_f represent the volume of composite, matrix and fibre respectively, then:

$$v_c = v_m + v_f \qquad \ldots(8.1)$$

Then, volume fraction of the matrix $$(V_m) = \frac{v_m}{v_m + v_f} = \frac{v_m}{v_c} \qquad \ldots(8.2)$$

Similarly, volume fraction of the fibres $$(V_f) = \frac{v_f}{v_c} \qquad \ldots(8.3)$$

Moreover, $$V_m + V_f = 1$$

Now let w_m and w_f represent the weight of matrix and fiber respectively,

the weight of composite material $$w_c = w_n + w_f \qquad \ldots(8.4)$$

Weight fraction of the matrix $$(W_m) = \frac{w_m}{w_m + w_f} = \frac{w_m}{w_c} \qquad \ldots(8.5)$$

Similarly, Weight fraction of the fibres

$$(W_f) = \frac{w_f}{w_m + w_f} = \frac{w_f}{w_c} \qquad \ldots(8.6)$$

Moreover, $W_m + W_y = 1$

Generally, *volume fractions* are used for the theoretical analysis of the composite materials. But *weight fractions* are easier to obtain during fabrication or by one of the experimental methods after fabrication.

Weight fractions and the volume fractions can be interconverted with the help of density of the composite material, which in turn can be easily obtained as follows:

From equation (8.1), $$v_c = v_m + v_f$$

or $$\frac{w_c}{d_c} = \frac{w_m}{d_m} + \frac{w_f}{d_f} \qquad \left(\text{since } d = \frac{w}{v} \text{ and hence } v = \frac{w}{d}\right)$$

Dividing by w_c on both sides of above equation, we get

$$\frac{1}{d_c} = \frac{w_m / w_c}{d_m} + \frac{w_f / w_c}{d_f}$$

Utilizing equations (8.5) and (8.6), we get

or $$\frac{1}{d_c} = \frac{W_m}{d_m} + \frac{W_f}{d_f}$$

or $$d_c = \frac{1}{W_m/d_m + W_f/d_f} \qquad \ldots(8.7)$$

i.e. $$\textbf{Density of composite} = \frac{1}{\left(\dfrac{\textbf{Wt. fraction of matrix}}{\textbf{Density of matrix}}\right) + \left(\dfrac{\textbf{Wt. fraction of fibre}}{\textbf{Density of fibre}}\right)}$$

We know that; $$w_c = w_m + w_f$$

$$d_c v_c = d_m v_m + d_f v_f \qquad \left(\text{since } d = \frac{w}{v}\right)$$

Dividing by v_c on both sides of the above equation, we get

$$d_c = d_m \frac{v_m}{v_c} + d_f \frac{v_f}{v_c}$$

Utilizing equations (8.2) and (8.3), we get

$$\boxed{\mathbf{d_c = d_m V_m + d_f V}} \qquad \ldots(8.8)$$

Conversion relations between the weight fractions and the volume fractions:

$$W_f = \frac{w_f}{w_c} = \frac{d_f v_f}{d_c v_c} = \frac{d_f}{d_c} \times V_f$$

Hence, $$\boxed{\mathbf{V_f = \frac{d_c}{d_f} W_f}} \qquad \ldots(8.9)$$

Similarly $$W_m = \frac{d_m}{d_c} \times V_m \text{ and}$$

Hence, $$\boxed{\mathbf{V_m = \frac{d_c}{d_m} W_m}} \qquad \ldots(8.10)$$

Void Content

Let us assume that there is some space in between the fiber material and the matrix of the composite, which may be called *void content*. For its evaluation;

Let d_{ct} = Theoretical composite density and

d_{cc} = Experimental composite density

Then $(d_{ct} - d_{cc})$ indicates the void content and *i.e.* volume fraction of voids (V_v) is given by

$$\boxed{V_v = \frac{d_{ct} - d_{cc}}{d_{ct}}} \qquad \ldots(8.11)$$

Importance of void content

The quality of composites can be estimated from the knowledge of void content. This is because the mechanical properties of the composite are significantly affected by its void content. Smaller void contents usually mean higher fatigue resistance, lower susceptibility to water penetration and weathering, and decreased variation in strength properties. A good composite should have less than 1% voids.

2. **Longitudinal Strength and Stiffness of Unidirectional Composites:** Longitudinal behaviour of a composite can be predicted by assuming that

(*a*) Fibers are uniform in diameter and properties, they are continuous and parallel throughout the composite *i.e.*

$$l_f = l_m = l \qquad \text{...(8.12)}$$

(*b*) A perfect bonding exists between the fibers and the matrix and

(*c*) Under applied stress, the deformation of matrix and fibers is the same, *i.e.*, the strains experienced by the fibre, matrix and composite are equal

$$\varepsilon_f = \varepsilon_m = \varepsilon_c \qquad \text{...(8.13)}$$

Under these conditions, the total load carried by the composite (P_c) is the sum of the loads carried by the matrix and the fibers.

$$P_c = P_m + P_f \qquad \text{...(8.14)}$$

Suppose, stress σ_m and σ_f act over the cross-sectional areas of the matrix A_m and the fibers A_j, respectively.

If the matrix and the fibres behave elastically, the stresses are

$$\sigma_m = E_m\varepsilon_m \text{ and } \sigma_f = E_f\varepsilon_f \qquad \text{...(8.15)}$$

Thus, the loads carried by the matrix (P_m) and the fibers (P_f) are

$$P_m = \sigma_m A_m = E_m \varepsilon_m A_m \qquad \text{...(8.16)}$$

$$P_j = \sigma_f A_f = E_f \varepsilon_f A_f \qquad \text{...(8.17)}$$

From equation (8.14), the resultant load carried by the composite (P_c) cause an average stress σ_c in the composite acting over the entire cross-sectional area A_c .

$$P_c = P_m + P_f$$

or $$\sigma_c A_c = \sigma_m A_m + \sigma_f A_f$$

or $$\sigma_c = \sigma_m \frac{A_m}{A_c} + \sigma_f \frac{A_f}{A_c}$$

or $$\sigma_c = \sigma_m \frac{A_m \times l}{A_c \times l} + \sigma_f \frac{A_f \times l}{A_c \times l}$$

or $$\sigma_c = \sigma_m \frac{v_m}{v_c} + \sigma_f \frac{v_f}{v_c}$$

Thus $$\boxed{\boldsymbol{\sigma_c = \sigma_m V_m + \sigma_m V_f}} \qquad \text{...(8.18)}$$

Differentiating the above equation with respect to strain, we get

$$\frac{d\sigma_c}{d\varepsilon} = \frac{d\sigma_m}{d\varepsilon} V_m + \frac{d\sigma_f}{d\varepsilon} V_f \qquad \text{...(8.19)}$$

In the above equation $\frac{d\sigma}{d\varepsilon}$ represents the slope of the corresponding stress-strain diagram at the given strain. These slopes constants for linear stress-strain curves and are replaced by the corresponding elastic modulii in equation.

$$\boxed{E_c = E_m V_m + E_f V_f} \quad \text{...(8.20)}$$

Relationships such as equations (8.18) and (8.20) are known as *rule of mixtures,* according to which the contributions of the matrix and the fibers to the average composite properties are proportional to their volume fractions. Equation (8.20) is primarily applicable for glass or ceramic-fiber-reinforced thermosetting plastics, when both the constituents deform elastically.

3. Relation Between Load Carried and Elastic Module of the Fiber

Let us now consider the stresses to which constituents of a composite are subjected and how the load is shared between then:

$$\frac{\sigma_f}{\sigma_m} = \frac{E_f \varepsilon_f}{E_m \varepsilon_m}$$

Since $\varepsilon_f = \varepsilon_m = \varepsilon_c$

Hence $$\frac{\sigma_f}{\sigma_m} = \frac{E_f}{E_m} \quad \text{...(8.21)}$$

Similarly, $$\frac{\sigma_f}{\sigma_c} = \frac{E_f}{E_c} \quad \text{...(8.22)}$$

Now $$\frac{P_f}{P_m} = \frac{\sigma_f A_f}{\sigma_m A_m} = \frac{\sigma_f V_f}{\sigma_m V_m} \quad (\text{since } l_f = l_m = l)$$

Using equation (8.21), the above equation becomes

$$\boxed{\frac{\mathbf{P_f}}{\mathbf{P_m}} = \frac{\mathbf{E_f V_f}}{\mathbf{E_m V_m}}} \quad \text{...(8.24)}$$

From the above equation (8.24), it can be concluded that *the ratio of fiber load to matrix load depends on the ratio of elastic moduli of the fibers and the matrix and also on the ratio of their volume fractions.*

That fraction of total load carried by the fibers becomes higher for a higher ratio of elastic moduli of fibres and matrix and a higher volume content of fibers. Thus, if the fibers are to carry a higher proportion of the composite load for a given fiber-matrix system, *the volume fraction of fibers in the composite must be maximized.*

Similarly it can written from equation (8.24) that:

$$\frac{P_f}{P_c} = \frac{E_f V_f}{E_m V_m + E_f V_f}$$

Dividing the numerator and denominator of equation (8.25) by $E_m V_f$, we get

$$\frac{P_f}{P_c} = \frac{E_f/E_m}{V_m/V_f + E_f/E_m}$$

The above equation is used to calculate fraction of the load carried by a fiber in a composite material. It is a relationship between fraction of load by a fiber and the elastic module of the fiber and the matrix.

8.7 SHORT FIBRE REINFORCED COMPOSITES

We know that unidirectional composites have higher strength and modulus in the direction of the fiber axis and are generally very weak in the transverse direction. In some application, it is required to impose equal stresses in all directions. Such stresses are called ***isotropic stresses***. In such cases undirected composites cannot be used.

Multilayered composites of unidirectional layers having different fiber orientations are essentially isotropic in a plane, but their surface layers are still very weak in the transverse direction. Failure of multilayer composites is often initiated from surface layers. So, when protection from a corrosive environment is desired as in case of storage tanks in the chemical industry, multilayered composites cannot be used. In such cases short fiber reinforced composites are used in which each layer of lamina is isotropic. An isotropic layer is obtained. when randomly oriented short fibers are used as the reinforcement. Short fiber reinforced composites have another advantage. These composites can be easily and economically molded into desired shapes by injection or compression molding. Thus, in designing equipments in which equal stresses are imposed in all directions, the short fiber reinforced composites are used.

8.7.1 Properties of Short Fiber Reinforced Composites

The strengthening in short fiber reinforced composites is achieved by restraining the deformation of the matrix between the fibers and also by sharing of load between the fibers and the matrix.

The properties of these composites are governed by the orientation of the short fibers, their interfacial bond strength and kind of the matrix used.

The key properties of short fiber reinforced composites are given below:

(*i*) **Fatigue:** As the weaker matrix is required to sustain a much greater proportion of the cyclic load. Short-fiber reinforced composites have much less resistance to fatigue damage. This damage is generally initiated by bonding of those fibers that lie perpendicular to the loading direction. But in aligned short-fiber composites the fiber ends and weak interfaces can becomes sites for fatigue crack initiation.

Thus, an important aspect of fatigue of short-fiber composites is that local failures in the matrix and at the weak interface can destroy the composite even though the fibers remain undamaged.

(*ii*) **Impact strength:** The tensile impact energy of inductile matrices, such as polystyrene and that of ductile matrixes such as polyethylene decreases. The explanation for this is that in ductile matrices the triaxial restraint of the matrix between fibers limits the elongation of the matrix, and thus addition of rigid fibers greatly reduces the toughness. But in brittle matrices addition of fibers increase toughness due to crack blunting, branching and arrest effects.

8.8 FAILURE OF FIBRE REINFORCED COMPOSITES

The fiber reinforced composites are generally used as structural materials of the composites, so whenever structural material ceases to perform satisfactorily, it is cosisedered as failure of the composite. In other words, *the composite is said to have failed when it fails to respond to the normal characteristics when subjected to stresses which it used to bear normally.*

In fact, prior to the macroscopic failure, the composite fails to respond normally on the microscopic level. The failure load of a unidirectional composite is that load at which composite shows deviation from linear stress-strain behavior or the load applied at the time of its fracture.

The fibre reinforced composites can fail due to the following factors:

(*a*) Breaking of the fibers;

(*b*) Microcracking of the matrix;

(*c*) Debonding *i.e.* separation of fibers from the matrix; and

(*d*) Delemination of laminated composite i.e. separation of laminae from each other.

The fracture modes of the unidirectional composites subjected to different loading conditions are discussed below:

(A) Failure modes under longitudinal tensile load: On increasing the longitudinal tensile load on unidirectional composite, failure of fibers at their weakest cross-sections leads to initiation of failure of the composite. As the number of broken fibers increase, complete rapture of the composite takes place because that cross-section of the composite becomes too weak to support an increased load. This is because the fibers are brittle depending upon the fiber volume fractions as under:

(*i*) When the fiber volume fraction (V_f) is less than 0.40, the unidirectional composite will undergo little failure represented in Fig. 8.5(*a*).

(*ii*) When (V_f) is less 0.65, the composite will undergo brittle failure with fiber pull out as represented in Fig. 8.5(*b*).

(*iii*) When the volume fraction (V_f) is greater than 0.65, the composite will undergo brittle failure with fiber pull out and debonding or matrix shear failure as represented in Fig. 8.5(*c*).

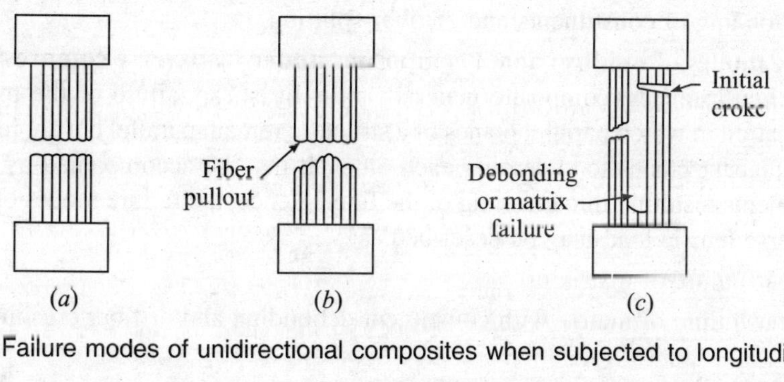

Fig. 8.5 Failure modes of unidirectional composites when subjected to longitudinal load

(B) Failure modes of unidirectional composites under longitudinal compressive load: When composites are subjected to compressive loads continuous fibers act as long columns. Failure modes of composites subjected to longitudinal compressive loads are summarized below:

1. Transverse tensile failure.	Generally, compressive failure of the composite is initiated by transverse splitting or fracture of the composite as represented in Fig. 8.6(*a*).
2. Fibre microbuckling in extensional mode.	Adjacent fibers in a composite buckle independent of each other and the resulting strains in the matrix are predominantly extensional. This generally occurs at low V_f. The modes is represented in Fig. 8.6(*b*).
3. Fibre microbuckling in shear mode.	Adjacent fibers in a composite buckle in a cooperative manner. The resulting strains in the matrix are predominantly shearing strains. This kind of failure generally occurs at practical V_f. This modes is represented in Fig. 8.6(*c*).

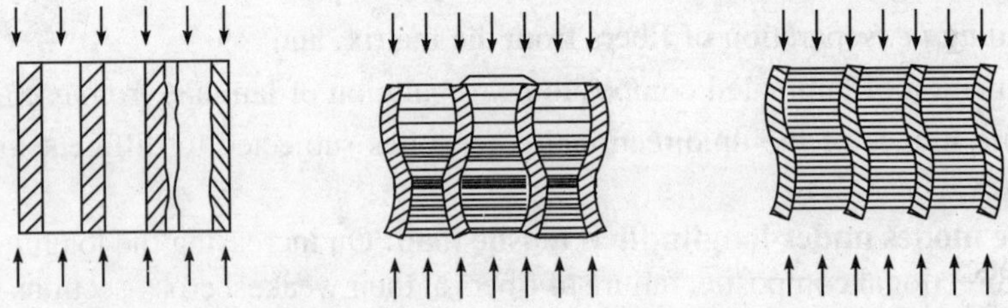

(*a*) (*b*) (*c*)

Fig. 8.6 Failure mode of composite under transverse tensile strength

(C) Failure modes of unidirectional composites under Transverse Tensile load: Unidirectional composites subjected to transverse tensile loads fail because of the tensile failure of either matrix or interface. It is because the fibers perpendicular to the loading direction generates stress concentration in the matrix and at the interface Sometimes, when the fibers are highly oriented and weak in the transverse direction, transverse tensile failure of fibers may occur leading to failure of composite. Hence, the failure modes of composite under transverse tensile loads may be described as:

(*i*) Tensile failure of matrix and

(*ii*) Debonding of constituents and/or fiber splitting.

(D) Failure modes of unidirectional composites under transverse compressive loads: Under these circumstances, composite generally fails by shear failure of the matrix. *i.e.* type of deformation in which parallel planes of a structure remain parallel but are relatively displaced from adjacent planes to slide over each other. It may be accompanied by debonding of the constituents resulting into crushing of the fiber. Hence, the failure modes of composite under transverse tensile load may be described as:

(*i*) Shear failure of matrix or

(*ii*) Shear failure of matrix with constituent debonding and/or fiber crushing.

8.9 PARTICLE REINFORCED COMPOSITES

In these composites, the dispersed phase is a particle which generally has nearly same dimensions in all the directions. In other words, it is equi-axed. The behavior of these composites is governed by:

(*i*) For effective reinforcement, distribution of particles should be uniform through out the matrix (*ii*) Greater the volume fraction of particulate reinforcement, more will be the improvement in the mechanical properties. It means that relative volume fraction of particles should be greater

(*iii*) Strength of the bond at the interface of particles and matrix *i.e.* stronger the bond, better will be the properties of composites

Depending upon the size of the particles reinforced composites belong to the following two categories:

(A) Large-particle Composites: In these composites the interactions between particle and matrix phases are not on the atomic or molecular level. Elastic modulus of particle reinforced composite fall between

$$E_c(u) = E_m(1 - V_p) + E_l V_p$$

and

$$E_c(l) = \frac{E_m E_p}{(1 - V_p)E_p + V_p E_m}$$

where $E_c(u)$ and $E_c(l)$ represent upper and lower elastic modulus of composite. E_m and E_p represents elastic modulus of matrix and particle phases and V_p represents volume fraction of particle phase.

Examples of large particle reinforced composites:

1. **Cermets (ceramic-metal composite):** In these composites ceramics like titanium carbide (TiC) and tungsten carbide (WC) are embedded in metals such as nickel or cobalt. These are used extensively as cutting tools. The cutting surface is being provided by hard ceramic carbide particles. Metal carbides being brittle are not used alone. For enhancing the toughness, these particles are incorporated in ductile metal matrix which prevents particle to particle crack propagation. Both the components of the composite used for cutting tools should be refractory materials. Because during cutting a very high temperature is generated. Ceramics and metal both can withstand high temperature being refractory materials.

2. **Vulcanised Rubber Composite:** In this case vulcanized rubber acts as a matrix which is reinforced by adding 15 30% by volume of carbon-black particles with 20-50 um diameter.

This composite has excellent tensile strength, toughness and abrasion resistance and is used for making automobile tyres.

(B) Fine Particle Composites: In this case reinforcement is done by the uniform dispersion of fine particles of very hard and inert material which may be metallic or non-metallic. The interaction between the fine particles and matrix phase is on the atomic or molecular level due to their small size. Strength is retained at elevated temperatures, because the inert fine particles do not react with matrix phase. Moreover, the resulting composites are durable.

Examples of these composites:

1. ***Thoria nickel composite:*** Here, about 3 volume% of thoria (ThO_2) in the form of fine particles are uniformly dispersed in nickel matrix for improving the high-temperature strength of nickel alloys.

2. ***Sintered aluminum powder:*** In this composite extremely fine (0.1 to 0.2*u* m thick) particles of aluminium oxide are reinforced in aluminium metal matrix. The resulting composites have excellent high temperature strength.

8.9.1 Important Types of Particulate Composites

There important types of particulate composites are given below:

(*i*) ***Metals in metallic composites:*** In these composites both the reinforcing phase and the matrix phase consist of metals. Some common examples are:

(*a*) Bearings made from Cu alloys utilize lead as a natural lubricant.

(*b*) Lead particles in copper alloys and steel, improves the machineabilty of the latter.

(*ii*) ***Metals in Non-metallic composites:*** Here the reinforcing phase is metallic, where as the matrix phase is non-metallic. A common examples is that of aluminium paints in which aluminium flakes act as reinforcing phase which is suspended in paints acting as a non-metallic matrix phase.

(*iii*) ***Non-metals in Non-metallic composites:*** In these composites both the reinforcing phase and the matrix phase consist of non-metals. Some common examples are:

(*a*) Concrete in which particles of sand and rock are dispersed in cement which then chemically react with water for hardening.

(*b*) Flakes of mica or glass in glass or plastic matrix respectively.

(*iv*) ***Non-metals in metallic composites:*** Here the reinforcing phase is non-mettalic, where as the matrix phase in metallic. A common example of this type is *cerments* in which ceramic particles (non-metallic) are suspended in metallic matrix. There are two types of commonly known cerements.

(*a*) ***Oxide based cerments:*** An important example is that of Al_2O_3 as a reinforcing phase dispered in chromium matrix. This composite possesses a very good thermal shock-resistance property and has a strength of 30 GPa.

(*b*) ***Carbide based cerments*:** Two common examples are:

(*i*) A composite in which tungsten carbide (WC) acts as reinforcing unit in cobalt metal which acts as a matrix. This composite is being used as a wire drawing die and other machine parts which require very high degree of hardness.

(*ii*) Another example is a composite consisting of chromium carbide (CrC) acting as reinforcing unit in cobalt metal acting as a matrix. This composite find application in the manufacture of valves, pump parts and spray nozzles because it has very high resistance to corrorion and abrasion.

8.10 ADVANTAGES OF COMPOSITES OVER CONVENTIONAL MATERIALS

The main advantages of the composites over conventional materials are given below:

(*i*) Weight of the composite articles are lesser than the weight of the conventional metallic materials. The weight saving contribute towards the minimization of fuel consumption in automobiles, space ships and other machinery. Therefore, parts of transportation industry used on earth, sea and air are made from composite materials new a days.

(*ii*) Composites show excellent mechanical properties *e.g.*

(*a*) Fiber-reinforced polymer composites show *high strength and stiffness.* They store greater energy per unit weight and so rotational components made out of them function faster than those made of steel alloys.

(*b*) High quality fiber reinforced composites show *excellent modulus strength and torsional stiffness.* These are required by various high speed vehicles and aircrafts.

(*c*) Composites also show *excellent impact and damage tolerance* characteristics.

(*d*) Carbon fiber reinforced polymers show extra ordinary *friction and wear resistance properties*. For instance, the coefficient of friction of a carbon composite is approximately 40% of that of steel.

(C) Composites are *more economical* than metals, ceramics and fine quality of wood.

(D) Composites also show *excellent chemical* properties, *e.g.*

(*a*) Glass fiber reinforced polymer composites show excellent corrosion resistance. Which mainly depends upon the nature of the matrix material.

(*b*) Fiber reinforced composites show better weather resistance and so they are more durable as compared to wood or metal. These composites are preferred in marine industry compared to wood or metal.

(E) Composites materials can *withstand higher thermal stresses* as compared to metals and alloys.

(F) Composites provide *improved insulation* because of less heat radiation losses from their surface as compared to metals.

(G) Due to better flexibility the products made from composites can be designed as per the needs of the market. Such products are more appealing and aesthetically better than those made from conventional materials.

8.11 APPLICATIONS

The main applications of composites are:

(*i*) ***In transportation industry,*** armid reinforced composites being light and tough are used in transportation industry for the manufacture of automobile bodies, racing vehicles and engine components. Not only in ground transportation, composites have replaced conventional materials in marine as well as aerospace transportation. For instance, composites are employed for the construction of racing boats, aircrafts and rockets.

(*ii*) ***As consumer products:*** Composites are being used in the manufacture of sport goods such as Lawn Tennis rackets, musical instruments and domestic equipments.

(*iii*) *In industrial and scientific equipments:* Composite materials are used in safety equipment including ballistic protection, high speed machinery and communication antennae etc.

(*iv*) ***In electronic circuit boards.*** Composite materials are extensively used in the manufacture of Printed Circuit Boards (PCB's) as well as in Printed Wiring Boards PWB's (In electronic equipments composites function primarily as an electrically insulating substrate on which various electrical or electronic components are mounted. PCB's eliminate bulky wiring, increase stability of circuitry and interconnections.)

REVISION QUESTIONS

1. Define composites. Why they should be made? How are composite materials classified?
2. Enlist the general characteristics of composites. Discuss the advantages and limitations of composite materials. Give the main components of composites.
3. What are fiber glass-reinforced composites? Discuss their properties and applications.
4. Discuss in brief, the role played by matrix, fiber and interface in fiber-reinforced composites.

5. Write short note on 'Laminar composites' and 'Sandwich panels'.
6. Briefly describe the failure modes of the fibre reinforced composites.
7. (*a*) List the factors controlling the quality of composites. What is the role of interface in composites?
   (*b*) Write a short note on various applications of composite materials. What are their advantages over the conventional materials ?
8. Name the two main constituents of composites. Discuss the functions of each constituent and the characterics of these constituents.
9. What do you understand by Matrix phase of a composite? What are the requirements of a good matrix material. Compare the different matrix phases used in composites.
10. What is meant by dispersed phase of a composite. Explain the important dispersed phases. What are fibres? Discuss the main types of fibres used in composites.
11. What are fiber reinforced composites. Give their main characteristics. On what factors these characteristics depend? Eulist the important applications of this type of composites.
12. What do you understand by structural composites. Explain in details the types of structural composites.
13. What do you know about unidirectional fibre-reinforced composites? Derive the relationship to evaluate the density of such composites.
14. Derive the relationship for longitudinal strength of unidirectional composites.
15. Derive an equation to calculate the fraction of load carried by a fiber and the elastic modulus of the fiber and the matrix.
16. Write a note on short fibre reinfored composites. Discuss the important properties.
17. What do you know about particle reinforced composites? Discuss the main types of these composites.
18. What do you mean by failure of fibre reinforced composites? Discuss the various fracture modes of the unidirectional composite subjected to different loading conditions.
19. What are composite materials? Discuss some important types of fibre-reinforced composites.
   (GGSIPU, B.Tech. 2nd Sem., May 2002)
20. Write short notes on:
   (*a*) Composites
   (*b*) Fibre-reinforced composites. *(Surat, June 01 KU, May 99)*
21. (*a*) What are composites? Give names of two natural composites. What are the advantageous characteristics of composites?
   (*b*) What are carbon fibres? How are they obtained? Mention their unique characteristics.
   (Delhi, June, 01)
22. (*a*) Discuss the potential applications of polymer reinforced composites in any three of the following fields: transportation, aircraft and aerospace, corrosion, resistance material and electric industries.
   (Anna, May, 01)
   (*b*) What are composite materials? Discuss some important types of fibre-reinforced composites.
   (GGSIP, May 02)
23. (*a*) Write short note on fibre-reinforced composites.
   (*b*) Define polymeric composite, giving examples. *(GGSIP, May 03)*

24. (*a*) How is a composite defined? Give a broad classification of composite materials.

(*b*) What are the factors on which the properties of a composite material depend?

(*c*) Distinguish between fibre and particle-reinforced composites. *(GGSIP, May 04)*

25. (*a*) What are composites? How are composite materials classified? Discuss the properties and applications of fibre reinforced polymeric composites. *(M.D.U., B.E, May, 2006)*

(*b*) Write a brief note on polymeric composite materials. *(M.D.U., B.E, May, 2007)*

(*c*) Define composites and write its applications over polymers. *(K.U. June, 2009)*

❑❑❑

PART II

LAB MANUAL PRACTICALS

(As per the Latest Syllabus of H.P. Tech. University)

Note: At least 10 to 12 Experiments to be Performed.

List of Experiments

1. To determine the surface tension of the given liquid by drop number method by using stalagmometer and identify the given liquid.
2. To determine the insoluble, soluble and total solids in given sample of sewage.
3. To determine the solid carbon, volatile matter, ash content and percentage of moisture in given sample of coal by proximate analysis method and classify the coal.
4. To determine the total alkalinity in a given sample of water using a standard acid.
5. To determine the percentage of Chlorine in a given sample of $CaOCl_2$ which has been dissolved in one litre of solution.
6. To determine the surface tension of the two given unknown liquids by using Stalagmometer and identify the given liquid.
7. To determine the coefficient of viscosity of the given unknown liquids by using Ostwald's Viscometer and identify the given liquid.
8. To determine the coefficient of viscosity of the given lubricating oil using Red Wood Viscometer.
9. To determine the coefficient of viscosity of the given lubricating oil using Seybolt Viscometer.
10. To determine the flash point and fire point of given sample of oil using Pens key Marten's apparatus.
11. To determine the amount of Chlorine in given sample of water approximate *N*/20 sodium Thiosulphate solution.
12. To determine the maximum wavelength of solution of cobalt chloride.
13. To determine the Beer's Law and apply it to find the concentration of given unknown solution by spectrophotometer.
14. To determine the chemical oxygen demand of waste water.
15. To determine the half-life period of given radioactive sample using GM counter.

EXPERIMENTS

Experiment 1

OBJECT

To determine the surface tension of the given liquid by drop number method by using stalagmometer and identify the given liquid.

THEORY

A stalagmometer (Drop pipette) consists of a pipette with a long capillary tube at its lower end and a wider tube at the upper end mounted with a piece of rubber tubing. The lower end of the capillary is flattened. Two marks A and B are etched on the drop pipette for a specific volume for drop formation.

When the drop of a liquid is formed and just falls, the force acting along its circumference due to surface tension of the liquid is equal to weight of the drop, *i.e.*, $mg = 2\pi r \times \gamma$

Where m is the mass of the drop, g is acceleration due to gravity,

r = radius of the drop and γ = the surface tension of the liquid.

Now, if we compare surface tension of two liquid, we shall get

$$m_1 g = 2\pi r \gamma_1 \quad \text{...}(i)$$

$$m_2 g = 2\pi r \gamma_2 \quad \text{...}(ii)$$

Dividing (*i*) by (*ii*), we get

$$\frac{m_1}{m_2} = \frac{\gamma_1}{\gamma_2} \quad \text{...}(iii)$$

But $m = v \times d$ (where v = Volume of a drop and d = density of the liquid)

$$\frac{r_1 \times d_1}{r_2 \times d_2} = \frac{\gamma_1}{\gamma_2} \quad \text{...}(iv)$$

Now, if the volume v of each liquid is changed in to n_1 and n_2 number of drops respectively, we shall have $n_1 \times r_1 = n_2 \times r_2$

$$\frac{n_1}{n_2} = \frac{r_1}{r_2} \quad \text{...}(v)$$

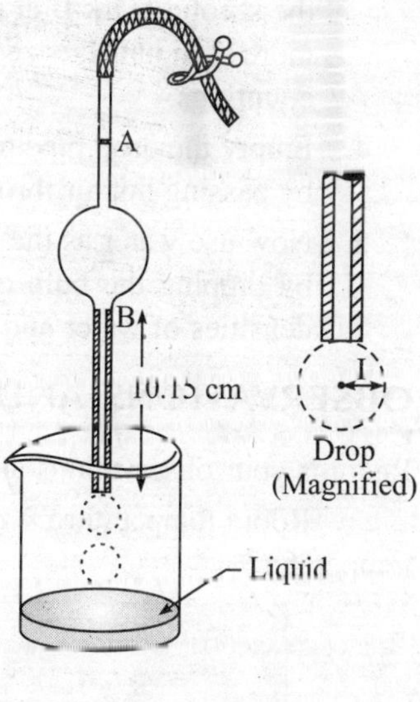

Substituting the value of $\frac{n_1}{n_2}$ from (*v*) in (*iv*)

$$\frac{\gamma_1}{\gamma_2} = \frac{n_2 d_1}{n_1 d_2}$$

Knowing the densities of both the liquids and surface tension of any one of them, the surface tension of the other can be calculated from the above expression.

For comparison, one of the liquid may be water with known density and surface tension and the other is an unknown liquid whose density is usually given or if need be, the same can be determined separately by mean of a specific gravity bottle.

The experimental liquid is placed in a breaker and it is sucked into a drop pipette. When it crosses the mark A it is released and allowed to fall in drops in the beaker below. The number of drops is counted for the liquid passing from A to B.

REQUIREMENTS

(*i*) Stalagmometer (Drop pipette) (*ii*) Beakers (*iii*) Measuring cylinder (*iv*) A rubber tube with a screw pinch cork (*v*) Burette stand.

PROCEDURE

1. Clean the stalagmometer by washing with water. If it is greasy clean it with chromic acid solution and then with water freely. Rinse it with alcohol and dry it in an electric oven. Put the rubber tube with a screw cock at its upper end and fix it on a burette stand.
2. Put the given liquid in a small dry beaker and dip the lower end of the drop pipette under it. Suck up the liquid. When its level crosses the point A, tighten the screw of the pinch cork. Then losen the pinch cork carefully and let the liquid fall slowly in drops. Receive the drops in a beaker containing the remaining liquid. Adjust the pinch cock so as to form drops at an easily countable speed.
3. When the level of the liquid reaches the mark A start counting the drops. When the liquid reaches the second mark B at once tighten the pinch cork. If the last drop hanging from the capillary is more than half in size count it as one. Repeat this process three times and take the mean of the countings.
4. Empty the drop pipette, clean and dry it. If the liquid used was volatile, you may dry it directly by passing hot air through it and there is no need of washing it with water.
5. Now use water as the known liquid and count the number of drops. Note the room temperature by dipping the bulb of a thermometer in a beaker of water for about five minutes. Read the densities of water and the given liquid at room temperature from the constant tables.

OBSERVATIONS AND CALCULATIONS

Records your observations and make calculation as follows:

Room temperature = d_w, Density of water at room temperature = d_1, Density of liquid at room temperature =

No. of drops of water, n_w	*No. of drops of the given liquid, n_1*	*Surface Tension of liquid*
(1) .... (2) .... (3) .... Mean = ...	(1) (2) (3) Mean = ...	$r_1 = \frac{n_w d_1}{n_1 d_w} \times r_w$

PRECAUTIONS

1. The length of capillary should be around 10–15 cm and diameter 1 mm or so. This will maintain proper pressure and given appropriate number of drops.
2. The rubber tube should be perfectly dry and the liquid sucked into the drop pipette should not touch the rubber tube.
3. Normally, the number of drops for a liquid/solution should be 50-100. The liquid should not flow from the drop pipette. The liquid drop should take sometime for its formation and then it should fall.
4. The stalagmometer should be cleaned properly and dried before use.
5. Fit the stalagmometer vertically.
6. The drops should fall from the tip of the stalagmometer under their own weight rather than pushing them by force.

VIVA-VOCE

1. Define the term Surface Tension.

Ans. Surface Tension of liquid is the force acting per centimeter along the surface of the liquid at right angle to any line on the surface of the liquid in any direction.

2. Give the CGS and SI units of Surface Tension.

Ans. The CGS units are dynes per cm and the SI units are Newton per meter. One SI unit is 1000 time of the CGS units.

3. What method do you use in laboratory for determination of surface tension?

Ans. Surface Tension can be measured by stalagmometer using drop number or drop weight method.

4. What is the shape of liquid drop falling from capillary?

Ans. Liquid drops are spherical in shape.

5. Why liquid drops are spherical in shape?

Ans. Due to surface tension, the surface of liquid tends to become minimum. The shape of sphere has minimum surface area.

6. Small insects are seen moving over surface of water tanks. Why?

Ans. The surface of a liquid behaves like a stretched membrane due to the property of surface tension. The molecules on the surface are experiencing a new downward pull due to molecules present in the bulk of the liquid.

7. What instrument is used for measuring the density of a liquid?

Ans. Density of a liquid can be measured by using specific gravity bottle or a Pyknometer.

8. What formula is used for calculating the surface tension of a liquid by drop number method?

Ans. $\gamma_1 = \frac{n_w d_1}{n_1 d_w} \times \gamma_w$, where γ_w represents the surface tension of the water at same temperature n_w and n_1 represents the no. of drops from the same volume of the water and liquid respectively d_w and d_1 represent the density of water and liquid.

9. Why liquids rise in a capillary tube?

Ans. The rise of liquid is due to the inward pull of surface tension acting on the surface which pushes the liquid into the capillary tube.

10. What is the effect of the rise in temperature on the surface tension of the liquids?

Ans. The surface tension of the liquid decreases with the increase in temperature and becomes zero at critical temperature, where the difference between the liquid and vapour disappears.

11. Why surface tension decreases with rise in temperature?

Ans. The decrease in surface tension of liquid with increase of temperature is due to the fact that with increase in temperature the kinetic energy of the molecules (and hence the speed of molecules) increase and therefore the intra-molecular attractions decrease.

12. What is meant by the term 'Surface Energy'.

Ans. It is defined as the amount of work required to increase the surface of a liquid by one sq cm.

13. What are the units of Surface Energy?

Ans. Ergs per cm^2.

14. What is the effect of soluble impurities on surface tension?

Ans. The presence of soluble impurities decreases surface tension.

15. What is angle of contact?

Ans. The angle between the tangent to the liquid surface at the point of contact and the wall of the vessel, inside the liquid, is called the angle of contact.

16. On what factors does the angle of contact depend?

Ans. It depends on the nature of the liquid and the solid constituting the vessel.

17. When and why does surface tension becomes zero?

Ans. The surface tension of a liquid becomes zero at its critical temperature since at critical temperature, surface of separation between the liquid and its vapour disappears.

18. Name different methods for determining surface tension.

Ans. Capillary rise method; Drop weight method, Drop count method; Tensiometric or Torsion balance method, Maximum bubble pressure method.

19. What are surface active agents?

Ans. Substances which decrease the surface tension of water when dissolved in it *e.g.,* soap, detergents, alcohol etc.

20. What is interfacial tension?

Ans. If two immiscible or partially miscible liquids are brought in contact with each other, it is observed that there exists a state of tension at their surface of separation *i.e.,* at the boundary between their layers. This is called *interfacial tension* and is defined as the force acting per unit length along the surface of separation of the immiscible or partially miscible liquids in contact with each other.

21. Why is the surface tension of water much higher than those of organic liquids?

Ans. It is because the intermolecular forces in water are much stronger than those in organic liquids.

Experiment 2

OBJECT

To determine the insoluble, soluble and total solids in a given sample of sewage.

THEORY

Analytically, the term total solids in waste water refers to the matter that remains as residue after evaporation and drying at 103–105°C. The total solids can be classified as filterable and suspended solids. The filterable solids are those that are able to pass through a specified filter and consist of colloidal and dissolved solids. The colloidal solids range in size from 10^{-3} to 10^{-6} mm, while dissolved solids are smaller than 10^{-6} mm. The suspended solids are retained on the filter and their minimum diameter is about 1 μm. The suspended and non-filterable solids put together are also termed as insoluble solids. The settleable solids are those suspended solids which settle in a cone shaped container (Imhoff cone) in a definite time period.

Suspended solids may be further classified on the basis of their volatility at 600°C. The organic matter that is oxidized at this temperature is known as **'volatile suspended solids' (VSS),** and the inorganic fraction that remains as ash is termed **'fixed suspended solids' or 'non-volatile suspended solids' (NVSS)**

Gravimetric methods are used for determination of solids in waste water.

REQUIREMENTS

Silica and porcelain crucibles, beakers, measuring cylinders, Whatmann filter paper no. 44, Imhoff cone, weighing balance etc.

PROCEDURE

(*i*) **Determination of total solids:** Take 50 or 100 ml of water sample in a weighed crucible or beaker. Place the sample in an oven and evaporate to dryness at 98°C. Then dry the evaporated sample for at least on hour at 103–105°C. Cool the crucible/beaker in a desiccator and weigh it.

(*ii*) **Dissolved and suspended solids:** Take 50 or 100 ml of water sample and filter it by using Whatmann filter paper no. 44. Place the filtrate in a dry, weighed crucible and evaporate to dryness in an oven at 98°C. Then dry for one hour at 103–105°C. Cool the crucible with residue in a desiccator and weigh it.

(*iii*) **Volatile and fixed solids:** Take a known quantity of total solids in a dry and weighed silica crucible. Ignite it in a muffle furnace at 500 ± 50°C for 15–20 minutes. Cool the crucible in a desiccator and weigh it.

(*iv*) **Settleable Solids:** Take one litre of sample in an Imhoff cone. Allow the sample to stand undisturbed for two hours. Read the quantity of settleable solids collected at the bottom of the Imhoff cone directly on the graduated scale.

OBSERVATIONS AND CALCULATIONS

(*i*) Total solids

Weight of crucible = W_1 g

Weight of crucible + residue = W_2 g

Weight of residue = $(W_2 - W_1)$ g

Volume of sample taken = V ml.

$$\text{Total solids (mg/L}^{-1}) = \frac{\text{Weight of residue (mg)} \times 100}{\text{Volume of sample (ml)}}$$

$$= \frac{(W_2 - W_1) \times 10^3}{V} \times 1000 = A \text{ mg/L}^{-1}$$

(*ii*) Dissolved and suspended solids

Weight of crucible = W_3 g

Weight of residue + crucible = W_4 g

Weight of residue = $(W_4 - W_3)$ g

$$\text{Dissolved solids (mg/l)} = \frac{\text{Weight of residue of filtrate (mg)}}{\text{Volume of sample (ml)}} \times 1000$$

$$= \frac{(W_4 - W_3) \times 10^3}{V} \times 1000 = B \text{ mg/L}^{-1}$$

Suspended or non-filterable residue or solids

= Total solids – Dissolved solids

$= A - B \text{ mg/L}^{-1} = C \text{ mg/L}^{-1}$

The suspended or filterable solids are nothing but total Insoluble solids.

Therefore insoluble solids = C mgL^{-1}.

(*iii*) Volatile and fixed solids

Weight of crucible = W_5 g

Weight of crucible + residue before heating = W_6 g

Weight of crucible + residue after heating = W_7 g

Weight of initial residue = $(W_6 - W_5)$ g

Weight of final residue = $(W_7 - W_5)$ g

$$\% \text{ Volatile matter} = \frac{\text{Loss in weight of residue}}{\text{Initial weight of residue}} \times 100$$

$$= \frac{W_6 - W_7 \text{ g}}{W_6 - W_5 \text{ g}} \times 100 = X$$

$$\text{Total volatile solids} = \frac{X}{100} \times \text{Total solids is mg/L}^{-1} = D \text{ mg/L}^{-1}$$

Total fixed solids = Total solids – Total volatile solids

$= A - D = E \text{ mg/L}^{-1}$

Settleable solids = F mg/L^{-1}

RESULT

Total solids = X mg/L^{-1}; Dissolved solids = Y mg/L^{-1};
Suspended solids = Z mg/L^{-1}; Volatile solids = L mg/L^{-1};
Fixed solids = M mg/L^{-1}; Settleable solids = N mg/L^{-1}.

VIVA-VOCE

1. What is the significance of solids in water?

Ans. Solids may adversely affect the quality of water. Waters containing suspended solids are not all acceptable for drawing purposes. Further, dissolved solids in drinking water may not be more than 500 mg/L^{-1}. Higher concentrations may cause laxative or sometimes the reverse effect. Suspended solids are also objectionable for bathing or swimming waters. Higher levels of suspended solids in lakes or streams inhibit the penetration of sunlight and promote conditions favourable for growth of pathogenic bacteria. Dissolved solids can also cause hardness in water.

2. What are the harmful effects of solids in boiler waters?

Ans. Both suspended and dissolved impurities are harmful for boiler waters. They cause priming and wet steaming and also deposite sludge and scale in the boiler.

3. What are the disadvantages of scales and sludges?

Ans. Scales and sludges are poor conductors of heat. They decrease the efficiency of the boiler.

4. What are the methods used for removing hardness from water?

Ans. Hardness causing soluble salts can be removed from water by following methods:
(*i*) Lime-soda process (*ii*) Ion-exchange process.

5. What do you mean by the term coagulation?

Ans. Finely divided particles of colloidal dimensions, suspended in water, settle at a very slow rate and readily pass through the pores of the filter. Coagulation is the process of adding chemical to water containing colloidal particles to enable them to settle. Coagulants destroy the charge on colloidal particles. The neutral particles coalese to form bigger particles which settle under the gravitational force. Most effective coagulants are aluminium sulphate [$Al_2(SO_4)_3$], sodium aluminate ($NaAlO_2$), ferrous sulphate ($FeSO_4$) etc.

6. What is the effect of silica in water?

Ans. Presence of silica results in priming in boilers. It also results in formation of calcium and magnesium silicate scales in boilers.

7. How silica can be removed from water?

Ans. Silica can be removed by following methods:

(*i*) By adding magnesia or dolomite lime to water. Magnesia acts as $Mg(OH)_2$. Silica is removed by magnesia by adsorption or by the formation of magnesium silicate.

(*ii*) By the use of ferrous sulphate or sodium aluminate as coagulants. They act as $Fe(OH)_2$ or $Al(OH)_3$ which enmesh finely suspended silica particles.

(*iii*) By passing the demineralised water through a strongly basic special anion exchanger.

8. What is demineralised water?

Ans. It is the water which is completely free from soluble salts. It is obtained by ion exchange process.

9. How will you differentiate between organic and inorganic solids?

Ans. Organic solids consist of carbohydrates (such as cellulose, starch, sugar, fibre etc.), fats and oils, nitrogenous compounds such as proteins and their decomposed products. Inorganic solids consist of minerals (such as sand, gravel etc.) and salts. Generally, the volatile solids represent the organic solids and fixed solids represent the inorganic solids.

10. Explain the importance of examination of solids in water.

Ans. (*i*) Knowledge of the organic or volatile fraction of solids in necessary for the design of biological treatment units and for disposal of sewage by dilution.

(*ii*) The quantity of inorganic and organic-settleable solids helps in designing the grit chamber and sedimentation tank, respectively.

Experiment 3

OBJECT

To determine the solid carbon, volatile matter, ash content and percentage of moisture in given sample of coal by proximate analysis method and classify the coal.

THEORY

Coal is an important fossil fuel. It occurs in layers under the earth crust. It is formed by the partial decay of plants burried under earth millions of years ago. It is highly carbonaceous matter and the process of conversion of vegetable matter into coal can be represented in the following sequence:

Plants $\rightarrow$ Peat $\rightarrow$ Lignite (Brown coal) $\rightarrow$ Bituminous coal $\rightarrow$ Anthracite $\rightarrow$ Graphite.

The composition of coal varies widely depending upon its type, source and age. The carbon content is lowest in Peat and highest in Anthracite. The composition of coal further depends upon the combined effect of heat and pressure to which it has been put. It is necessary to analyse the coal sample to determine its quality and thus the industrial use to which it can be put. The quality of coal is ascertained by the following two types of analysis:

Proximate Analysis: It tells us about the partical utilization of coal. It includes the determination of (*a*) Moisture content (*b*) Volatile matter (*c*) Ash content and (*d*) Fixed carbon as percentages of the original weight of the coal sample.

Significance of Proximate Analysis: This analysis inform us about the relative qualities of the coal samples with regard to its cost and the probability of its use for a particular industry.

1. **Moisture:** Coal contains two types of moisture

(*i*) **Free, Adsorbed or Surface moisture:** The moisture which is lost on air drying. It depends upon the treatment given to coal after mining, the rank, size and surface nature of the coal. It reduces the calorific value of the sample. A lot of heat is wasted in evaporating the free moisture during combustion. It also increases the cost on transport of the coal.

(*ii*) **Inherent or Internal Moisture:** It is that type of moisture which is obtained from air dried coal. It is an indicator of the rank of the coal *i.e.,* the degree of decomposition of vegetable matter to carbon content of coal. It decreases with increasing rank of the coal. In other words the lower the moisture content, the better is the quality of coal. The sum of the weight loss on air drying and Inherent moisture content gives the total moisture content of the coal sample.

2. **Volatile Matter:** It consists of a complex mixture of gaseous and liquid products resulting from the thermal decomposition of coal. It is not a constituent of coal as such. Its magnitude depends upon the conditions of heating particularly the temperature. So the experimental conditions must be maintained during its determination.
   (*i*) High volatile matter content of a coal gives high flames, greater smoke and relatively low heating quality. Thus, it helps in deciding the shapes of furnaces. Hence the higher the volatile matter content, the larger is the combustion space required.
   (*ii*) Moreover, it determines the proportion of coal which would be converted into tarry matter on heating. It includes the combustible gases like CO, H_2, CH_4 etc. and incombustible gases like N_2, CO_2, etc. The presence of incombustible gases is undesirable as they do not have any heating value.
   (*iii*) The volatile matter content in Anthracite coal is lower than in bituminous coal. The lower the volatile matter content the better would be the quality of coal.
   (*iv*) Coking quality of coal also depends upon its volatile matter content. Coking power decreases with increase of volatile matter.
3. **Ash Content:** The non-combustible inorganic matter which is left behind after burning off the organic matter in coal is called ash. It is of two types:
   Free of Extraneous Ash: The ash which occurs in different layers of the coal is known as free ash or extraneous ash. It can be removed by washing.
   Fixed or Inherent Ash: The ash which is intimately interspersed within the mass of coal is called fixed or inherent ash.
4. **Fixed Carbon:** The sum total of percentages of moisture, volatile matter and ash subtracted from 100 gives the percentage of fixed carbon. The fixed carbon content increases from low ranking coals such as lignite to high ranking coal such as anthracite. More the amount of fixed carbon in coal more will be its calorific value. The volume of furnace required for the combustion of such a coal sample will be low. Fixed carbon amount varies inversely to volatile matter in the coal.

REQUIREMENTS

(*i*) Powdered and air dried coal/coke sample
(*ii*) Electric oven (*iii*) Analytical balance (*iv*) Muffle Furnace
(*v*) Desicator (*vi*) Silica crucible with lid (*vii*) Long legged tongs
(*viii*) Normal size tongs

PROCEDURE

(*i*) **Determination of Moisture Content:** Moisture is generally determined by heating a known quantity of air-dried coal from 105°C to 110°C for one hour and calculating the loss in weight as percentage.

If the coal/coke sample is wet, spread it on a porcelain plate or a dried steel pan for 5 to 6 hours to dry it up at room temperature. Usually an air-dried sample is provided in the laboratory. Heat a silica crucible with lid, cool it in a desicator and weigh. Repeat it again until constant weight is obtained. Take about 1 g of the air-dried sample in it and again weigh it accurately on an analytical balance. Now remove the lid of the crucible and place it in the desicator. Place the crucible without lid in an air oven. Now heat the sample and maintain the temperature of the oven between 105°C to 110°C for one hour. Cool the crucible in a desicator and weigh it again along with the lid. The loss in weight corresponds to moisture.

(ii) **Determination of Volatile Matter Content:** After determining moisture content, the dried sample of coal taken in the crucible with lid is heated in a muffle furnace maintained as 925°C ± 25°C for exactly seven minutes. The crucible and the lid are removed from the furnace with the help of long legged tongs carefully. It is first placed on a cooled iron plate and the warm crucible is then transferred in a desicator. Allow it to cool till it attains room temperature. Weight it finally. Loss in weight is reported as volatile matter content on percentage basis.

(iii) **Determination of Ash Content:** The residual coal in the crucible after volatile matter determination is now heated in a muffle furnace maintained as 725 ± 25°C for nearly half-an-hour or until a constant weight of the crucible is obtained. The silica crucible and the lid are placed separately in the furnace. The crucible is then cooled first in air on an iron plate and then in a desicator. The amount of unburnt residue in the crucible is ash, which is reported on percentage basis.

(iv) **Determination of Fixed Carbon:** The sum total of percentages of moisture, volatile matter and ash subtracted from 100 gives the percentage of fixed carbon.

OBSERVATIONS

Weight of empty and previously ignited crucible with lid $= W_1$ g
Weight of crucible with air dried sample (with lid) $= W_2$ g
Weight of crucible with oven died sample (with lid) $= W_3$ g
Weight of crucible after ignition in muffle furnace at 925 ± 25°C $= W_4$ g
Weight of crucible after ignition of residual sample at 725°C ± 25°C $= W_5$ g

CALCULATIONS

1. Moisture Percentage $= \dfrac{\text{Loss in weight of coal}}{\text{Weight of coal taken}} \times 100 = \dfrac{W_2 - W_3}{W_2 - W_1} \times 100$

2. Volatile matter percentage $= \dfrac{\text{Loss in weight due to volatilematter}}{\text{Weight of moisture free coal laken}} \times 100 = \dfrac{W_3 - W_4}{W_2 - W_1} \times 100$

3. Ash content percentage (Air dry basis) $= \dfrac{W_5 - W_1}{W_2 - W_1} \times 100$

   Ash content percentage (oven dry basis) $= \dfrac{W_5 - W_1}{W_2 - W_1} \times 100$

4. Fixed carbon percentage = 100 – % of (Moisture + Volatile matter + Ash)

PRECAUTIONS

(i) The empty crucible with lid should be ignited and cooled in a desicator before weighing.

(ii) The process of ignition and cooling should be repeated until a constant weight is obtained.

(iii) The crucible after washing and ignition should never be placed in the open. It is always placed in a desicator.

(iv) The crucible should never be held with fingers. The tongs should be used for putting the crucible from desicator to the oven or muffle furnace.

(v) The weighings should be made up to fourth decimal places.

VIVA-VOCE

1. **Define a fuel.**

Ans. A combustible substance containing carbon as the main constituent, which on proper burning liberates large amount of heat, which can be used economically for domestic as well as industrial purposes is called a fuel.

2. **What are fossil fuels and how are they formed?**

Ans. The fossil fuels are formed by the decomposition of dead plants and animals under earth, which are subjected to heat and high pressures.

3. **Name three fossil fuels.**

Ans. Coal, Petroleum and Natural gas.

4. **What is meant by calorific value of a fuel?**

Ans. It is the total quantity of heat liberated when a unit mass (or volume) of the fuel is burnt completely in presence of sufficient quantity of air/oxygen.

5. **Define gross calorific value of a fuel.**

Ans. Gross calorific value is the total quantity of heat liberated, when one unit of the fuel has been burnt completely and the products of combustion have been cooled to room temperature.

6. **Define net calorific value of a fuel.**

Ans. Net calorific value is the net heat evolved when one unit of the fuel is burnt completely and the hot products are allowed to escape as such.

7. **What are the different varieties of coal formed inside the earth?**

Ans. Peat, Lignite, Bituminous and Anthracite.

8. **Arrange the different varieties of coal in the decreasing order of their calorific values.**

Ans. Anthracite > Bituminous > Lignite > peat.

9. **Arrange different varieties of coal in decreasing order of their moisture content.**

Ans. Anthracite < bituminous > lignite > peat.

10. **Name the first stage in the coalification of wood.**

Ans. Peat.

11. **What is meant by ignition temperature?**

Ans. The lowest temperature at which the fuel must be pre-heated so that it starts burning smoothly.

12. **What do you mean by Proximate analysis?**

Ans. It tells us about the partical utilization of coal. It includes the determination of (*a*) Moisture content (*b*) Volatile matter (*c*) Ash content and (*d*) Fixed carbon as percentages of the original weight of the coal sample.

13. **What do you mean by Ultimate Analysis?**

Ans. It includes the estimation of carbon, hydrogen, nitrogen, sulphur, and oxygen. This analysis is useful in calculating the calorific value of the fuel.

14. **What is the significance of Proximate analysis?**

Ans. This analysis inform us about the relative qualities of the coal samples with regard to its cost and the probability of its use for a particular industry.

15. **Name the highest ranking coal.**

Ans. Anthracite.

16. Mention four factors taken into consideration for selecting coal for different uses.

Ans. (*i*) High calorific value (*ii*) low moisture content (*iii*) low ash content, and (*iv*) high calorific intensity.

17. What is the main function of proximate analysis of coal?

Ans. It provides valuable information regarding practical utility of coal.

18. What are the characteristics of a good quality coal?

Ans. A good quality coal possesses:

(*i*) Low moisture (*ii*) Low volatile matter,

(*iii*) Low ash content and (*iv*) High fixed carbon.

19. Why a good fuel must have low ash content?

Ans. Ash is a useless non-combustible matter present in a fuel. It reduces the calorific value, causes hinderance in the flow of air and heat, thereby reducing the temperature attained. Moreover, it causes rouble during firing by forming clinkers. Hence, lower the ash, better the quality of fuel.

20. Why a good fuel must have low volatile matter content?

Ans. High volatile content of a fuel gives greater smoke, high flames and low heating quality. So good fuel should have low volatile matter content.

21. During this experiment, what is the use of desicator?

Ans. In desicator, crucible is attaining room temperature without gaining moisture as it contains anhydrous calcium chloride.

22. What is the significance of fixed carbon?

Ans. Higher the value of fixed carbon in a given coal high is the calorific value and better is the quality of coal.

Experiment 4

OBJECT

To determine the total alkalinity in a given sample of water using a standard acid.

THEORY

The determination of alkalinity of irrigation water is particularly important because the extent of alkalinity in water affects the growth of plants. The effect of fertilizers which are being used for the growth of a particular class of plants is dependent on the alkalinity of water. Besides agriculture, the knowledge of alkalinity of water is necessary for controlling the corrosion of the boilers for steam generation, for calculating the amount of lime and soda needed for water softening. High alkalivity of water may lead to the precipitation of sludges and deposition of scales.

Alkalinity of water means the total content of those substances in it which cause an increased $[OH^-]$ ions concentration upon dissociation or due to hydrolysis. Although many substances can contribute towards alkalinity, but it is mainly due to the presence of hydroxide ion (OH^-), carbonate ion (CO_3^{2-}) and bicarbonate ion (HCO_3^-) present in the given sample of water.

The alkalinity due to different kind of ions responsible for it can be estimated separately by titration against standard acid using phenolphthalein and methyl orange as indicators. A known volume of water sample is titrated against standard H_2SO_4 using phenolphthalein as indicator. The phenolphthalein end point (P) marks the neutralization of OH^- and half neutralization of CO_3^{2-} to HCO_3^- ions. The sample is further titrated with standard H_2SO_4 acid in presence of methyl orange as indicator. The volume of acid used after phenolphthalein end point is for neutralization of all the bicarbonate ions *i.e.*, present originally or obtained from half neutralization of carbonate ions. The titration of water upto methyl orange end point (M) marks completion of all three reactions.

$$\left.\begin{array}{l}\left.\begin{array}{l}(i)\ OH^- + H^+ \rightarrow H_2O \\ (ii)\ CO_3^{2-} + H^+ \rightarrow HCO_3^-\end{array}\right] P \\ (iii)\ HCO_3^- + H^+ \rightarrow H_2O + CO_2\end{array}\right] M$$

Phenolphthalein alkalinity $P = OH^- + \frac{1}{2}CO_3^{2-}$

Total or Methyl orange alkalinity $M = OH^- + CO_3^{2-} + HCO_3^-$

With respect to the constituents causing alkalinity in water, the following situations may arise:

(*a*) $P = 0$; Only HCO_3^- is present. Alkalinity due to $HCO_3^- = M$.

(*b*) $P = M$; Only HO^- is present. Alkalinity due to $OH^- = P = M$.

(*c*) $P = \frac{1}{2}M$; Only CO_3^{2-} is present. Alkalinity due to $CO_3^{2-} = 2P$

(*d*) $P > \frac{1}{2}M$; CO_3^{2-} and OH^- are present; Alkalinity due to $CO_3^{2-} = 2\ (M - P)$

Alkalinity due to $OH^- = 2P - M$

(*e*) $P < \frac{1}{2}M$; CO_3^{2-} and HCO_3^- are present; Alkalinity due to $CO_3^{2-} = 2P$

Alkalinity due to $HCO_3^- = M - 2P$

It is important to note that OH^- and HCO_3^- can not be together due to following reaction:

$$OH^- + HCO_3^- \rightarrow CO_3^{2-} + H_2O$$

REQUIREMENTS

(*i*) Standard $H_2SO_4\left(\frac{N}{10}\right)$ (*ii*) Phenolphthalein indicator solution.

(*iii*) Methyl orange indicator solution.

INDICATORS

(*i*) Phenolphthalein (*ii*) Methyl orange.

END POINTS

Colour change from pink to colourless with phenolphthalein. From light yellow to reddish orange using methyl orange.

PROCEDURE

Rinse and fill the burette with $\frac{N}{10}$ H_2SO_4. Note initial burette reading. Pipette out 20 ml of the water sample into a conical flask. Add 1-2 drops of phenolphthalein indicator and titrate this sample against

H_2SO_4 until pink colour just disappears. Note down this reading as phenolphthalein end point. Add 2-3 drops of methyl orange in the same solution. The solution turns light yellow. Titrate further against the acid until the light yellow colour changes to orange red. Note down the final burette reading as methyl orange end point. Repeat the whole procedure to get concordant readings.

OBSERVATIONS AND CALCULATIONS

Volume of water sample taken for each titration = 20 ml.

Let volume of $\frac{N}{10}$ acid used upto phenolphthalein end point = A ml.

Total volume of acid used upto methyl orange end point = B ml.

OBSERVATION TABLE

S.No.	*Initial burette reading (x)*	*Final burette reading using phenolphthalein (y)*	*Final burette reading using methyl orange (z)*	*A = y − x (ml)*	*B = z − x (ml)*
1.					
2.					
3.					
4.					
5.					

Phenolphthalein Alkalinity

$$\underset{\text{(Sample)}}{N_1V_1} = \underset{\text{(Acid)}}{N_2V_2}$$

$$N_1 \times 20 = \frac{1}{10} \times A \text{ or } N_1 = \frac{A}{200} N$$

$$\text{Strength (in terms of } CaCO_3) = \text{Normality} \times \text{Eq. wt. of } CaCO_3 = \frac{A}{200} \times 50 \text{ gL}^{-1}$$

or $$\text{Phenolphthalein Alkalinity (P)} = \frac{A}{200} \times 50 \times 1000 \text{ mgL}^{-1} = 250 \times A \text{ ppm.}$$

Total or Methyl Orange Alkalinity

$$\underset{\text{(Sample)}}{N_3V_3} = \underset{\text{(Acid)}}{N_4V_4} \text{ or } N_3 \times 20 = \frac{1}{10} \times B \text{ or } N_3 = \frac{B}{200} N$$

$$\text{Strength (in terms of } CaCO_3) = \frac{B}{200} \times 50 \text{ gL}^{-1}$$

or $$\text{Methly orange Alkalinity (M)} = \frac{B}{200} \times 50 \times 1000 \text{ mgL}^{-1} = 250 \times B \text{ ppm.}$$

Types of Alkalinity

From 5 cases *i.e.,* $P = O, P = M, P = \frac{1}{2}M$; $P > \frac{1}{2}M$ and $P < \frac{1}{2}M$

Select the type of alkalinity in the given sample and calculate the extent of alkalinity accordingly.

RESULTS

The various forms of alkalinity present in the given sample of water are as follows:

Total or Methyl orange alkalinity = ... ppm
Phenolphthalein Alkalinity = ... ppm
Hydroxide (OH^-) Alkalinity = ... ppm
Carbonate (CO_3^{2-}) Alkalinity = ... ppm
Bicarbonate (HCO_3^-) Alkalinity = ... ppm.
(All in terms of $CaCO_3$ equivalent).

Instructions to the Lab. Technician

1. $N/10H_2SO_4$ or N/10HCl: Dilute 2.8 ml. of conc. H_2SO_4 or 8.6 ml of conc. HCl to 1 litre with distilled water.
2. **Phenolphthalein:** Dissolve 1 g of phenolphthalein in 100 ml of ethyl alcohol and mix 100 ml of distilled water. Filter, if necessary.
3. **Methyl orange:** Dissolve 50 mg of sodium salt of methyl orange solid in 100 ml of distilled water. Add 1 ml of N/10 HCl and filter, if necessary. Addition of HCl is not required in case of methyl orange solid.

VIVA-VOCE

1. What is alkalinity of water?

Ans. The alkalinity of water is a measure of its capacity to neutralise acids.

2. What materials are responsible for the alkalinity of water?

Ans. The alkalinity of water is mainly due to the hydroxides, carbonates and bicarbonates of alkali and alkaline earth metals.

3. What do you understand by Phenolphthalein alkalinity?

Ans. The alkalinity due to the neutralisation of standards acid by hydroxyl (OH^-) ions and half of the carbonate (CO_3^{2-}) ions present in water.

4. What do you understand by Methyl orange alkalinity?

Ans. The alkalinity due to the neutralisation of one-half of normal carbonate (CO_3^{2-}) plus all the bicarbonate (HCO_3^-) ions present in water.

5. What do you understand by total alkalinity?

Ans. The total alkalinity corresponds to the neutralisation of standard acid by hydroxyl (OH^-) ions, carbonate (CO_3^{2-}) ions and bicarbonate (HCl) ions in water.

6. What are the possible combination of ions causing alkalinity in water?

Ans. (*i*) OH^- ions only (*ii*) CO_3^{2-} only (*iii*) HCO_3^- ions only (*iv*) OH^- and CO_3^{2-} and (*v*) HCO_3^- and CO_3^{2-}.

7. Which combination of ions is not possible for alkalinity in water and why?

Ans. The possibility of OH^- and HCO_3^- ions together is ruled out, but they combine to form carbonate ions. $OH^- + HCO_3^- \longrightarrow CO_3^{2-} + H_2O$

8. Show the various forms of alkalinity on pH scale.

Ans. The various forms of alkalinity present in water are shown on the pH scale as under:

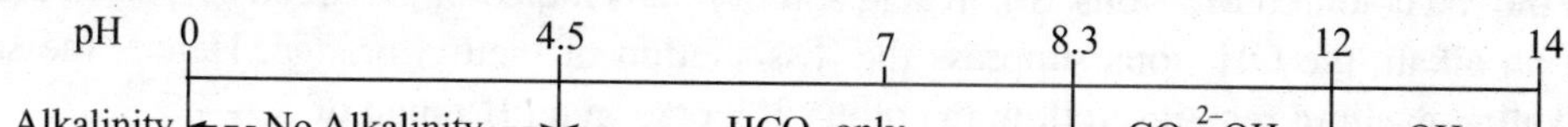

9. What is the significance of determing alkalinity in water?

Ans. The knowledge of alkalinity of water is necessary due to following:

(*a*) For controlling corrosion of boiler

(*b*) In conditioning boiler feed water as highly alkaline water which may lead to caustic embrittlement and deposition of precipitates and sludges in boiler tubes and pipes.

(*c*) In calculating amounts of lime and soda required for water softening.

(*d*) In neutralizing acidic solutions produced by hydrolysis of salts.

(*e*) The amount of coagulants like aluminium sulphate and ferrous sulphate to be added depends upon the amount of alkalinity in water.

(*f*) To control the alkalinity of water being used for irrigation purposes.

10. For which type of the acid base titration phenolphthalein is used as an indicator? Why does it change colour at the end point?

Ans. Phenolphthalein is used as indicator for strong base titrations against a strong acid or a weak acid. It is a weak acid and may be represented as Hph. It ionises as under:

$$\underset{\text{Colourless}}{\text{Hph}} \rightleftharpoons \text{H}^+ + \underset{\text{Pink}}{\text{Ph}^-} \quad ...(i)$$

The strong base (NaOH) will ionese fully as under:

$$\text{NaOH} \longrightarrow \text{OH}^- + \text{Na} \quad ...(ii)$$

The OH^- ions from the strong base take away the H^+ ions and equilibrium (*i*) is favoured in the forward direction. The colour changes to pink due to the Ph^- ion which exists in the quinoid form responsible for colour.

11. Why phenolphthalein is not a suitable indicator for titrating a weak base like, NH_4OH, against a strong acid?

Ans. Phenolphthalein is an acid-base indicator which works in pH range of 8.0–9.8. It is colourless in acidic solution and pink in alkaline solution. When a weak base like NH_4OH is titrated with a strong acid, the OH^- ions furnished by the weak base at the end point of the titration are too few to shift the equilibrium sufficiently towards the right to raise the pH value of 8.3, at least and, therefore, the pink colour does not appear just at the end point. A sufficient excess of weak base has to be added to get the pink colour.

12. How does methyl orange function as an indicator?

Ans. Methyl orange is a weak base which may be represented as MeOH. It dissolves in water and undergoes dissociation to a small extent. The undissociated molecules are yellow while the cation Me^+ are red in colour. The dissociation may be written as:

$$\underset{\text{Yellow}}{\text{MeOH}} \longrightarrow \underset{\text{Red}}{\text{Me}^+} + \underset{\text{Colourless}}{\text{OH}^-}$$

If the solution is acidic, the H^+ ions furnished by the acid combine with OH^- ions furnished by the indicator to form unionised water. This shifts the equilibrium towards the right giving more of the red coloured Me^+ ions. So, in acid solution, this indicator gives red colour. In the presence of an alkali, the OH^- ions suppress the dissociation of methyl orange. Hence, the solution in alkaline medium remains yellow in colour. It works in a pH range of 3.1–4.5.

13. Why methyl orange can not be used as indicator for titrating a weak acid like acetic acid against a strong base?

Ans. Methyl orange can not be used for titrating a weak acid against strong base because the H^+ ions furnished by the weak acid at the end point are too few to combine with a sufficient number of OH^- ions of methyl orange to shift the equilibrium in favour of red coloured Me^+ ions. A sufficient excess of the weak acid has to be added to get the end point.

14. Why the alkalinity of irrigation water is determined?

Ans. It is because of the following two reasons:

(*i*) The extent of alkalinity controls the growth and development of different kind of plants.

(*ii*) Alkalinity also effects the fertilizers being used. Different fertilizers are employed for different types of vegetation and so it is significant that alkalinity is due to which ion and the amount of that ion in irrigation water.

(*iii*) It determines the salinity of the soil. High salinity is not suitable for irrigation.

15. On what factor the choice of indicator for a particular acid-base titration depends?

Ans. It depends on the pH value of the solution at the end point.

16. Tell whether the pH value at the end point is always 7 or not. Explain your answer.

Ans. The pH value is not always seven at the end point. The pH value at the end point depends upon the nature of the acid and the base used in the titration. The salt formed gets hydrolysed and pH value depends upon the nature of salt formed.

17. Can the pH value at the end point be 7. Give reason for your answer.

Ans. Yes, it can be 7 only when both the acid and the base are strong (say NaOH and HCl). The salt formed in this case would be NaCl and it does not get hydrolysed to make any change in the pH value.

18. Give an example of the acid-base titration, where pH value is more than 7 at, the end point.

Ans. The pH value is more than 7 when strong base is titrated against a weak acid. For example, in the titration of CH_3COOH against NaOH. The salt formed would be $H_3CCOONa$, which will get hydrolysed to give a basic solution near the end point.

19. Give an example of the acid-base titration when the pH value is less than 7 at the end point.

Ans. The pH value is less than 7, when a strong acid (say HCl) is titrated against a week base (say NH_4OH). The salt formed NH_4Cl in this case would get hydrolysed to give and acid solution which must have pH value less than 7.

20. How do you infer that the alkalinity is due to OH^- ions?

Ans. When the titre value is of phenolphthalein end point (P) comes out to be equal to that of methyl orange end point (M).

21. In which case the alkalinity is due to OH^- and CO_3^{2-} ions both?

Ans. When $P > \frac{1}{2}M$; Both CO_3^{2-} and OH^- ions are present. Alkalinity due to CO_3^{2-} ions = 2 (M – P).

Alkalinity due to OH^{-3} ions = 2P – M.

Experiment 5

OBJECT

To determine the percentage of Chlorine in a given sample of $CaOCl_2$ which has been dissolved in one litre of solution.

THEORY

Bleaching powder is used as a bleaching agent and also as a disinfectant. The main constituent of bleaching powder calcium hypochlorite supplies chlorine with dilute acids as under.

$$\underset{\text{Calcium hypochlorite}}{2Ca(OCl)_2} + 8HCl \longrightarrow 2CaOCl_2 + 4H_2O + 4Cl_2 \text{ (Available Chlorine)}$$

So available chlorine is defined as the percentage of chlorine made available by bleaching powder when treated with dilute acid.

The available chlorine present in bleaching powder ($CaOCl_2$) sample is determined iodometrically by treating its solution with an excess of KI solution in the acidic medium.

$$OCl^- + 2H^+ + 2e^- \longrightarrow Cl^- + H_2O$$
$$2I^- \longrightarrow I_2 + 2e^-$$
$$OCl^- + 2H^+ + 2e^- \longrightarrow Cl^- + H_2O$$

The liberated iodine is titrated with Hypo solution using freshly prepared starch solution to be added near the end point.

$$I_2 + 2e^- \longrightarrow 2I^-$$
$$2S_2O_3^{2-} \longrightarrow S_4O_6^{2} + 2e^-$$
$$I_2 + \underset{\text{Thiosulphate ion}}{2S_2O_3^{2-}} \longrightarrow 2I^- + \underset{\text{Tetrathionate ion}}{S_4O_6^{2-}}$$

$$I_2 + \text{Starch Solution} \longrightarrow \text{Blue Coloured Complex}$$

REQUIREMENTS

(*i*) Standard sodium thiosulphate solution $\left(\frac{N}{20} Na_2S_2O_3\right)$

(*ii*) 10% KI solution.

(*iii*) dil H_2SO_4

(*iv*) Freshly prepared starch solution.

INDICATOR

Freshly prepared starch solution to be added near the end point.

END POINT

Disappearance of blue colour.

PROCEDURE

Weigh accurately 1-2 g of bleaching powder sample in a weighing bottle. Transfer it to a porcelain mortar and weigh the bottle again. Difference in the weights will be the exact weight of sample taken.

Put some distilled water on the sample and rub it with a pestle to form a smooth paste. Add more water to it grind it with pestle, keep it undisturb for a while and decant milky solution into the 250 ml measuring flask. Grind the residue with pestle, add more water and repeat the same process, until whole of the sample is transferred to the flask. Make up the volume upto the mark by adding distilled water and shake it well.

Now pipette out 20 ml of the above prepared solution to a titration flask. Add 10 ml of 10% KI solution and 10 ml dil H_2SO_4. Titrate the liberated I_2 with standard starch as indicator, again titrate it with same Hypo solution until the disappearance of blue colour. Repeat the same titration 3-4 times so as to get the concordant reading.

OBSERVATIONS

Weight of (bleaching powder sample + weighing bottle) = W_1 g
Weight of weighing bottle alone = W_2 g
Weight of sample = $W_1 - W_2$ = Wg
Volume of solution prepared = 250 ml
Volume of solution taken for each titration = 20 ml
Record the burette reading in a Tabular form
Concordant volume = A ml

CALCULATIONS

(*i*) Normality of bleaching powder sample solution.

$$N_1V_1 = N_2V_2$$

$$N_1 \times 20 = \frac{1}{20} \times A$$

$$N_1 = \frac{A}{400} N$$

(*ii*) Amount of chlorine present in 1 litre of the solution = $N_1 \times$ Eq. wt. of chlorine

$$= \frac{A}{400} \times 35.5 \text{ g} = \text{BgL}^{-1}$$

(*iii*) Amount of chlorine present in 250 ml. of the solution = $\frac{B}{1000} \times 250$ g = CgL^{-1}

(*iv*) % of available chlorine = $\frac{C}{W} \times 100\%$

$$= \frac{A \times 35.5 \times 250}{400 \times 1000} \times \frac{100}{W} \%$$

PRECAUTIONS

1. Bleaching powder sample should be weighed into a weighing bottle otherwise there may be corrosion of the balance material due to decomposition of same.
2. The sample solution should be pipetted out carefully as chlorine vapours are harmful.
3. The prepared sample solution should be titrated immediately after its preparation.
4. Freshly prepared starch solution should be added as indicator near the end point.
5. The sample solution should be shaken well before pipetting out for each titration.

VIVA-VOCE

1. What do you mean by 'Available Chlorine' in given bleaching powder sample?

Ans. Available chlorine in bleaching chlorine means the grams of chlorine liberated from 100 g of the sample when treated with dil. acid.

2. What is the need of determinging the available chlorine in bleaching powder sample?

Ans. During storage bleaching powder gets decomposed by absorbing moisture from atmosphere and evolves chlorine

$$OCl^- + Cl^- + Ca^{2+} + H_2O \longrightarrow Cl_2 + Ca(OH)_2$$

Due to this, it always contains less amount of chlorine than the expected amount. Therefore, the available chlorine must be determined in a sample before its use.

3. Give the chemical formula of bleaching powder.

Ans. Bleaching powder is the mixture of calcium Hypochlorite [$Ca(OCl)_2 . 4H_2O$] and basic calcium chloride [$CaCl_2 . Ca(OH)_2 . H_2O$] with some free slaked lime. Conveniently it is assumed to be a mixed calcium Hypochlorite-chloride [$Ca(OCl)Cl$] or $CaOCl_2$.

4. Give another name of bleaching powder.

Ans. Bleaching powder is also called chloride of lime.

5. In determination of available chlorine, what happens when HCl is used instead of H_2SO_4 for acidification.

Ans. When HCl is used for acidification then, calcium chlorate present in bleaching powder sample reacts slowly with KI liberating additional I_2 and thus higher results are obtained.

$$ClO_3^- + 6H^+ + 6e^- \longrightarrow Cl^- + 3H_2O$$

$$2I \longrightarrow I_2 + 2e^-] \times 3$$

$$ClO_3^- + 6H^+ + 6I^- \longrightarrow Cl^- + 2I_2 + 3H_2O$$

6. Give some other commonly used bleaching agents.

Ans. (*i*) Hydrogen Peroxide (H_2O_2)

(*ii*) Benzoyl peroxide ($(C_6H_5COO)_2$)

(*iii*) Ozone (O_3)

(*iv*) Chlorine gas (Cl_2 gas).

7. Explain why starch-iodine blue colour reappears after the end point if acetic acid is used instead of H_2SO_4 for acidification.

Ans. In presence of acetic acid, chlorite (ClO_2^-) which is formed from moist bleaching powder, reacts slowly with KI to liberate I_2.

$$ClO_2^- + 4H^+ + 4e^- \longrightarrow Cl^- + 2H_2O$$

$$2I^- \longrightarrow I_2 + 2e^-] \times 4$$

$$ClO_2^- + 4H^+ + 4I^- \longrightarrow Cl^- + 4I_2 + 2H_2O$$

This I_2 combines with starch to give blue colour even after the end point, while in presence of H_2SO_4, a stable end point is obtained.

Experiment 6

OBJECT

To determine the surface tension of the two given unknown liquids by using Stalagmometer and identify the given liquid.

THEORY

(*i*) Determination of surface tension with a stalagmometer has already been discussed in Experiment No. 1.

(*ii*) **Determination of percentage composition by volume of two liquids:** The given two liquids should be miscible in all proportions. We can prepare mixtures of known composition by volume. Then surface tensions of these mixture and of the unknown given mixtures are measured. A graph is plotted between surface tension and composition values and the composition of the known mixture is read off, from the graph by intrapolation.

Instead of determining the surface tension of liquid mixtures, the graph can be plotted even between the number of drops for a fixed volume of the mixtures and their respective compositions. From this graph the composition of the given mixture with unknown composition can be determined by intera polation.

REQUIREMENTS

(*i*) Stalagmometer (Drop pipette) (*ii*) Beakers (*iii*) Measuring cylinder (*iv*) A rubber tube with a below pinch cock (*v*) Burette stand.

PROCEDURE

Let the two unknown liquids be Ethanol and Benzene which are miscible in each other. We can prepare mixtures of known composition of Ethanol and Benzene (90%, 70%, 50%, 30% and 10%) by mixing in the following proportions by volume so as to get 50 ml of solution in each case.

Composition	*90%*	*70%*	*50%*	*30%*	*10%*
Ethanol (ml)	45	35	25	15	5
Benzene (ml)	5	15	25	35	45

Now, the number of drops formed by allowing the same volume (between A and B of the stalagmometer) solution having different composition is determined using a stalagmometer in the following steps.

1. Clean the stalagmometer by washing with water. If it is greasy clean it which chromic acid solution and then with water freely. Rinse it with alcohol and dry it in an electric oven. Put the rubber tube with a screw cock at upper end and fix it on a burette stand.
2. Take the solutions prepared above one-by-one in a small dry beaker and dip and the lower end of the drop pipette upper it. Suck up the liquid. When its level crosses the point A, tighten the screw of the punch cock. Then losen punch cock carefully and let the liquid fall slowly in drops. Receive the drops in a beaker containing the remaining liquid. Adjust the punch cock so as to form drops at an easily countable speed.

3. When the level of the liquid reaches the mark A start counting the drops. When the liquid reaches the second mark B at once tighten the punch cock. If the last drop hanging from the capillary is more than half in size count it as full. Repeat this process three times and take the mean of the countings.
4. Empty the drop pipette, clean and dry it. If the liquid used was volatile, you may dry it directly by passing hot air through it and there is no need of washing it with water.
5. Now repeat the experiment with other solutions of different compositions one-by-one. Count the number of the drops formed while passing the same volume of different solutions flowing through the stalagmometer between the end. Perform the experiment with the given solution of unknown composition.

OBSERVATIONS AND CALCULATIONS

Records your observations in the following table.

S.No.	*% Composition*	*Number of Drops formed*
1.	90% Ethanol 10% Benzene	(*i*) … (*ii*) … Mean = …
2.	70% Ethanol 30% Benzene	(*i*) … (*ii*) … Mean = …
3.	50% Ethanol 50% Benzene	(*i*) … (*ii*) … Mean = …
4.	30% Ethanol 70% Benzene	(*i*) … (*ii*) … Mean = …
5.	10% Ethanol 90% Benzene	(*i*) … (*ii*) … Mean = …
6.	Given solution	(*i*) … (*ii*) … Mean = …

Now plot a graph between % composition by volume of the solutions and the number of drops formed from the same volume of different solutions. Take % composition along X axis and number of drops formed along Y axis of the graph as shown in Fig. 6.1.

Knowing the number of drops formed from the same volume of the given solution, its composition by volume can be determined from the graph plotted. If AB is the plot obtained the point C corresponds to the given solution. The best situation for finding composition of mixture by such physical methods is when a straight line graph is available. But this is usually not so and analysis by interpolation is only possible.

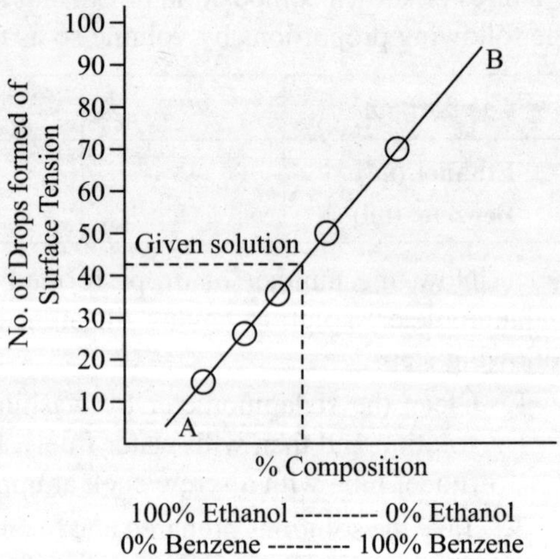

Fig. 6.1

For Surface Tension Determination

Room temperature =

d_w, Density of water at room temperature =

d_l, Density of liquid at room temperature =

No. of drops of water, n_w	*No. of drops of the given liquid n_l*	*Surface Tension of liquid*
(1) .... (2) .... (3) .... Mean = ...	(1) (2) (3) Mean = ...	$\gamma_1 = \frac{n_w d_l}{n_l d_w} \times y_w$

The graph can also be plotted against surface tension and % composition and from the graph % composition of the given solution can be determined in a similar way. But it will involve lot of calculations. More over the experiment becomes lengthy. Therefore the method discussed earlier is preferred.

Note: In the given two liquids are unknown (say liquid A and B) and the two liquids are to be identified. Their surface tension can be determined as in Experiment No. 1. They can be identified by comparing their values determined experimentally with the given values in the table.

PRECAUTIONS

1. The length of capillary should be around 10.15 cm and diameter 1 mm or so. This will maintain proper pressure and give appropriate number of drops.
2. The rubber tube should be perfectly dry and the liquid sucked into the drop pipette should not touch the rubber tube.
3. Normally, the number of drops for a liquid/solution should be 50–100. The liquid should not flow from the drop pipette. The liquid drop should take sometime for its formation and than it should fall.
4. The stalagmometer should be cleaned properly and dried before use.
5. Fit the stalagmometer vertically.
6. The drops should fall from the tip of the stalagmometer under their own weight rather than pushing them by force.

VIVA-VOCE

1. How do you define the term Surface Tension?

Ans. Surface Tension of a liquid is the force acting per centimeter along the surface of the liquid at right angle to any line on the surface of the liquid in any direction.

2. Give the CGS and SI units of Surface Tension.

Ans. The CGS units are dynes per cm and the SI units are Newton per meter. One SI unit is 1000 time of the CGS units.

3. What do you understand by surface energy of liquid?

Ans. Surface Energy of a liquid is the work required to extend the surface of the liquid by one unit area.

4. What are units of Surface energy?

Ans. Its unit are ergs/cm^2 in the CGS system and joules/meter^2 in SI units.

5. How Surface tension and Surface energy are related to each other?

Ans. The value of Surface Energy for any liquid surface is numerically equal to the value of Surface Tension of liquid in either set of units.

Thus if for a liquid, surface tension equals γ dynes cm^{-1} or $\gamma \times 10^3$ Nm^{-1}. Surface Energy will equals γ ergs cm^{-2} or $\gamma \times 1000$ jm^{-2}.

6. What method do you use in laboratory for determination of surfaces tension?

Ans. Surface Tension can be measured by using stalagmometer or a drop pipette method.

7. What is the shape of liquid drop falling from capillary?

Ans. Liquid drops are spherical in shape.

8. Why liquid drops are spherical in shape?

Ans. Due to surface tension, the surface of liquid tends to become minimum. The shape of sphere has minimum surface area.

9. Small insects are seen moving over surfaces of water tanks. Why?

Ans. The surface of a liquid behaves like a streched membrane due to the property of surface tension. The molecules on the surface are experiencing a new downward pull due to molecules present in the bulk of the liquid.

10. What instrument is used for measuring the density of a liquid?

Ans. Density of a liquid can be measured by using specific gravity bottle or a Pyknometer.

11. What formulas is used for calculating the surface tension of a liquid?

Ans.
$$\gamma_1 = \frac{n_w d_l}{n_l d_w} \times \gamma_w$$

where γ_w represents the surface tension of the water at same temperature n_w and n_l represents the no. of drops from the same volume of the water and liquid respectively. d_w and d_l represent the density of water and liquid.

12. What do you understand by the interpolation of graph?

Ans. The graph is plotted by taking two physical quantities on the two axis of the graph. Knowing the value of either of the two values that of the other can be found by interapolation. Knowing the numerical value of the property on Y-axis can be known by interapolation.

13. What do you understand by extrapolation of graph?

Ans. By extrapolation we mean extension of the curve obtained on the graph to either of two axis. The curve is extended towards the particular axis on which that physical quantity to be determined has been taken. The corresponding point at which the plot touches the axis gives the required value.

14. Why liquids rise in a capillary tube?

Ans. The rise of liquid is due to the inward pull of surfaces tension acting on the surface which pushes the liquid into the capillary tube.

15. Why mercury level falls and that of other liquids rise in capillary tube?

Ans. Liquids like mercury which do not wet the glass, their level inside fall below the level outside. Moreover the upper surface of liquids which wet the glass in concave and that of mercury is convex. Surface tension becomes operationl pulls up the liquids which wet the glass. The liquid rise in capillary and it is pulled up due to surface tension acting all along the circumference of the tube along with the wall of the tube.

16. What is the effect of the rise in temperature on the surface tension of the liquids?

Ans. The surface tension of the liquids decreases with the increase in temperature and becomes zero at critical temperature. Where the difference between the liquid and vapour disappears.

17. Why surface tension decreases with rise in temperature?

Ans. The decrease in surface tension of liquid with increase of temperature is due to the fact that with increase in temperature the kinetic energy of the molecules (and hence the speed of molecules) increase and therefore the intramolecular attractions decrease.

Experiment 7

OBJECT

To determine the coefficient of viscosity of the given unknown liquids by using Ostwald's Viscometer and identify the given liquid.

APPARATUS

Distilled water, stop watch, viscometer (Ostwald's), unknown liquid, pyknometer or R.D. bottle, pipette etc.

DESCRIPTION

Ostwald's viscometer is a U-shaped glass tube as shown in the Fig. 7.1. In one arm, the bulb A is connected with a fine capillary. The lower end of the capillary is connected with a U-tube provided with a bulb B in the second arm. The bulbs are necessary to maintain the hydrostatic pressure during flow of liquid. Through the capillary tube, the liquid flows with measurable speed. There are two marks C and D and the below the bulb A. The upper end of the bulb A is attached with a rubber tube. The liquid flows under its own weight.

The density of liquid is determined either by pyknometer or by R.D. bottle.

THEORY

The measurement of viscosity by viscometer is based on the Poiseulle's equation, which is

$$\eta = \frac{\pi P r^4 \cdot t}{8V \cdot l} \quad \text{...(1)}$$

where V – volume of liquid viscosity, η flowing in time t, through the capillary tube of radius r and length l.

P = Pressure (hydrostatic pressure) of liquid.

If t_1 and t_2 are times required to flow for equal volumes of two liquids through the same length of a capillary tube, then from equation (1), we have

$$\frac{\eta_1}{\eta_2} = \frac{P_1 \cdot t_1}{P_2 t_2} \quad \text{...(2)}$$

The pressure of liquid = $h \cdot d \cdot g$.

Since for two liquids h and g are same, hence

$$\frac{P_1}{P_2} = \frac{d_1}{d_2} \quad \text{...(3)}$$

Therefore $$\frac{\eta_1}{\eta_2} = \frac{d_1 \cdot t_1}{d_2 \cdot t_2} \qquad \ldots(4)$$

where d_1 and d_2 are the densities of two liquids

If η_1 = Viscosity of the liquid

η_2 = Viscosity of water

Then $\frac{\eta_1}{\eta_2}$ = Relative viscosity of liquid with respect to water

Relative viscosity has no unit.

PROCEDURE

Clean the viscometer with chromic acid and then wash thoroughly with distilled water. It is finally washed with alcohol or acetone and then dried. A sufficient volume of distilled water is introduced by pipette in bulb B so that the bend portion of tube and half or a little more than half of bulb B are filled up. Clamp the viscometer in quite vertical position. Through the rubber tube attached to upper arm of a bulb A, suck up water until it rises above the upper mark C and allow it to flow under its own weight. The time of flow of water from C to D is counted by starting the stop watch as the meniscus just reaches upper mark C and stopping the watch as the meniscus just passes the lower mark D. Take at least 3 or 4 readings with water and then take the mean of them.

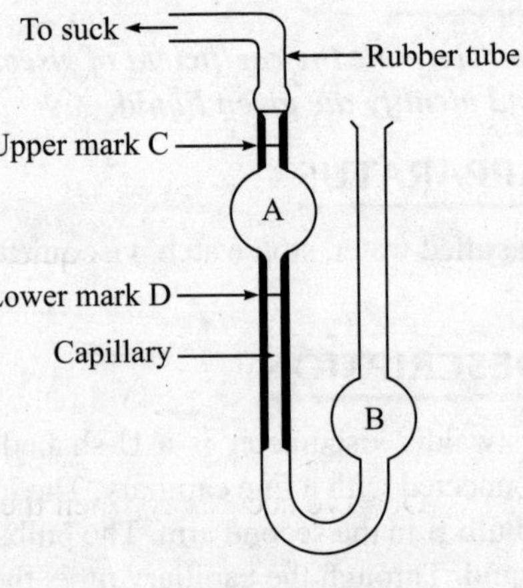

Fig. 7.1 Ostwald's viscometer.

Now remove the water from viscometer and dry it. Introduce in bulb B the same volume of liquid and measure the time of flow of liquid as before. Take at least 3 or 4 readings and then take the mean value.

Now, wash and dry the pyknometer (or R.D. bottle) and then weight it empty. Then fill it with distilled water and weigh. Remove water and dry. Then fill it with experimental liquid and weigh. Note the room temperature.

OBSERVATIONS

1. Room Temperature.......
2. Time of flow between C and D:

S.No.	Water		Liquid under experiment	
	Time of flow	*Mean (t_2)*	*Time of flow*	*Mean (t_1)*
1.	......			......
2.	......			......
3.	......			......
4.	......			......

3. Density of liquid (d_1) and of water (d_2)

Mass of empty of R.D. bottle (or pyknometer) = ... gm.

Mass of R.D. bottle (or pyknometer) + water = ... gm.

∴ Wt. of water = ... gm.

Mass of R.D. bottle (or pyknometer) + liquid = ... gm.

∴ Mass of liquid = ... gm.

CALCULATIONS

$$\frac{\text{Density of liquid } (d_1)}{\text{Density of water } (d_2)} = \frac{\text{wt. of liquid}}{\text{wt. of water}}$$

$$\text{Relative viscosity of liquid} = \frac{\eta_1}{\eta_2} = \frac{d_1 t_1}{d_2 t_2}$$

By inserting the respective values, the relative viscosity of liquid is calculated.

RESULT

The relative viscosity of this liquid with respect to water, at room temperature is ....

PRECAUTIONS

1. The viscometer should be thoroughly cleaned before use.
2. Viscometer must be strictly kept in vertical position.
3. Observe accurately when the meniscus of liquid and water are just passing the upper and lower marks.
4. Same volumes of water and liquid should be taken in viscometer.

VIVA-VOCE

1. Define viscosity.

Ans. It may be defined as the property of a liquid by virtue of which if offers resistance is flow.

2. What is meant by coefficient of viscosity.

Ans. It is defined as the force per unit area required to maintain velocity is difference of cm. per sec between two parallel layers one cm. apart.

3. What are the units of coefficient of viscosity?

Ans. It is expressed in units of dynes-sec-cm^{-2} or poise. Lower values of viscosity however, generally expressed in centipoise (= .01 poise) or millipose (= 0.001 poise).

4. What are the dimensions of viscosity coefficients?

Ans. It has dimensions of $ML^{-1}T^{-1}$.

5. What is fluidity?

Ans. The reciprocal of the coefficient of viscosity is called fluidity. Fluidity is a measure of the ease with which the liquid flows.

6. What is the effect of temperature on viscosity?

Ans. Viscosity of a liquid decreases rapidly with rise in temperature.

7. How does viscosity change when a solid melts or a liquid vaporises?

Ans. It decreases in both the cases.

8. What is the effect of hydrogen bonding on viscosity?

Ans. The presence of hydrogen bonds results in an increase in viscosity.

9. Glass is a super cooled liquid. What do you think its viscosity should be?

Ans. It should have high viscosity.

10. How is viscosity related to molecular weight?

Ans. For similar substances, it increases with increase in molecular weight.

11. How can you increase the viscosity of a solution?

Ans. By adding lyophilic colloids and substances as such sugar, gum etc.

12. Define density.

Ans. Density of a substance is defined as its mass per unit volume.

13. What is the effect of temperature on the density of a substance?

Ans. The density of a substance decreases with increase in temperature.

14. Define relative density.

Ans. It is defined as the ratio of the weight of a given volume of the substance to the right of the same volume of water at the same temperature.

15. Define specific gravity.

Ans. It is defined as the ratio of the weight of a given volume of the substance at a given temperature to the weight of the same volume of water at 4°C.

Or

It is defined as the ratio of the density of a substance to the density of water at the same temperature.

16. What are the units of specific gravity.

Ans. It has no units since it is a pure ratio.

17. What is the maximum value of density of water and at what temperature?

Ans. The maximum value of density of water is 1 at 4°C.

Experiment 8

OBJECT

To determine the coefficient of viscosity of the given lubricating oil using Redwood Viscometer.

THEORY

Viscosity is a measure of flowability at definite temperature. It is the resistance offered by a liquid towards. In other words, viscosity is reverse of fluidity. Glycerine or honey is more viscous than water. In other words, water has greater mobility as compared to glycerine.

Viscosity determines the performance of an oil under the operating conditions. A lubricating oil should have sufficient viscosity to enable it to stay in position. In case of machines moving at slow speeds, a heavy oil (high viscosity) should be used as lubricant and in case of machines, moving at high speeds, a light oil (low viscosity) should be used as lubricant.

The absolute viscosity or dynamic viscosity of a lubricating oil is determined by measuring the rate of flow of the oil through a capillary tube kept at a uniform temperature. But in case of lubricating oils, specific viscosity is generally determined by measuring the time taken for a given quantity of oil to flow through jet of standard size under standard conditions.

The viscosity is usually measured with Redwood, viscometer.

Redwood viscometers are available in two sizes No. 1 (Universal) and No. 2 (Admirality). The two viscometers, RW_1 and RW_2 are identical in principle, shape and mode of testing a sample. The difference is in the dimensions of the jet.

Viscometer	*Diameter*	*Length of the Jet*
RW_1	1.62 mm	10 mm
RW_2	3.8 mm	50 mm

The RW_1 will correctly indicate the viscosity for a liquid having flow time between 30 seconds to 2000 seconds. If flow time measured with this apparatus for any oil exceeds 2000 seconds the test should be repeated with RW_2, which will give the correct value of Viscosity.

The rate of discharge of the oil through RW_2 is nearly 10 times the discharge through RW_1. Therefore, RW_1 is commonly used for light or thin oils *e.g.*, Kerosene oil, mustard oil etc. and RW_2 is used for heavy or highly viscous oils *e.g.*, mobile oil, glycerol etc.

APPARATUS REQUIRED

(*i*) Red wood Viscometer (*ii*) Stop watch.

DESCRIPTION OF VISCOMETER

The Redwood Viscometer consists of the standard oil cup fitted with an agate jet in the base and opened at the upper end Fig. 8.1. RW_1 is commonly used for determining viscosities of thin lubricating oils. The jet can be opened or closed by a ball valve which permits the flow of oil. The cup is provided with a pointer to indicate the level of oil to be filled in the cup. The lid of the cup is provided with a thermometer to note the temperature of oil. The oil cup is placed in a cylinderical copper vessel containing water which serves as a water bath to

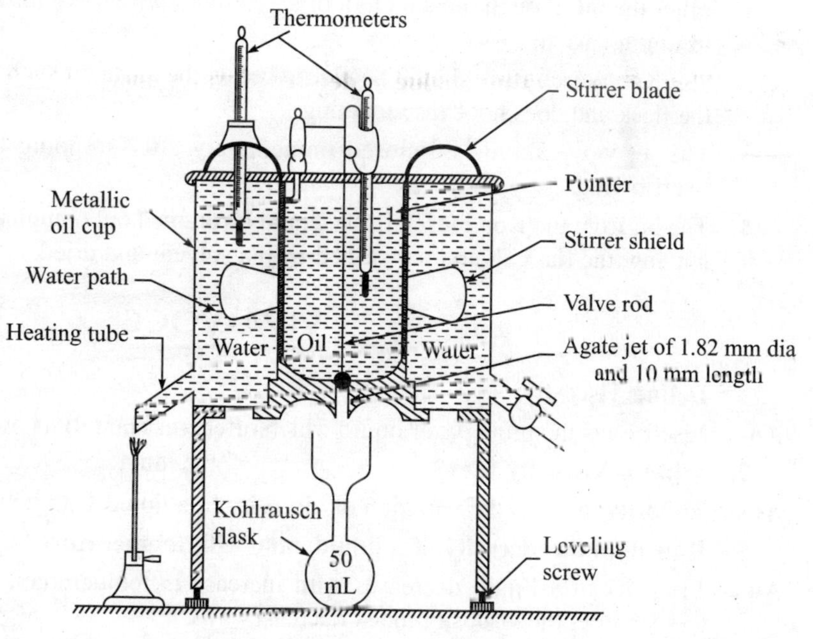

Fig. 8.1 Redwood Viscometer No. 1

maintain the desired oil temperature with the help of heating coils. This water bath is provided with a stirrer having four blades to maintain the uniform temperature and a thermometer to know the temperature of water. The water bath is provided with an outlet for removing water. At the base levelling screws are provided to level the apparatus and a 50 ml. Kohlrausch flask for receiving the oil from the jet outlet.

PROCEDURE

Level the viscometer with the help of levelling screws. Fill the water bath with water and connect to the electric mains. Clean the oil cup with a suitable solvent like CCl_4, benzene and properly dry it. Place the ball valve on the agate jet to close it and pour the oil under test carefully upto the pointer. Insert a thermometer and stirrer and cover the lid. Adjust the temperature of water both until the oil attains the desired constant temperature and keep stirring the water in water bath and the oil in the oil cup. Keep the Kohlrausch flask below the agate jet. When the oil attains a constant desired temperature, lift the ball valve and simultaneously start the stop watch, allow the oil to fill in the flask upto 50 ml mark. Stop the stop watch and note down the time in seconds. Repeat the same experiment 3–4 times to get the concordent reading. Report the viscosity of oil as redwood seconds.

OBSERVATIONS

S.No.	*Temperature*	*Viscosity (Redwood seconds)*
1.		
2.		
3.		

PRECAUTIONS

1. Filter the oil through muslin cloth or a 100 mesh wire sieve to remove the dust particles before testing for its viscosity.
2. The Kohlrausch flask should be placed below the agate jet such that the oil touches the neck of the flask and does not cause foaming.
3. The jet valve should be closed immediately after stopping the stop watch to prevent the overflowing.
4. The oil from the Kohlrausch flask should be drained out completely after each reading. For next reading, the flask should be washed with a solvent and dried.

VIVA-VOCE

1. Define Viscosity.

Ans. Viscosity is the property of liquid which offers resistant to its own flow. It is reverse of fluidity.

2. What is Viscosity Idex?

Ans. Viscosity Index is the variation of viscosity of a liquid with temperature.

3. How does the viscosity of a liquid vary with temperature?

Ans. Viscosity of a liquid decreases with increase in temperature. Generally for every 1° rise in temperature, the viscosity index decreases by 2%.

4. How is Viscosity index of a lubricating oil improved?

Ans. Viscosity Index of a lubricating oil is improved by adding linear polymers like polystyrenes, polymetharylates etc.

5. **What do you mean by the term 'Viscous-Static'?**

Ans. Viscous-Static is the state of a liquid when its viscosity does not change with rise in temperature.

6. **How is a viscous-static lubricant prepared?**

Ans. Viscous-Static lubricant is prepared by adding proper amount of a suitable linear polymer to increase the viscosity index.

7. **What is the Significance of viscosity?**

Ans. Viscosity is the only property of the lubricating oils which determines their performance under the operating conditions. A lubricating oil should have sufficient viscosity to enable it to stay in position.

8. **What do you understand by Viscosity Index (V.I.)?**

Ans. The rate at which the viscosity of an oil changes with temperature is measured by an arbitrary scale, known as the Viscosity Index (V.I.)

9. **What do you understand by low and high Viscosity Index of an oil? How are Viscosity and V.I. interrelated?**

Ans. If the viscosity of an oil falls rapidly as the temperature is raised, it has a low viscosity index on the other hand, if viscosity of an oil is only slightly affected on raising the temperature, its V.I. is high.

10. **What is meant by the term oiliness of a lubricating oil?**

Ans. It is a measure of capacity of the lubricating oil to stick to the surfaces of machine parts under conditions of heavy pressure or load.

11. **How can oiliness of a petroleum oil be increased?**

Ans. By adding vegetable oils and higher fatty acids.

Experiment 9

OBJECT

To determine the coefficient of viscosity of the given lubricating oil using Seybolt Viscometer.

THEORY

Viscosity is the property of a liquid or a fluid by virtue of which it offers resistance to its own flow. If two layers of a liquid separated by a distance '*d*' and moving with a relative velocity difference '*v*' then force per unit area (F) required to maintain this velocity difference is given by

$$F = \frac{\eta v}{d}$$

If $d = 1$ cm and $v = 1$ cm/s

Then $F = \eta$

where η is the coefficient of viscosity.

Hence coefficient of viscosity may be defined *as the force per unit area required to maintain a unit velocity gradient between two parallel layers.*

The unit of viscosity is poise and poise may be defined as: *"A force of 1 dyne is required to maintain a relative velocity difference of 1 cm/sec between two parallel layers 1 cm apart".*

Viscosity is the most important single property of any lubricating oil. If the viscosity of the oil is too low, a liquid oil film cannot be maintained between two moving surfaces and excessive wearing takes place.

Absolute viscosity *is the tangential force in dynes per square centimeter required to maintain a velocity difference of 1 cm/sec in two parallel layers of fluid 1 cm apart.*

Determination of viscosity: In case of heavy pressure and low speed machines, less viscous oils will be easily squeezed out, so thick viscous oils are used. Similarly in light machines less viscous oils are used. The apparatus which is used to determine the viscosity is known as viscometer. Viscosity is determined by following methods:

(*a*) Ostwalds Viscometer Method (*b*) Redwood Viscometer Method

(*c*) Saybolt Viscometer Method (*d*) Engler Viscometer Method.

In industry viscosity of lubricating oil is determined by Redwood. Saybolt and Engler viscometers. In the *Redwood viscometer* the measure of viscosity of an oil is the time in seconds for 50 ml of an oil to flow through standard orific under a given set of conditions. This viscometer is mostly used in Common Wealth Countries. In case of *Saybolt viscometer* the viscosity is expressed as the number of seconds required for 60 ml of oil to pass through a standard orifice under a given set of conditions. This instrument is used mostly in USA. In the case of *Engler viscometer* the viscosity is expressed as the ratio of the ratio of the time for oil to the time taken by water at 20°C. This viscometer is mostly used in Europe.

APPARATUS REQUIRED

Saybolt Viscometer and Stop watch.

DESCRIPTION OF THE SAYBOLT VISCOMETER

A single-unit Saybolt Universal Viscometer is shown in Fig. 9.1.

In the multiple-unit viscometer a number of oil-cups can be accommodated in the same bath, thus enabling tests on a number of oils to proceed at the same time. Instruments can be fitted with an electric immersion heater–a U-tube for steam heating or water cooling, and a gas ring which is placed inside the air-jacket surrounding the water-bath. The bath liquid is stirred by rotating the cover by means of the two handles as a turn-table arrangement. Its temperature can be regulated by running cold or warm water through the U-tube, whatever may be the heating arrangement used. The jet is made of a hard non-corrodible metal such as Monel or stainless steel. The lower end of the jet opens into a larger tube. This tube, when stoppered by a cork, becomes a closed air chamber preventing the oil flowing out.

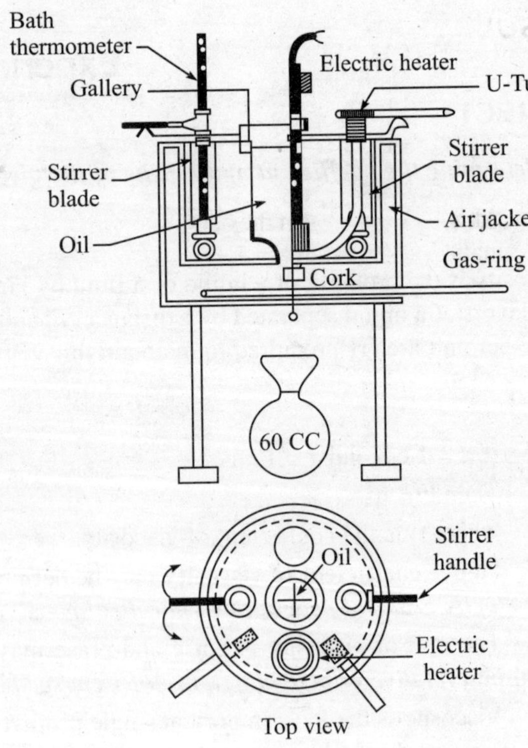

Fig. 9.1 Saybolt Viscometer.

PROCEDURE

(*i*) To start with the experiment, the bath is brought to the desired temperature and the oil is heated to the same temperature in a separate vessel.

(ii) The oil is then poured into the oil cylinder and stirred with the oil thermometer, any excess oil flowing over into the surrounding gallery.

(iii) When the oil and the bath are at the same temperature, the oil thermometer is removed, he excess oil drawn off from the gallery with a pipette.

(iv) The cork is withdrawn and the stop-watch started.

(v) The collecting flask is so arranged that the oil stream will strike its neck and thus avoiding the formation of foam.

(vi) The time of outflow of 60 ml of the oil is the viscosity in second Saybolt Universal at the test temperature.

For very viscous fuels, a viscometer with a larger jet known as the Saybolt Furol Viscometer is used. The Saybolt Universal viscometer can be used for oils having flow times of more than 32 seconds. There is no maximum limit, but in general, for liquids having flow times over 1000 seconds. Saybolt Furol Viscometer is better.

OBSERVATIONS

Record the observations in the following table

S.No.	*Temperature*	*Viscosity in seconds*
1.		
2.		
3.		

RESULT

The mean value of the above readings gives the viscosity expressed in seconds.

VICE-VOCE

Already discussed at the end of Experiment 7 and 8. All these three experiment deal with viscosity determination with different viscometers.

Experiment 10

OBJECT

To determine the flash point and fire point of given sample of oil using Pens key Marten's apparatus.

THEORY

Flash point is the minimum temperature at which an oil gives enough vapours which burn for a moment when a small flame is brought near it.

Fire point is the minimum temperature at which oil given enough vapour which burns for at least 5 seconds when a small flame in brought near to it. Generally, fire point is 5 to 40°C higher than the Flash point.

Significance of Flash and Fire Point

(*i*) Flash and fire point tells us the maximum temperature upto which a lubricant can be used. A good lubricating oil should flash and fire point above its working temperature. This ensure safety against fire hazard during storage, transport and use of the lubricating oil. A good lubricating oil should not volatalise under the working temperature. If some volatialisation takes place the vapours formed should not form inflammable mixture with air.

(*ii*) Flash and fire points are also used to detect the solvent contamination and to determine the approximate extent of dilution of the lubricating oils.

Flash point and fire point of an oil is determined by either. Abel's apparatus of Pensky-Marten's apparatus. For flash point less than 120ºF Abels's closed cap apparatus is used, while flash point more than 120°F Pensky Marten's closed cup apparatus is used.

DESCRIPTION OF THE APPARATUS

The apparatus consists of a brass cylinderical cup with a lid having four openings of standard dimensions for a stirrer, thermometer, an air inlet and a device *i.e.*, shutter for introducing the standard flame. The shutter has a lever mechanism when it is opened, a test flame comes in contact with the oil vapour to check the flash and fire point

The difference in the two apparatus is that in Pensky-Marten's apparatus, the oil cup is surrounded by air bath and in case of Abel's apparatus the oil cup is first surrounded by copper air jacket and then placed in water bath

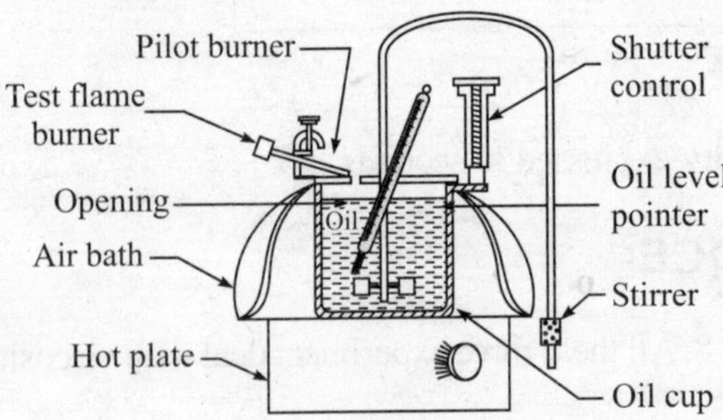

Pensky-Marten's flash point Apparatus.

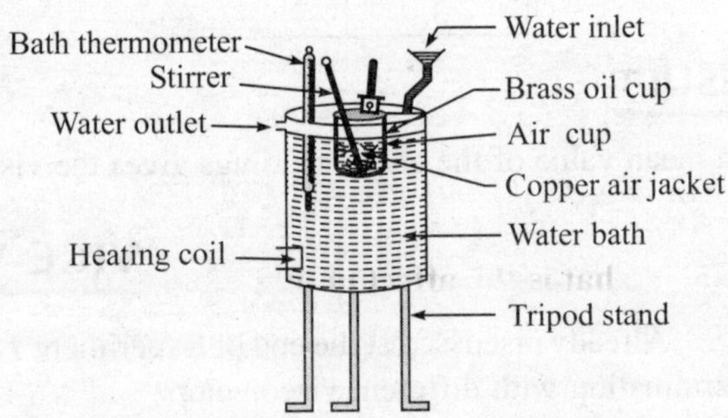

Abel's flash point apparatus.

Fig. 10.1

REQUIREMENT

Flash point apparatus.

PROCEDURE

Thoroughly clean and dry all the parts of the apparatus. Fill the oil cup with given lubricating oil sample upto the mark. Cover it with lid. The apparatus is heated so that the oil temperature increases by about 5 to 6°C per minute. Keep stirring the oil in the oil cup. When the temperature of the oil reaches within 15°C of the probable, the test flame is dipped into the oil vapours to check the flash point. The test flame is applied at every 1°C, rise in temperature. When test flame produces a distinct flash in the oil cup, the temperature is noted and recorded as flash point (say T_1°C). The heating is continued after every 1°C rise in temperature the oil vapours are tested for fire point. The temperature at which oil vapours given a clear and distinct flame atleast for 5 seconds, is noted and recorded as fire point (say T_2°C).

PRECAUTIONS

1. Clean and dry the apparatus thoroughly before use as moisture affects the flash and fire point.
2. The bulb of thermometer should dip inside the oil sample.
3. The shutter should be open slowly for applying test flame and closed quickly.
4. Stirring of the oil should be stopped at the time of application of test flame.
5. For each reading, fresh oil sample should be filled in the oil cup.
6. Oil should be filled carefully in the oil cup upto the pointer.

VIVA-VOCE

1. Define flash point of an oil.

Ans. Flash point is the minimum temperature at which an oil give enough vapours which burn for a moment when a small flame is brought near it.

2. Define Fire point of an oil.

Ans. Fire point is the minimum temperature at which an oil gives enough vapours which burn for at least five seconds when a flame is brought near it.

3. What is the main significance of determining the flash and fire point of the lubricating oil sample?

Ans. Flash and fire point tells us the maximum temperature upto which a lubricating oil can be used. This ensures safety against fire hazard during its use.

4. What are the factors affecting the flash and fire point?

Ans. The main factors affecting the flash and fire point are:

(*i*) Presence of moisture.

(*ii*) Vapour pressure of the oil—Higher vapour pressure means lower flash and fire points.

5. What is the affect of moisture on the flash and fire point?

Ans. Presence of moisture in the lubricant may increase or decrease the flash point. It is increased when steam prevents vapour from igniting and decreased when steam distillation of low molecular weight constituents present in oil takes place.

6. Give the main difference between Abel's apparatus and Pensky-Marten's apparatus.

Ans. In the Abel's apparatus, the oil cup is surrounded by an air jacket and then heated by placing in a water bath whereas in Pensky Marten's apparatus the oil cup is surrounded by air jacket and heated directly. In this apparatus the rate of heating. Therefore it is most suitable and commonly used for determing the flash points of oils above 120°F. Whereas Abel's apparatus is used for oils flashing upto 120°F.

7. Give the main difference between open cup and closed cup flash point apparatus.

Ans. In open cup flash point apparatus, the flash point is determined by heating the oil in a cup without cover. In this case, there is loss of vapour of oil into the atomosphere during heating. So, the results are not reproducible and are higher.

In closed cup flash point apparatus the flash point is determined by heating the oil in a closed cup and test flame is introduced into the cup, therefore, Ignition of vapour takes place inside the cup. Therefore, accurate and reproducible results are obtained.

8. How water is removed from the oil before determining the flash point?

Ans. Water is removed by:

(*i*) Filteration through suitable filter paper.

(*ii*) Absorption by anhydrous $CaCl_2$.

(*iii*) Centrifugal action.

(*iv*) Settlement and decantation.

9. What is meant by freaky flash?

Ans. Freaky flash is the production of irregular flashes below the true flash point of an oil due to presence of contamination of oil with small amount of volatile organic substances.

Experiment 11

OBJECT

To determine the amount of Chlorine in given sample of water approximate N/20 sodium Thiosulphate solution.

THEORY

For disinfection of potable water and municipal water supplies chlorination is done with the help of bleaching powder, chlorine gas or with chloramines as under:

$$CaOCl_2 + H_2O \longrightarrow Ca(OH)_2 + Cl_5$$

$$Cl_2 + H_2O \longrightarrow HOCl + HCl$$

$$\underset{\text{(Hypochlorous acid)}}{HOCl} \longrightarrow \underset{\text{(Nascent Oxygen)}}{HCl + [O]}$$

$$HOCl + \text{Germs} \longrightarrow \text{Germs are killed by oxidation}$$

Hypochlorous acid and nascent oxygen both have powerful germicidal properties. However, excess of free chlorine in drinking water produces unpleasant taste and effect the digestive system. Hence, the amount of free chlorine in municipal water is estimated before the domestic supply so as to make necessary adjustment in dose rate accordingly.

The amount of free chlorine in water is determined Iodometrically by treating a known amount of water with excess of KI solution and then titrating the liberated iodine against stardard Hypo solution, using freshly prepared starch as indicator.

$$Cl_2 + 2KI \longrightarrow 2KCl + I_2$$

$$I_2 + \underset{\text{(Sodium thiosulphate)}}{Na_2S_2O_3} \longrightarrow \underset{\text{Sodium tetra thionate}}{Na_2S_4O_6 + 2NaI}$$

The ionic equation can be written as:

$$Cl_2 + 2I^- \longrightarrow Cl^- + I_2$$

$$I_2 + S_2O_3^{2-} \longrightarrow S_4O_6^{2-} + 2I^-$$

$$I_2 + \text{Starch solution} \longrightarrow \text{Blue coloured complex}$$

CHEMICALS REQUIRED

(*i*) Standard $K_2Cr_2O_7$ solution (N/20)
(*ii*) Hypo solution = N/10
(*iii*) $NaHCO_3$ solution
(*iv*) 10% KI
(*v*) Conc. HCl.

INDICATOR

Freshly prepared starch solution

END POINT

Disappearance of blue colour (Hypo solution in the burette)

PROCEDURE

(*i*) **Standardisation of Hypo Solution:** Pipette out 20 ml of standard $K_2Cr_2O_7$ solution into a conical flask. Add half test tube dil. H_2CO_4 and 10 ml of 10% KI solution. Stopper the flask and shake it vigrously. Keep it undisturbed for about 2 minutes. Rinse the stopper with 5-10 ml. of distilled water into the flask. Titrate the solution with Hypo solution taken in the burette until the solution is pale yellow. Then add 2 ml. (10-12) drops of freshly prepared starch, the solution would become bluish in colour. Continue titration until the solution becomes light green at the end point. Repeat the same titration 3-4 times so as to get the concordent reading.

(*ii*) **Determination of Free Chlorine in the Water Sample:** Pipet out 20 ml. of water sample into a conical flask. Add 10 ml. of 10% KI solution. Shake it vigrously by placing stopper on it. Rinse the stopper with 5-10 ml of distilled water into the flask. Titrate the solution with stardardised Hypo solution taken in the burette until the solution is pale yellow. Then, add 2 ml. (10-12 drops) of freshly prepared starch indicator and continue the titration until the solution becomes just colourless at the end point. Repeat the same titration 3-4 times so as to get the concordent reading.

CALCULATIONS

Let the first titre value be A ml. and second titre value be B ml. against 20 ml of the solution taken in the titration flask.

I. Standardisation of Hypo Solution:

$$\underset{\text{Hypo}}{N_1V_1} = \underset{K_2Cr_2O_7}{N_2V_2}$$

$$N_1 \times A = \frac{1}{20} \times 20 \quad \text{or} \quad N_1 \text{ (Normality w.r.t. Hypo)} = \frac{1}{A} N$$

II. For Normality of free chlorine present in the water sample

$$\underset{\text{(water sample)}}{N_3V_3} = \underset{\text{Hypo}}{N_4V_4}$$

$$N_3 \times 20 = \frac{1}{A} \times B \quad \text{or} \quad N_3 = \frac{B}{A \times 20} \text{ N}$$

III. Strength of free chlorine present in the water sample

$$= N_3 \times (\text{Eq. wt. of } Cl_2) = \frac{B}{A \times 20} \times 35.5 \text{ g/L}$$

$$= \frac{B \times 35.5 \times 1000}{A \times 20} \text{ mg/L or ppm.}$$

VIVA-VOCE

1. **Give the principle of the method employed for employed for determined residual chlorine in water.**

Ans. Chlorine being an oxidising agent liberates iodine from KI solution. The liberated iodine can be titrated against hypo.

$$KI + Cl_2 \longrightarrow 2KCl + I_2$$

$$\underset{\text{Hypo}}{2Na_2S_2O_3} \longrightarrow \underset{\text{Sodium Tetrathionate}}{Na_2S_4O_6} + 2NaI$$

2. **What is Iodometry?**

Ans. The process of titration of the liberated iodine against hypo ($Na_2S_2O_3$) by an oxidising agent from KI solution is called iodometry.

3. **What is iodimetric titration?**

Ans. The direct titration of iodine solution against hypo solution using starch as indicator is called iodimetric titration.

4. **Give the name of oxidising and reducing agents in iodine titration.**

Ans. Iodine acts as an oxidising agent, whereas Hypo acts as a reducing agent.

$$2Na_2S_2O_3 + I_2 \longrightarrow Na_2S_4O_6 + 2NaI$$

5. **Which indicator is used in Iodometric titrations? Explain in the colour change at the end point.**

Ans. Starch is used as an indicator and disappearance of blue colour is the end point. The liberated I_2 form a surface complex with starch which is blue in colour. The iodine from the complex reacts with hypo and the original colour of starch (colourless) appears.

6. **Why freshly prepared starch solution is used as indicator?**

Ans. Because gelatinous solution is appeared if starch is kept for longer time so for practical purpose, it is to be freshly prepared.

7. **What is meant by stabilizing the chlorine?**

Ans. Free chlorine is readily dissipated in the distribution system and does not give complete protection against bacteria due to short contact period. In order to prolong bacterial action, free chlorine residuals are converted into more stable chloramines by adding a little NH_3 to chlorinated water. This process is known as stabilizing the chlorine.

8. **Give the mechanism of chlorination of water.**

Ans. Cl_2 reacts with water to produce Hypo-chlorous acid which reacts with germs and the germs are killed.

$$Cl_2 + H_2O \longrightarrow HOCl + HCl \qquad HOCl + \text{Bacterias} \longrightarrow \text{Bacterias are killed}$$

9. What is the effect of overdose of chlorine added to the water for chlorination?

Ans. Excess of chlorine produces an unpleasant taste and odour. Moreover, its excess produces an irritation on mucus membrane.

10. What are the advantages of the process of chlorination?

Ans. (*i*) It is effective and economical

(*ii*) It required very small space for storage.

(*iii*) It is stable and does not decompose on storage.

(*iv*) It can be used at low as well as high temperatures.

(*v*) It introduces no impurities in water.

11. Name three substances used for sterilization of water?

Ans. (*i*) Bleaching powder (*ii*) Liquid chlorine and

(*iii*) Chloramine

12. Chloramines is preferred to bleaching power or chlorine for sterilization of drinking water. Give reasons.

Ans. Chloramine is preferred over bleaching powder or chlorine because it.

(*i*) causes no irritation odour even if used in excess and

(*ii*) imparts good taste to drinking water

13. Why is boiled water not 100% safe for drinking purposes?

Ans. Boiling kills only the existing germs in water, but does not provide any protection against further possible contamination during storage etc.

14. What is the mechanism of disinfecting action of chlorination?

Ans. It was believed earlier that the disinfecting action of chlorination is due to the nascent oxygen liberated, which oxidises harmful bacterias etc. However, it is now believed that the death of micro-organisms results from chemical reaction of hypochlorous acid with the enzymes in the cells of the organisms. Since, enzymes are essential for the metabolic processes of the micro-organisms, so their death results due to inactivation of enzymes by hypochlorous acid.

15. Why chlorine is more effective disinfactent at pH value lower than 6.5?

Ans. This is due to the fact the HOCl is about 80 times more destructive to bacterias than OCl^-ion. At higher pH values than 6.5, HOCl gives OCl^- ion which becomes less effective.

$$HOCl \longrightarrow H^+ + OCl.$$

Experiment 12

OBJECT

(a) To determine the maximum wavelength of solution of cobalt chloride.

(b) To determine the Beer's law and apply it to find the concentration of the given unknown solution by spectrophotometer.

THEORY

When a beam of monochromatic light falls on a homogeneous medium, a portion of the incident light is reflected, a portion is absorbed and the remainder is transmitted.

$$I_0 = I_a + I_r + I_t$$

It case of aqueous solution, I_r is negligible as compared to I_a and I_r.

$$\therefore \quad I_0 = I_a + I_t$$

According to Beers Lambert's law, *the rate of decrease in intensity of a beam of monochromatic radiation with thickness of the solution is proportional to the intensity of incident radiation as well as concentration of the solution.*

Mathematically,

$$-\frac{dI}{dx} \propto CI \quad \text{or} \quad \frac{dI}{I} = -\in C\,dx$$

It can also be stated as:

$$I_t = I_0\, 10^{-\in CL}$$

Where C is the concentration of solute expressed in mole/liter, L is the length of the cell and is the constant characteristic of the solute called molar extinction coefficient or molar absorptivity. Taking logarithm of both sides.

$$\log I_t = \log I_0 - e^{\in CL} \quad \text{or} \quad \log \frac{I_0}{I_t} = e^{\in CL} = A$$

Where A is called absorbance or optical density (O.D.)

Transmittance 'T' of a solution is the ratio of $\frac{I_t}{I_{ot}}$ *i.e.*, the fraction of incident light transmitted by solution.

$$T = \frac{I_t}{I_0} \quad \text{or} \quad -\log T = \log \frac{I_0}{I_t} = A \quad \text{or} \quad A = -\log T = e^{\in CL}$$

The study of variation in intensity of a given coloured solution with the change in concentration of the given coloured component is termed as colorimetric analysis.

Spectrophotometer: A spectrophotometer is a device which detects the percentage transmittance of light radiation when light of certain intensity and frequency range is passed through the sample. This instrument compares the intensity of the transmitted light with that of the incident light. An ultra-violet visible spectrophotometer consist of light source (LS), monochromater (M), detector (D), amplifier (A) and the recording device (R).

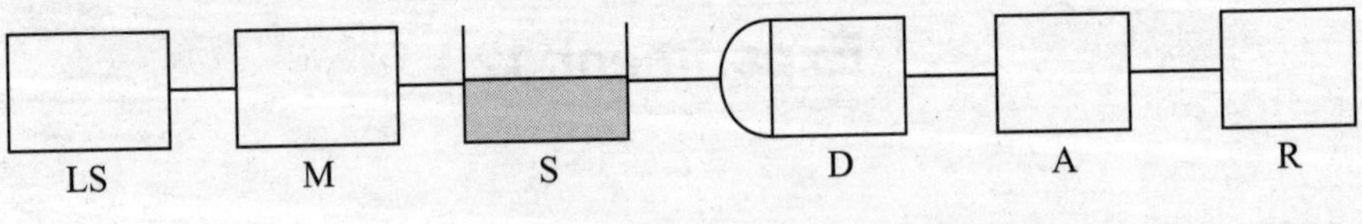

Sample block diagram of photospectorphotometer

REQUIREMENTS

Spectrophotometer, Beaker, Distilled water, Standard Cobalt Chloride ($COCl_2 \cdot 5H_2O$) solution, Tissue paper.

PROCEDURE

Step I. Initial setting of the spectrophotometer

(*i*) Ensure that the meter initially reads zero on transmittance scale (T) before starting the experiment.

(*ii*) Connect the instrument to the power supply. Set the switch to the ON position and ensure that the red light indicator glows.

(*iii*) Adjust the wavelength knob to the required wavelength region on scale. Choose the position of gain selector switch corresponding either to 340-400 mm or 400-960 nm wavelength.

(*iv*) Adjust the "set zero" knob so that meter needle reads zero on T scale and 100 on O.D. scale.

Step II. Final setting of spectrophotometer

(*i*) Open the lid of the sample housing and insert a cuvette containing blank solution (distilled water). Close the lid so that it fits properly.

(*ii*) Adjust the control knob (set 100) in appropriate direction to bring the needle to 100 transmittance or zero optical density (O.D.).

(*iii*) Open the lid and remove the cuvette. Close the lid tightly again.

(*iv*) Check zero on the meter. Adjust zero if disturbed.

(*v*) Repeat again for zero and 100% transmittance

Step III. Determination of λ_{max}.

(*i*) Insert the cuvette containing standard solution in the similar fashion as explained in (*i*) above.

(*ii*) Note the reading of optical density at zero-wave length.

(*iii*) Change the wavelength by 20 nm every time and note down corresponding optical densities.

(*iv*) Plot a graph between wavelength on the *x*-axis and O.D. on the *y*-axis.

(*v*) Determine the value of λ_{max} by extra-ploation of the curve towards *x*-axis.

(*vi*) The maxima in the curve (*i.e.*, maximum value of optical density, gives the value of λ_{max})

Step IV. Verification of Beer's law

(*i*) Fix the wavelength of λ_{max} positions.

(*ii*) Prepare the standard solution of KM_nO_4 or $CoCl_2$ in distilled water. Now prepare the solution of different concentration from stock solution by mixing 2, 4, 6, 8, 10 ml of standard KM_nO_4 and 8, 6, 4, 2, 0 ml. of distilled water to get 20, 40, 60, 80 and 100 percent concentration of solution.

(*iii*) Note down optical density of series of solutions of cobalt chloride prepared above.

(*iv*) Plot the O.D. against concentration. It should be a straight line.

(*v*) Now take a solution of the unknown concentration and note down optical density. Find out the concentration of the unknowns solution from the graph corresponding to the O.D. of the solution.

OBSERVATION AND CALCULATIONS

Record the observation in the following table:

S.No.	*Concentration*	*Absorbance (or optical density)*
1.	20%	....
2.	40%	....
3.	60%	....
4.	80%	....
5.	100%	....
6.	unknown (let it be C_1)	*y*

We know $A = \log \frac{I_0}{I} = \in cx.$

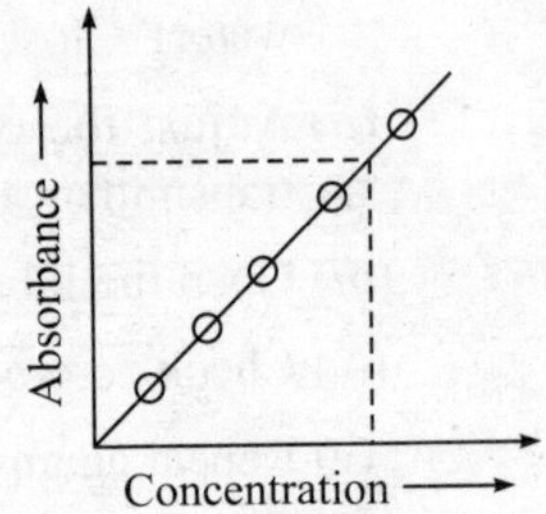

It is clear that absorbance A is directly proportional to concentration of the solution.

Plot the graph between concentration and absorbance *i.e.*, optical density by taking concentration on X-axis absorbance on Y-axis.

A linear plot as shown in the figure verify the Lamberta Beers law.

From the graph the concentration of unknown solution can be determined. Corresponding to the observed absorbance (*y*) the Y-axis, the unknown concentration (cl) on the X-axis can be noted from the graph.

PRECAUTIONS

(*i*) For preparing the standard calibration curve, dilute solution of known concentration should be used.

(*ii*) As the cuvette is very fragile, it should to used with great care.

(*iii*) λ_{max} value should be carefully observed.

(*iv*) Cuvette should be properly cleaned before use and should be wiped with tissue paper gently.

Experiment 13

OBJECT

To determine the Beer's Law and apply it to find the concentration of given unknown solution by spectraphotometer.

Note: The experiment has already been discussed in the previous Experiment 12.

VIVA-VOCE

1. Define spectroscopy.

Ans. Spectroscopy is defined as the interaction between the matter and electromagnetic radiations.

2. What are electromagnetic radiations?

Ans. Different forms of energy which are propagated in space in the form of waves and are associated with electric and magnetic fields perpendicular to each other and to the direction of propagation.

3. What are the characteristics of electromagnetic radiations?

Ans. All forms of electromagnetic radiations travel with same velocity (3.0×10^{10} cm/sec) and require no medium for their transmission. All the radiations are characterised in terms of their wave-lengh, frequency, velocity, wave number and amplitude.

4. How an electromagnetic radiation is produced?

Ans. An electromagnetic radiation is produced by the oscillation of a magnet in an electric field or by oscillation of a charged body in a magnetic field.

5. Name some important electromagnetic radiations.

Ans. Light, Ultra-violet, Infrared, Radio waves, Television waves, Micro waves, X-rays and cosmic rays etc.

6. What is electromagnetic spectrum?

Ans. The arrangement of different types of electromagnetic radiations in the order of increasing wave length (or decreasing frequency) is known as electromagnetic spectrum as shown below:

Increasing wave length ⟶

Cosmic rays	γ-rays	x-rays	u.v.	Visible	I.R.	Micro waves	Radar waves	Television waves	Radio waves

⟵ Increasing frequency

7. Define wavelength and frequency of a wave. How are they related?

Ans. The shortest distance between any two successive crests or troughs or any two similar points on the wave is called wavelength. The number of waves that pass through a fixed point per second is called frequency of the wave. Frequency (ν) = $\frac{C}{\lambda}$, where C is the velocity and λ is the wave-length. As the wavelength of a radiation decrease, its frequency increases and energy also increases.

8. How is the energy of a wave related to frequency or wavelength?

Ans. $E = h\nu = \frac{hc}{\lambda}$, where E is the energy, ν is the frequency, λ is the wavelength and c is the velocity of the radiation h is plank's constant (6.626×10^{-34} Joule second)

9. What is visible region of the electromagnetic radiations?

Ans. Visible light is only a small portion of electromagnetic radiations, which we can see. Wavelength of visible region is approx. 3800Å to 7600Å. It ranges from violet (λ = 380 nm) to red (λ = 760 nm).

10. What are ultra-violet and infra-red radiations?

Ans. Radiations with frequencies higher and wavelength lower than those of violet radiation are known as ultra-violet radiation. Radiation with frequencies lower than red radiations of the visible region are called infra-red radiations.

11. What is a spectrophotometer?

Ans. It is an instrument which compares the intensity of the transmitted light with that of the incident light. It detects the percentage transmittance of light when light of certain intensity and frequency range is passed through the sample solution.

12. State Beer's Law.

Ans. The rate of decrease of intensity of a beam of monochromatic radiation with thickness of the solution is proportional to the intensity of incidence radiation as well as concentration of the solution.

13. Which region of electromagnetic spectrum is taken into consideration during electronic transitions?

Ans. UV-visible range of electromagnetic spectrum.

Experiment 14

OBJECT

To determine the chemical oxygen demand of waste water.

THEORY

COD is a measure of any kind of oxidisable impurities present in the sewage. COD is a measure of both the biologically oxidisable and biologically inert organic matter present in the sewage sample. It is an important and quickly measured parameter for steam and industrial waste water analysis and water treatment plants.

Most types of orgainc matter are completely oxidized by a boiling mixture of chromic acid and sulphuric acid to produce CO_2 and H_2O. A measured quantity of the sample is refluxed with a known excess of standard potassium dichromate solution and dil. H_2SO_4 in the presence of catalyst. The excess dichromate remaining unreacted is titrated with ferrous Ammonium Sulphate solution using Ferroin as a indicator. The $K_2Cr_2O_7$ consumed is proportional to the amount of oxidizable impurities present in the given water sample.

APPARATUS/REAGENTS REQUIRED

Reflux apparatus consisting of a 500 ml round bottomed flask with ground glass joint and a condenser. 0.25 N potassium dichromate solution.

Sulfuric acid-silver sulphate reagent.

Standard 0.25 N Ferrous Ammonium Sulphate (FAS)

Ferroin Indicator

Mercuric Sulphate ($HgSO_4$) A.R. Grade.

PROCEDURE

To 50 ml of the sample taken in a 500 ml reflux flask, add 1 g $HgSO_4$ and some broken porcelain pieces or glass beads. Immerse the flask in cold water and slowly add 75 ml of $Ag_2SO_2 - HgSO_4$ reagent while continuously shaking the flask. Add 25 ml of 0.25 N $K_2Cr_2O_7$ Solution mix the contents of the flask thoroughly. Attach the reflux condenser and reflux the flask for 2 hours. Wash the condensor with distilled water into the flask. Cool and dilute to about 300 ml with distilled water. Add 2 to 3 drops of ferroin indicator and titrate against of 0.25 N FAS. The colour changes at the end point is from blue green to wine red. A blank titration is done by taking a volume of distilled water, equal to that of the sample in another round bottom flask. Add the same amounts of reagents, reflux for 2 hours titrate in the same way as in the sample titration. Record volume of FAS used for titration as V_2 ml. COD of the sample corresponds to $(V_2 - V_1)$ ml of standard 0.25 N FAS.

OBSERVATIONS AND CALCULATIONS

Volume of Sample taken = 50 ml

Volume of 0.25 N FAS used in the sample titration = V_1 ml

Volume of 0.25 N FAS used in blank titration = V_2 ml

Therefore, volume of 0.25 N FAS equivalent to $K_2Cr_2O_7$ used for COD determination

COD Determination = $(V_2 - V_1)$ ml

$$N_1V_1 = N_2V_2$$

Sample FAS

$$N_1 \times 50 = 025 \times (V_2 - V_1)$$

$$\text{or } N_1 = \frac{(V_2 - V_1) \times 0.1}{50}$$

Therefore,

$$\text{COD} = \frac{(V_2 - V_1)}{50} \times 0.25 \times 8 \text{ g/lit}$$

$$= \frac{(V_2 - V_1) \times 0.25 \times 8 \times 1000}{50} \text{ mg/lit}$$

$$= (V_2 - V_1) \times 40 \text{ mg/lit.}$$

PRECAUTIONS

1. Adding of sulfuric acid reagent to reflux flask should be done slowly and by shaking the flask. As it is an exothermic reaction the flask should be cooled during mixing.
2. Smaller volume of sample should be taken in the flask, if it has high COD.
3. End point of the titration should be carefully observed.

VIVA-VOCE

1. What are water pollutants?

Ans. Sewage and other oxygen demanding waste are known as water pollutants. They are so called due to following reasons.

(*a*) They promote the growth of pathogenic bacteria.

(*b*) They decay causes a decrease in dissolved oxygen which has a damaging affect on aquatic life.

(*c*) They give bad taste, foul smell and colour to water.

2. What is B.O.D.?

Ans. Biological Oxygen Demand (B.O.D.) is defined as the amount of dissolved oxygen required by bacteria while oxidizing organic matter under aerobic conditions. It measures the oxygen consumed by living organisms while assimilating organic matter present in the waste water.

3. Differentiate between B.O.D. and C.O.D. of waste water?

Ans. B.O.D. measures the oxygen consumed by living organisms while assimilating organic matter present in the waste water. Whereas, C.O.D. is a measure of biologically oxidisable and biologically inert organic matter such as cellulose.

C.O.D. values are higher than B.O.D. values, further C.O.D. determination is over in 2-3 hours compared to 5 days required for the B.O.D. determination.

4. What are the major sources of error in the above method of determining C.O.D.?

Ans. (*i*) Cl^- ions, usually present in sewage in appreciable amounts interfere as they also get oxidised under experimental conditions.

(*ii*) Certain compounds like pyridine, aromatic hydrocarbons, low molecular weight fatty acids and straight chain aliphatic hydrocarbons resist oxidation.

5. How are above errors minimised?

Ans. The above mentioned errors can be minimised as follows:

(*a*) Interference due to Cl^- is removed by using $HgSO_4$,

$$Hg^{+2} + 4Cl^- \longrightarrow \underset{\text{(very loss dissociated)}}{[HgCl_4]^{2-}}$$

(*b*) Oxidation of low molecular weight fatty acids and straight chain aliphatic hydrocarbons is made effective by using Ag_2SO_4 as catalyst. Pyridine and aromatic hydrocarbons are still not oxidised.

6. What is sewage?

Ans. The liquid waste, which includes human and house hold waters, industrial wastes, ground wastes, street washings and storm waters is called sewage.

7. What is stable sewage?

Ans. When aerobic decomposition of sewage is continuing, the sewage is called stale. It gives offensive smell.

8. What is a septic sewage?

Ans. When the anaerobic decomposition of sewage is continuing, the sewage is called septic.

9. Is COD the key test for determining the pollution load in water samples? Justify your answer.

Ans. Yes! COD is the key test for determining the pollution load in water samples because it is a measure of all oxidisable impurities including cellulose etc. in the sewage water.

10. Distinguish between primary and secondary sewage treatment.

Ans. Primary treatment removes large and small suspended particles by filtration and sedimentation, respectively. Secondary treatment decomposes orgainc matter by the action os aerobic bacteria and sedimentation.

11. Correlate OD and BOD.

Ans. BOD measure the oxygen consumed by living organisms while assimilating organic matter present in the waste water. While COD is a measure of biologically oxidisable and biologically inert (chemically oxidisable) matter such as cellulose.

12. Why COD values are higher than BOD values?

Ans. In COD, the oxygen is required for oxidation of biologically oxidisable organic matter as well as chemically oxidisable substances. But in case of BOD, oxygen required for only biologically oxidisable substances is taken into consideration.

Experiment 15

OBJECT

To determine the half-life period of given radioactive sample using GM counter.

Note: The experiment has not been dealt with in this book because the equipment to be used (GM Counter) is not available in most of chemistry laboratories meant for B.Tech standard. Moreover, as per instructions of the university at least 10–12 experiments are to be performed out of the 15 experiments given.

❑❑❑

PART III

SOLVED QUESTION PAPER (December-2012)

(H.P. Tech. University)

and

MODEL QUESTION PAPERS (I–V)

Alongwith Solution of the Compulsory Question No. 9 (Section E)

SOLVED QUESTION PAPER

H.P. TECH. UNIVERSITY, December, 2012

B. Tech. 1st Semester Examination

ENGINEERING CHEMISTRY (NS-103)

Times: 3 Hours **Max. Marks: 100**

The candidates shall limit their answers precisely within the answer-book (40 pages) issued to them and no supplementary/continuation sheets will be issued.

Note: Attempt five questions in all, select one question from each section A, B, C and D. Section E (questions 9) is compulsory.

Section A

1. (*i*) What are reference electrodes? Discuss the construction and limitations of glass electrode? (8)

Ans. Reference Electrodes

An electrode whose electrode potential is accurately known or whose electrode potential has been arbitrarily fixed is called a reference electrode. In other words it is an electrode of standard potential with which we can compare the potentials of other electrodes.

It is combined with any electrode whose electrode potential has to be determined to from an electrochemical chemical cell through a salt bridge. The e.m.f. of the cell thus formed is noted and the electrode potential of the electrode under test can be known. In case of normal hydrogen electrode (NHE), the electrode potential is taken as zero. So, the e.m.f. of the cell itself becomes the electrode potential of the electrode concerned. Commonly used reference electrodes include. Normal Hydrogen Electrode, Calomal Electrode and Glass Electrode.

Glass electrode: When two solutions of different pH are separated by a glass membrane, a potential is established across the membrane whose magnitude depends on the difference in pH of the two solutions, a potential is established between the glass and the solution. The value of the potential is a function of the particular glass used and is given by the equation.

$$E_g = E_g^\circ + \frac{RT}{F} \text{In}\,(H^+) = E_g^\circ - \left(\frac{2.303\,RT}{F}\right)pH$$

The electrode reaction is assumed to involve the reduction of H^+ ions. The glass electrode thus functions in the same manner as a reversible hydrogen electrode.

Constructions of glass electrode: It consists of a thin walled glass bulb with a long neck and is made up of a special type of glass of low melting point and comparatively high conductivity [Fig. 1 (*a*) (*b*)]. A solution of known pH (usually 1 N-HCl) is taken in the bulb. A platinum electrode is introduced into the solution and a pinch of quinhydrone is added to it. A glass electrode is used for determining pH of solution, especially for coloured solutions and solutions containing oxidizing and reducing agents. The combination representing the half cell for glass electrode is represented as

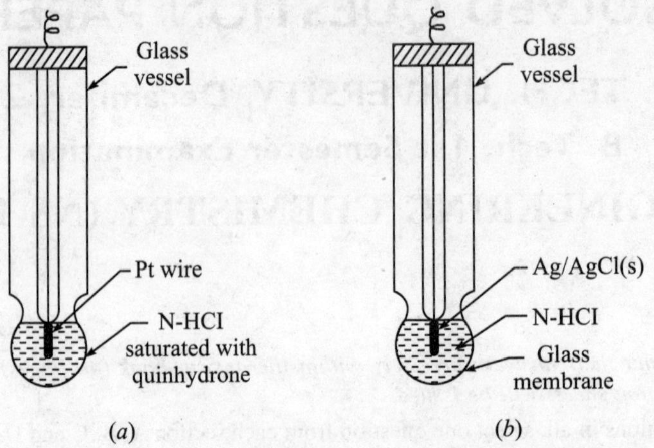

Fig. 1 Glass electrode.

Pt | 1.0 N-HCl | Glass

Sometimes a silver-silver chloride electrode in place of platinum is also employed. It is then represented as : Ag/AgCl(s) | 0.1 MHCl | glass.

Limitations of Glass Electrode

(*i*) For e.m.f. measurement using glass electrodes ordinary potentiometers can not be employed because of high resistance glass. Either quadrant electrometers or vacuum tube voltameters (VTVM) which require practically no current for their operations must be employed.

(*ii*) It can be used upto a pH of 13 but being sensitive to Na^+ ions above a pH of 9, the linear relationship between glass electrode and pH is altered. So an alkaline error is introduced beyond a pH of 9, which should be taken into consideration.

(*iii*) In strongly acidic solutions of pH < 1, the salt effect due to anions is observed.

(*iv*) It does not function properly in some organic solvents like pure alcohol.

1. (*ii*) Derive Nernst equation and give its significance? **(7)**

Ans. For a general cell reaction: $aA + bB \rightarrow cC + dD$

The free energy (ΔG) accompanying the above equation is given as:

$$\Delta G = \Delta G^\circ + RT \ln \frac{[C]^c [D]^d}{[A]^a [B]^b} \quad ...(i)$$

Since $\Delta G = -nFE$ and $\Delta G^\circ = -nFE^\circ$

∴ Equation (*i*) becomes,

$$-nFE = -nFE^\circ + RT \ln \frac{[C]^c [D]^d}{[A]^a [B]^b}$$

or

$$E = E^\circ - \frac{RT}{nF} \ln \frac{[C]^c [D]^d}{[A]^d [B]^b}$$

$$E = E^\circ - \frac{2.303\,RT}{nF} \log \frac{[C]^c [D]^d}{[A]^a [B]^b} \quad ...(ii)$$

where E is e.m.f. of cell, E° is the standard e.m.f., [A], [B], [C] are the molar concentration of species, A, B, C, and D respectively. Equation (*ii*) is Nernst equation.

Significance of Nernst Equation

(*i*) It helps in calculating the e.m.f. of a cell.

(*ii*) It also helps in calculating the equilibrium constant.

1. (*iii*) Calculate the e.m.f. of Daniel cell at 25°C, when the concentration of $ZnSO_4$ and $CuSO_4$ are 0.01 M and 0.1 M respectively. The standard potential of cell is 1.2 volts. (5)

Ans. According to Nerust equation;

$$E_{cell} = E^{\circ}_{cell} + \frac{0.591}{2} \log \frac{[Zn^{2+}]}{[Cu^{2+}]}$$

Putting the given values, we get

$$E_{cell} = 1.2V + 0.0296 \log \frac{0.01}{0.1}$$

$$= 1.2V + 0.0296 \log \frac{1}{10}$$

$$= 1.2V + 0.0296\,(-1) = \mathbf{1.1814\ V}$$

2. (*i*) Explain Phase rule with the help of two component system. Give the practical applications of this rule. (8)

Ans. Let us discuss a common example of two component system

Lead-Silver System

It is the simplest case of two component solid-liquid system involving eutectic formation. The two components are lead and silver. The four phases involved in the system are: (*i*) Solid lead (*ii*) Solid silver (*iii*) Solutions of lead and silver in the molten state and (*iv*) Vapour.

In this system pressure will not have considerable effect because the vapour pressure of solids and molten metals are very small and can be neglected. The phase diagram, therefore, is drawn at atmospheric pressure so that no gaseous phase is present. The system can be studied by using *Reduced. Phase rule equation.*

According to the reduced phase rule equation when one phase is present, F = 2 – 1 + 1 = 2. Since F = 2, only two variables (excluding pressure which is considered constant) must be specified in order to describe the systems and that the equilibrium can be represented in two dimensions by means of temperature-composition diagram. (Fig. 2)

(*i*) **Point A** represents the *melting or freezing point of pure lead* (327°C) and **point B** represents the *melting or freezing point of pure silver* (961°C). Thus at *point A* liquid lead is in equilibrium with solid lead and at *point* B liquid silver is an equilibrium with solid silver.

(*ii*) **Curve AO** represents the *freezing point of lead* to which successive small amounts of silver are added. It is seen that addition of silver to lead lowers the melting point of lead. All along this curve, the silver, which is added, goes into the solution while the separation of solid lead takes place. *Thus along this curve solid lead is in equilibrium with liquid melt* (solution of silver in liquid lead) *of varying composition.* The curve AO represents the freezing point curve of lead. Along this curve two phases solid lead and liquid melt are present in equilibrium. Therefore, F = C – P + 1 = 2 – 2 + 1 = 1, *i.e.*, the system is univariant.

(*iii*) **Curve BO** represents the freezing point of silver to which successive small amounts of lead are added. It is observed that the melting point of silver is lowered, gradually by the addition of lead into it. All along this curve, the added lead goes into the solution and the separations of solid silver takes place. Thus *along this curve, solid silver is in equilibrium with liquid melt* (solution of lead in liquid silver) of *varying compositions.* The curve BO represents the freezing point of silver separated from the melt containing varying amounts of lead. Along the curve BO two phases – solid silver and melt are present, and the system is univariant as shown below:

$$F = C - P + 1 = 2 - 2 + 1 = 1$$

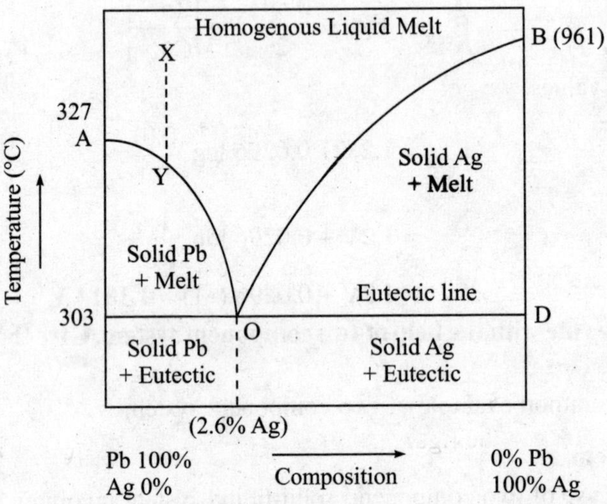

Fig. 2 Phase diagram for Lead-Silver System

At the point O in the phase diagram lead and silver freeze out simultaneously. At the point, the composition of the melt is the same as that of the solid mixture of lead and silver. The liquid melt freezes out as if it was a pure compound. The whole of the melt freezes out at constant temperature. This point O is called **eutectic point** (Greek – low melting) *and the* solid mixture *is called eutectic* which has a characteristic composition for each system.

The eutectic even though melts at a constant temperature, is not a chemical compound of lead and silver as (*i*) *its* composition does not correspond to any simple chemical formula (*ii*) its composition depends on temperature.

It may be made clear here that the phase diagram 2 will not be exactly similar at different pressures and the eutectic composition will also vary with pressure.

Below point O in Fig. 2 only solid phase exists and the cooling of this solid eutectic is represented by a dotted line.

Eutectic Point O is the point where two curves AO and BO meet. At this point solid lead and solid silver are in equilibrium with the melt, *i.e.*, three phases are present. Therefore, there is no degree of freedom $F = C - P + 1 = 2 - 3 + 1 = 0$.

This point O in Fig. 2 represents the lowest possible temperature (303°C) below which a liquid phase cannot exist and beyond which the liquid phase cannot be enriched in either composition by freezing out the other component. *Such a liquid mixture of two components Pb and Ag. which has the lowest freezing point compared to all other liquid mixtures is called* **eutectic mixture.** The mixture freezes out completely at a constant temperature and on freezing given the two components as solid phase in the same proportion in which they are present in the liquid phase. The temperature corresponding to the eutectic point O is called the **eutectic temperature**. In the above system, we see that the alloy of silver 2.6.% with lead 97.4% melts at 303°C, a temperature lower than the melting point of silver (961°C) and of lead (327°C).

A practical Applications of the System is Pattinson's Process For Desilverisation of Lead

Galena (PbS, an ore of lead) is usually associated with silver. When galena is worked out for the extraction of lead, some traces of silver pass into it because silver is soluble in lead to some extent. Such a lead is known as *argentiferous* lead. The desilverisation of lead can be easily understood from the phase diagram of lead silver system (shown in Fig. 3).

The agentiferous lead consisting of very small percentage of silver is first of all heated to temperature well above its melting point, so that the system consists only of liquid phase represented by the point *x* in the figure. It is then allowed to cool, gradually along the line *xy* without any change in composition till the point *y* is reached. As soon as this point is attained, lead starts separating out and the solution will contain respectively increasing amount of silver. Further cooling will shift the system along the line YO. Lead continues to separate out and the melt continues to be richer and richer in silver, till the eutectic point O is reached where an alloy containing about 2.6% of silver is obtained.

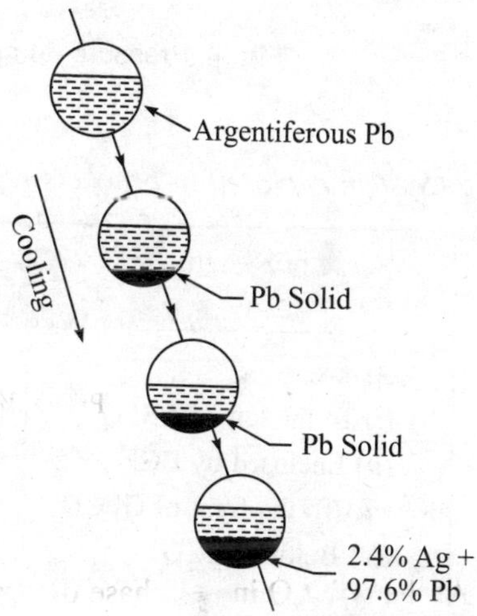

Fig. 3 Cooling of aregentiferous lead in pattison process

The increase in the relative amount of Ag during the cooling process of molten argentiferous lead is shown in Fig. 3. The amount of Ag present in traces in the alloy increases upto 2.4 percent by gradually cooling the molten mass.

This alloy is then further refined by the process of cupellation. This above principle is used in Pattison's Process (*i.e.*, the process of increasing the relative proportions of silver in the alloy) for the desilverisation of lead.

2. (*ii*) Discuss the application of phase rule to sulphur system. Draw a labelled diagram. (8)

Ans. Refer to Textbook.

However, the labelled diagram of Sulphuric System along with summary in the form of table is given below:

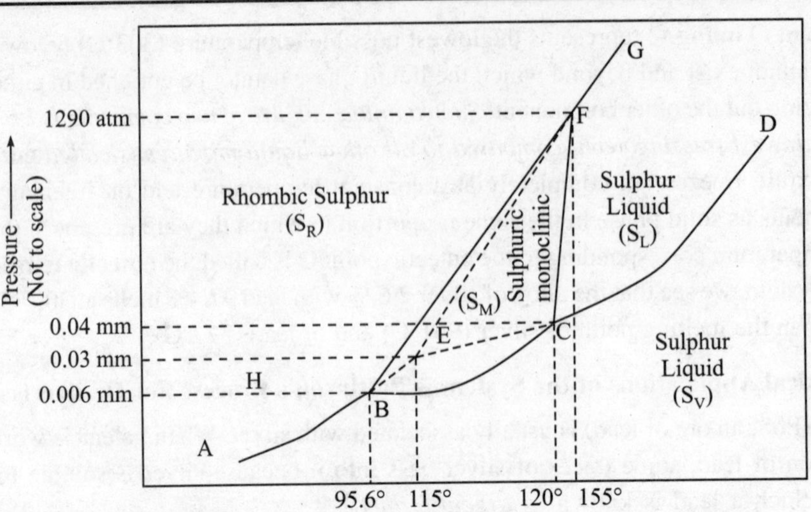

Fig. 4 Pressure–temperature phase diagram of the sulphur system

The Sulphur System

(System enclosed in brackets represents metastable curves)

Name of System	Phases is Equilibrium	Degrees Freedom $F = C - P + 2$
Areas		$F = 1 - 1 + 2 = 2$ (Bivariant
(*i*) To the left of ABFG	S_R	Two
(*ii*) Enclosed by BCF	S_M	Two
(*iii*) To the right of GFCD	S_L	Two
(*iv*) Below ABCD	S_V	Two
Curves		$F = 1 - 2 + 2 = 1$ (Univariant)
(*i*) AB (BE)	$S_R \rightleftharpoons S_V$	One
(*ii*) BC (BH)	$S_M \rightleftharpoons S_V$	One
(*iii*) CD (CE)	$S_L \rightleftharpoons S_V$	One
(*iv*) FG (FE)	$S_R \rightleftharpoons S_V$	One
(*v*) BF	$S_R \rightleftharpoons S_M$	One
(*vi*) CF	$S_M \rightleftharpoons S_L$	One
Triple Points		$F = 1 - 3 + 2 = 0$ (Invariant)
(*i*) B	$S_R \rightleftharpoons S_M \rightleftharpoons S_V$	Zero
(*ii*) C	$S_M \rightleftharpoons S_L \rightleftharpoons S_V$	Zero
(*iii*) F	$S_R \rightleftharpoons S_L \rightleftharpoons S_V$	Zero
(*iv*) E (Metastable)	$S_M \rightleftharpoons S_L \rightleftharpoons S_V$	Zero

2. (*iii*) What are the differences between phase diagrams of one- and two-component systems? (4)

Ans. Phase Diagram of One-component Systems: The minimum number of phases that can be possible in any system is one. Therefore, according to the phase rule equation, a one-component system should have a maximum of two degrees as is clear from the following:

We know, $C = 1 \quad P = 1$

$\therefore \quad F = C - P + 2 = 1 - 1 + 2 = 2$

Thus, *a one-component system requires a maximum of two variables to be fixed in order to define the system completely.* The two variables are *temperature* and *pressure*. Hence, a one-component system can conveniently be studied by plotting the conditions of pressure against the corresponding temperatures. The diagrams thus obtained are called **P – T diagrams.**

A phase diagram consists of *areas*, *curves* or *lines* and *points*. In case of a one-component system, an *area represents a bivariant system* whereas a *curve* or a *line represents a univariant system.* A *point* on a phase diagram *represents a non-variant system.*

The general shape of the phase diagram for one component system is shown below in Fig. 5(*a*) and that in case of water system in Fig. 5(*b*).

The inclination of curve OB representing solid $\rightleftharpoons$ liquid equilibrium in Fig. 5(*a*) in general is always away from the Y-axis, whereas for water it is inclined towards Y-axis. This explains why the effect of pressure on melting point of ice is different from that of other solids in general. It is clear from the curves that melting point of ice decreases, while that of other solids increases with increase of temperature.

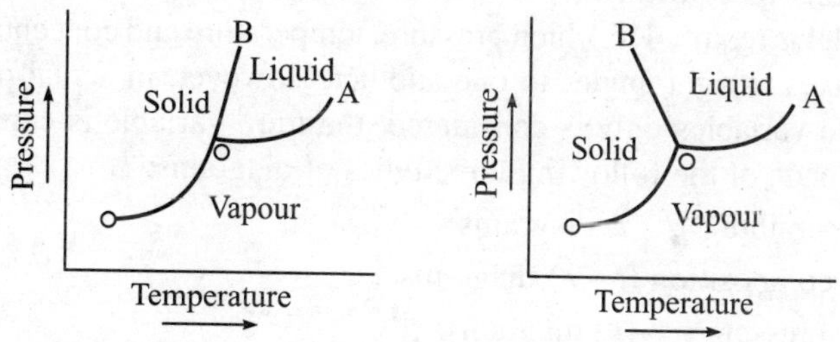

(*a*) (*b*)

Fig. 5 Phase diagram of one component

Let us consider one-component systems with different number of phases:

(*a*) **Single Phase:** When a pure substance is in single phase represented by a particular area, it is evident from the phase rule equations, that a maximum of two degree of freedom are required to be fixed in order to define the system.

$$F = C - P + 2 \qquad [\because \; C = 1, P = 1]$$
$$= 1 - 1 + 2 = 2$$

All such systems can be completely described by stating two variables. The two variables are temperature and pressure, concentration being out of question as three is only one component. Such systems are referred to a *bivariant systems.*

(*b*) **Two Phase:** When we are considering a pure substance, *i.e.*, one-component system in two phases in contact with one another, it is evident that only one degree of freedom is required to be fixed in order to define the system.

$$F = C - P + 2 \qquad [\because \; C = 1, P = 2\,]$$
$$= 1 - 2 + 2 = 1$$

Such systems are represented by a curve or a line. All such systems can be completely described by stating only one variable—either temperature or pressure. If temperature is specified, pressure of the system is automatically fixed and, vice-versa. Such systems are referred to as *univariant systems.*

(*c*) **Three Phases:** When there phases in equilibrium in one-component system, it is evident that the system has no degree of freedom.

$$F = C - P + 2 = 1 - 3 + 2 = 0$$

Such a system is represented by a point and is referred to as non-variant or invariant system.

In phase diagrams of one component system, bivariant system are represented by 'areas', univariant systems by 'lines' or curves' and non-*variant systems by 'points'*.

Phase Diagrams of Two-Component System

A two-component system can be defined as one in which it is possible to express the composition of all the phases, in terms of two substances. A two-component system may have as many as three degrees of freedom. This is because when P = 1, then

$$F = C - P + 2 = 2 - 1 + 2 = 3$$

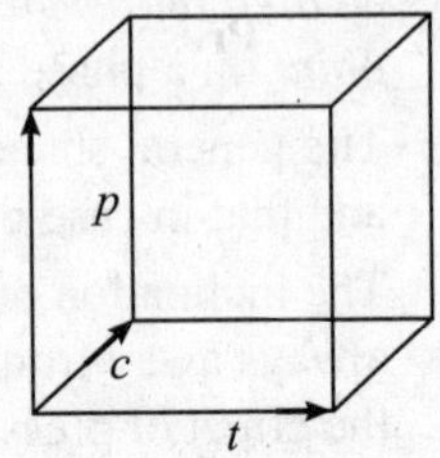

Fig. 6 Space Model for a Two-Component System

Thus the system will have, besides temperature and pressure variables, one concentration variable for each phase present at equilibrium. Naturally, for the graphic representation of all the possible conditions of equilibrium of such a system, a three-dimensional model is required in which pressure, temperature and concentration would be plotted along the three axes at right angles to one another. However, in actual practice, a simple plain diagram with two variables only is considered; the third variable is considered as constant. In this way, use is made of the following three types of diagrams:

(*a*) pressure-temperature ($p - t$) diagrams.

(*b*) Temperature-composition ($t - c$) diagrams.

(*c*) Composition-pressure ($c - p$) diagrams.

But in most of the cases, it is usually convenient to keep pressure constant at the atmosphere value and thus a $t - c$ diagram is often met with. In case of only solid-liquid phases, small changes of pressures have very little effect on the system. In this equilibrium, pressure will not have considerable effect because only gaseous systems, which involve large changes of volume, are affected by pressure. Such systems are called **condensed systems.** The degree of freedom in such a case is reduced by one as the vapour pressures of solids are very small and experiments are performed at atmospheric pressure. Thus we may use the **reduced phase rule equation** in such cases: $F' = C - P + 1$.

F' in the above equation represents the number of degrees of freedom which the system can possess in addition to pressure, namely, temperature and concentration (or composition) of the constituent. When the two components are miscible in the liquid state, we have following three types of two-component systems.

(*i*) **Simple Eutectic Systems:** In such systems, the two-components although miscible in the liquid state are not miscible in the solid state and form a *eutectic mixture* (Greek: *eutectos* meaning easy melting).

(*ii*) **Systems Forming Compounds with Congruent Melting Points:** In this type of systems, the two-components form a stable compund which possesses a *congruent melting point* (the temperature corresponding to the maxima on the freezing point curve at which the solid compound and the liquid phase have identical composition).

(*iii*) **Systems Forming Compounds with Incongruent Melting Points:** In this type of systems, the two-components form a compound which possesses an *incongruent melting point.*

Section B

3. (*i*) What are the factors which cause boiler corrosion? How can it be minimised?

Ans. Boiler Corrosion

The decay of boiler material by its environment is termed as ***boiler corrosion.***

Factors Causing Boiler Corrosion: The main factor responsible for boiler corrosion are as follows.

(*i*) **Presence of Dissolved oxygen:** The presence of dissolved oxygen in water is the main cause of boiler corrosion. When water containing dissolved oxygen is heated in the boiler, the dissolved oxygen gets liberated and attacks the boiler material as follows:

$$4Fe + 3O_2 + 2xH_2O \longrightarrow \underset{\text{(Rust)}}{2Fe_2O_3.\ xH_2O}$$

(*ii*) **Presence of Dissolved Carbon Dioxide:** Carbon dioxide usually remains dissolved in natural waters. It may also be released inside the boiler if water used contains impurities of bicarbonates. Carbon dioxide dissolves in water to form carbonic acid

$$CO_2 + H_2O \longrightarrow H_2CO_3$$

Carbonic acid thus formed causes intense local corrosion called **pitting.**

(*iii*) **Presence of Acid Forming Salts:** Certain inorganic salts such as $MgCl_2$ and $CaCl_2$ hydrolyse in water to from acids which have a corrosive action on the boiler material.

Magnesium chloride hydrolyses completely at 200°C to from hydrochloric acid.

$$MgCl_2 + 2H_2O \longrightarrow Mg(OH)_2 \downarrow + 2HCl$$

The liberated acid reacts with the boiler material (iron) in *chain-like reactions* producing HCl again and again.

$$Fe + 2HCl \longrightarrow FeCl_2 + H_2 \uparrow$$

$$FeCl_2 + 2H_2O \longrightarrow Fe(OH)_2 \downarrow + 2HCl$$

Calcium chloride also undergoes hydrolysis but to a lesser extent. It hydrolyses to an extent of 25% at 600°C. Silicic acid catalyses the reaction. Therefor, if water contains silica, appreciable quantities of HCl may be formed even at lower temperatures.

(*iv*) **Galvanic Cells:** Corrosion can also be due to simple galvanic cells which iron forms with some of the boiler fittings made of other materials or with impurities. It can be avoided by suspending zinc plates. Since zinc is more electoro positive than iron, it gets corroded and iron is saved.

Minimisation of Boiler Corrosion: The boiler corrosion can be prevented by the following methods.

(*i*) **By Mechanical Deaeration:** The dissolved oxygen as well as CO_2 can be removed by this method. The apparatus used is called a *deaerator* (Fig. 7).

In a deaerator, water is sprayed in a perforated plate fitted tower which is heated by an external heater surrounding it and is connected to a vacuum pump. The water flowing down through perforated plates undergoes deaeration at high temperature and low pressure. This is because the solubility of a gas in water is directly proportional to pressure and inversely proportional to temperature.

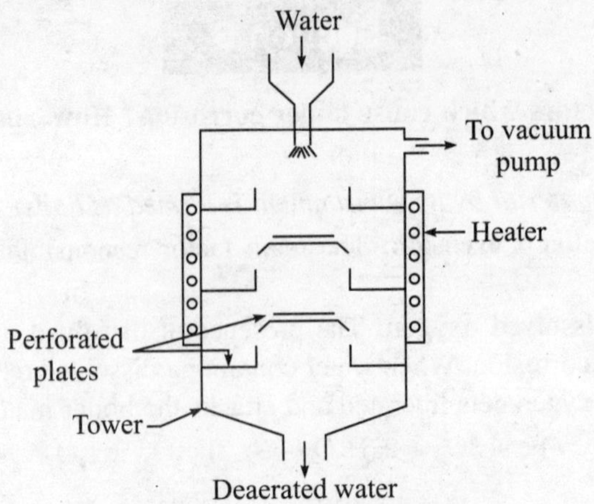

Fig. 7. Mechanical deaerator.

(ii) **By the Addition of Oxygen Removing Substances:** Certain substances such a hydrazine, sodium sulphide and sodium sulphite when added in calculated amounts to water remove dissolved oxygen.

$$\underset{\text{Hydrazine}}{N_2H_4} + O_2 \longrightarrow N_2 + 2H_2O$$

$$Na_2S + 2O_2 \longrightarrow Na_2SO_4$$

$$2Na_2SO_3 + O_2 \longrightarrow 2Na_2SO_4$$

The use of hydrazine is preferred over to that of Na_2S or Na_2SO_3 because nitrogen formed in the process is harmless.

(iii) **By the Addition of Alkali:** Corrosion by acids can be prevented by addition of calculated amount of alkalies. It in suggested that except in high pressure boilers, the boiler feed water should have a pH value of about 10.5 so that acid may not attack iron.

3. ***(ii)* Write notes on**

***(a)* Zeolite process for Softening Water.**

***(b)* Disinfection of Drinking Water.**

***(c)* Disadvantages of Hard Water.** **(12)**

Ans. *(a)* **Zeolite or Permutit Process for Softening Water:** Permutit is the trade name given to sodium zeolites *i.e.*, $Na_2OAl_2O_3 \times SiO_2 \times H_2O$. When $y = 2$, we get $Na_2Al_2Sl_2O_8 \cdot xH_2O$. For simplicity it may be written as Na_2Z, where $Z = Al_2Si_2O_8 \cdot xH_2O$.

In this method hard water is passed through a bed of permutit contained in a cylindrical vessel (Fig. 8). The water percolates at a specified rate through the bed. The loose sodium ions are exchanged for Ca^{2+} and Mg^{2+} ions. Thus calcium and magnesium salts get removed in the from of the insoluble zeolites and soft water is collected.

$$CaSO_4 + Na_2Z \longrightarrow CaZ\downarrow + Na_2SO_4$$

$$MgSO_4 + Na_2Z \longrightarrow MgZ\downarrow + Na_2SO_4$$

$$CaCl_2 + Na_2Z \longrightarrow CaZ\downarrow + 2NaCl$$

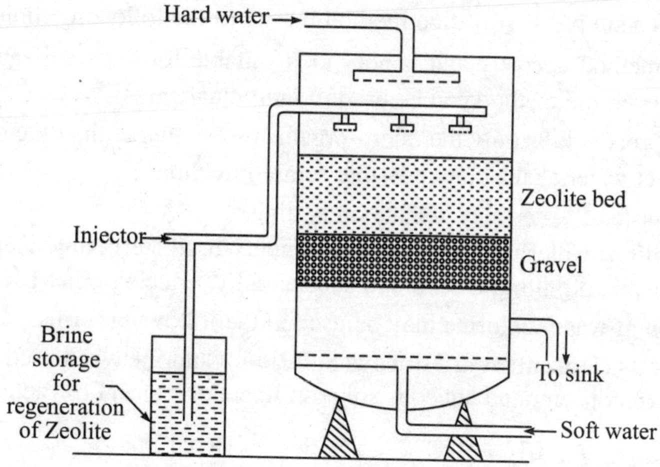

Fig. 8 Zeolite Softener

This process removes both temporary and permanent hardness. After long use, the zeolite bed gets exhausted. It can be regenerated by using chemicals, such as brine solution, NaCl or sodium nitrate or sodium sulphate. However, NaCl is preferred on account of its cheapness, easy availability and low molecular weight. Moreover, the products calcium chloride and magnesium chloride are highly soluble in water and can be easy washed out. The softening and regeneration process can be represented as follows.

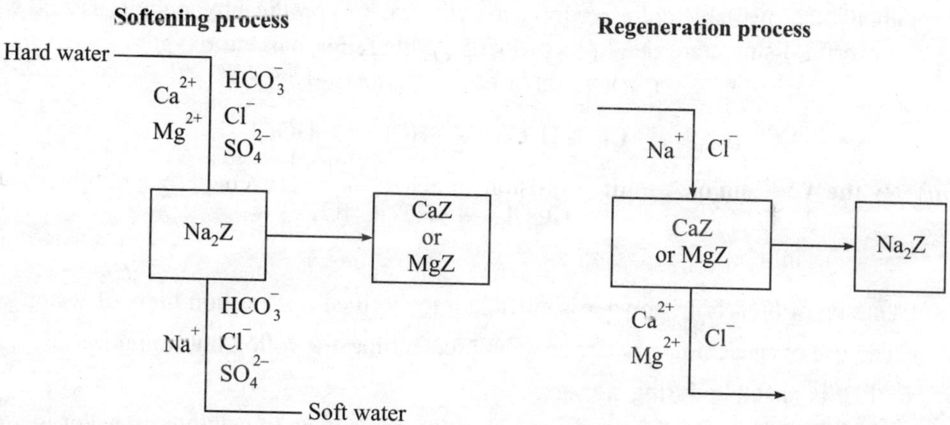

(*b*) **Disinfection of Water:** Water may contain small amounts of pathogenic (disease producing) bacterias. In order to use this water for drinking purpose, all the pathogenic bacterias and other micro-organisms present in it must be killed. *The process of destroying killing of pathogenic bacterias and other micro-organisms is called disinfection and the chemicals used for this purpose are called disinfectants.*

Several methods are available for the sterilization of water. Some of the important ones are as follows:

(*i*) **Boiling:** The harmful disease producing bacteria can be killed by boiling water for 15-20 minutes.

Boiling is a simple sterilization method but it has the following limitations:

(*a*) The method is costly and is not much suitable for the sterilization of water in bulk. However, the method can be used in individual cases.

(*b*) The process kills only the micro-organisms existing at the time of boiling. It does not protect water against any possible future infection.

(*c*) The taste of water may change on boiling.

(ii) **Chlorination:** Chlorine is the most commonly used sterilizing agent in water treatment and is capable of killing B. coli and substantially reducing other bacteria.

For sterilization of water, chlorine may be used in the following forms:

(i) **As gas or as Concentrated Aqueous Solution:** Chlorine when used directly as a gas or in the form of concentrated aqueous solution liberates nascent oxygen as shown below.

$$Cl_2 + H_2O \longrightarrow HCl + \underset{\text{Hypochlorous acid}}{HOCl}$$

$$HOCl \longrightarrow HCl + \underset{\substack{\text{Nascent}\\\text{oxygen}}}{[O]}$$

The nascent oxygen so liberated destroys the harmful germs and bacteria by oxidation. The OCl^- ions given by hypochlorous acid on ionisation also act as disinfectants as they are capable of rupturing the cell membranes of pathogenic micro-organisms.

(ii) **As Bleaching Powder:** Bleaching powder is commonly represented as $CaOCl_2$. It contains about 30% available chlorine. It reacts with water to produce hypochlorous acid which is a powerful disinfectant due to its property of liberating nascent oxygen.

$$CaOCl_2 + H_2O \longrightarrow Ca(OH)_2 + Cl_2$$

$$Cl_2 + H_2O \longrightarrow HCl + \underset{\text{Hypochlorous acid}}{HOCl}$$

$$HOCl \longrightarrow HCl + \underset{\substack{\text{Nascent}\\\text{oxygen}}}{[O]}$$

One kg of bleaching powder is sufficient to sterilize one million liters of water.

The use of bleaching powder as a disinfectant has the following limitations:

1. It is unstable during storage.
2. It introduces calcium in water. Thus the percentage of calcium in water increases and water becomes more hard.
3. When used in excess, it produces disagreeable odour in water. Too much excess may cause irritation to mucous membranes. Hence, it needs to be used only in calculated quantities.

(iii) **As Chloramine:** The use of excess of chlorine gas or bleaching powder as disinfectants often produces disagreeable odour and unpleasant taste in water.

The unpleasant taste can be avoided by the use of chloramine (NH_2Cl) which may be obtained by treating chlorine with ammonia in the ratio 2 : 1 by volume.

$$Cl_2 + NH_3 \longrightarrow NH_2Cl + HCl$$

Chloramine provides a greater lasting effect than that of chlorine.

$$NH_2Cl + H_2O \longrightarrow HCl + \underset{\text{Hypochlorous acid}}{HOCl} + NH_3$$

$$HOCl \longrightarrow HCl + \underset{\text{Nascent oxygen}}{[O]}$$

Since chloramine decomposes slowly to gives chlorine, it is particularly useful when the treated water has to be stored.

Superchlorination: *The sterilization proess involving a large excess of chlorine is called superchlorination.* Superchlorination not only destroys the micro-organisms but also the other organic impurities present in water. The process is usually followed by *dechlorination by* NH_3 or SO_2.

Break-point Chlorination: Break-point chlorinations is a carefully controlled chlorination process in which water is treated with such an amount of chlorine which is just sufficient.

(*a*) to destroy bacteria;

(*b*) to oxidise organic matter;

(*c*) to react with any ammonia, if present and

(*d*) to leave behind slight excess of free chlorine which could act as disinfectant during the storage of water.

For details refer Textbook.

(*c*) Disadvantages of Hard Water: Hard water contains large amounts of bicarbonates, sulfates and chlorides of calcium and magnesium dissolved in it. These dissolved salts cause undesirable effects both in houses and industries. Hard water causes number of problems in domestic use, industrial use and in boilers. These problems are explained below:

1. Problems in Domestic Use: Hard Water causes many undesirable effects when used for domestic purposes. Some instances are given below:

(*a*) Wastage of Soap: A large amount of soap is wated when hard water is used for washing purposes. It is because hard water does not form lather with soap readily. Most of the soap is wasted in reacting with calcium and magnesium salts present in water and only after this soap forms lather hard water.

(*b*) Spoiling clothes: Soap reacts with hard water to from insoluble calcium and magnesium soap, which sticks to the clothes in the form of scum. These scums are not easy to remove and make dirty marks on the clothes. The insoluble soaps also damage silk and nylon fabrics.

(*c*) Deposition in Kettles: Calcium and magnesium salts form an incrustation inside the kettles, if hard water is bouled in it. It is due to the formation of carbonates and hydroxides in boiling water. It wastes fuel, this incrustation can be removed by adding a little vinegar.

(*d*) Cooking: Pulses and other vegetable do not cook well in hard water.

2. Problems in Industrial Use

(*a*) Paper Industry: In paper industry, water is used along with soap for sizing the paper. If hard water is used, Ca and Mg salts present in it waste a good amount of soap by forming insoluble precipitates

(*b*) Textile Industry: Water is used in textile industry for cleaning. washing and whitening of yarn. For such purposes, soap is required. If hard water is used much soap is wasted.

(*c*) **Dyeing:** The hard water is used for dyeing purposes, salts of Ca and Mg spoil the desired shade and thus the dyed clothes do not give a homogeneous look.

(*d*) **Chocking and Clogging Problem:** Water pipes are blocked due to precipitation of salts during boiling of hard water. This involves extra cost for maintenance and repair of water pipes.

3. **Problems in Boilers Use:** Huge quantity of water is used in boilers for generation of steam. If hard water is used in the boiler, a number of difficulties arise which are listed as follows:
   (*a*) Scale and sludge formation
   (*b*) Boiler corrosion
   (*c*) Priming and foaming
   (*d*) Caustic embrittlement

4. **(*i*) Explain differential aeration corrosion, Waterline Corrosion and pitting corrosion.**

Ans. Differential Aeration Corrosion: This type of corrosion occurs when the metal surface is exposed to an electrolyte of varying concentration or of varying aeration. It also occurs when one part of the metal is exposed to a different air concentration from the other part. It has been observed that poorly oxygenated parts become anodic and are corroded, where more oxygenated portion acts as cathode.

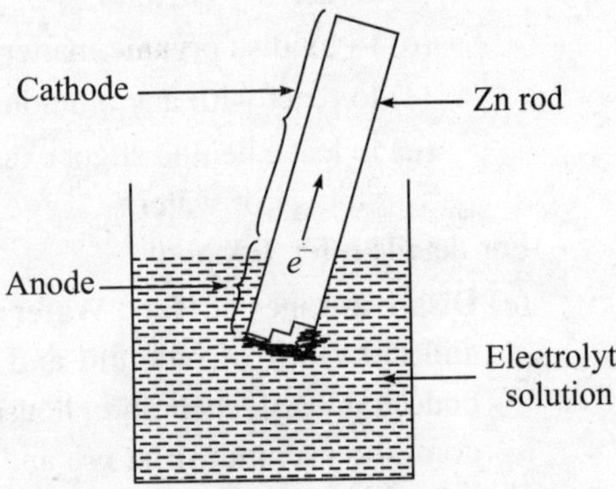

Fig. 9 Differential aeration corrosion

It occurs when a metallic surface is partially immersed in a solution and partially exposed to the atmosphere. Consider a metal (say Zn) is partially immersed in a dilute solution of NaCl and solution is not agitated for long. The part of the metal immersed in solution has less access to oxygen and act as anode, whereas the part above and closely adjacent to the waterline is more strongly aerated and act as cathode (Fig. 9). So a difference of potential is created which causes flow of electrons between two differentially aerated areas of the same metal . Zinc change to Zn^{2+} ions at the anodic area and oxygen takes up electrons at the cathodic area of the metal to form hydroxial ions. The Zn^{2+} and OH^- ions interact to form $Zn(OH)_2$.

$$\text{At anode } Zn \rightarrow Zn^{2+} + 2e^- \text{ (oxidation)}$$

$$\text{At cathode } \tfrac{1}{2}O_2 + H_2O + 2e \rightarrow 2OH^- \text{ (reduction)}$$

$$Zn^{2+} + 2OH^- \rightarrow Zn(OH)_2.$$

Similarly a crevice (joint) between different metallic objects *i.e.*, bolts, nuts, rivets etc. also cause corrosion. The joint area has lack of oxygen and becomes anodic and the exposed area becomes cathodic. The corrosion is accelerated at the joint area acting as anode. This type of corrosion acting at the junction of two metals exposed to corrosive environments is called *crevice corrosion* (Fig. 10).

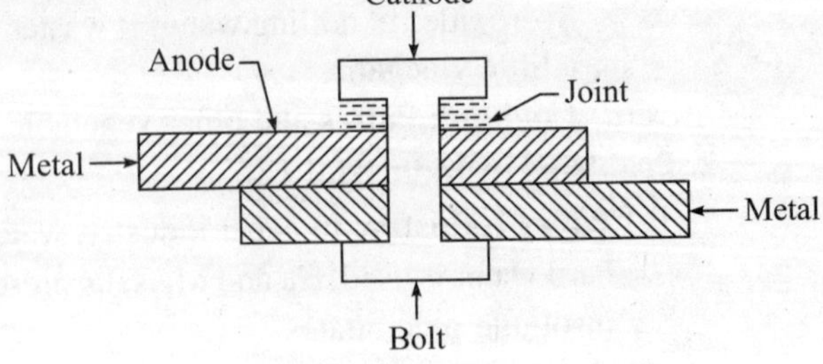

Fig. 10 Crevice

Pitting Corrosion: A small part of the metal surface covered by an impurity or water has a low oxygen concentration and it acts as an anode, whereas the uncovered metal surface acts as a cathode due to high oxygen concentration. The small covered area acting as anode gets corroded giving rise to localized spots (pits), pinholes and cavities in the metal. Once a small pit is formed, the rate of corrosion increases. The electrochemical reaction may be represented as:

At anode: $$Fe \longrightarrow Fe^{2+} + 2e^-$$

At cathode: $$\frac{1}{2}O_2 + H_2O + 2e^- \longrightarrow 2OH^-$$

$$Fe^{2+}\ 2OH^- \longrightarrow Fe(HO)_3 \xrightarrow{\text{Oxidation}} Fe(OH)_3$$

Pitting corrosion is usually caused due to cracking of the protective film on a metal at specific points. The cracks serve as initiators of corrosion by creating two areas, one fairly accessible to oxygen and the remaining poorly accessible. The presence of impurities such as dust, sand, scale etc. also cause pitting corrosion.

(*i*) **Water line Corrosion:** When water is stored in a metallic tank, it is observed that corrosion takes place along a line just below the level of water in the container. The area below the level of water is poorly oxygenated and acts as anode whereas area above the water level is highly oxygenated and acts as a cathode. The part of the steel tank just below the water level is corroded, whereas the area above the water level remains protected from corrosion (Fig. 11).

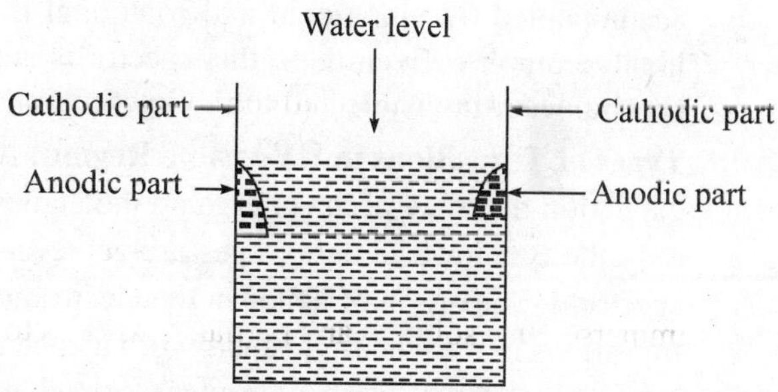

Fig. 11 Water level corrosion

At Anode: $$Fe \longrightarrow Fe^{2+} + 2e^- \text{(oxidation)}$$

At Cathode: $$2H^+ + \frac{1}{2}O_2 + 2e^- \longrightarrow 2OH^- \text{(reduction)}$$

$$Fe^{2+} + 2OH^- \longrightarrow Fe(OH_2) \xrightarrow{\text{Oxidation}} Fe(OH_3)$$

Waterline corrosion is a problem for the marine engineers when marine plants are attached to the ships in ocean, waterline corrosion comes into action and it can be checked by anti-fouling paints to some extent.

(*ii*) **Discuss Various Corrosion Control Methods:** Some of the important methods controlling corrosion are:

(*a*) Material Selection and Proper Designing
(*b*) Barrier Protection
(*c*) Sacrificial Protection
(*d*) Cathodic Protection
(*e*) Anodic Protection

For details refer to Texbook.

Section C

5. ***(i)*** **What do you understand by UV-visible spectroscopy? What are different types of transitions that take place on absorption of energy in UV-visible regions? Explain the significance of these transition with examples.** **(8)**

Ans. UV-Visible Spectroscopy: The visible region of the spectrum comprises photon energies of 36 to 72 kcal/mole, and the near ultraviolet region, out to 200 nm, extends this energy range to 143 kcal/mole. Ultraviolet radiation having wavelengths less than 200 nm is difficult to handle, and is seldom used as a routine tool for structural analysis.

The energies noted above are sufficient to promote or excite a molecular electron to a higher energy orbital. Consequently, absorption spectroscopy carried out in this region is sometimes called **"electronic spectroscopy"**. The spectra arise in a molecule from electronic transitions due to absorption of radiations falling in the visible and ultra violet regions. In the visible region the spectra occur in the spectral range of 12,500 to 25,000 cm^{-1}, whereas in the ultra violet region, the spectral range is 25,000-70,000 cm^{-1}. Since electronic transitions in a molecule are accompanied by vibrational and rotational transitions, the electronic spectra of molecules is highly complex. Even then, this spectra is studied in preference to other types because these spectra occur in visible and ultra violet region and can be examined easily.

Types of Transitions in UV-visible Region: A diagram showing the various kinds of electronic excitation that may occur in organic molecules is shown below. Of the six transitions outlined, only the two lowest energy ones are achieved by the energies available in the 200 to 800 nm spectrum. The origin of bands in the electronic spectra (ultraviolet and visible spectra) is due to transitions in the electronic energy of valence electrons. These valence electrons exist in atomic orbitals in atoms. When a molecule is formed, then valence electrons taking part in bond formation exist in *bonding orbitals*. Electrons which forms bonds are called σ-electrons and those which form π-electrons are called π-electrons. Some unshared electrons in a molecule do not take part in any bond formation and exist in *non-bonding orbitals*. Such electrons are called *n*-electrons. Apart from these, there are higher energy levels termed as antibonding orbitals by molecular orbitals theory. The anti-bonding orbitals associated orbitals associated with σ and π-bonds are called σ* (sigma star) and π* (pi star) orbitals. As *n*-electrons do not take part in bonding orbital, no anti-bonding orbital is associated with them. Anti-bonding orbitals are vacant in the ground state of molecule, but they are filled in excited states.

As a rule, energetically favored electrons promotion will be from the highest occupied molecular orbital (HOMO) to the lowest unoccupied molecular orbital (LUMO), and the resulting species is called an excited state.

(*i*) ***Transition between bonding orbitals and anti-bonding orbitals,*** *e.g.*, $\sigma \rightarrow \sigma^*$ or $\pi \rightarrow \pi^*$. This excited state closely resembles the polar character.

(*a*) **σ → σ* transition:** Compounds in which all the valence electrons from ∇ bonds absorb in infra red region and show this transition, e.g. CH absorbs at 125 mμ.

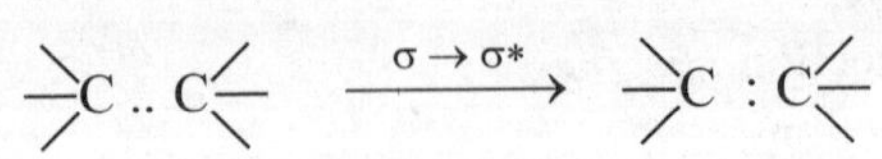

$\sigma \rightarrow \sigma^*$ transition requires highest energy. In *u.v*, this energy is not of much importance as oxygen $(O)_2$ starts absorbing radiation.

(*b*) **σ → σ* transition:** Molecules having unsatination show this kind of transition and absorb in ultraviolet or visible region.

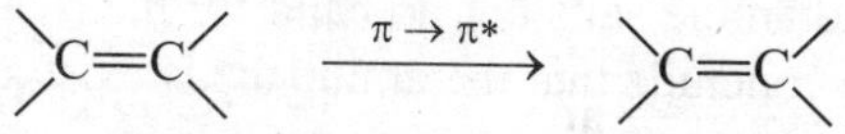

(*ii*) ***Promotion of non-bonding electron*** (*i.e.*, unshared pair, *n*) into antibonding sigma or antibonding pi, *i.e.*, $n \rightarrow \sigma^*$ or $n \rightarrow \pi^*$. These involve transitions of non-bonding lone pair of hetro atom (*e.g.*, O, N) to an empty anti-bonding molecular orbital. The absorption bands for such transitions occur in longer wavelength with low intensity.

(*iii*) ***Transition between bonding (occupied) σ and π orbitals to anti-bonding (unoccupied) σ* and π* orbitals*** *i.e.*, $\sigma \rightarrow \pi^*$ and $\pi \rightarrow \sigma^*$ can also take place. According to selection rules these radiations are forbidden and therefore weak in intensities.

All the above six type of electronic transitions have been summed up Fig. 12. The relative energies required for these transitions have also been shown. The energy required is maximum for $\sigma \rightarrow \sigma^*$ transition and minimum for $n \rightarrow \pi^*$ transitions. The energy requirements in the decreasing order of these transitions is given below:

$$\sigma \rightarrow \sigma^* > \sigma \rightarrow \pi^* > \pi \rightarrow \sigma^* > n \rightarrow \sigma^* > \pi \rightarrow \pi^* > \nu \rightarrow \pi^*$$

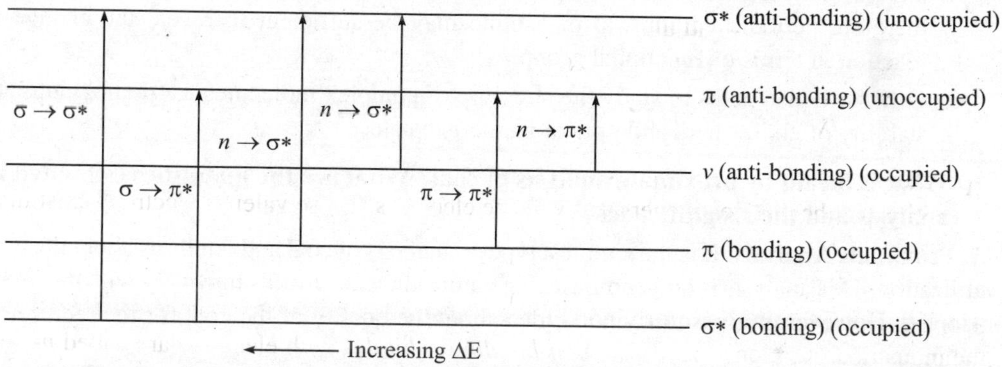

Fig. 12 Relative energies required for various types of electrons transitions.

5. (*ii*) **In relation to NMR spectra, explain the following (*a*) Shielding-deshielding effects (*b*) Coupling constant.**

Ans. (*a*) **Shielding-deshielding Effects:** Refer to Textbook.

(*b*) **Coupling Constant:** Refer to Textbook.

5. (*iii*) **Explain Application of IR spectroscopy**

Ans. Applications of Infrared spectroscopy

1. **Identification of an Organic Compound:** The infrared spectrum of an organic compound is the most reliable method to establish its identity. The region of the spectrum from 1400 cm^{-1} to 700 cm^{-1} is complex because it contains, apart from fundamental streching and bending vibrations, many bands resulting their vibrational frequencies. Thus, this part of the spectrum is characteristic of a compound and is called *fingerprint region*. The molecules containing the same functional groups may have similar absorptions above 1500 cm^{-1} but their spectra differ significantly in the fingerprint region.

2. **Structure Determinations:** Infrared spectroscopy is used to establish the structure of a compound. For example, the infrared spectra of amino acids exhibits bands for ionised carboxylic acids and amine salts but no band for the free amino and carboxylic groups is observed. Thus, we conclude that the amino acids exist as zwitter ions shown below:

$$H_3N^+—CH_2—COO^-$$

For the same molecule formula, different isomers can be proposed and their identify can be established with the help of IR spectra.

3. **Detection of Impurities:** The use of IR spectroscopy for the detection of impurities is also used although IR bands/peaks are not very sharp and intense even for pure sample. The presence of impurities can be detected if they absorb in the region where the sample does not show any absorption.
4. **Distinction between Inter and Intramolecular Hydrogen Bonding (Effect of Dilution):** IR spectroscopy offers a simple method to establish inter-molecular as well as intra-molecular hydrogen bonding shows the change in absorption pattern of the sample if diluted with non-polar solvents.
5. **Kinetic Study of Reactions:** It helps to follow the progress of chemical reaction. Since most of the chemical reactions involve changes in the functional groups of the reactants, therefore periodic withdrawal of sample may be sufficient to reveal the change in the reaction in terms of functional group.
6. **Others:** It also helps to study the structure of complex molecules, keto-enol tautomerism, stability of conformers and optical isomerisms etc.

6. (*i*) What is meant by proximate analysis of coal? What are the quantities estimated in this analysis and their significance?

Ans. **A. Proximate Analysis:** It is the simplest type of analysis of coal and it tells us about the practical utilization. This analysis is quite empirical in nature since the results may vary with the procedure adopted. However, it gives fairly good idea about the quality of the coal for the requirement of the industries.

Significance of Proximate Analysis: The results of proximate analysis have no absolute significance. However, if the observations are made in accordance with standard specifications, reproducible results can be obtained. This analysis informs about the relative qualities of the coal samples with regard to its cost and the probability of its use for a particular industry.

Quantities Estimated their Significance: The proximate analysis involves the determinations of four quantities, which are (1) Moisture content (2) Volatile matter (3) Ash content (4) Fixed carbon.

1. **Moisture:** Coal contains two types of moisture:

   (*i*) **Free Moisture Content, Adsorbed or Surface Moisture:** The moisture of this type is lost on air drying. It depends upon the treatment given to coal after mining the rank, size and surface, nature of the coal. It reduces the calorific value of the sample. A lot of heat is wasted in evaporating the free moisture during combustion. It also increases the cost on transport of the coal.

   (*ii*) **Inherent or Internal Moisture:** It is that type of moisture which is obtained form air dried coal. It is an indicator of the rank of the cal *i.e.,* the degree of decomposition of vegetable matter to carbon content of coal. It decreases with increasing rank of the coal. In other words the lower the moisture content, the better is the quality of coal.

The sum of the weight loss on air drying and Inherent moisture content gives the total moisture content of the coal sample.

Significance of Moisture in Coal: Moisture present in coal evaporates during burning of the coal and takes away some portion of heat liberated, in the form of latent heat of evaporation. Therefore, higher the % of moisture in a coal sample, lower is the effective calorific value of the coal. Moreover, it quenches the fire and increases the cost since water is brought and transported at the fuel price. However, coal containing 5-10% free moisture improve the yield and quality of metallurgical coke formed. It helps in forming uniform fuel bed and reduces the amount of "fly ash".

2. **Volatile Matter:** It consists of a complex mixture of gaseous and liquid products resulting from the thermal decomposition of coal. It is not a constituent of coal as such. Its magnitude depends upon the condition of heating particularly the temperature. So the experimental conditions must be maintained during its determination. High volatile matter content of a coal gives high flames, greater smoke and relatively low heat content, the larger is the combustion space required. Moreover, it determines the proportion of coal, which would be converted into tarry matter on heating. It includes the combustible gases like CO, H_2, CH_4 etc. and incombustibe gases like N_2, CO_2, etc. The presence of incombustible gases is undesirable as they do not have any heating value. The lower the volatile matter content, the better would be the quality of coal. Coking quality of coal also depends upon its volatile matter content. Coking power decreases with increases of volatile matter.

   Significance of Volatile Matter: The volatile matter present in coal distills over as gas or vapour, as the fuel is burnt in air, without adding to the heat value of the coal. Therefore, higher the % of volatile matter in coal, lower is the relative calorific value.

   A high volatile matter in the coal sample means high % of combustible or non-combustible gases in it. The presence of non-combustible gases is always undesirable, as they increase the volume of the furnace without adding to the heat liberated. Therefore, furnace with small volume is not suitable for burning fuels with high volatile matter content since large portion of fuel goes unburnt. On the other hand, burning of low volatile coals requires forced drought. As the % of volatile matter decreases, the intensity of drought needs to be increased.

   Coals containing high volatile matter burn with long flame and high smoke. Therefore, they are not suitable for domestic purpose. Since the heat produced by such coals is dissipated over large space (calorific intensity is low). Such coals are not suitable for purposes like steam generation.

   However, coals with high volatile matter content are suitable for production of coal gas. Moreover, coals with medium amount of volatile matter (around 20%) are suitable for producing hard and strong coke. Coals with very high or very low volatile matter content are not suitable for the purpose.

3. **Ash Content:** The non-combustible inorganic matter which is left behind after burning off the organic matter in coal is called ash. It is of two types:

   Free or Extraneous Ash: The ash which occurs in different layer of the coal is known as free ash or extraneous ash. It can be removed by washing.

   Fixed or Inherent Ash: The ash which is intimately interspersed with in the mass of coal is called fixed or inherent ash.

Significance: Ash is a totally useless, non-combustible matter present in the coal, which does not add to the heating value of the coal. Therefore, more the % of ash in coal sample lesser is the relative calorific value. Moreover, it also hinders the flow of air and heat. Therefore, it decreases the flame temperature and causes trouble during firing by forming clinkers which blocks the inter spaces of the grating on which the coal is burnt. As a result the combustion becomes irregular. The presence of ash causes early wear of furnace and also increases the cost of transportation, handling and storage of the coal. It also increases the cost in ash disposal. The presence of ash also demands increase in furnace size, which is also undesirable.

4. **Fixed Carbon:** The percentage of fixed carbon can be obtained by subtracting the percentages of moisture, ash and volatile matter from 100.

$$\% \text{ of fixed carbon} = 100 - \% \text{ (moisture + volatile matter + ash)}$$

$$\text{Alternatively, \% of ash} = \frac{W_4 - W_5}{W_2 - W_1} \times 100$$

Significance: Carbon present in the coal burns in air to give carbon dioxide.

$$C + O_2 \longrightarrow CO_2 + \text{Heat}$$

It is highly exothermic process and hence greater the percentage of fixed carbon in a coal sample, greater is the heat liberated and hence greater is its calorific value. Moreover, greater the % age of fixed C, smaller is the percentage of volatile matter. Moreover, the percentage of fixed C helps in deciding the shape and size of the furnace to be used.

6. ***(ii)* What is meant by cracking? What are the advantages of catalytic cracking?**

Ans. **Cracking:** The process of decomposition of less volatile higher hydrocarbons into more volatile lower hydrocarbons with the application of heat and catalysts is known as cracking.

$$\underset{\substack{\text{Dodecane} \\ b.p.\ 216°\text{C}}}{C_{12}H_{26}} \xrightarrow[\Delta,\ \text{Pressure, Catalyst}]{\text{Cracking}} \left\{\begin{matrix} C_7H_{16} + C_5H_{10} \\ C_6H_{14} + C_6H_{12} \\ C_5H_{12} + C_7H_{14} \end{matrix}\right\} \underset{30\text{–}150°\text{C}}{\text{Boiling Range}}$$

$$\underset{\text{Hexadecane}}{C_{16}H_{34}} \xrightarrow[\text{Catalyst}]{\text{Heat, Pressure}} \left\{\begin{matrix} C_8H_{18} + C_8H_{16} \\ C_9H_{20} + C_7H_{14} \\ C_{10}H_{22} + C_6H_{12} \end{matrix}\right\} \underset{30\text{–}150°\text{C}}{\text{Boiling Range}}$$

The nature of products obtained depends upon the following:

(*i*) Structure of the hydrocarbon.

(*ii*) Temperature and pressure applied during cracking

(*iii*) Presence or absence of the catalyst.

Changes Occurring During Cracking: Besides decrease in the size of the molecule which results in increased volatility or lower boiling fractions, the other important changes taking place are as follows:

(*i*) Cracking results in formation of mixture of large number of hydrocarbons—all having lower b.pts than initial alkane.

(*ii*) Straight chain alkanes are converted branched chain hydrocarbons *i.e.*, isomerisation may take place.

(*iii*) Saturated hydrocarbons are converted to unsaturated hydrocarbons.

(*iv*) Ring closure *i.e.,* cyclization of aliphatic alkanes may occur.

(*v*) The breakage of C—C as well as C—H bonds may take place with the formation of free C and H_2 gas.

Types of Cracking: There are two main processes used for cracking (1) Thermal cracking (2) Catalytic cracking.

Thermal Cracking: Thermal cracking is carried out simply by heat and applying suitable pressure and is mainly divided into two catagories namely (*i*) vapour phase cracking and liquid phase cracking.

Catalytic Cracking: When cracking is carried out in the presence of catalysts, high pressure and temperature are usually not necessary. The feed stock used is generally in the boiling range of 350-550°C. The catalytic cracker may be fixed bed, moving bed or fluid bed type. Now-a-days generally the latter types are used. The product mix obtained from a catalytic cracker depends upon a number of variables, *viz.* nature of feed stock, catalyst, cracking conditions. etc.

Advantages of Catalytic Cracking

(*i*) Most of the gasoline (petrol) that is being used is obtained by catalytic cracking rather than by direct refining of petroleum refining.

(*ii*) It also gives several gaseous hydrocarbons which are employed in the manufacture of petrochemicals. These basic petrochemicals give rise to second and third generation petrochemicals which are used in industry for the manufacture of industrial products such as polymers, synthetic rubbers and composite materials.

(*iii*) The petrol obtained by catalytic cracking is of better quality because of its higher octane number.

(*iv*) A comparison of thermal cracking with catalytic cracking reveals that the temperature and pressure required in catalytic cracking is less as compared to thermal cracking. Moreover, the quality of petrol obtained by using catalytic cracking is much better.

6. (*iii*) Write notes on (*a*) Water Gas (*b*) Producer Gas **(8)**

Ans. (*a*) For water gas, refer to Textbook.

(*b*) For producer gas, refer to Textbook.

Section D

7. (*i*) What are polymers? Give classification of polymers based upon (*i*) structure (*ii*) configuration (*iii*) Type of monomers involved.

Ans. Polymers: *Polymers are macro-molecules of high molecular masses formed by the combination of large number of simple molecules by covalent bonds to form long chain compounds.* The simple molecules which combine to form polymers are called **monomers**. The process by which monomers combine to form polymers is called polymerisation. For example, a commonly known polymer, polyethene is obtain from the monomer, ethane, by the process of polymerisation.

$$nCH_2{=}CH_2 \xrightarrow{\text{Polymerisation}} (-CH_2{-}CH_2)_n$$

Ethylene (Monomer) Polyethylene (Polymer)

(*i*) Classification of Polymers on the Basis of Structure

1. **Linear Polymers:** *These are polymers in which monomeric units are linked together to form linear chains*. These linear polymers are well packed and therefore, have high densities, high tensile strength and high melting points. Some common examples of linear polymers are: **polyethylene, nylons, polyesters etc.**
2. **Branched Chain Polymers:** *These are polymers in which the monomers are joined to form long chains with side chains or branches of different lengths*. These branched chain polymers are irregularly packed and therefore, they have low density, tensile strength and melting points than linear polymers. Some common examples are: **low density polythene, glycogen, starch** etc.
3. **Cross-linked Polymers:** *These are polymers in which monomer units are cross-linked together to form a three-dimensional network*. These are also called three-dimensional network polymers. These polymers are hard, rigid and brittle because of network structure. For example, **bakelite, melamine formaldehyde resin** etc.

(*ii*) Classification of Polymers Based Upon Configuration: Configuration of a molecule means that position of certain groups or atoms in space is different in the two isomers of a molecule. A type of space isomerism is geometrical isomerism also called *cis*-trans isomerism. When the two different groups attached to carbon atoms joined by a double bond are on the same side, it is called *cis*-form. But when different groups lie on opposite sides, it is called trans from of the molecule. Polymers also exist as *cis*-trans geometric isomerism. For example, natural rubber is isomorphous, while gutta percha is crystalline, although both polymers are made up of the same monomeric unit *i.e.,* 2-methybuta-1, 3-diene. In natural rubber, all the double bonds are *cis*-with the result it is amorphous, possesses elasticity and melts comparatively at lower temperature than its trans-isomer gutta percha. In the latter, the double bonds are *trans*-with the result it is crystalline, hard substance and melts at higher temperature than natural rubber. The structures of both gutta percha and natural rubber are shown below:

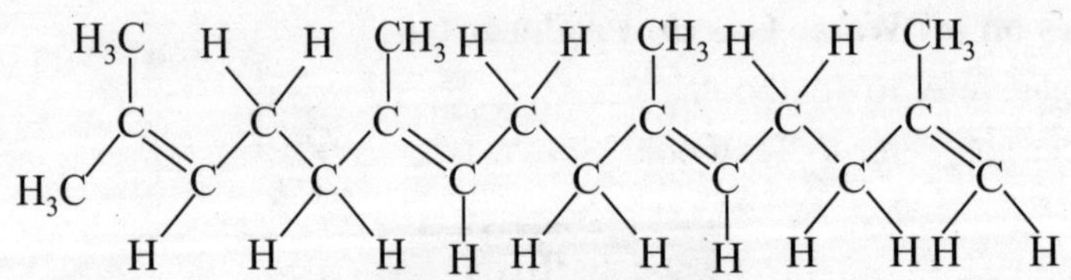

Gutta percha (all *trans*-configuration)

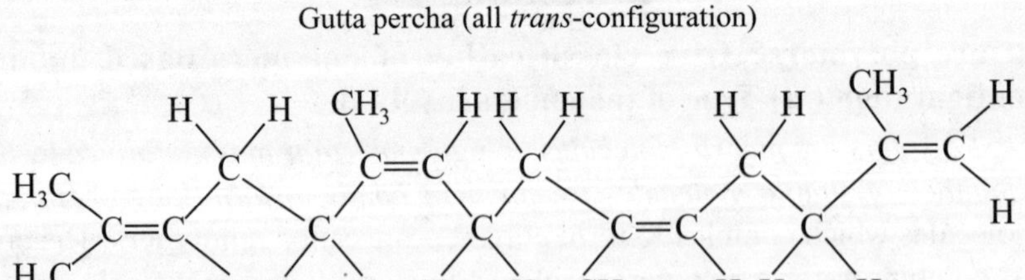

Natural rubbers (all *cis*-configuration)

In natural rubber the $-CH_3$ group and H atom attached to the cation joined by double bond lie on the same side, whereas in gutta percha the groups lie on the opposite sides.

(iii) **Classification of Polymers Based on Types of Monmers Involved:** On this basis the polymers may be classified as:

(i) *Addition Polymers* (ii) *Condensation polymers*

This may also be referred to as classification based on mode of synthesis

1. **Addition Polymers:** *A polymers formed by direct addition of repeated monomers without the elimination of by product molecules is called* **addition polymer.**

   In this type, the monomers are unsaturated compounds and are generally derivatives of ethene. The addition polymers have the same empirical formula as their monomers.

   Very often the molecular weight of the polymer is an integral multiple of the monomer. The polymers thus obtained are termed as addition polymer. For a monomer, to undergo the addition polymerisation, the basic requirement possessed by the monomer is the multiplicity in it *i.e.,* the presence of double or triple bond. For example.

$$nCH_2{=}CH_2 \longrightarrow (\text{–}CH_2\text{—}CH_2\text{–})_n$$

ethene (monomer) polyethene (polymer)

$$nCH_2{=}CH\text{—}Cl \longrightarrow (\text{–}CH_2\text{—}CHCl\text{–})_n$$

vinyl chloride (monomer) polyvinyl chloride (polymer)

$$nCH_2{=}CHOCOCH_3 \longrightarrow (\text{–}CH_2\text{—}CHOCOCH_3\text{–})_n$$

vinyl acetate (monomer) polyvinyl acetate (polymer)

$$nF_2C{=}CF_2 \longrightarrow (\text{–}CF_2\text{—}CF_2\text{–})_n \text{ etc.}$$

Tetrafluoroethylene teflon

Condensation Polymers: *A polymer formed by the condensation of two or more than two monomers with the elimination of simple molecules like water, ammonia, hydrogen, chloride, alcohol etc. is called* condensation polymer.

For a monomer, to undergo the condensation polymerisation, the basic requirement possessed by the monomers is the presence of some reactive functional group like —OH, —COOH, —NH_2, etc. For example,

$$nH_2N\text{—}(CH_2)_6\text{—}NH_2 + nHOOC\text{—}(CH_2)_4\text{—}COOH \xrightarrow[-n\,H_2O]{\Delta} [\text{–}NH\text{—}(CH_2)_6\text{—}NH\text{—}CO\text{—}(CH_2)_4\text{—}CO\text{—}]_n$$

hexamethylenediamine adipic acid Nylon 66

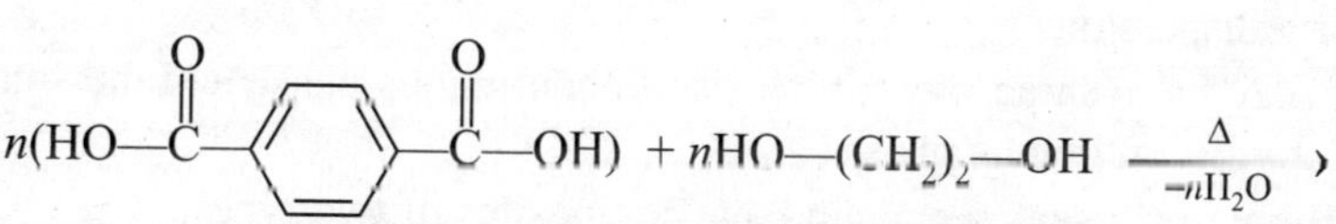

terephthalic acid ethylene glycol

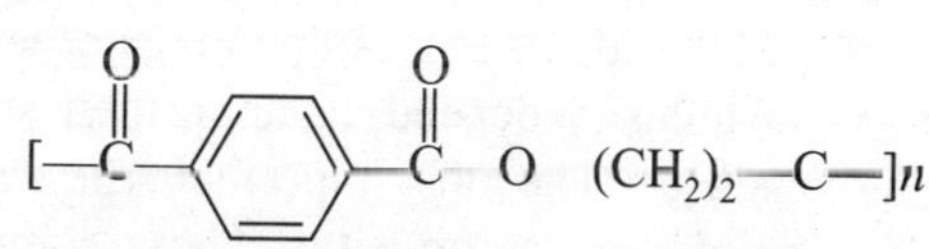

terylene or dacron

There are many other examples, such Nylon-6, Nylon-11 etc.

7. (*ii*) What are the important thermosetting plastics/resins? Describe the preparation and application of (*a*) Phenol-formaldehyde resins and (*b*) Urea-formaldehyde resins.

Ans. A thermosetting plastic can be heated only once because it undergoes permanent change on melting and sets into a new solid which cannot be remelted. These materials have 3-dimensional cross-linked structure which are formed during moulding. Prolonged heating chars these materials. Due to the presence of strong covalent bonds and cross-linking, these materials are insoluble in all organic solvents. They are usually hard, strong and more brittle than the thermoplastic materials. They cannot be reclaimed from waste materials.

Examples: Bakelite, urea formaldehyde resin, polyesters, alkyd resins, epoxy resins, polyurethanes etc.

(*a*) For preparation and applications of phenol formaldehyde resius, refer to Textbook.

(*b*) For urea-formaldehyde resius, refer to Textbook.

7. (*iii*) What is plexiglass? Discuss its preparation, properties and important applications.

Ans. Refer to article page of Textbook.

8. (*i*) How will you define composite materials? Briefly describe the important types of composites. What is role of matrix in the composites?

Ans. Composite Materials: *A composite is defined as a multiphase materials made up of two or more components, which are physically and chemically different phases separated by an interface.* For example, concrete is a composite. It is made up of cement, gravel and sand and has iron rods inside it to reinforce it. Wood is another example, which is a composite of cellulose and lignin.

Important Types of Composites

There are two main parts *i.e.,* the constituents of any composite material:

1. Body constituents (also called as matrix).
2. Structural constituents (also called as reinforcing phase).

1. On the basis of type of matrix, we have the following three types of composites:
   (*i*) Polymer matrix composites (PMC),
   (*ii*) Metal matrix composites (MMC), and
   (*iii*) Ceramic matrix composites (CMC).
2. On the basis of type of reinforcement used in the matrix, we have the following two types:
   (*i*) Particles which may be large or small in size. The small size particle may be dispersed in the form of a dispersion.
   (*ii*) Fibers which may be of large size giving rise to continuous phase and the small size may give rise to discontinuous phase.
3. On the basis of number of layers, we again have composites of two types:
   (*i*) Unilayer or single ply composites and (*ii*) Multiple layer composites.

All the above types of composites are very useful materials and are being extensively used in industry depending upon their specific characteristics and suitability for a particular industry. The most important type among all the various types mentioned above are **Fiber reinforced composites**. These composites are strong and light. They are often stronger than steel but they weigh much less. So, these composites can be used to make automobiles lighter and thus much more fuel efficient. They pollute the environment to much lesser extent. The main reinforcing fibers are discussed in the next part of this question.

Role of Matrix Phases in the Composites

(*i*) It binds the fibers together by virtue of its cohesive and adhesive forces.

(*ii*) It acts as a medium to transmit and distribute an externally applied load.

(*iii*) It protects the individual fibers from chemical reaction with the environment and from surface damage due to mechanical abrasion.

(*iv*) It keeps the reinforcing fibers in the proper position and orientation.

(*v*) It prevents the propagation of brittle cracks due to its relative softness and plasticity.

(*vi*) It provides interlaminar shear strength to the composite.

Requirements of a Good Matrix Materials are:

(*i*) It should be ductile.

(*ii*) The bonding strength between the matrix and fibre must be high.

(*iii*) It should have good corrosion resistance.

The Properties Controlled by Matrix Phase: Generally matrix determines the overall service temperature limitations of the composite. It also controls the environmental resistance of the composite. For example, composites based on polymers are limited to service temperature below 300°C.

8. (*ii*) What are fiber reinforced composites? Discuss unidirectional fiber reinforced composites. (8)

Ans. Fiber reinforced composites

The main reinforcing fibers are discussed below:

(*i*) **Glass Fibers** are obtained by forcing glass melt though an instrument having a number of small openings followed by cooling the emerging continuous filaments. Individual fibre (of diameter about 10 N m) is called '*monofilament*'. Usually, a "sizing agent" in liquid form is applied to glass monofilaments so as to help in handling strands and protect them from abrasion.

Glass fibres are considered as high performance materials due to the following points: (*i*) ready availability, (*ii*) low cost, (*iii*) easy to draw from molten melt, (*iv*) high specific strength, (*v*) can be economically used to produce glass-reinforced polymer composite, (*vi*) can be coupled with polymer matrix to get chemical inertness. But they can be used only at lower temperature only and do not possess the desired stiffness and rigidity for making structural components. They mainly find applications in plastic pipes, storage tanks and automobile industries to reduce vehicle weight and boost fuel efficiency.

(*ii*) **Carbon Fibres** are obtained as continuous filament by the pyrolysis in an inert atmosphere of organic fibres such as cellulose, but usually of polyacrylonitrile. The monofilaments have diameter of about 8 Nm. They are used as reinforcing materials with epoxy or polyester resins to form composites, which have higher strength/weight ratio than metals. These are high performance fibres. However, these fibres are quite costly.

Reasons for popularity of carbon fibres as reinforcing material are:

(*i*) High modulus, specific strength and stiffness even at elevated temperatures,

(*ii*) resistant to moisture, acids, bases and a number of solvents. These materials find limited applications because of their high costs. However, these are used for aircraft structural components such as wings, body and stabiliser components of helicopters and equipments like fishing rods etc.

(*iii*) **Aramid Fibers:** They are basically aromatic polyamides (aramids) like Nomex and Kevlar. For instance, Kevlar is *p*-phenylene terephthalamide. It is spun from a nematic liquid crystalline solution using sulphuric acid as the solvent and cold water as coagulant.

The Characteristic Properties of Kevlar Fibers are:

(*a*) Higher tensile strengths and tensile moduli compared to other polymeric fiber materials.
(*b*) Resistant to combustion and stability even at relatively high temperature.
(*c*) Retain mechanical properties over a long temperature range.
(*d*) Excellent toughness, impact resistance and resistance to creep and fatigue failure.
(*e*) Weak in compression and
(*f*) Susceptible to degradation by strong bases and acids.

However, the limitations of these fibers are: Aramid fibers are commonly used in polymer matrix composites. The common matrix materials are polyesters and epoxy glass.

8. (*iii*) Briefly describe the failure modes of fibre reinforced composites. (4)

Ans. Failure of Fibre Reinforced Composites: The fiber reinforced composites are generally used as structural materials of the composites, so whenever structural material ceases to perform satisfactorily, it is considered as failure of the composite. In other words, *the composites is said to have failed when it fails to respond to the normal characteristics when subject to stresses which it used to bear normally.*

In fact, prior to the macroscopic failure, the composite fails to respond normally on the microscopic level. *The failure load of a unidirectional composite is that load at which composite shows deviation from linear stress-strain behaviour or the load applied at the time of its fracture.*

The fibre reinforced composites can fail due to the following factors:

(*a*) Braking of the fibers;
(*b*) Microcracking of the matrix;
(*c*) Debonding *i.e.,* separation of fibers from the matrix and
(*d*) Delemination of laminated composite *i.e.,* separation of laminae from each other.

The fracture modes of the unidirectional composites subjected to different loading conditions are discussed below:

A. Failure Modes Under Longitudinal Tensile Load: On increasing the longitudinal tensil load on unidirectional composite, failure of fibers at their weakest cross-sectional leads to initiation of failure of the composite. As the number of broken fibers increase, complete rupture of the composite takes place because that cross-section of the composite becomes too weak to support an increase load. This is because the fibers are brittle depending upon the fiber volume fractions as under:

(*i*) When the fiber volume fraction (V_f) is less than 0.40 the unidirectional composite will undergo little failure.
(*ii*) When (V_f) is less 0.65, the composite will undergo brittle failure with fiber pull out.
(*iii*) When the volume fraction (V_f) is greater than 0.65, the composite will undergo brittle failure with fiber pull out and debonding of matrix shear failure.

B. Failure Modes of Unidirectional Composites Under Longitudinal Compressive Load: When composites are subjected to compressive loads. Continuous fibers act as long columns.

Failure modes of composites subjected to longitudinal compressive loads are summarized below:

(*i*) Transverse tensile failure

(*ii*) Fibre microbuckling in extensional mode

(*iii*) Fibre micro buckling in shear mode

C. Failure Modes of Unidirectional Composites Under Transverse Tensile Load: Unidirectional composites subjected of transverse tensile loads fail because of the tensile failure of either matrix or interface. It is because the fibers perpendicular to the loading direction generate stress concentration in the matrix and at the interface. Sometimes, when the fibers are highly oriented and weak in the transverse direction, transverse tensile failure of fibers may occur leading to failure of composite. Hence, the failure modes of composite under transverse tensile loads may be described as:

(*i*) Tensile failure of matrix and

(*ii*) Debonding of constituents and/or *fiber* spliting.

D. Failure Modes of Unidirectional Composites Under Transverse Compressive Loads: Under these circumstances, composite generally fails by shear failure of the matrix *i.e.,* type of deformation in which parallel planes of a structure remain parallel but are relatively displaced from adjacent planes to slide over each other. It may be accompanied by debonding of the constituents resulting into crushing of the fiber.

Hence, the failure modes of composite under transverse tensile load may be described as:

(*i*) Shear failure of matrix or

(*ii*) Shear failure of matrix with constituent debonding and/or fiber crushing.

Section E

9. (*i*) Why does dry cell become dead after sometime even if it has not been used?

Ans. In a dry cell, NH_4Cl is packed in a zinc container. The salt being acidic in nature get hydrolysed by water in the paste and forms hydrochloric, which corrodes the zinc container, even if the cell is not functioning.

$$NH_4Cl + H_2O \xrightarrow{\text{Hydrolysis}} NH_4OH + HCl$$

$$Zn + HCl \longrightarrow ZnCl_2 + H_2$$

9. (*ii*) Can a system have more than one triple point? Illustrate your answer with an example.

Ans. Yes, there can be more than one triple points in a system. Let us take an example of sulphur system in which there are three stable and one meta-stable triple points. A triple point in any one component system, there are three phases in equilibrium and degree of freedom is zero. According to phase rule, F = C – P + 2 or F = 1 – 3 + 2 = zero. The three phases present in equilibrium at each of the four triple points in case of sulphur system are given below.

1. The point B corresponding to transition temperature (95.6°C) of S_R to S_M, the three phases in equilibrium are: $S_R \rightleftharpoons S_M \rightleftharpoons S_V$.
2. The point C corresponding to melting point (120°C) of S_M to S_L, the three phases in equilibrium are: $S_M \rightleftharpoons S_L \rightleftharpoons S_V$.

3. The point F corresponding to melting point (155°C) of S_R to S_L, the three phases in equilibrium are: $S_R \rightleftharpoons S_L \rightleftharpoons S_V$.
4. The point E(Metastable) corresponding to metastable melting point (115°C), the three phases in equilibrium are: $S_R \rightleftharpoons S_L \rightleftharpoons S_V$. Here the extended curves meet.

9. (*iii*) What are the hardness causing constituents?

Ans. Temporary hardness is caused due to the presence of bicarbonates of calcium and magnesium. Permanent hardness is due to the presence of sulphates and chlorides of calcium and magnesium. Ca and Mg are not the only salts responsible for hardness. They are mentioned because they are most commonly present. However, all soluble salts of heavy metals such as $FeCO_3$, $Fe(HCO_3)_2$, $FeSO_4$, $FeCl_3$ etc. cause hardness.

9. (*iv*) Define COD and its significance.

Ans. COD is defined as *the amount of oxygen used for chemical oxidation of organic matter content of a sample with a strong chemical oxidising agent under acidic conditions.* Most types of organic matter are completely oxidised by boiling mixture of chromic acid and sulphuric acid to produce CO_2 and H_2O.

Significance:

(*i*) COD gives a measure of any kind of oxidisable impurities present in waste water. It is a measure of both the biologically oxidisable (like glucose) and biologically inert organic matter (like cellulose) present in waste water.

(*ii*) It is an important and quickly measurable parameter for stream and industrial waste water analysis and control of waste water plants.

(*iii*) It indicates the magnitude of the biologically inert organic matter, when used in conjugation with BOD test.

(*iv*) COD test is widely used for measuring the pollutional strength of domestic and industrial wastes.

(*v*) COD is important in management and design of plants because of its rapid determination. It is used in calculating the efficiency of treatment plants and proposing standards for discharging domestic waste water from various types of water streams.

9. (*v*) Why impure metals are more susceptical to corrosion than pure metals?

Ans. Impurities in metals generally cause heterogeneity and form minute electro-chemical cells at the exposed parts. Thus the anodic parts get easily corroded. In pure metals there is no chance of formation of electrochemical cell and therefore less chances of corrosion of pure metals.

9. (*vi*) Explain the role of pH on the nature of corrosion.

Ans. Acidic media are generally more corrosive than alkaline or neutral media. So lower the pH, greater is the corrosion. However, metals like Al, Zn, Pb etc. dissolve and are corrodes in alkalinge media. The corrosion rate of iron in oxygen free water is slow, until pH is below 5. The corresponding corrosion rate in presence of oxygen is much higher. Zinc is rapidly corroded even in weakly acidic solutions such as that of carbonic acid.

9. (*vii*) How can you distinguish $CH_3CH_2CH_2CHO$ from $(H_3C)_2CHCHO$ by proton NMR spictroscopy?

Ans. The two structures under consideration are

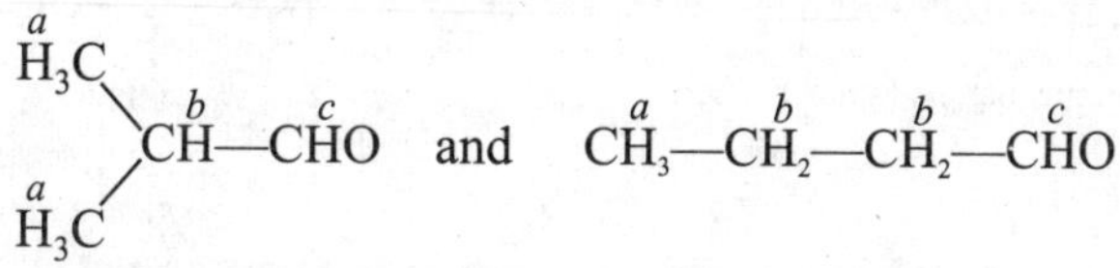

(I) (II)

For structure (I), the NMR spectrum will consist of a singlet for the two methyl groups since they contain six equivalent $\overset{a}{H_3}$ atoms, two triplets one for the proton $(\overset{b}{H})$ and the other for proton $(\overset{c}{H})$ in the aldehyde group.

For structure (II), the NMR spectrum will consist of three signals corresponding to $\overset{a}{H}, \overset{b}{H}$ and $\overset{c}{H}$. We will get a triplet for $\overset{a}{H}$, a complex multiplet (8 lines) for $\overset{b}{H}$ and a triplet for $\overset{c}{H}$ present in the aldehyde group similar to the one in structure I. Thus the proton NMR signals for the two structures will be different and the two structures can be distinguished.

9. (*viii*) Explain the terms hypsochromic (blue) and bathochromic (red) shift.

Ans. **Hypsochromic (Blue) Shift:** It is an effect due to which the absorption maximum is shifted to a *shorter wavelength.* If an electron acceptor group like acetyl group is substituted in the compound, then the absorption band is shifted to a shorter wavelength. If methyl group is introduced in an unsaturated chain of an organic compound, then conjugation is destroyed and absorption band is shifted towards a shorter wavelength. Hydrogen bonding also produces this effect as more energy is required for breaking H-bonds, hence less wavelength. Even change in solvent can also bring this effect.

Bathochromic (Red) Shift: It is an effect due to which the absorption maximum is shifted to a *longer wavelength* due to presence of a anxochrome or by change of solvent. If an electron donor group like —OH, —CH_3, —NH_2 etc. is substituted in a compound, the absorption band in shifted to longer wavelength. For example, the λ_{max} for C_6H_6 (Benzene) and C_6H_5—CH_3(Toluene) are 254 mμ and 261 mμ respectively. Conjugation of double bonds also cause this effect. For example, ethylene ($H_2C{=}C_2H$) absorbs at I_{max} = 175 nm in comparison to λ-butene ($H_2C{=}CH$—CH—CH_3) at I_{max} = 185 nm. The shift is more pronounced by increase in the number for alkyl groups. For instance in case of isobutylene, the I_{max} = 188 nm.

9. (*ix*) What is octane number? How does it effect the quality of petrol?

Ans. The octane number of a fuel is a measure of its tendency to knock, when burnt in a spark-ignition engine. It may be defined *as the percentage of iso-octane in a mixture of iso-octane and n heptane which has the same anti-knocking qualities as the fuel being tested under the similar experimental conditions.*

For example, if the octane number of a fuel is 80, it means this fuel behaves like a mixture of 80 percent iso-octane and 20 percent *n*-heptane. It has been observed that the hydrocarbon *n*-heptane produces maximum knocking when used as a fuel. Due to its very poor anti-knock quality, it has been arbitrarily assigned an octane number of zero. On the other hand 2, 2, 4 trimethyl pentane (incorrectly called iso-octane) produces minimum knocking and has been assigned an octane number of 100.

$H_3C—CH_2—CH_2—CH_2—CH_2—CH_2—CH_3$,

n-heptane, octane number = zero

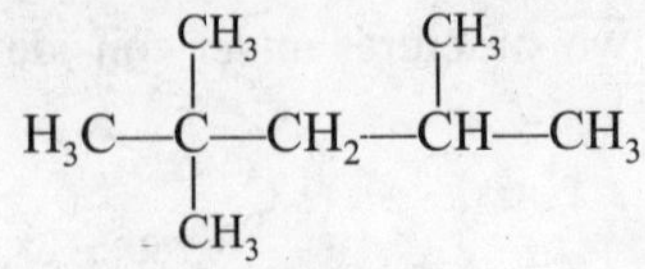

2,2,4 Trimethylpentane,
octane number = 100

It means that *n*-heptane is the worst and iso-octane the best fuel. All other fuels are rated between zero and 100.

Hence, it is clear that higher the octane number of fuel, the lesser will be its knocking tendency and better is the quality of fuel for an internal combustion engine.

9. (*x*) Why is teflon highly chemical resistant in nature? **(1)**

Ans. Teflon is polytetrafluoroethylene (PTFE). Fluorine is the most electronegative element and teflon has very large number of C—F bonds, which are highly polar. There are very strong attractive forces between the different chains of teflon. Hence, it possesses very high chemical resistance towards any type of chemicals.

9. (*xi*) Name different varieties of coal.

Ans. There are four main varieties of coal, namely; Peat, Lignite, Bituminous and Anthracite.

The bituminus type is further divided into three types namely:

Sub-bituminous, Bituminous and semi-bituminous.

MODEL QUESTION PAPERS

MODEL QUESTION PAPER–I

Times: 3 Hours **Max. Marks: 100**

The candidates shall limit their answers precisely within the answer-book (40 pages) issued to them and no supplementary/continuation sheets will be issued.

Note: Attempt five questions in all, select one question from each section A, B, C and D. Section E (Questions 9) is compulsory.

Section A

1. (*i*) Derive Nenst equation for single electrode. Explain the determination of standard electrode potential using standard hydrogen electrode. 8

(*ii*) Write a short note on ion selective electrode. 7

(*iii*) Calculate the voltage generated in the following cell.

$$Fe \mid Fe^{2+} \parallel Mn^{2+} \mid Mn$$

When iron rod is immersed in 1.0 M $FeSO_4$ solution and Mn rod is immersed in 0.1 M $MnSO_4$ solution. Given E° for $Fe^{2+} \mid Fe$ is –0.40 V and $Mn^{2+} \mid Mn$ is –1.18 V. 5

2. (*i*) Define and explain the term components of a system with suitable examples. 5

(*ii*) Draw a labelled phase diagram of water system and discuss the metastable curves and triple point. 6

(*iii*) The melting point of water system has a negative slope whereas the melting point curve of sulphur has a positive slope. Explains. 5

(*iv*) Explain the terms : stable equilibrium and metastable equilibrium. 4

Section B

3. (*i*) Explain the process of galvanisation of iron. How does it prevent the corrosion of iron? 5

(*ii*) How material selection and design can prevent corrosion? Which protective coatings can be used for corrosion prevention? 3

(*iii*) Comment on the use of Al in place of Zn for the cathodic protection of iron from rusting. 6

(*iv*) Write short notes on (a) stress corrosion (h) water line corrosion. 6

4. (*i*) What do you mean by hardness of water? How is it determined by complexometric titration (EDTA Method) 9

(*ii*) Distinguish between carbonate and bicarbonate hardness of water with examples. 3

(*iii*) Differentiate between BOD and COD. 5

(*iv*) Why does hard water consumes a lot of soap? 3

Section C

5. (*i*) Explain the following in respect of UV-visible spectroscopy

(*a*) Red shift or bathochromic effect

(*b*) Hyperchromic effect

(*c*) Blue shift or hypochromic effect

(*d*) Isobestic point 8

(*ii*) State and explain the principle of NMR spectroscopy. 4

(*iii*) Describe a UV-visible double spectrophotometric. 4

(*iv*) On the basis of Infra Red spectroscopy how can you distinguish among 1-hexyne, 1-hexene and hexane? 4

6. (*i*) What is the significance of octane number and cetane number and for which these terms are used? How can these be improved? 8

(*ii*) What is meant by calorific value of a fuel? Distinguish between Gross and Net calorific values of a fuel. 3

(*iii*) Differentiate between chemical fuel and nuclear fuel. Give complete classification of chemical fuels with example. 4

(*iv*) Give in detail the composition and manufacture of water gas. What is its calorific value? 5

Section D

7. (*i*) Write the preparation, properties and uses any one thermoplastic and thermosetting material. Distinguish between the thermoplastic and thermosetting polymers. 8

(*ii*) Write a short note on Elastomers. What is vulcanisation of rubber? Mention its uses. 5

(*iii*) What are polyurethanes? What is the principal linkage is Polyurethane? 3

(*iv*) How is Neoprene prepared? Mention its properties and uses. 4

8. (*i*) What are composites? How are the composite materials classified? Discuss the properties and applications of Fibre reinforced polymeric composites. 8

(*ii*) Distinguish between fibre and particle reinforced composites. 5

(*iii*) State the factors on which the properties of a composite material depend. Enlist the general characteristics of composites. 5

(*iv*) Give names of two natural and two synthetic composites. 2

Section E with Solutions

9. (*i*) What is a reference electrode? Why use of calomel electrode as a reference electrode preferred? 2

Ans. (*i*) The electrode which is used for measuring the electrode potential of another electrode by combining with it to form a cell is called a reference electrode. The electrode potential of NHE electrode is taken as zero and it can be used as a reference electrode. NHE is obtained by dipping platinum foil in 1 M- HCl solution through which hydrogen gas is passed at 298 K under 1 atm. pressure. The use of NHE as a reference electrode is not convenient because of the following:

(*i*) It is difficult to maintain a pressure of H_2 gas at 1 atm. pressure and $[H^+]$ ion concentration of solution at 1 M.

(*ii*) The platinum electrode used get poisoned by the traces of impurities.

(*iii*) Handling of NHE is difficult.

It is, therefore, replaced by other reference electrodes. The most commonly used is calomel electrode represented as Hg; $Hg_2Cl_2(s)$ KCl solution.

Its reduction potential depends upon the concentration of KCl solution used.

When saturated KCl solution is used, it is referred as SCE (Saturated Calomel Electrode). It has a reduction potential of 0.5415 V. With 1.0 N KCl solution (known as NCE), the potential is 0.3338 V. The calomel electrode, once set, can be used for long..

(ii) Why is salt bridge used in the construction of a cell? **2**

Ans. (*ii*) Salt bridge completes the electrical circuit and permits the ions to move from one solution to another without direct contact of the two. It avoids mixing of the two solutions. Moreover, it maintains electrical neutrality of the solutions in the two half cells. A wire can not be used to connect the two electrodes internally because it produces a voltage drop. Consequently, accurate measurement of the emf can not be made.

(iii) What is condensed Phase Rule? When is it applied? **2**

Ans. (*iii*) Phase rule equation, F = C – P + 2; where F, C and P represents degree of freedom, number of components and number of phases in a system respectively. When this phase rule equation is used for two component system and with minimum number of phases *i.e.,* one, the above equation gives F as

$$F = C - P + 2 = 2 - 1 + 2 = 3$$

When F = 3 for two component systems, the phase behaviour of a binary system shall have to be represented by three dimensional diagrams representing pressure, temperature and composition along three axis. This can not be conveniently drawn on paper, since a three dimensional picture would give a model and not a diagram.

However, in practice, a solid-liquid equilibrium has no gaseous phase and effect of pressure is almost negligible. Therefore, keeping the pressure constant, the phase rule can be revised as :

$$F' = C - P + 1.$$

This equation is known as *condensed phase rule equation*. Here the maximum value of F = 2 for a two component system. So solid-liquid phase equilibria are represented on temperature composition (*t-c*) diagrams. For such systems, pressure is taken as constant because vapour is neglected and condensed phase rule is applied.

(iv) Why does part of nail inside the wood undergoes corrosion easily?

Ans. (*iv*) The part of the nail inside the wood is not exposed to atmospheric conditions; whereas the remainder part of it is exposed directly to atmospheric air. Due to differential aeration, the nail inside the wood (not oxygenated part) becomes anodic, while the upper part of nail (well-oxygenated) becomes cathodic. Cosequently, a differential current starts flowing, whereby the anodic part (nail inside the wood) gets corroded easily.

(v) Why do we express hardness of water in terms of $CaCO_3$ equivalent? **2**

Ans. (*v*) The hardness of water is expressed in terms of $CaCO_3$ equivalent because $CaCO_3$ is the most insoluble compound in water treatment. Moreover, its molecular mass is 100 and it makes the calculations of hardness comparatively easier. Although hardness of water is never due to $CaCO_3$ because, being insoluble, it is not present in water as such. The hardness of water is due to soluble salts like bicarbonates, chlorides and sulphates of calcium and magnesium, but it is a convenient way of expressing the hardness in terms of $CaCO_3$ equivalents of the salts causing hardness.

(vi) Most of absorption bands in the UV-visible spectra are very broad. Give reasons. **2**

Ans. (*vi*) UV-visible spectra involve transitions between electronic energy level which are associated with vibrational and rotational energy levels. The electronic transition may occur from any of the several vibrational and rotational states of one electronic level to any of the various corresponding states of higher level. Since a large number of transitions are possible, so a large number of absorptions will take place corresponding to different wavelengths. Hence, spectra are very broad.

(vii) Predict the NMR spectrum of $\underset{c}{CH_3}\underset{b}{CH_2}\underset{a}{OH}$. **2**

Ans. (*vii*) CH_3CH_2OH has three kinds of protons *viz.* $\underset{c}{CH_3}\underset{b}{CH_2}\underset{a}{OH}$

for protons *a*, $n = 2$ so triplet is observed due to coupling with —CH_2–group at downfield.

for protons *b*, $n = 3$ and $n' = 1$, so $(3 + 1)(1 + 1) = 8$ lines (multiplet) is observed at in between.

for protons *c*, $n = 2$ so triplet is observed due to coupling with —CH_2–group at up-field.

(viii) Why is Net calorific value less than Gross calorific value? When are these same in magnitude? **2**

Ans. (*viii*) Gross calorific value is the total amount of heat liberated, including latent heat of steam when one unit of the fuel is completely burnt. On the other hand, Net calorific value in the net amount of useful heat liberated excluding the latent heat of steam, when one unit of the fuel is completely burnt. So, net calorific value is less than gross calorific value by an amount equal to the energy equiualant of latent heat of steam.

When the fuel does not contain hydrogen (*e.g.*, pure carbon, sulphur etc.), then the latent heat of steam for changing H_2O (*l*) to H_2O (vapour) does not come into operation, because no water is formed. So the two types of calorific values are same is magnitude.

(ix) Why do all simple organic molecules not produce polymers? **2**

Ans. (*ix*) The simple organic molecules like C_2H_5Cl, C_2H_5OH, H_3C COOH etc. do not form polymers, because they are mono functional molecules. To form a polymer, the monomer molecule must be at least bifunctional For example, ethylene forms polythylene. The double bond in ethylene provides two bonding sites per molecule. So a large number of ethylene molecules can join with one another to form polythylene. Similarly, ethylene glycol (HO—CH_2—CH_2—OH) is a bifunctional molecule and can act as a monomer for the formation of a polymer (polyethylene terephthalate).

(x) Write a short note on : Fibre Reinforced Polymer Composites.

Ans. (*x*) Fiber-reinforced polymer composites materials in which a fiber made from a suitable material such a glass, is embedded in another material, called matrix. Polymer composites are usually made of two components:

(*i*) Fiber (*ii*) Matrix

(*i*) **Fiber:** The fiber is normally glass but sometimes carbon fiber or polyethylene are also used. Glass is the most popular fiber because it is really cheap. Glass fibers are quite strong and flexible. There are still stronger fibres available like carbon fiber and polyethylene. Polyethylene fibres are even stronger than carbon fibers.

(*ii*) **Matrix:** The matrix is usually a thermosetting polymer like epoxy resin or polimide. Epoxyresins do not absorb water and they do not shrink very much when they are cured. They can be used upto a temperature of 160°C but for higher temperature polimides are good, but they absorb a lot of water which causes them to breakdown.

The fiber is embedded in the matrix to make the matrix stronger. These composites are strong and light. They are often stronger than steel but weigh much less. This means that composites can be used to make automobiles lighter, and thus much more fuel efficient. This means they pollute less, too.

❑❑❑

MODEL QUESTION PAPER–II

Times: 3 Hours **Max. Marks: 100**

The candidates shall limit their answer precisely within the answer-book (40 pages) issued to them and no supplementary/continuation sheets will be issued.

Note: Attempt five questions in all, select one question from each section A, B, C and D. Section E (Questions 9) is compulsory.

Section A

1. (*i*) Define fuel cell. Explain the construction and working of hydrogen-oxygen fuel cell. What are the advantages and limitations of fuel cell? Why is water formed in this cell removed continuously. 10

(*ii*) How does fuel cell differ from battery? 5

(*iii*) Write short note on fuel cell. How it is different from commercial cell. 5

2. (*i*) Describe zinc-magnesium alloy system, with the help of a neat phase diagram. 8

(*ii*) Mention the merits of phase rule. 7

(*iii*) Differentiate between true equilibrium and metastable equilibrium. 5

Section B

3. (*i*) How hardness is removed by soap-titration method? 8

(*ii*) 50 ml of a sample water consumed 15 ml of 0.01 M-EDTA after boiling. Calculate total hardness, permanent hardness and temporary hardness. 7

(*iii*) Mention various units of hardness of water. 5

4. (*i*) What is Stress Corrosion? Give its causes. Explain the mechanism of Stress Corrosion. Give types of Stress Corrosion. 10

(*ii*) Write short notes on :

(*a*) Sacrificial anodic protection 3

(*b*) Cathodic protection 4

(*c*) Pilling Bedworth's rule 3

Section C

5. (*i*) In relation to NMR spectra, explain the following terms: 12

(*a*) Chemical Shift

(*b*) Spin-spin Coupling

(*c*) Shielding-deshielding effect.

(*ii*) Why molecules absorb in UV-visible region? 4

(*iii*) What are the types of electronic transitions that can occur in a molecule? Discuss giving examples. 4

6. (*i*) Name the parameters, which are studied under proximate and ultimate analysis of coal. How ultimate analysis is carried out? Give its significance. 12

(*ii*) Write short notes on:

(*a*) Oil gas (*b*) Coal gas 8

Section D

7. (*i*) Differentiate between addition polymerisation and condensation polymerisation with suitable examples. 7

(*ii*) What is Bakelite? How is it prepared? Give its properties and uses. 8

(*iii*) How will you synthesize nylon 6 : 6 from 1, 3-butadiene? 5

8. (*i*) What are carbon fibres? How are they obtained? Mention their unique characteristics. 8

(*ii*) Write short note on fibre reinforced composites. 7

(*iii*) What are composites? Give their applications. 5

Section E with Solutions

9. (*i*) Why does an electrochemical cell stop giving current after sometime? 2

Ans. (*i*) The flow of electric current is due to the potential difference between the two electrodes. With the passage of time, the reduction potential of anodic half cell increases while that of the cathodic half cell decreases and a stage is reached when the electrode potentials of the two electrodes become equal and hence the flow of current stops. With the passage of time the concentration of Zn^{2+} ions in the anodic half cell of Daniel cell increases resulting in the decrease of reduction potential. On the other hand the concentration of Cu^{2+} ions in the cathodic half cell decreases resulting into the increase of reduction potential. When the two reduction potentials become equal, there is no tendency for the electrons to move from one electrode to the other and the cell stops working.

(*ii*) Can we measure the absolute value of the electrode potential? Explain your answer. 2

Ans. (*ii*) No, the absolute value of electrode potential can not be measured. An electrode has to be coupled with a reference electrode to set up an electrochemical cell. Hence the absolute value of electrode potential can not be measured. It is only the relative value of electrode potential with reference to a standard electrode such as NHE or calomal electrode. The standard electrode potential of NHE is taken to be zero. So, the measured voltage of the cell set up with NHE gives the standard electrode potential of the other electrode. No current flows from an isolated half cell, the current flows only when it is joined with some other electrode.

(*iii*) Define and explain degree of freedom. 2

Ans. (*iii*) *The number of degrees of freedom is the smallest number of variable factors i.e., temperature, pressure and concentration of the components which must be arbitrarily fixed in order that the condition of the system may be completely defined.* For example, in ice $\rightleftharpoons$ water $\rightleftharpoons$ vapour system, there are three phases of one component (H_2O). So for three phases to exist simultaneous by in equilibrium.

∴ According to Gibb's Phases rule:

$$F = C - P + 2$$

$$F = 1 - 3 + 2 = 0 \qquad \therefore F = 0$$

Thus three phases can co-exist in equilibrium only at one particular temperature and under one particular pressure. Any variation of these factors will lead to disappearance of one or more phases. Hence, this system has no degrees of freedom or it is non-variant.

(*iv*) A copper equipment should not possess a small steel bolt. Give reason. 2

Ans. (*iv*) Iron (higher in electrochemical series) acts as anodic end is attached and corroded; while copper equipment (lower in electochemical series) acts as cathode and is not attacked at all. Since anodic area is very small. So it is competely corroded in due course of time and the action of bolt is finished, *i.e.*, equipment loses its bolt after sometimes.

(*v*) What do you understand by an isobestic point? 2

Ans. (*v*) An isobestic point for different absorption spectra is that point where the wavelength corresponding to these spectra is equal and the corresponding wavelength at this point is called *isobestic wavelength*. The absorbance (ε_{max}) and λ_{max} values of different species are, therefore, equal at this point. Although λ_{max} is the characteristic properly of a compounds, yet different compounds may possess same value of λ_{max} by changing the solvent or pH values of the solutions.

Some substances exists in two different forms at diferent pH values of the solution. For instance, the acid base indicators such as phenolphthalein and methyl orange show a colour change at the end point due to change of benzenoid form into the quinoid form with change in pH value of the solution. At the end point both the forms co-exists in equilibrium. On comparing the absorption spectra of the two forms, which coexists and have the same value ε_{max} at isobestic wavelength. Thus, the appearance of an isobestic point confirms that two forms of the same substance co-exists in equilibrium.

(*vi*) How many NMR signals do you expect from each of the following compounds, indicate also the splitting pattern of the various signals.

(*a*) CH_3OCH_3 **(*b*) $CH_3OCH_2CH_3$** **(*c*) CH_3CH_2OH** 2

Ans. (*vi*) (*a*) One signal (singlet); (*b*) Three signals (singlet, quarlet, triplet); and (*c*) Three signals (singlet, quartlet, triplet).

(*vii*) Why a good fuel must have low ash content? 2

Ans. (*vii*) Ash is useless non-combustible matter present in a fuel. It reduces the calorific value, causes hindrances in the flow of air and heat, thereby reducing the temperature attained. Moreover, it causes trouble during firing by forming clinkers. Hence, lower the ash content, better the quality of fuel.

(*viii*) What are eutectics? Give an example. 2

Ans. (*viii*) A binary system consisting of two substances, which are miscible in all proportions in the liquid phase, but do not react chemically is known as eutectic system. Eutectic means easy to melt. For example, a mixture of Pb-Ag, Zn–Cd, and Bi–Cd etc. are common Eutectic systems, so these two substances capable of forming solid solution with each other and have the property of lowering each other's freezing point. The minimum freezing point is obtained at a definite composition and lowest possible freezing point obtained from this composition is known as Eutectic point.

Low melting alloys *i.e.*, Eutectics are used in safety devices (as plugs in automobiles). Fire sprinklers and as safety device in boilers and also in fuse plugs at home.

(*ix*) Name the first and last stage in the calorification of wood. Arrangte different varieties of coal in the decreasing order of their calorific value. **2**

Ans. (*ix*) The first and the last stage of calorification of wood are peat and anthracite sespectively. The decreasing order of calorific values of different varieties of coal are:

Anthracite < Bituminous < Lignite < Peat

(*x*) Why is Teflon highly chemical resistant? Point out its one importance? **2**

Ans. (*x*) Due to presence of most electronegative element, F, in teflon, there are very strong attractive forces between its different chains. Hence, it possesses extremely high chemical resistance towards most of the chemical. Its most important application is that it is used for making nonstick utensils.

❑❑❑

MODEL QUESTION PAPER–III

Times: 3 Hours **Max. Marks: 100**

The candidates shall limit their answers precisely within the answer-book (40 pages) issued to them and no supplementary/continuation sheets will be issued.

Note: Attempt five questions in all, select one question from each section A, B, C and D. Section E (Questions 9) is compulsory.

Section A

1. (*i*) What is a concentration cell? Explain with example. 5
   (*ii*) Discuss concentration cell with and without transference. 8
   (*iii*) What is electrochemical series? Give its applications with suitable examples. 7
2. (*i*) State the reduced phase rule and discuss its application to lead-silver system. 10
   (*ii*) What is meant by phase diagram? With the help of phase diagram, explain the following terms. (*a*) Triple point (*b*) Eutectic point 5
   (*iii*) Explain the applications of phase diagrams. 5

Section B

3. (*i*) Explain Boiler corrosion. What are its causes? How it is removed. 8
   (*ii*) Calculate the temporary and total hardness of a sample of water containing $Mg(HCO_3)_2$ = 73 mg/l, $Ca(HCO_3)_2$ = 162 mg/l, $MgCl_2$ = 95 mg/l, $CaSO_4$ = 136 mg/l. 5
   (*iii*) What is hardness of water? How is it determined by O.Hehner's method. 7
4. (*i*) Explain preventive measure of corrosion. 8
   (*ii*) Why impure metals are more susceptible to corrosion than pure metals. 4
   (*iii*) Pitting corrosion is dangerous, explain. 4
   (*iv*) Corrosion is reverse to extraction. Justify. 4

Section C

5. (*i*) Define Infrared Spectroscopy. Describe the various molecular vibrations in this technique. 7
   (*ii*) Give different types of electronic excitations
   (*iii*) Write short notes on: (*a*) Chromophore (*b*) Auxochrome 4
6. (*i*) Explain proximate analysis of coal? How is it carried out? What is it significate. 8
   (*ii*) What are the advantages of gaseous fuels over solid and liquid fuels. 7
   (*iii*) What is producer gas? How it is prepared? Give its properties and uses. 5

Section D

7. (*i*) Distinguish between thermoplastic and thermosetting polymers. 8

(*ii*) Describe the method of preparation, properties and applications of the following: 12

(*a*) Teflon

(*b*) Polyvinyl chloride

(*c*) Polystyrene

8. (*i*) What are composites? Give constituents of composites. 8

(*ii*) Distinguish between fibre and particle reinforced composites? 7

(*iii*) Write short note on particulate composites. 5

Section E with Solutions

9. (*i*) What are the salts responsible for the temporary and permanent hardness of water? 2

Ans. (*i*) **Temporary Hardness:** $Ca(HCO_3)_2$ and $Mg(HCO_3)_2$

Permanent Hardness: $CaCl_2$, $MgCl_2$, $CaSO_4$, $MgSO_4$, $FeSO_4$, $Al_2(SO_4)_3$ etc.

(*ii*) Why thermosetting plastics can not be reused and reshaped? 2

Ans. (*ii*) Thermosetting polymers have three dimensional, cross linked net worked structure. Nighbouring polymeric chains in them are held together by cross links, which are strong covalent bonds. Heating does not soften them, since softeing would require breaking of covalent bonds. On strong heating they coould burn without getting softened. So, they can not be reused or reshaped.

(*iii*) Define Octane number and cetane number. 2

Ans. (*iii*) **Octane Number:** The percentage of iso-octane in a mixture of iso-octane and *n*-heptane, which matches the gasoline under consideration in knocking characteristics.

Cetane Number: The percentage of hexadecane in a mixture of hexadecane and 2-methylnaphthalene, which possesses the same ignition characteristics as the diesel fuel under consideration.

(*iv*) What is the effect of temperature on molar conductivity? 2

Ans. (*iv*) On increasing the temperature, the molar conductivity increases, because at a higher temperature, the ionic mobilities increases.

(*v*) Corrosion of water filled steel tanks occur below the waterline. Give reason. 2

Ans. (*v*) This is because the area above the waterline is highly oxygenated and acts as cathode while the fully immersed part is poorly oxygenated and acts as anode. So the anodic part gets corroded, while the cathodic part remains completely unaffected by corrosion.

(*vi*) What is meant by polymorphism? Given an example. 2

Ans. (*vi*) The existence of a solid substance in more than one crystalline forms is called polymorphism. For example, sulphur exist in two different polymorphic forms which differ in their physical properties, which are Sulphur Rhombic (S_R) and Sulphur Monoclinic (S_M).

(*vii*) Predict the proton NMR spectra of H_2, CH_4, C_2H_6 and C_6H_6. 2

Ans. (*vii*) All the four molecules contain equivalent protons which do not show spin-spin interaction. Hence, their proton NMR spectra will consist of one signal only is a singlet.

(viii) Name any two particulate–composites. **2**

Ans. (*viii*) Cermet and concrete.

(xi) Small anodic area results in intense corrosion. Give reason. **2**

Ans. (*ix*) Corrosion is more rapid, intense and highly localised, if the anodic area is small, because the current density at a smaller anodic area is very large. As a result, the demand for electrons by the large cathodic area can be met by the smaller area only by under going corrosion more rapidly.

(x) What is reforming? **2**

Ans. (*x*) A process of bringing about structural modifications in the component of straight run gasoline, with the object of improving its anti-knock properties is called reforming. If this process the straight chain hydrocarbons are convertcd to their branched chain isomers, cyclisation and aromatisation may also take place.

❑❑❑

MODEL QUESTION PAPER–IV

Times: 3 Hours **Max. Marks: 100**

The candidates shall limit their answers precisely within the answer-book (40 pages) issued to them and no supplementary/continuation sheets will be issued.

Note: Attempt five questions in all, select one question from each section A, B, C and D. Section E (Questions 9) is compulsory.

Section A

1. (*i*) What is meant by electrochemical series? Discuss briefly the various uses to which the series can be put. Explain why Zn reacts with H_2SO_4 to give H_2 gas but Ag does not react. 8

(*ii*) Describe the construction of Zn-MnO_2 battery with relevant reactions occurring during discharge. 7

(*iii*) Calculate the EMF of the given cell at 298°K.

$$Ag(s) \mid AgNO_3\ (0.018\ m) \parallel AgNO_3(1.2\ m) \mid Ag(s)$$ 5

2. (*i*) What is condensed phase rule? When it is applied. Explain lead-silver system on this basis. Explain the terms

(*a*) Eutectic point

(*b*) Pattinson's process of desilverisation. 12

(*ii*) Justify the statement, "The eutectic is a mixture and not a compound." 3

(*iii*) Define the terms:

Phase, Component, Degrees of freedom. 5

Section B

3. (*i*) Describe the process of sewage treatment in detail. 8

(*ii*) Discuss, in brief, the boiler corrosion. How is it controlled? 8

(*iii*) Define degree of hardness. Why it is determined in terms of $CaCO_3$ equivalent. 4

4. (*i*) Explain various factors which are responsible for corrosion. 8

(*ii*) Explain how corrosion can be controlled by proper designing? 4

(*iii*) Write short note on

(*a*) Waterline corrosion 4

(*b*) Differential aeration corrosion 4

Section C

5. (*i*) Explain Absorption Laws in detail 5

(*ii*) Indicate diagrammatically the splitting of signals in NMR spectra of

(*a*) CH_3CH_2Br (*b*) CH_3CHBr_2

(*c*) $CH_2Br—CHBr_2$ 7

(*iii*) Write short notes on:

(*a*) Fluorescence

(*b*) Phosphorescence 8

6. (*i*) What is reforming of petrol? How does reforming increase Octane number? Give any two reforming reactions. 5

(*ii*) Differentiate proximate and ultimate analysis of coal. 10

(*iii*) Discuss in brief the proximate analysis of coal. 5

Section D

7. (*i*) Explain classification of polymers, on the basis of:

(*a*) Source (*b*) Structure

(*c*) Synthesis (*d*) Molecular forces 10

(*ii*) What is plexiglass? How is it prepared? Give its properties and applications. 5

(*iii*) Give the synthesis of epoxyresin. Mention its important uses.

8. (*i*) What are different types of composites? 10

(*ii*) Explain the failure of fibre-reinforced composites. 7

(*iii*) Write short note on Layered composites. 3

Section E with Solutions

9. (*i*) The presence of CO_2 in boiler feed-water should be avoided. Why? 2

Ans. (*i*) CO_2 forms carbonic acid (H_2CO_3), which slowly attacks the boiler materials, thereby making the boiler weaker and weaker progressively.

(*ii*) What is an elastomer? Explain the cause of their eleastic character. 2

Ans. (*ii*) Any rubber–like elastic polymer, which can be stretched by applying force, but it returns to its original shape and size as soon as stretching force is removed. They have long flexible chains with weak intramolecular forces and occasional cross-links between the chains. Moreover, there are no polar groups or sites for hydrogen bonding. For elongation and recovery the long chains of an elastomer must be connected to each other by a few cross links. These cross-links prevent the shipping of molecular across each other, but at the sometime the flexibility of the chain is preserved.

(*iii*) Why are gaseous fuels more advantageous than solid fuels. 2

Ans. (*iii*) Gaseous fuels possess advantages over solid fuels.

(*i*) They can be distributed and transported easily through pipelines.

(*ii*) They can be ignited at moment's notice.

(*iii*) Their combustion can be controlled easily.

(*iv*) They burn without producing smoke and ash.

(*v*) They possess high calorific values.

(*vi*) They do not affect the quality of metal produced, when used in metallurgy as fuel.

(iv) Specific conductance decreases with dilution, while the equivalent conductance increases. Give reason. **2**

Ans. (*iv*) When an electrolytic solution is diluted, two things happens, namely degree of dissociation increases, but the total volume increases. The number of ions per cm^3 decreases, because the effect of increased volume in decreasing the number of ions cm^3 out weighs the minor increase in conductance with increase of dilution. On the other hand, equivalent conductance increases, because the total number of ions per gram equivalent increases with dilution, due to increased degree of ionization.

(v) Bolt and nut made of the same metal is preferred in practice. Why? **2**

Ans. (*v*) Bolt and nut made of the same metal is preferred, because such a combination will not permit galvanic corrosion to takes place.

(vi) What is metastable state? Explain with an example **2**

Ans. (*vi*) The state of supercooled or supersaturated solution in which the phase, which is normally stable under the given conditions, does not form is called metastable state. For example, in water system the vapour pressure curve of water can be extended by super cooling water to minus zero °C carefully. But on slight disturbance, the supercooled phase at once changes to the stable phase ice. The special feature of the vapour pressure curve of supercooled water *i.e.*, metastable curve is that its vapour pressure is greater than that of solid ice, which is the stable phase at that temperature.

(vii) Predict the NMR spectrum of $CH_2Cl—CHCl_2$. **2**

Ans. (*vii*) $\overset{a}{C}H_2Cl—\overset{b}{C}HCl_2$ has two sets of equivalent protons. So number of signals = 2. For proton '*a*', $n = 1$ so it gives a doublet as only one hydrogen atom is attached to neighbouring carbon. For proton '*b*', $n = 2$ so it gives a triplet as two hydrogen atoms are attached to neighbouring carbon.

(viii) How are carbon fibres formed? **2**

Ans.(*viii*) They are obtained as continuous filament by the pyrolysis in an inert atmosphere of organic fibres such as cellulose, but usually of polyacrylonitrile.

(xi) What is the composition of the salt bridge used between two half cells? Which type of salt is used? **2**

Ans. (*ix*) It is an inverted U-tube containing a paste of KCl or NH_4NO_3 and agar gel. The salt such as KCl or NH_4NO_3, in which the transport number of cation and anion is almost equal are used. So that equal amount of current is carried out towards the anode and the cathode.

(x) Arrange the chromophores present in CH_3OH, CH_3Br, $CH_2{=}CH_2$ and $CH{=}CH$ in increasing order of their λ_{max} values. **2**

Ans. (*x*) $CH_3OH < CH_3Br < CH_2{=}CH_2 < CH{\equiv}CH$.

MODEL QUESTION PAPER–V

Times: 3 Hours **Max. Marks: 100**

The candidates shall limit their answers precisely within the answer-book (40 pages) issued to them and no supplementary/continuation sheets will be issued.

Note: Attempt five questions in all, select one question from each section A, B, C and D. Section E (Questions 9) is compulsory.

Section A

1. (*i*) What are reference electrodes. Discuss primary and secondary reference electrodes. Describe the construction of calomel electrode. 8

(*ii*) Describe the construction of Zn—MnO_2 battery with working principle. 7

(*iii*) What is single electrode potential? Derive Nernst equation for single electrode potential. 5

2. (*i*) Define and explain the terms involved in phase rule. 10

(*ii*) If there is no triple point in the diagram of one-component system, then what inference do you draw. 3

(*iii*) State phase rule and draw sulphur system on this basis 7

Section B

3. (*i*) What is the principle of EDTA method? Describe the estimation of hardness of water by EDTA method. 10

(*ii*) Distinguish between COD and BOD. 5

(*iii*) What happens when temporary hard water is boiled (give equations)? 5

4. (*i*) Differentiate between chemical corrosion and electrochemical corrosion. 8

(*ii*) Explain the mechanism of hydrogen evolution and oxygen absorption in electrochemical corrosion. Illustrate your answer with figure. 10

(*iii*) Aluminium corrodes in alkaline medium but iron does not, explain. 5

Section C

5. (*i*) What do you mean by shielding and deshielding effects involved in NMR spectroscopy? 8

(*ii*) Briefly explain the principle of IR spectroscopy. 5

(*iii*) Describe a UV-visible double-beam spectrophotometer. 7

6 (*i*) What are fuels? Give their classification. What are the characteristics of good fuel? 10

(*ii*) What is meant by cracking of petroleum? Explain fluidized-bed catalytic method of obtaining gasoline. Give its mechanism. 5

(*iii*) Write short note on watergas. 5

Section D

7. (*i*) What are plastics. Mention their properties and uses. 5

(*ii*) Describe the method of preparation, properties and applications of urea-formaldehyde resin. 8

(*iii*) Write short note on elastomers. Differentiate between elastomer and plastic. 7

8. (*i*) What are composites? Give names of natural composites. What are the advantageous characteristics of composites? 8

(*ii*) Explain the failure of Fibre-reinforced composites. 8

(*iii*) Write short note on Layered composites. 4

Section E with Solutions

9. (*i*) What is degree of hardness of water? Classify water on its basis. 2

Ans. (*i*) It is the parts of calcium carbonate equivalent hardness present in a definite number of parts of water, depending upon the unit employed. It is usually expressed in ppm or mgL^{-1} units. Water can be classified into the following types the basis of degree of harduess.

(*i*) **Softwater:** When degree of harduess is less than 75 ppm.

(*ii*) **Moderately hard water:**When degree of hardness is between 75–150 ppm.

(*iii*) **Hard water:** When degree of hardness is between 150–300 ppm.

(*iv*) **Strongly hard water:** When degree of hardness is above 300 ppm.

(*ii*) What is vulcanisation? In what respects vulcanised rubber differ from natural rubber? 2

Ans. (*ii*) The process of heating natural rubber with sulphur to improve its properties is called vulcanisation. During this process, sulphur cross-links are formed. The formation of cross-links makes ruffer hard, tough with greater tansile strength. The vulcanised rubber has excellent elasticity, low wtaer alsorption tendercy, ressestance to exidation and solvents. The vulcansied rubber has high wear and tear resistance and can be used over wide rangs of temperatures. In vulcanised rubber, the polymer chain are held toghther by polysulphide bridges, or cross-links. When the streching force is applied, the chains can be straightened out, but they can not slip accross again and the rubber returns to its original shape.

(*iii*) Is catalytic reforming an exothermic or endothermic process? 2

Ans. (*iii*) Endothermic, since the strong C—H bonds require sufficient amounts of energy to break them. This energy is not retrieved by making the bonds in double bonds, created during reforming.

(*iv*) Why do we use a very small current during the determination of transport number? 2

Ans. (*iv*) We use a very small current during determination of transport number in order to minimise thermal and resulting diffusion effects. This inturn causes change in concentration only around the two electrode.

(*v*) What is galvanic corrosion? 2

Ans. (*v*) When two dissimilar metals are electrically connected and immersed in an electrolyte, the metal higher in electrochemical series undergoes corrosion. This is called galvanic corrosion.

(vi) What is triple point of water ? How does it differ from the Freering point? 2

Ans. (*vi*) A point at which the gaseous, liquid and solid phases of the system co-exist in equilibrium. At this point degree of freedom (F) is zero. For a one component system, F = C – P + 2 = 1 – 3 + 2 = Zero. Triple point is a self defined point corresponding to definite temperature 0.0098°C and definite pressure, *viz.*, 4.58 mm. The three phases of water can co-exist only at this temperature and pressure. On making any change in these, one or two phases would disappear. It is different from ordinary freezing point *i.e.,* 0°C at 760 mm, where only two phases (solid and liquid) can co-exist.

(vii) "IR spectra is often characterized as molecular finger points." Justify statement. 2

Ans. (*vii*) Different functional groups produce recognizable peaks at nearly definite positions in IR spectra. Most of the peaks can be assigned due to the presence of specific groups in the molecule. So IR spectra can be used in establishing the identity of the compounds. It can be characterised as finger prints of the molecule.

(viii) What are advantages of composites over conventional materials? 2

Ans.(*viii*) The main advantages of composites over conventional materials namely wood, metals, alloys, ceramics etc. are:

(*a*) Composites are more economical than conventional materials.

(*b*) Weights of the composites are lesser as compared to conventional materials. The weight saving contribute towards minization of fuel consumption in automobiles, space ships, aircrafts and other machinery.

(*c*) Composite show excellent mechanical properties such high strength and stiffness, excellent impact and damage tolerance characteristics, outstanding friction and wear properties etc.

(*d*) They also show excellent chemical properties such as higher corrosion resistance and higher thermal stresses as compared to metals and alloys.

(*e*) They can provide improved insulation because of less heat radiation losses from their surface as compared to metals.

(xi) ^{13}C is NMR active while ^{12}C is not. Explain. 2

Ans. (*ix*) ^{12}C does not have a nuclear spin and cannot be used in an NMR experiment. In contrast, its isotope ^{13}C, whose natural abundance is only 1.1 %, has a nuclear spin of 1/2.

(x) What is a fuel cell? Why water is removed from H_2–O_2 fuel cell? 2

Ans. (*x*) A device for converting the energy of a fuel directly into electrical energy is called fuel cell. In H_2–O_2 fuel cell, the only product discharged by the cell is water. Water should be removed at the same rate at which it is formed. Otherwise, it will dilute the KOH and the cell becomes inoperative. Therefore, the electrolyte is kept hot, so that water evaporates as fast as it is formed.

❑❑❑

Index

❑❑❑

Win Prizes

ATTENTION READERS

Win Prizes

We request for your frank assessment regarding various aspects of the book as given below:

NEW COURSE ENGINEERING CHEMISTRY

(Theory and Practicals)

BY : S.K. Bhasin; Punit Bhardwat

Please Fill up the Sheet in Neat and Clean Writing

(i) What topics of your syllabus, important for your examination point of view, are not covered in this book ?

....

....

....

....

(ii) What are the chapters wherein the treatment of the subject matter is not systematic or organised or updated ?

....

....

....

....

(iii) Have you come across misprints/mistakes/factual inaccuracies in the book ? Please specify the chapters and the page numbers.

....

....

....

....

(iv) What other books on the same subject have you found/heard better than the present book ? Please specify in terms of price and quality.

....

....

....

....

(v) Further suggestions and comments for the improvement of this book.

....

....

....

....

(PS : If need be please attach a separate sheet to write your views)

OTHER DETAILS

(i) Who recommended you this book ? Please tick (✓) in the box near the option relevant to you.

☐ Teacher ☐ Friends ☐ Bookseller

....

....

....

....

(ii) Name of the recommending teacher, his designation and address:

....

....

....

....

(iii) Name and address of the bookseller you purchased the book from :

....

....

....

....

(iv) Name and address of your institution (Please mention the University or Board as the case may be)

....

....

....

....

(v) Your name and complete address :

....

....

....

....

(vi) Course you are enrolled in :

....

....

....

....

The best assessment will be awarded each month : The award will be in the form of our publications as decided by the Editorial Board.

Please mail the filled up coupon at your earliest to :

S.K. KATARIA & SONS®
4885/109, Prakash Mahal, Dr. Subhash Bhargav Lane,
Opposite Delhi Medical Association, Daryaganj, New Delhi-110002
☎ +91-11-2324389, 23269324; Telefax: +91-11-23243489
E-mail: katariabooks@yahoo.com
website: www.skkatariaandsons.com